Designing Audio Effect Plugins in C++

Designing Audio Effect Plugins in C++ presents everything you need to know about digital signal processing in an accessible way. Not just another theory-heavy digital signal processing book, nor another dull build-a-generic-database programming book, this book includes fully worked, downloadable code for dozens of professional audio effect plugins and practically presented algorithms.

Sections include the basics of audio signal processing; the anatomy of a plugin; AAX, AU, and VST3 programming guides; implementation details; and actual projects and code. More than 50 fully coded C++ audio signal-processing objects are included. Start with an intuitive and practical introduction to the digital signal processing theory behind audio plugins, and quickly move on to plugin implementation, gain knowledge of algorithms on classical, virtual analog, and wave digital filters, delay, reverb, modulated effects, dynamics processing, pitch shifting, nonlinear processing, sample rate conversion, and more. You will then be ready to design and implement your own unique plugins on any platform and within almost any host program.

This new edition is fully updated and improved and presents a plugin core that allows readers to move freely between application programming interfaces and platforms. Readers are expected to have some knowledge of C++ and high school math.

Will C. Pirkle is Associate Professor and Program Director of Music Engineering Technology at the University of Miami Frost School of Music. He teaches classes in C++ Audio Programming, Signal Processing, Audio Synthesis, and Mobile App Programming. In addition to 14 years of teaching at the University of Miami, Will has 20 years of experience in the audio industry working and consulting for such names as Korg Research and Development, SiriusXM Radio, Diamond Multimedia, Gibson Musical Instruments, and National Semiconductor Corporation. An avid guitarist and studio owner, Will still seeks projects that combine all his skills.

Designing Audio Effect Plugins in C++

For AAX, AU, and VST3 With DSP Theory

Second Edition

Will C. Pirkle

Routledge
Taylor & Francis Group

NEW YORK AND LONDON

Second edition published 2019
by Routledge
605 Third Avenue, New York, NY 10017

and by Routledge
2 Park Square, Milton Park, Abingdon, Oxon, OX14 4RN

Routledge is an imprint of the Taylor & Francis Group, an informa business

First edition published by Focal Press 2013

Library of Congress Cataloging-in-Publication Data
Names: Pirkle, William C. author
Title: Designing audio effect plugins in C++ : for AAX, AU, and VST3 with DSP theory / Will C. Pirkle.
Description: Second edition. | New York, NY : Routledge, 2019.
Identifiers: LCCN 2018056123 (print) | LCCN 2018057004 (ebook) | ISBN 9780429954320 (pdf) |
   ISBN 9780429954313 (epub) | ISBN 9780429954306 (mobi) | ISBN 9781138591899 (hardback : alk. paper) |
   ISBN 9781138591936 (pbk. : alk. paper) | ISBN 9780429490248 (ebook)
Subjects: LCSH: Computer sound processing. | Plug-ins (Computer programs) | C++ (Computer program language)
Classification: LCC MT723 (ebook) | LCC MT723 .P57 2019 (print) | DDC 006.5—dc23
LC record available at https://lccn.loc.gov/2018056123

Typeset in Times New Roman
by Apex CoVantage, LLC

ISBN 13: 978-1-138-59189-9 (hbk)
ISBN 13: 978-1-138-59193-6 (pbk)
ISBN 13: 978-0-429-49024-8 (ebk)

DOI: 10.4324/9780429490248

Dedicated to

John Robert Pirkle
and
Emily Grace Pirkle

Contents

Figures

Tables

Preface

There are a lot of things to love about teaching at the university level—the periodicity of the semesters (even when Hurricane Irma comes for a brief visit), the revival and rebirth as each graduating class of seniors gives way to a new batch of freshmen, and of course there's summer break. But one of the coolest parts is the fact that you get to reshape, revitalize, and reinvent your classes every year. You get a perpetual "do-over" to fix the mistakes, the off-topic tangents, and the badly timed pop quizzes with each new school year—and each year on the day of spring graduation, I make a list of the stuff to do differently the next year.

The day I received my first copy of *Designing Audio Effect Plug-Ins in C++* I began a list of the stuff I'd do differently if given another chance to write a second edition. That happened in the fall of 2016 when I met with my Focal Press editor at the Audio Engineering Society (AES) Convention and I bounced the idea off of her for a new edition that fixed the issues with the first edition.

I wrote *Designing Audio Effect Plug-Ins in C++* from a notebook of class notes I had made and used during the 1990s in the Music Engineering Technology program at the University of Miami. There were few sources on the practical implementation of audio effects in C++ or digital signal processing (DSP) assembly. We wrote flangers, choruses, long delays, and even a plucked string model, right out of those original class notes.

When I put the first book together, the idea was to combine DSP theory and C++ practice to show how those algorithms could be implemented in code. I used my RackAFX platform and application programming interface (API) for two reasons: the API was stunningly lightweight and simple to understand, and it was not tied to any other software. I didn't have to worry about it becoming obsolete. But some of the readers didn't like the fact that the plugins didn't run in Virtual Studio Technology (VST®) or Audio Unit (AU®), and at first I didn't understand that. The book wasn't supposed to be about those things; it was about the algorithms and the C++. RackAFX was an example API to use as a conduit to demonstrate the ideas. After all, the other *API's are all basically the same and do the same exact stuff*.

The need to generate plugins for these APIs caused a ripple effect in the RackAFX software and the original API, now named *RAFX1* and gracefully retired. The last versions of RAFX1 could be exported to AAX® (Avid Audio eXtension), AU, and VST2/3 on Windows or MacOS in Microsoft Visual Studio and Apple Xcode projects, and they ran directly as VST2 and VST3 plugins on Windows 32-bit digital audio workstations (DAWs). This all came with a price: there was a lot of code dating back to 2004 or earlier, much of it now considered "illegal" in C++11. Back then, a lot of programmers didn't use the *std::* library, preferring to hand-code the linked lists and hash tables. In addition, the original RAFX1 was never intended to need to display a custom graphical user interface (GUI). The original API was written for a consulting project that involved a hardwired DSP chip—there was no GUI required. So when I approached my Focal Press editor that fall in 2016, I had a new premise to work on: first, the algorithms should be implemented in C++ objects that are fully portable across all APIs. After all, the title does say "... *in C++*."

Second, and more importantly, the second edition could not be RackAFX-centric, and there needed to be a platform that anyone could use to write and design plugins on their DAW and operating system of choice (Windows or MacOS). On top of that, this platform needed to be free from third parties for the same reasons as before—no worrying about someone making the product obsolete, and no worrying about someone deciding to charge fees or royalties for its use. I would have to write that new platform myself, independently of RackAFX or any software development kit (SDK) or operating system (OS).

So, after saying goodbye to my editor at the AES show, I got on the airplane and began the notes for what is now called ASPiK (Audio Specific Plugin Kernel). Two years later, the new platform is complete, along with a new book to show off dozens of algorithms in more than 55 C++ objects for you to use in your own projects. These objects are written for any API and may easily be used across platforms and SDKs. If you are already committed to using another framework, like *JUCE*, you will see that the C++ objects will snap into any API or framework—"Abstraction 101" as my grad student Madhur used to say. The book is not about ASPiK (although it is there if you need it) and it doesn't rely on any platform. Those 55 C++ objects are not ASPiK objects either: they are straight C++ and ready to rock right out of the box.

Chapters 1–8 are about plugins and the various plugin APIs. Instead of trying to re-print complete code for all of the plugins across all of the APIs (something I won't try again), each API gets its own "Programming Guide" chapter. If you look at the Table of Contents you can see that each chapter's section lists are virtually identical. These programming guides are to help you navigate the various API documentation and sample code. As you move through them, you will start to notice something: *the API's are all basically the same and do the same exact stuff,* only the implementation details are different. If you are new to plugin programming, you can use the free ASPiK platform. It is completely self-contained, including a built-in drag-and-drop GUI designer that launches from any plugin window (AAX, AU, or VST) on any platform (Windows or MacOS). Chapters 6 and 7 cover ASPiK basics, and how to use RackAFX to generate ASPiK projects—this is optional, and you don't need RackAFX to use ASPiK or generate new blank projects. Chapter 8 explains how the C++ book objects are packaged and how an example plugin GUI parameter list is shown. There are DSP objects for low-level signal processing chores, effect objects that encapsulate complete effect algorithms and are designed to interface with a GUI in any plugin framework, and finally an entire library of wave digital filter (WDF) components and adaptors for generating countless types of digital ladder filters.

Chapters 9 through 22 encompass the DSP theory, effect algorithms, and C++ objects and their use in audio plugins. Chapters 9 and 10 are the basic DSP theory chapters and remain almost unchanged from the first edition, as they seemed to be reader favorites. However, Chapter 10 has been beefed up with the rest of the common biquadratic filter structures, along with a bonus structure in the Homework section. Chapters 11–22 cover everything else—all of the original effects plus new chapters on modulation, WDF and virtual analog filters, a fantastic new reverb, nonlinear processing, FFTs (fast Fourier transforms), the phase vocoder, and finishing with sample rate conversion. The effects have all been re-coded from scratch, and they sound fantastic and professional across the board.

If you look at the flow of the sections of each half of the book in the table of contents, you can see that there are built-in "through lines" that run parallel across the chapters. This is most evident by the ordering and naming of the sections, which repeat in sets, at first along with the various SDKs, and later around the different DSP and effect objects. You can find the information on setting the plugin latency for AAX, AU, or VST in the same spot in each chapter. Likewise, once you get the hang of doing the first few C++ projects, then the rest will be easy because they all follow very consistent design and coding patterns. Using ASPiK is extremely simple, and there are numerous tutorial videos at www.willpirkle.com/support/video-tutorials/ to get you started. ASPiK is also very deep, including very complex GUI and drawing options, GUI zooming, and other topics too advanced for this text—all of this is also documented along with the ASPiK product and at my website as well.

Thanks to Lara Zoble, my Focal Press editor who encouraged me and believed in the premise for the second edition. In addition, I need to say thanks to all those at the University of Miami's Frost School of Music who've both supported me and entrusted me with the Music Engineering Technology program. They include Dean Shelly Berg, Rey Sanchez, and Serona Elton, along with my faculty colleagues Joe Abbati, Chris Bennett, Dana Salminen and of course my old boss, Ken Pohlmann, who is now surfing somewhere in north Florida, and having finally attained the "higher plane" we used to dream about, is blissfully ignorant of the yearly SACS reports that need filing. Again. Thanks also to my students who helped greatly with the RackAFX software and helped guide the directions of my classes all these years. Most recently that would include Madhur Murli, Lucas Pylypczak, Luke Habermehl, Akhil Singh, Jay Coggin, Hunter Garcia and Sanket Kulkarni, plus many more who also played a role. I must include a massive shout-out to my former student Alex Zinn, who translated various parts of the book *Digitale Signalverarbeitung* by Kammeyer and Kroschel—the original source of the combined WDF *RL*, *LC* and *RL* models in Chapter 12—from German.

As with the original book, I look forward to hearing what you've cooked up in your own plugin labs. You can find me at www.willpirkle.com. Please send videos and recordings of your plugin audio madness!

Will Pirkle
August 1, 2018

CHAPTER 1

Introduction

The first affordable digital audio devices began appearing in the mid-1980s. Digital signal processing (DSP) mathematics had been around since the 1960s. Commercial digital recordings first appeared in the early 1970s, but the technology did not become available for widespread distribution until about 15 years later when the advent of the compact disc ushered in the age of digital audio. Digital sampling refers to the acquisition of data points from a continuous analog signal. The data points are sampled on a regular interval known as the "sample period" or "sample interval." The inverse of the sample period is the sampling frequency, which we denote as f_s. A compact disc uses a sampling frequency of 44,100 Hz, producing 44,100 discrete samples per channel each second, with a sample interval of about 22.7 microseconds (μS). While digital sampling applies to many different systems, this book is focused on only one of those applications: audio. Prior to the early 1990s digital audio effects and synthesizers were implemented in hardware devices that consisted of microprocessors, microcontrollers, DSP chips, and other components. These devices almost exclusively featured analog input and output connections at first, and then later began to incorporate digital I/O. Digital audio workstations (DAWs) appeared in the late 1980s and audio software began to bloom. Microsoft released a software development kit (SDK) called DirectX® aimed at what was originally termed *multimedia*—a phrase many used at the time, but few really understood. Within DirectX were signal processing specifications to process audio and video data streams, and in short time the first DAWs appeared that supported DirectX "plugins." Around the same time, Steinberg Media Technology published a specification for its Virtual Studio Technology (VST), which became arguably the most popular plugin format ever conceived.

During the course of this book, you will learn both DSP theory and C++ software applications. You have instant access to dozens of C++ objects that implement every signal processing algorithm in the book, and Chapters 9 through 22 will specifically reveal the conversion of the algorithms into C++ code. Each of these chapters will include one to three of these C++ objects. These objects are not tied to any plugin format or operating system. You may easily combine them with other signal processing objects or frameworks, and you have license to use them as you wish for commercial or noncommercial plugins. The resulting products sound fantastic. You will get great satisfaction when you begin inventing your own unique plugins, or implementing cutting-edge audio algorithms as soon as they are published.

1.1 Using This Book

Included are individual chapters for the three major commercially available effects specifications AAX (Avid Audio eXtension), Audio Unit (AU), and VST3 (Steinberg Media Technology GmbH). Plugin specifications are also known as APIs. The most important aspect of the book design is that you may use whatever platform you wish for writing and testing your plugins on MacOS or Windows and in the DAW of your choice. The DSP algorithms encapsulated in the aforementioned C++ objects are platform and API neutral. As you study the API-specific chapters, you will learn that all three of these APIs are really very similar in their internal functionality, but the implementation details for the different APIs can be overwhelming. This has lead to the use of third party plugin development frameworks.

You may already be using one of these frameworks, such as *JUCE*—if so, that is fine. You can easily incorporate the C++ signal processing objects into these frameworks because the objects are not API or platform centric. Make sure you adhere to any licensing restrictions that these frameworks require and follow their rules for software release or code publication. You will need to know how to get audio in and out of the frameworks as well as how to design your graphical user interface (GUI) and handle parameter changes from the GUI or DAW automation so please refer to documentation for those frameworks for that information.

If you are like most readers you probably have not developed a plugin using one of those frameworks—or perhaps you tried, but got bogged down in the steep learning curves and the new implementation details that the frameworks were supposed to help alleviate. Or you may have downloaded one of the SDKs and were bewildered by its contents. That is also fine because we've spent the last three years developing a fully functional, truly professional, license-free platform for you to use called ASPiK™ (which stands for "Audio Specific PlugIn Kernel"). ASPiK consists of a set of C++ objects that are custom tailored to interface directly with AAX, AU, and VST3 plugins, along with a mechanism for easily generating new plugin projects.

The projects for the book are packaged as ASPiK plugin projects and are simple to integrate into other third party frameworks if you would rather use them instead. All of the plugin projects in the book are written in a way that will make the code simple to transport to other development systems. In addition, the code in the individual chapters does not focus on any API, or on ASPiK itself. The fundamental goal of the book is to help you understand audio signal processing algorithms and how to convert them into C++ objects; it tries to be as platform, framework, and API neutral as possible.

Included in the ASPiK framework is a complete, platform-independent GUI design system called the *PluginGUI* object, with a drag-and-drop interface for creating beautiful and professional looking GUIs without writing a single line of code. If you want to experiment with GUI programming, you can find out about custom GUI design in Chapter 21.

If you are new to plugin programming, or you want to develop new plugins or test algorithms in the most efficient manner possible, you might want to use the RackAFX™ software, however this book is not RackAFX-centric nor requires it for use. There is absolutely no RackAFX-specific code in ASPiK or its accompanying objects. However, RackAFX simplifies plugin programming and design by allowing you to quickly define the plugin parameters and user interface, and it includes tools for testing your plugin algorithm such as an oscillator, oscilloscope, spectrum analyzer, and more. You may also use RackAFX to generate ASPiK projects based off of your designs, and these ASPiK projects do not rely upon, nor include anything RackAFX-specific. Almost all of the real-time response plots you see in this book—including frequency, impulse, step, and phase responses—were done using RackAFX's plugin analyzer, which can make analytical measurements on your plugin's algorithm while it is running.

1.2 Fundamentals of Audio Signal Processing

Because plugins are software variations on hardware designs, it's worth examining how the hardware systems operate: acquisition of data, audio reconstruction, and numerical representation of the audio signal. Then we define our analytical test signals and the basic signal processing blocks.

1.2.1 Acquisition of Audio Samples

The incoming analog audio signal is sampled with an analog to digital converter (ADC or A/D). Analog to digital converters must accurately sample and convert an incoming signal, producing a valid digitally coded number that represents the analog voltage within the sampling interval. This means that for CD audio, a converter must produce an output every 22.7 μS. Figure 1.1a shows the block diagram of the input conversion system with low-pass filter, A/D, and encoder while Figure 1.1b demonstrates the Nyquist frequency's encoding consisting of only two points.

> The sampling theorem states that a continuous analog signal can be sampled into discrete data points and then reconstructed into the original analog signal without any loss of information—including inter-sample fluctuations—if and only if the input signal has been band-limited so that it contains no frequencies higher than half the sample rate, also known as the Nyquist frequency or Nyquist rate. Band-limited means low-pass filtered (LPF). Band-limiting the input signal prior to sampling is known as adhering to the Nyquist criteria.

Violating the Nyquist criteria will create audible errors in the signal in the form of an erroneously encoded signal. We will refer to the Nyquist frequency simply as "Nyquist" from now on. Frequencies higher than Nyquist will fold back into the spectrum. This effect is called aliasing since the higher-than-Nyquist frequencies are encoded "in disguise" or as an "alias" of the actual frequency. This is easiest explained with a picture of an aliased signal shown in Figure 1.1c where the encoding error is shown in the dotted waveform—it is clearly not the correct frequency.

Once the aliased signal is created it will remain as a permanent error in the signal. The LPF that band-limits the signal at the input is called the "anti-aliasing filter." Another form of aliasing occurs at the movies. An analog movie camera takes 30 pictures (frames) per second. However, it must often film objects that are rotating at much higher rates than 30 per second, like helicopter blades or car wheels.

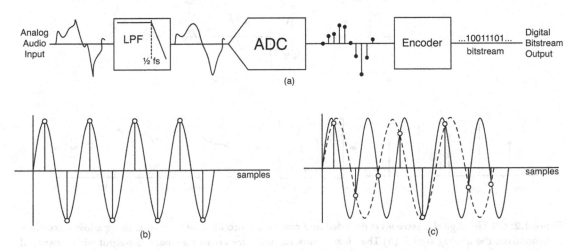

Figure 1.1: (a) An analog-to-digital encoding system. (b) The Nyquist frequency is the highest frequency that can be encoded with two samples per period. (c) Increasing the frequency above Nyquist but keeping the sampling interval the same results in an obvious coding error—the aliased signal (dotted line) is the result.

The result is visually confusing: the helicopter blades or car wheels appear to slow down and stop, then reverse directions and speed up, then slow down and stop, reverse, etc. This is the visual result of the high frequency rotation aliasing back into the movie as an erroneous encoding of the actual event.

1.3 Reconstruction of the Analog Signal

The digital to analog converter (DAC or D/A) first decodes the bitstream, then takes the sampled data points or impulses and converts them into analog versions of those impulses. The DAC output is then low-pass filtered to produce the final analog output signal, complete with all the inter-sample fluctuations. As with the ADC diagram, the decoder and D/A are both inside the same device (chip). Figure 1.2a shows the conceptual block diagram of the decoding end of the system. The output filter is called the *reconstruction filter* and is responsible for re-creating the seemingly missing information that the original sampling operation did not acquire—all the inter-sample fluctuations. The reason it works is that the low-pass filtering converts impulses into smeared out, wiggly versions. The smeared out versions have the shape *f(x) = sin(x)/x,* which is also called the *sinc()* function; it resembles the profile of a sombrero, as in Figure 1.2b. When a train of impulses is filtered, the resulting set of *sin(x)/x* pulses overlap with each other and their responses add up linearly. The addition of all the smaller curves and damped oscillations reconstructs the inter-sample curves and fluctuations in Figure 1.2c.

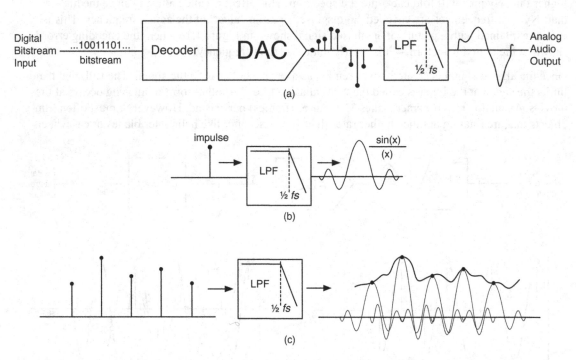

Figure 1.2: (a) The digital bitstream is decoded and converted into an analog output using a low-pass filter to reconstruct the analog signal. (b) The ideal reconstruction filter creates a smeared output with a damped oscillatory shape, while the amplitude of the *sin(x)/x* shape is proportional to the amplitude of the input pulse. (c) The *sin(x)/x* outputs of the LPF are summed together to reconstruct the original band-limited input waveform (the inter-sample information has been reconstructed).

1.4 Numerical Representation of Audio Data

The audio data can be numerically coded in several different ways. Basic digital audio theory reveals that the number of quantization levels available for coding is $q = 2^N$ where N is the bit depth of the signal. Thus an 8-bit system can encode 2^8 values or 256 quantization levels. A 16-bit system can encode 65,536 different values. Figure 1.3a shows the hierarchy of encoded audio data. As a hardware system designer, you must first decide if you are going to deal with unipolar (unsigned) or bipolar (signed) data. After that, you need to decide on the data types. Figure 1.3b reveals how the quantization levels are encoded. The dashed lines represent the minimum and maximum values.

- Unipolar or unsigned data are on the range of zero to +max, or −min to zero and only has one polarity (+ or −) of data, plus the number zero.
- Bipolar or signed data varies from −min to +max and is the most common form today; it also includes the number zero.
- Integer data are represented with integers and no decimal place; unsigned integer audio varies from 0 to +65,535 for 16-bit systems while signed integer audio varies from −32,768 to +32,767 for the same 16-bit audio—in both cases, there are 65,536 quantization levels.
- Fractional data are encoded with an integer and fractional portion, combined as *int.frac* with the decimal place separating the parts (e.g. 12.09).
- Within the fractional data subset are fixed and floating point types: fixed point data fixes the number of significant digits on each side of the decimal point and is combined as Int-Sig-Digits. Frac-Sig-Digits. For example, "8.16" data would have eight significant digits before the decimal place and 16 afterwards. Floating point data has a moving mantissa and exponent that code the values across a range predetermined by the IEEE. The positive and negative portions are encoded in 2's complement so that addition of exactly opposite values (e.g. −0.5 and + 0.5) always results in zero.

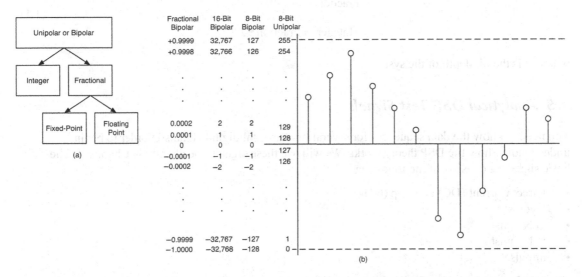

Figure 1.3: (a) The hierarchy of numerical encoding. (b) A comparison of several different types of data representations. The floating point version is fixed for a range of −1.0 to +0.9999, though any range can be used.

A fundamental problem is that the number of quantization levels will always be an even number since 2^N is always even. In bipolar systems, we would very much like to have the number zero encoded as 0. If we do that, then we use up one of our quantization levels for it. This leaves an odd number of levels that cannot be split exactly in half. That creates the anomaly you see in Figure 1.3b: there are always more negative (−) quantization levels than positive ones for bipolar coding. For the unipolar case, there is no value that exactly codes the zero level; in Figure 1.3b it is midway between the values 127 and 128.

This slight skewing of the data range is unavoidable if we intend on using the number zero in our algorithms, and that is almost guaranteed. In some systems the algorithm limits the negative audio data to the second-most negative value. This is because phase inversion is common in processing algorithms, either on purpose or in the form of a negative-valued coefficient or multiplier. If a sample came in with a value of −32,768 and it was inverted, there would be no positive version of that value to encode. To protect against that, the −32,768 value is treated as −32,767. The audio data that travel from the audio hardware adaptor (DSP and sound card) as well as that stored in *.wav* files is often signed integer–based. However, for audio signal processing, we prefer to use floating point representation. Fortunately, most audio file decoders are set up to deal with fixed and floating point files, and will automatically produce floating point audio data on the range [−1.0, +1.0] for us; note that this simplifies the actual range of −1.0 to +0.9999. In fact the plugins you code in this book will all use data that is on that same range. The reason has to do with overflow. In audio algorithms, addition and multiplication are both commonplace. With integer-based numbers, you can get into trouble quickly if you mathematically combine two numbers that result in a value that is outside the range of known numbers. Consider the 16-bit integer bipolar format on the range of −32,768 to +32,767. When multiplied together, most of the values on this range will result in a product that is outside these limits. Addition and subtraction can cause this as well, but for only half the possible values. However, numbers between −1.0 and +1.0 have the interesting property that their product is always a number in that range. Converting an integer value to a fractional value along the normalized range of −1.0 to +1.0 in an *N*-bit digital audio system is easy, as is the reverse conversion.

$$\text{Fraction} = \frac{\text{Integer}}{2^N}$$

$$\text{Integer} = \text{Fraction} * 2^N$$

(1.1)

where N is the bit depth of the system.

1.5 Analytical DSP Test Signals

You need to know the data sequences for several fundamental digital signals in order to begin understanding how the DSP theory works. We will use these signals specifically in Chapter 2. The basic signal set consists of the following:

- Direct Current (DC) and step (0 Hz)
- Nyquist
- ½ Nyquist
- ¼ Nyquist
- impulse

The first four of these signals are all you need to get a ballpark idea of the frequency response of some basic DSP filters—and the good news is that all the sequences are simple to remember. The figures that follow show samples on an *x*-axis labeled *x(n)*. In these figures, *n* is the sample number, starting at time = 0 with *n* = 0.

1.5.1   DC and Step (0 Hz)

The DC or 0 Hz and step responses can both be found with the DC/step input sequence { . . . 0, 0, 1, 1, 1, 1, 1, 1, 1, 1 . . . } as shown in Figure 1.4a. This signal contains two parts: the step portion where the input changes from 0 to 1 and the DC portion where the signal remains at the constant level of 1.0

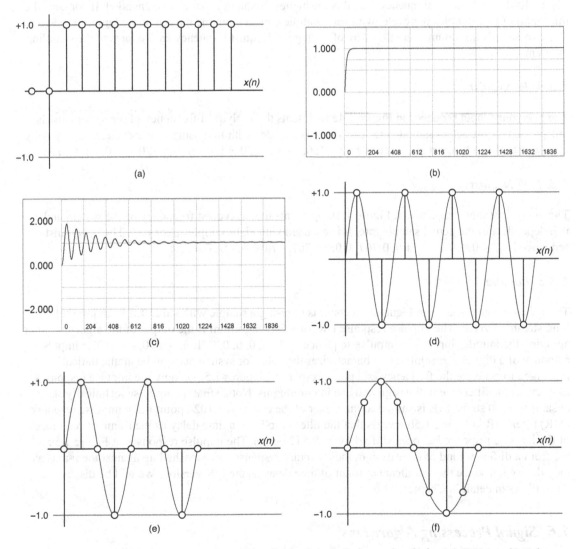

Figure 1.4: A graphical representation of the digitized analytical test signals for (a) DC/step (0 Hz); (b) the step-response of a 1st order low-pass filter with f_c = 500 Hz; (c) the step-response of a 2nd order low-pass filter with f_c = 1 kHz and Q = 8, along with the digitized analytical test signals for (d) Nyquist, (e) ½ Nyquist, and (f) ¼ Nyquist.

forever. When you apply this signal to your DSP filter and examine the output you will get two pieces of information; the step portion will tell you the transient attack time and the DC portion will give you the response at DC or 0 Hz. Figure 1.4b and 1.4c show the step responses of two different low-pass filters.

1.5.2 Nyquist

The Nyquist input sequence represents the Nyquist frequency of the system and is independent of the actual sample rate. The Nyquist sequence is {. . . −1, +1, −1, +1, −1, +1, −1, +1 . . .} and is shown in Figure 1.4d. The Nyquist frequency signal is the highest frequency that can be encoded. It contains the minimum of two samples per cycle, with each sample representing the maximum and minimum values. The two-sample minimum is another way of stating the Nyquist frequency as it relates to the sampling theorem.

1.5.3 ½ Nyquist

The ½ Nyquist input sequence in Figure 1.4e represents the ½ Nyquist frequency of the system and is independent of the actual sample rate. The signal is encoded with four samples per cycle, twice as many as Nyquist. The ½ Nyquist sequence is { . . . −1, 0, +1, 0, −1, 0, +1, 0, −1, 0, +1, 0, −1, 0, +1, 0 . . .}.

1.5.4 ¼ Nyquist

The ¼ Nyquist input sequence in Figure 1.4f represents the ¼ Nyquist frequency of the system and is independent of the actual sample rate. It is encoded with eight samples per cycle. The ¼ Nyquist sequence is {. . . 0.0, 0.707, +1.0, 0.707, 0.0, −0.707, −1.0, −0.707, 0.0 . . .}.

1.5.5 Impulse

The impulse signal shown in Figure 1.5a consists of a single sample with value 1.0 in an infinitely long stream of zeros. The impulse response (IR) of a DSP algorithm is the output of the algorithm after applying the impulse input. The impulse sequence is: { . . . 0, 0, 0, 0, 1, 0, 0, 0, 0, . . . }. The impulse response of a digital algorithm fully characterizes the behavior system and can be mathematically massaged to generate the frequency and phase responses. Figures 1.5b–f show the impulse responses of a few of the filters we will design and use in our plugins. Notice that the impulse technically starts at sample $n = 0$ so the axis is shifted to the center of the event. The 1024 point finite impulse response (FIR) filter's IR in Figure 1.5b reveals that the filter introduces a time delay as the frame of reference of the impulse response has been shifted over by 512 points. The impulse responses in Figure 1.5c–f are from a different kind of filter design: their impulse responses are smashed up against the y-axis as they do not introduce the significant amount of time delay as the FIR version. We will be discussing these filters in detail in Chapters 11 and 16.

1.6 Signal Processing Algorithms

In the broadest sense, an algorithm is a set of instructions that complete a predefined task. In the specialized case of audio signal processing, an algorithm is a set of instructions that operates on audio input data to produce an audio output bitstream. For real-time processing, the algorithm must accept

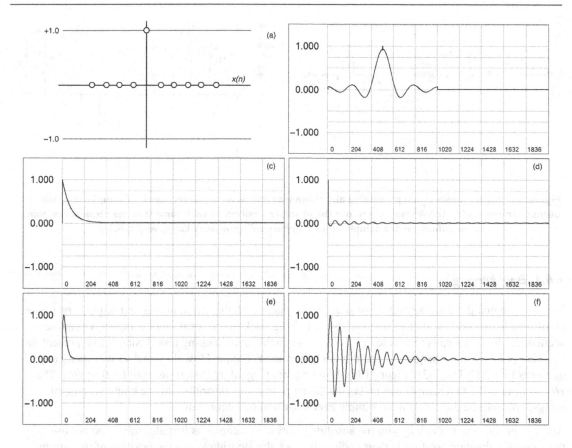

Figure 1.5: (a) An impulse signal consisting of a single sample with value 1.0; (b) 1024 point FIR low-pass filter with f_c = 1 kHz; (c) 1st order low-pass filter with f_c = 100 Hz; (d) 2nd order high-pass filter with f_c = 500 Hz and Q = 10; (e) 2nd order low-pass filter with f_c = 500 Hz and Q = 0.2; (f) 2nd order low-pass filter with f_c = 500 Hz and Q = 10. For all impulse responses, the sample rate f_s = 44.1 kHz.

a new input sample (or set of samples), do the processing, then have the output sample(s) available before the next input arrives—if the processing takes too long, clicks, pops, glitches, and noise will be the real-time result. Most of the exercises in this book involve processing incoming audio data and transforming it into a processed output. However, synthesizing a waveform to output also qualifies. In this special case, there is no real-time audio input to process. Most of the plugins in this book use the effects model where an input sequence of samples is processed to create an output sequence, as shown in Figure 1.6a. Throughout the book, I will use the following terms to represent the input, output and impulse response signals:

- $x(n)$ is always the input sequence; the variable n represents the location of the nth sample of the x-sequence.
- $y(n)$ is always the output sequence; the variable n represents the location of the nth sample of the y-sequence.
- $h(n)$ is the impulse response of the algorithm and is a special sequence that represents the algorithm output for a single sample input or impulse.

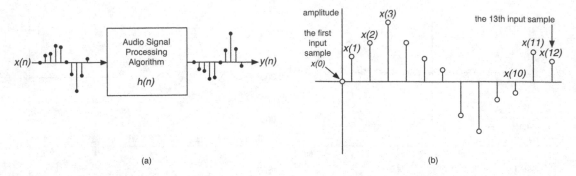

Figure 1.6: (a) An audio signal processing algorithm transforms an input sequence of samples *x(n)* into an output sequence *y(n)*. (b) Attempting to use the absolute position of each sample in the sequence is going to be difficult as the sample index numbers will grow very large, very quickly.

1.6.1 Bookkeeping

You can see that there are already three sequences to deal with the input, output and impulse response, all of which are coded with the same variable *n* to keep track of the location of samples within the sequence. The first step is to decide how to use *n* to do this bookkeeping task. Using it to represent the absolute position in the sequence would quickly become tiresome. How do you deal with indexing numbers like $x(12,354,233)$? Figure 1.6b shows an input signal, *x(n)* starting from $t = 0$ or *x(0)*. The *x(0)* sample is the first sample that enters the signal processing algorithm. In the grand scheme of things, *x(0)* will be the oldest input sample ever known to the algorithm.

Another problem with dealing with the absolute position of samples is that algorithms do not use the sample's absolute position in their coding. Instead, the algorithms use the position of the current input sample as their reference frame and make everything relative to that sample. On the next sample period, everything gets reorganized in relation to the (new) current input sample again. This might sound confusing at first, but it is a much better method of keeping track of the samples—and more importantly, defining the input/output characteristics of the algorithm, called the *transfer function*. Figure 1.7a shows the input signal frozen at the current time; *x(n)* and the other samples are indexed based on its location. One sample period later, in Figure 1.7b, you can see the frame of reference has shifted to the right and that the original *x(n)* has now become *x(n − 1)*.

Bookkeeping Rules

- The current sample is labeled "*n*."
- Previous samples are negative, so *x(n − 1)* would be the previous input sample.
- Future samples are positive, so *x(n + 1)* would be the next input sample relative to the current one.
- On the *next* sample interval, everything is shuffled and referenced to the new current sample, *x(n)*.

1.6.2 The One-Sample Delay

Whereas analog processing circuits like tone controls use capacitors and inductors to alter the phase and delay of the analog signal, digital algorithms use time delay instead. You will uncover the math and science behind this fact later on when you start to use time delay blocks in your signal processing

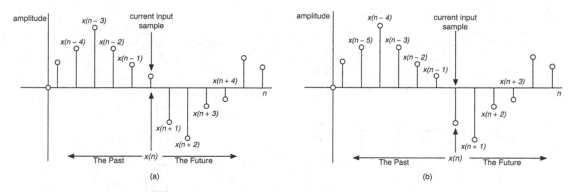

Figure 1.7: (a) DSP algorithms use the current sample location as the reference location. All other samples are indexed based on that sample, while the input sequence is frozen in time at the current input sample *x(n)*. (b) One sample period later, everything has shifted and the previous *x(n)* is now indexed as *x(n − 1)* while what was the next sample, *x(n + 1)*, now becomes the current input sample *x(n)*.

structures. In our algorithm diagrams, a delay is represented by a box with the letter z inside. The z-term will have an exponent such as z^{-5} or z^{+2} or z^0; the exponent codes the delay in samples following the same bookkeeping rules, with negative (−) exponents representing a delay in time backwards (past samples) and positive (+) representing delay in forward time (future samples). We call z the *delay operator* and as it turns out, it will be treated as a mathematical operation.

You are probably asking yourself how you can have a positive delay towards the future. The answer is that you can't for real-time signal processing. In real-time processing you never know what sample is going to arrive next. However, in non-real-time processing (for example, an audio file that you are processing offline) you do know the future samples because they are in the file. Figure 1.8a shows two common orientations to denote a one-sample delay in an algorithm block diagram.

Delay Rules

- Each time a sample goes into the delay register (memory location), the previously stored sample is ejected.
- The ejected sample can be used for processing or deleted.
- The delay elements can be cascaded together with the output of one feeding the input of the next to create more delay time.

If a sample *x(n)* goes into the one-sample delay element, then what do you call the sample that is ejected? It's the previous sample that came in, one sample interval in the past. So, the output sample is *x(n − 1)* as shown in Figure 1.8b. In Figure 1.8c and 1.8d you can see how delay elements can be cascaded with outputs taken at multiple locations generating multiple samples to use in the algorithm.

1.6.3 Multiplication With a Scalar Value

The next algorithm building block is the scalar multiplication operation. It is a sample-by-sample operator that simply multiplies the input samples by a coefficient. The multiplication operator is used in just about every DSP algorithm. Figure 1.8e shows the multiplication operator in action. The outputs are simply scaled versions of the inputs.

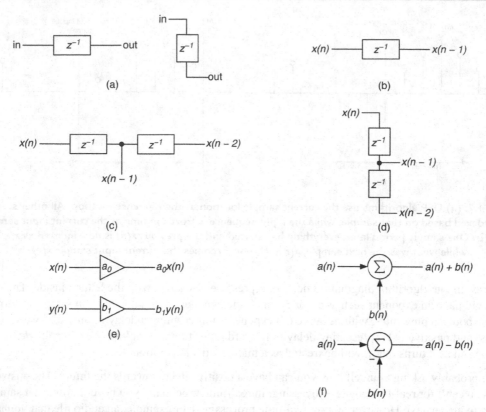

Figure 1.8: Common block diagram components for audio signal processing algorithms include the following: (a) the one-sample delay in two orientations; (b) the one-sample delay processing an input $x(n)$ and delaying it to produce $x(n - 1)$; (c) an example of using two delays to create two delayed signals $x(n - 1)$ and $x(n - 2)$; (d) an alternative way to use two delays to create $x(n - 1)$ and $x(n - 2)$; (e) scalar multiplier blocks with coefficients a_0 and b_1; and (f) summing and subtraction algorithms.

1.6.4 Addition and Subtraction

Addition and subtraction are really the same kind of operation, because subtracting is the addition of a negative number. There are several different algorithm symbols to denote addition and subtraction. The operation of mixing signals is really the mathematical operation of addition. Figure 1.8f shows the addition and subtraction operations in block diagram form. Notice that the subtraction operation is shown with the sigma (Σ) symbol and a small ($-$) sign is next to the branch being subtracted.

1.6.5 Some Algorithm Examples and Difference Equations

By convention, the output sequence of a DSP algorithm is named $y(n)$ and the mathematical equation that relates it to the input is called the *difference equation*. The difference equation does not necessarily require a subtraction (difference) operation. Its name corresponds to the analog equivalent version called the *differential equation*. Combining the operations will give you a better idea of what the difference equation looks like. Figure 1.9a shows the difference equations for several combinations of algorithm building blocks. The output $y(n)$ is a mathematical combination of inputs.

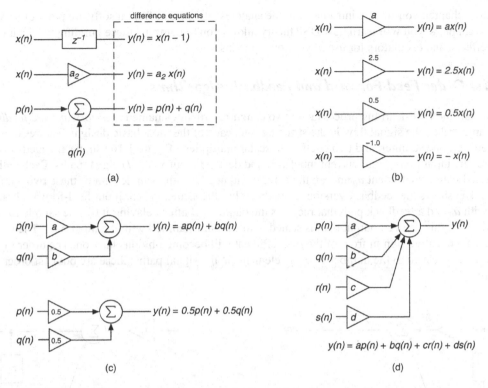

difference equations

$y(n) = x(n - 1)$

$y(n) = a_2\, x(n)$

$y(n) = p(n) + q(n)$

(a)

$y(n) = ax(n)$

$y(n) = 2.5x(n)$

$y(n) = 0.5x(n)$

$y(n) = -\,x(n)$

(b)

$y(n) = ap(n) + bq(n)$

$y(n) = 0.5p(n) + 0.5q(n)$

(c)

$y(n) = ap(n) + bq(n) + cr(n) + ds(n)$

(d)

Figure 1.9: (a) These difference equations relate the input(s) to the output via mathematical operations. (b) A simple coefficient multiplier will handle the three basic audio processing functions. (c) A generalized mixer/summer with a separate coefficient on each line (top) and a normalized mixer that will not overflow or clip (bottom). (d) An example of a weighted-sum algorithm (each sample from each channel has its own weighting coefficient, $a - d$).

1.6.5.1 Gain, Attenuation and Phase Inversion

As shown in Figure 1.9b, a simple coefficient multiplier will handle the three basic audio processing functions of gain, attenuation, and inversion. If the coefficient is a negative number, phase inversion will be the result. If the coefficient has a magnitude less than 1.0, attenuation will occur, while amplification occurs if the magnitude is greater than 1.0. Notice the different notation with the coefficient placed outside the triangle; this is another common way to designate it. A problem with mixing multiple channels of digital audio is the possibility of overflow or creating a sample value that is outside the range of the system. You saw that by limiting the bipolar fractional system to the bounds of -1.0 to $+1.0$, multiplication of any of these numbers always results in a number that is smaller than either, and always within the same range of $[-1.0, +1.0]$. However, addition of signals can easily generate values outside the ± 1 limits. In order to get around this problem, N-channel mixing circuits incorporate attenuators to reduce the size of the inputs, where the attenuation value is $1/N$. When mixing two channels, the attenuators each have a value of one-half while a three-channel mixer would have attenuators with a value of one-third on each mixing branch. If all channels happen to have a maximum or minimum value at the same sample time, their sum or difference will still be limited to the range of $[-1.0, +1.0]$. Figure 1.9c and 1.9d show the generalized mixing and weighted-sum algorithms.

In the next chapter, you will be introduced to the anatomy of a plugin from a software point of view. In future chapters, you will learn how DSP theory allows you to combine these building blocks into filters, effects, and oscillators for use in your own plugins.

1.7 1st Order Feed-Forward and Feedback Algorithms

We will start the DSP-learning process with two common structures named *feed-forward* and *feedback*. Their names relate the signal flow in the structures and each of the most basic designs features one delay register, one summer, and two coefficient scalar multipliers. Figure 1.10a shows the feed-forward structure; the input is split into current input $x(n)$ and delayed input $x(n - 1)$ signal paths. Each path is weighted with a coefficient a_0 and a_1; the current output $y(n)$ is the simple sum of these two paths. Figure 1.10b shows the feedback version. Note that it has the same, current input feed-forward branch scaled with a_0 and a feedback path that recycles the output $y(n)$ after delaying it by one sample to form $y(n - 1)$; this delayed output path is scaled with $-b_1$ and summed with the input path. The reason for placing a negative sign in front of the b_1 coefficient will become obvious later on. The order of the structure is related to the number of delay elements in the signal paths; these are both 1st order structures.

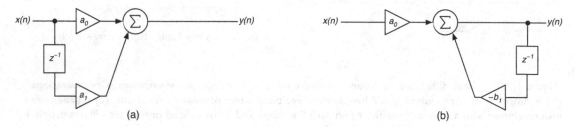

Figure 1.10: (a) A 1st order feed-forward structure. (b) A 1st order feedback algorithm.

1.8 Bibliography

Ballou, G., ed. 1987. *Handbook for Sound Engineers*. Carmel, IN: Howard W. Sams & Co., 1987, pp. 898–906.

IEEE 754–2008 Standard for Floating-Point Arithmetic. 2008. New York: Institute of Electrical and Electronics Engineers. http://www.dsc.ufcg.edu.br/~cnum/modulos/Modulo2/IEEE754_2008.pdf

Jurgens, R. K., ed. 1997. *Digital Consumer Electronics Handbook*, Chap. 2. New York: McGraw-Hill.

Kirk, R. and Hunt, A. 1999. *Digital Sound Processing for Music and Multimedia*, Chap. 1. Burlington, MA: Focal Press.

KORG Wavestation SR Service Manual. Tokyo, Japan: KORG, Inc.

Limberis, A. and Bryan, J. 1993. *An Architecture for a Multiple Digital-Signal Processor Based Music Synthesizer with Dynamic Voice Allocation*, Journal of the Audio Engineering Society, Convention Paper 3699, Audio Engineering Society, 95th Convention, October, San Francisco, CA.

Pohlmann, K. C. 2011. *Principles of Digital Audio*. New York: McGraw-Hill, pp. 16–30.

Stearns, S. D. and Hush, D. R. 1990. *Digital Signal Analysis*, Eaglewood Cliffs, NJ: Prentice-Hall, pp. 44–52.

CHAPTER 2

Anatomy of an Audio Plugin

The first thing to understand about the audio plugin specifications for AAX, AU, and VST3 is this: they are all fundamentally the same and they all implement the same sets of attributes and behaviors tailored specifically to the problem of packaging audio signal processing software in a pseudo-generic way. Plugins are encapsulated in C++ objects, derived from a specified base class, or set of base classes, that the manufacturer provides. The plugin host instantiates the plugin object and receives a base class pointer to the newly created object. The host may call only those functions that it has defined in the API, and the plugin is required to implement those functions correctly in order to be considered a proper plugin to the DAW—if required functions are missing or implemented incorrectly, the plugin will fail. This forms a contract between the manufacturer and the plugin developer. The deeper you delve into the different APIs, the more you realize how similar they really are internally. This chapter is all about the similarities between the APIs—understanding these underlying structural details will help your programming, even if you are using a framework like *JUCE* or our new ASPiK. In the rest of the book I will refer to the plugin and the host. The plugin of course is our little signal processing gem that is going to do something interesting with the audio, packaged as C++ objects, and the host is the DAW or other software that loads the plugin, obtains a single pointer to its fundamental base class, and interacts with it.

Each plugin API is packaged in an SDK, which is really just a bunch of folders full of C++ code files and other plugin components. None of the APIs come with precompiled libraries to link against so all of the code is there for you to see. To use any of the frameworks, like *JUCE* or ASPiK, you will need to download the SDK for each API you wish to support. The SDKs always come with sample code; the first thing you want to do is open some of that code and check it out, even if you are using a framework that hides the implementation details. Remember: **do not be afraid**. All of the APIs do the same basic thing. Yes, there is a lot of API-specific stuff, but you really need to get past that and look for the similarities. This chapter will help you with that, and the following four chapters will provide many more details and insights. A good exercise is to open up the most basic sample project that the SDK provides (usually this is a volume plugin) and then try to find each of the pieces in this chapter somewhere in the code.

2.1 Plugin Packaging: Dynamic-Link Libraries (DLLs)

All plugins are packaged as dynamic-link libraries (DLLs). You will sometimes see these called *dynamic linked* or *dynamically linked* but they all refer to the same thing. Technically, *DLL* is a Microsoft-specific term for a shared library, but the term has become so common that it seems to have lost its attachment to its parent company. We are going to use it universally to mean a precompiled library that the executable links with at runtime rather than at compile time. The executable is the DAW and the precompiled library is the DLL, which is your plugin. If you already understand how

DLLs work then you can safely skip this section, but if they are new to you, then it is important to understand this concept regardless of how you plan on writing the plugin.

To understand DLLs, first consider static-link libraries, which you have probably already used, perhaps unknowingly. C++ compilers include sets of precompiled libraries of functions for you to use in your projects. Perhaps the most common of these is the math library. If you try to use the *sin()* method you will typically get an error when you compile stating "sin() is not defined." In order to use this function you must link to the library that contains it. The way you do this is by placing *#include <math.h>* at the top of your file. Depending on your compiler, you might also need to tell it to link to *math.lib*. When you do this, you are statically linking to the *math.h* library, a precompiled set of math functions in a *.lib* file (the file is suffixed with *.a* on MacOS). Static linking is also called *implicit linking*. When the compiler comes across a math function, it replaces the function call with the precompiled code from the library. In this way, the extra code is compiled *into* your executable. You cannot un-compile the math functions. Why would you want to do this? Suppose a bug is found in the *sin()* function and the *math.h* library has to be recompiled and redistributed. You would then have to recompile your software with the new *math.h* library to get the bug fix. Static linking is shown figuratively in Figure 2.1a where the *math.lib* code is *inside* of the host code.

A solution to our math bug problem would be linking to the functions at runtime. This means that these precompiled functions will exist in a separate file that our executable (DAW) will know about and communicate with but only after it starts running. This kind of linking is called *dynamic linking* or *explicit linking* and is shown in Figure 2.1b. The file that contains the precompiled functions is the DLL. The advantage is that if a bug is found in the library, you only need to redistribute the newly compiled DLL file rather than recompiling your executable with a fixed static library. The other advantage is that the way this system is set up—a host that connects to a component at runtime—works perfectly as a way to extend the functionality of a host without the host knowing anything about the component when the host is compiled. It also makes an ideal way to set up a plugin processing system. The host becomes the DAW and your plugin is the DLL, as shown in Figure 2.1c.

When the host is first started, it usually goes through a process of trying to load all of the plugin DLLs it can find in the folder or folders that you specify (Windows) or in well defined API-specific folders (MacOS). This initial phase is done to test each plugin to ensure compatibility if the user decides to

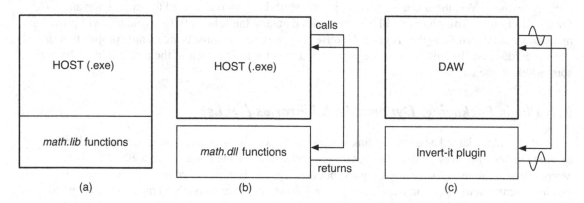

Figure 2.1: (a) A host with a statically linked math library compiled into it. (b) A host communicating with a DLL version. (c) A plugin version.

load it. This includes retrieving information such as the plugin's name, which the host will need to populate its own plugin menus or screens, and may also include basic or even rigorous testing of the plugin. We will call this the *validation* phase. For a plugin developer, this can be a very frustrating phase: if your plugin has a runtime bug in the loading code that the compiler did not catch, it will likely crash the host application. (Reaper® is one exception to this; it will tell you if your plugin failed to load, but it usually won't crash as a result.) In addition, the host DAW may forbid itself from ever loading your plugin again in the future. This is particularly true with Apple Logic®, which will permanently forbid your plugin if it fails validation. This can make plugin development confounding. You can find some strategies for dealing with validation phase errors at www.willpirkle.com but the best thing you can do is prevalidate your plugins prior to running them in the host. AU and VST provide mechanisms to do this.

Usually, the host unloads all of the plugins after the validation phase and then waits for the user to load the plugin onto an active track during a session. We will call this the *loading* phase. The validation phase already did at least part (if not all) of this once already; if you have bugs during validation, they are often caused during the loading operation. When your plugin is loaded, the DLL is pulled into the executable's process address space, so that the host may retrieve a pointer from it that has an address within the DAW's own address space. For most people, this is the only way they have ever used a plugin. There is a less common way of doing this between processes, meaning your DAW would load a plugin from another executable's address space. The loading phase is where most of the plugin description occurs. The description is the first aspect of plugin design you need to understand, and it is mostly fairly simple stuff.

After the loading phase is complete, the plugin enters the *processing* phase, which is often implemented as an infinite loop on the host. During this phase, the host sends audio to your plugin, your cool DSP algorithm processes it, and then your plugin sends the altered data back to the host. This is where most of your mathematical work happens, and this is the focus of the book. All of the plugin projects have been designed to make this part easy and universally transportable across APIs and platforms. While processing, the user may save the state of the DAW session and your plugin's state information must also be saved. They may later load that session, and your plugin's state must be recalled.

Eventually, your plugin will be terminated either because the user unloads it or because the DAW or its session is closed. During this *unloading* phase, your plugin will need to destroy any allocated resources and free up memory. This is another potential place your plugin can crash the host, so be sure to test repeated loading and unloading.

2.2 The Plugin Description: Simple Strings

The first place to start is with the plugin description. Each API specifies a slightly different mechanism for the plugin to let the host know information about it. Some of this is really simple, with only a few strings to fill in: for example, the text string with its name (such as "Golden Flanger") that the plugin wants the host to show the user. The simplest part of the plugin description involves defining some basic information about the plugin. This is usually done with simple string settings in the appropriate location in code, or sometimes within the compiler itself. These include the most basic types of information:

- plugin name
- plugin short name (AAX only)

- plugin type: synth or FX, and variations within these depending on API
- plugin developer's company name (vendor)
- vendor email address and website URL

These descriptions are set with simple strings or flags; there isn't much more to discuss. If you use ASPiK, you will define this information in a text file—and it will take you all of about 30 seconds to do so. There are some four-character codes that need to be set for AU, AAX, and VST3, which date back to as early as VST1. The codes are made of four ASCII characters. Across the APIs, there are two of these four-character codes, plus one AAX-specific version, as follows here.

- product code: must be unique for each plugin your company sells
- vendor code: for our company it is "WILL"

AAX requires another code called the AAX Plugin ID that is a multi-character string, while VST3 requires yet another plugin code. This 128-bit code is called a *globally unique identifier* (GUID, which is pronounced "goo-id"); it is also called a *universally unique identifier* (UUID). Steinberg refers to it as a FUID as it identifies their *FUnknown* object (*FUnknown IDentifier*) but it is identical to the standard GUID. The GUID is generated programmatically with a special piece of software; there are freely available versions for most operating systems. The software aims to create a genuinely unique number based on the moment in time when you request the value from it, in addition to other numbers it gets from your network adaptor or other internal locations. The moment in time is the number of 100-nanosecond intervals that have elapsed since midnight of October 15, 1582. If you use ASPiK, this value is generated for you when you create the project.

2.2.1 The Plugin Description: Features and Options

In addition to these simple strings and flags, there are some more complicated things that the plugin needs to define for the host at load-time. These are listed here and described in the chapters ahead.

- plugin wants a side-chain input
- plugin creates a latency (delay) in the signal processing and needs to inform the host
- plugin creates a reverb or delay "tail" after playback is stopped and needs to inform the host that it wants to include this reverb tail; the plugin sets the tail time for the host
- plugin has a custom GUI that it wants to display for the user
- plugin wants to show a Pro Tools gain reduction meter (AAX)
- plugin factory presets

Lastly, there are some MacOS-specific codes that must be generated called *bundle identifiers* or *bundle IDs*. These are also supposed to be unique codes that you create for each of your plugin designs and are part of the MacOS component handler system. If you have used Xcode, then you are already aware of these values that are usually entered in the compiler itself or in the *Info.plist* file. As with the VST3 GUID, if you use ASPiK, these bundle ID values are generated for you and you'll never need to concern yourself with them.

2.3 Initialization: Defining the Plugin Parameter Interface

Almost all plugins we design today require some kind of GUI (also called the UI). The user interacts with the GUI to change the functionality of the plugin. Each control on the GUI is connected to a plugin *parameter*. The plugin parameters might be the most complicated and important aspects of each API; much of the next few chapters will refer to these parameters. The plugin must declare and describe the parameters to the host during the plugin-loading phase.

There are three fundamental reasons that these parameters need to be exposed:

1. If the user creates a DAW session that includes parameter automation, then the host is going to need to know how to alter these parameters on the plugin object during playback.
2. All APIs allow the plugin to declare "I don't have a GUI" and will instead provide a simple GUI for the user. The appearance of this *default UI* is left to the DAW programmers; usually it is very plain and sparse, just a bunch of sliders or knobs.
3. When the user saves a DAW session, the parameter states are saved; likewise when the user loads a session, the parameter states need to be recalled.

The implementation details are described in the next chapters, but the relationship between the GUI, the plugin parameters, the host, and the plugin are all fundamentally the same across all of the APIs— so we can discuss that relationship in more abstract terms. Remember that the APIs are all basically the same! Let's imagine a simple plugin named *VBT* that has some parameters for the user to adjust:

*   volume
*   bass boost/cut (dB)
*   treble boost/cut (dB)
*   channel I/O: left, right, stereo

Figure 2.2a shows a simple GUI that we've designed for VBT. Each GUI control corresponds to an underlying plugin parameter. The plugin must declare these parameters during the loading phase,

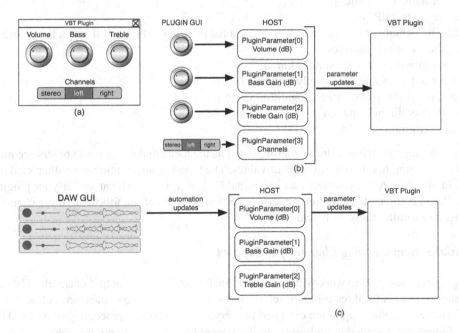

Figure 2.2: (a) A simple plugin GUI. (b) In this GUI, each GUI control connects to an underlying parameter object; the host gathers control changes and they are delivered to the plugin as parameter updates. (c) DAW automation works the same way, except the host is interacting with the parameters rather than the user/GUI (note that the Channel parameter is not automatable).

with one parameter per GUI control. Without exception, these parameters are implemented as C++ objects, and the result of the declaration is a list of these objects. The actual location of this list of parameters depends on the API, but it is generally stored in a plugin base class object, and the DAW has a mechanism to get and set parameter values in the list. This is how it stores and loads the plugin state when the user stores and loads a session that includes it. **In these diagrams I will show this parameter list as being part of the host regardless of where it is actually located.** We'll discuss the mechanisms that the host uses to update the parameters in a thread-safe way in the API programming guide chapters.

Figure 2.2b shows the concept, with each GUI control connecting to a parameter. The parameter updates are sent to the plugin during the processing phase. Figure 2.2c shows the same thing—only it is track automation that is updating the parameters rather than the user. The implementation details are different for each API, but the information being stored about the parameters is fundamentally the same. There are two distinct types of parameters: continuous and string-list. Continuous parameters usually connect to knobs or sliders that can transmit continuous values. This includes *float*, *double*, and *int* data types. String-list parameters present the user with a list of strings to choose from, and the GUI controls may be simple (a drop-down list box) or complex (a rendering of switches with graphic symbols). The list of strings is sometimes provided in a comma-separated list, and other times packed into arrays of string objects, depending on the API. For the Channels parameter in Figure 2.2a, the comma-separated string list would be:

```
"stereo, left, right"
```

The parameter attributes include:

- parameter name ("Volume")
- parameter units ("dB")
- whether the parameter is a numerical value, or a list of string values and if it is a list of string values, the strings themselves
- parameter minimum and maximum limits
- parameter default value for a newly created plugin
- control taper information (linear, logarithmic, etc.)
- other API-specific information
- auxiliary information

Some of the parameter attributes are text strings such as the name and units, while others are numerical such as the minimum, maximum, and default values. The tapering information is another kind of data that is set. In all of the APIs, you create an individual C++ object for each parameter your plugin wants to expose. If you compare the parameter interface sections of each of the following three chapters, you will see just how similar these APIs really are.

2.3.1 Initialization: Defining Channel I/O Support

An audio plugin is designed to work with some combination of input and output channels. The most basic channel support is for three common setups: mono in/mono out, mono in/stereo out, and stereo in/ stereo out. However, some plugins are designed for very specific channel processing (e.g. 7.1 DTS® vs. 7.1 Sony). Others are designed to fold-down multi-channel formats into mono or stereo outputs (e.g. 5.1 surround to stereo fold-down). Each API has a different way of specifying channel I/O support; it may be implemented as part of the plugin description, or the initialization, or a combination of both. You can find the exact mechanisms for each API in the plugin programming guide chapters that follow.

2.3.2 Initialization: Sample Rate Dependency

Almost all of the plugin algorithms we study are sensitive to the current DAW sample rate, and its value is usually part of a set of equations we use to process the audio signal. Many plugins will support just about any sample rate, and its value is treated similarly to plugin parameters—it is just another number in a calculation. A few plugins may be able to process only signals with certain sample rates. In all APIs, there is at least one mechanism that allows the plugin to obtain the current sample rate for use in its calculations. In all cases, the plugin retrieves the sample rate from a function call or an API member variable. This is usually very straightforward and simple to implement.

2.4 Processing: Preparing for Audio Streaming

The plugin must prepare itself for each audio streaming session. This usually involves resetting the algorithm, clearing old data out of memory structures, resetting buffer indexes and pointers, and preparing the algorithm for a fresh batch of audio. Each API includes a function that is called *prior to audio streaming,* and this is where you will perform these operations. In ASPiK, we call this the *reset( )* operation; if you used RackAFX v1 (RAFX1), the function was called *prepareForPlay( )*—a secret nod to the Korg 1212 I/O product that I designed, whose driver API used a function with the identical name. In each of the programming guide chapters, there is a section dedicated to this operation that explains the function name and how to initialize your algorithm. One thing to note: trying to use transport information from the host (i.e. information about the host's transport state: playing, stopped, looping, etc.) is usually going to be problematic, unless the plugin specifically requires information about the process—for example, a pattern looping algorithm that needs to know how many bars are looping and what the current bar includes. In addition, some hosts are notoriously incorrect when they report their transport states (e.g. Apple Logic) so they may report that audio is streaming when it is not.

2.4.1 Processing: Audio Signal Processing (DSP)

Ultimately the effects plugin must process audio information, transforming it in some way. The audio data come either from audio files loaded into a session, or from real-time audio streaming into the audio adaptor. These days, and across all APIs, the audio input and output samples are formatted as floating point data on the range of [−1.0, +1.0] as described in Chapter 1. AU, AAX (native), and RAFX use *float* data types while VST3 allows either *float* or *double* data types, though these will likely change over time. In either case we will perform floating point math and we will encode all of our C++ objects to operate internally with *double* data types. In an interesting—but not unexpected—twist, all of the APIs send and receive multi-channel audio in the exact same manner. From the plugin's point of view, it doesn't matter whether the audio data comes from an audio file or from live input into hardware, or whether the final destination for the data is a file or audio hardware. In Figure 2.3 we'll assume we have live audio streaming into the DAW.

On the left of Figure 2.3 is the Hardware Layer with the physical audio adaptor. It is connected to the audio driver via the hardware bus. The driver resides in the Hardware Abstraction Layer (HAL). Regardless of how the audio arrives from the adaptor, the driver formats it according to its OS-defined specification and that is usually in an interleaved format with one sample from each channel in succession. For stereo data, the first sample is from the left channel and then it alternates as left, right, left, right, etc. Note that this is the same way data are encoded in most audio files including the *.wav* format. So, if the audio originated from a file, it would also be delivered as interleaved data. The driver

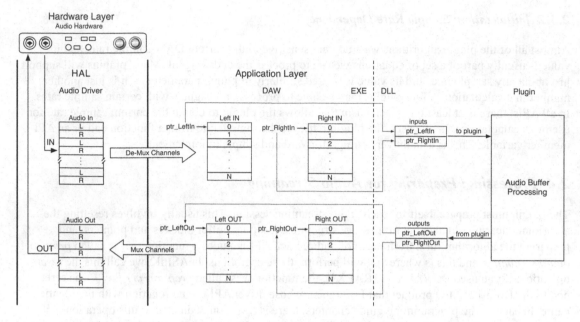

Figure 2.3: The complete audio path from hardware to plugin and back again.

delivers the interleaved data to the application in chunks called *buffers*. The size of the buffers is set in the DAW, the driver preferences, or by the OS.

The DAW then takes this data and de-interleaves it, placing each channel in its own buffer called the *input channel buffer*. The DAW also prepares (or reserves) buffers that the plugin will write the processed audio data into called the *output channel buffers*. The DAW then delivers the audio data to the plugin by sending it a pointer to each of the buffers. The DAW also sends the lengths of the buffers along with channel count information. For our stereo example here, there are four buffers: left input, right input, left output, and right output. The two input buffer pointers are delivered to the plugin in a two-slot array and the output buffers are delivered the same way, in their own two-slot array. For a specialized fold-down plugin that converts 5.1-surround audio into a stereo mix, there would be six pointers for the six input channels and two pointers for the pair of output channels. It should be obvious that sending and receiving the buffer data using pointers to the buffers along with buffer sample counts is massively more efficient than trying to copy the data into and out of the plugin, or coming up with some difficult-to-implement shared memory scheme. When the plugin returns the processed audio by returning the pointers to the output buffers, the DAW then re-interleaves the data and ships it off to the audio driver, or alternatively writes it into an audio file. This gives us some stuff to think about:

- The plugin receives, processes and outputs buffers of data.
- Each channel arrives in its own buffer.
- Each buffer is transmitted using a single pointer that points to the first address location of the buffer.
- The length of each buffer must also be transmitted.
- The number of channels and their configuration must also be transmitted.

With our audio data packaged as buffers, we have an important decision to make about processing the buffers. We can process each buffer of data independently, which we call *buffer processing,* or we can process one sample from each buffer at a time, called *frame processing.* Buffer processing is fine for multi-channel plugins whose channels may be processed independently—for example, an equalizer. But many algorithms require that all of the information to each channel is known at the time when each channel is processed. Examples would include a stereo-to-mono fold-down plugin, a ping-pong delay, many reverb algorithms, and a stereo-linked dynamics processor. Since frame processing is a subset of buffer processing and is the most generic, we will be doing all of the book processing in frames.

2.5 Mixing Parameter Changes With Audio Processing

Our plugin needs to deal with the parameter changes that occur in parallel with the audio processing and this is where things become tricky. The incoming parameter changes may originate with the user's GUI interaction, or they may occur as a result of automation. Either way we have an interesting issue to deal with: there are two distinct activities happening during our plugin operation. In one case, the DAW is sending audio data to the plugin for processing, and in the other the user is interacting with the GUI—or running automation, or doing both—and this requires the plugin to periodically re-configure its internal operation and algorithm parameters. The user does not really experience the plugin as two activities: they simply listen to the audio and adjust the GUI as needed. For the user, this is all one unified operation. We may also have output parameters to send back to the GUI; this usually takes the form of volume unit (VU) metering. We might also have a custom graph view that requires us to send it information. All of these things come together during the buffer processing cycle. We've already established that in all APIs the audio is processed in buffers, and the mechanism is a function call into your plugin. For AU and VST the function call has a predefined name, and in AAX you can name the function as you wish. So in all cases, the entire buffer processing cycle happens within a single function call. We can break the buffer processing cycle into three phases of operation.

1. **Pre-processing**: transfer GUI/automation parameter changes into the plugin and update the plugin's internal variables as needed.
2. **Processing**: perform the DSP operations on the audio using the newly adjusted internal variables.
3. **Post-processing**: send parameter information back to the GUI; for plugins that support MIDI output, this is where they would generate these MIDI messages as well.

2.5.1 Plugin Variables and Plugin Parameters

It is important to understand the difference between the plugin's internal variables and the plugin parameters. Each plugin parameter holds a value that relates to its current state—that value may be originating with a user–GUI interaction or automation, or it may be an output value that represents the visual state of an audio VU meter. The parameter object stores this value in an internal variable. However, the parameter's variable is not what the plugin will use during processing, and in some cases (AU), the plugin can never really access that variable directly. Instead, the plugin defines its own set of internal variables that it uses during the buffer processing cycle, and *copies* information from the parameters into them. Let's think about a volume, bass and treble control plugin. The user adjusts the volume with a GUI control that is marked in dB. That value in dB is delivered to the plugin at the start of the buffer processing cycle. The plugin needs to convert the value in dB to a scalar multiplier coefficient, and then it needs to multiply the incoming audio samples by this coefficient. There are a few strategies for handling this process. We are going to adhere to a simple paradigm for all of our plugin projects depicted in Figure 2.4. Notice that we name the input parameters as inbound and the output parameters as outbound.

- Each GUI control or plugin parameter (inbound or outbound) will ultimately connect to a member variable on the plugin object.
- During the pre-processing phase, the plugin transfers information from each of the inbound parameters into each of its corresponding member variables; if the plugin needs to further process the parameter values (e.g. convert dB to a scalar multiplier) then that is done here as well.
- During the processing phase, the DSP algorithm uses the plugin's internal variables for calculations; for output parameters like metering data, the plugin assembles that during this phase as well.
- During the post-processing phase, the plugin transfers output information from its internal variables into the associated outbound parameters.

Figure 2.4 shows how the DAW and plugin are connected. The audio is delivered to the plugin with a buffer processing function call (*processAudioBuffers*). During the buffer processing cycle, input parameters are read, audio processing happens, and then output parameters are written. The outbound audio is sent to the speakers or an audio file and the outbound parameters are sent to the GUI.

Concentrating on the pre-processing phase, consider two different ways of handling of handling the volume control on our plugin. In one naïve case, we could let the user adjust the volume with a GUI control that transmits a value between 0.0 and 1.0. This makes for a poor volume control because we hear audio amplitude logarithmically. Nevertheless, consider that arrangement of parameter and plugin variables. The GUI would transmit a value between 0.0 and 1.0. We would transfer that data into an

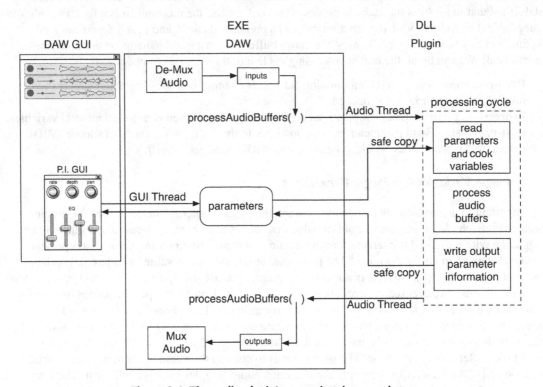

Figure 2.4: The audio plugin's operational connections.

internal variable named *volume*, then we would multiply the incoming audio data by the *volume* value. This is shown in Figure 2.5a. The raw GUI value is used directly in the calculation of the audio output.

In Figure 2.5b, the improved plugin transfers values in dB on a range of −60.0 to 0.0 into the plugin, which then cooks the raw data into a new variable also named *volume* that the plugin can use in its calculations. In this case, the cooking operation requires only one line of code to convert the dB value into a scalar multiplier. Figure 2.5c shows a panning control that transmits a value between −1.0 (hard left) and +1.0 (hard right). The plugin must convert this information into two scalar multipliers, one for each channel. Since this operation is more complicated, we can create cooking functions to do the work and produce nicer code. For almost all of our plugin parameters, we will need to cook the data coming from the GUI before using it in our calculations. To keep things as simple and consistent across plugins, we will adhere to the following guidelines:

- Each inbound plugin parameter value will be transferred into an internal plugin member variable that corresponds directly to its raw data; e.g. a dB-based parameter value is transferred into a dB-based member variable.
- If the plugin requires cooking the data, it may do so but it will need to create a second variable to store the cooked version.
- Each outbound plugin parameter will likewise be connected to an internal variable.
- All VU meter data must be on the range of 0.0 (meter off) to 1.0 (meter fully on) so any conversion to that range must be done before transferring the plugin variable to the outbound parameter.

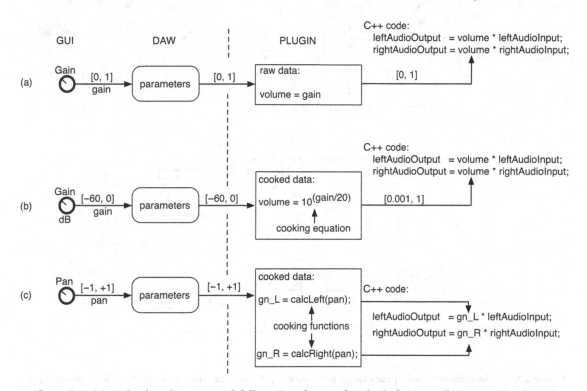

Figure 2.5: (a) A simple volume control delivers raw data to the plugin for immediate use. (b) A better version delivers raw dB values to the plugin, which it must first cook before using. (c) A panning control requires more complex cooking.

If you want to combine the first two guidelines as a single conceptual block in your own plugin versions, feel free to do so. Our experience has been that when learning, consistency can be very helpful so we will adhere to these policies throughout. Another reason to keep raw and cooked variables separate involves parameter smoothing.

2.5.2 Parameter Smoothing

You can see in Figure 2.6a that the plugin parameters are transferred into the plugin only at the beginning of the buffer processing cycle. If the user makes drastic adjustments on the GUI control, then the parameter changes arriving at the plugin may be coarsely quantized, which can cause clicks while the GUI control is being altered (which is sometimes called *zipper noise)*. If the user has adjusted the buffer size to be relatively large, then this problem is only exacerbated, as the incoming parameter control changes will be even more spread out in time. To alleviate this issue, we may employ parameter smoothing. It is important to understand that if this smoothing is to occur, it must

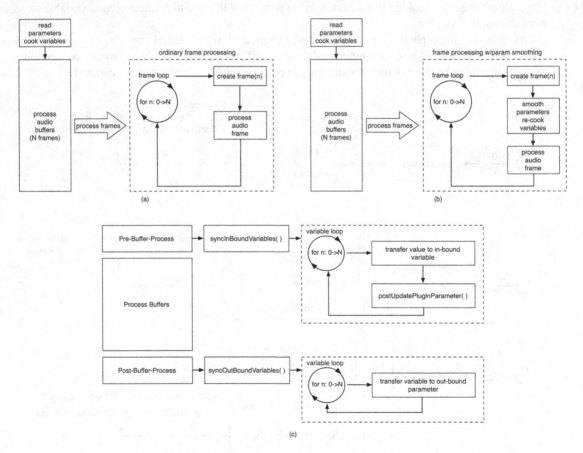

Figure 2.6: (a) In ordinary frame processing, parameter updates affect the whole buffer. (b) In processing with parameter smoothing, each frame uses a new and slightly different parameter value. (c) The complete buffer processing cycle includes the transfers.

be done on the plugin side, not in the GUI or DAW. With parameter smoothing, we use interpolation to gradually adjust parameter changes so that they do not jump around so quickly. Our parameter smoothing operations will default to do a new parameter smoothing update on each sample period within the buffer processing cycle. This means that the smoothed internal plugin variable will be updated during every *frame* of processing, as shown in Figure 2.6b. The good news is that this will eliminate clicks or zipper noise. The bad news is that if the parameter change requires a complex cooking equation or function, we can burn up a lot of CPU cycles. The parameter smoothing object we use may also be programmed to perform the smoothing operations only during every other *N* sample periods (called *granularity*)—but of course the more granularity we add, the more likely the control changes will produce clicks and noise. We'll discuss the parameter smoothing object in more detail in Chapter 6.

Since parameter smoothing will incur some CPU expense, you should use it only when absolutely necessary. For a GUI control that transmits a discrete value—for example, a three-position switch that transmits a value of 0, 1, or 2 corresponding to each position—there is usually no need to try to smooth this value to slow down the movement between positions. We will smooth only those parameters that are linked to continuous (*float* or *double*) plugin variables.

2.5.3 Pre and Post-Processing Updates

Our ASPiK kernel object (or your chosen plugin management framework) should provide a mechanism for performing the operations of transferring parameter data into and out of the plugin. To break this operation up further, we will adhere to another convention that separates the transfer of parameter information from the cooking functions that may be required; this will produce code that is easier to read and understand, and that is more compartmentalized. In Figure 2.6c you can see that the pre-processing block calls a function we will call *syncInboundVariables*. During this function call, each parameter's value is transferred into an associated plugin variable. After each transfer, the function *postUpdatePluginParameter* will be called. During this function call, the plugin can then cook incoming parameter information as it needs to. During the post-processing phase, the function *syncOutboundVariables* is called: this copies the outbound plugin variables into their corresponding plugin parameters. There is no need for a post-cooking function in this block as the plugin can format the information as needed easily during the processing phase.

2.5.4 VST3 Sample Accurate Updates

The VST3 specification differs from the others in that it defines an optional parameter smoothing capability that may be implemented when the user is running automation in the DAW session. This involves only automation and is entirely separate from the normal user–GUI interactions that produce the same per buffer parameter updates as the other APIs. In the case of automation, the VST3 host may deliver intermediate values in addition to the last parameter update received. The exact mechanism is covered in Chapter 3; it is not a perfect solution to the parameter smoothing problem with automation, but it is definitely a step in the right direction. ASPiK is already designed to handle these parameter updates with a feature that may be optionally enabled. As with normal parameter smoothing, care must be taken when the updated parameter values must be cooked with a CPU intensive function. The same issues will exist as with normal parameter smoothing.

At this point, you should have a nice big-picture view of how the plugin operates at a functional block level. You can see how the plugin parameter updates and the audio signal processing are interleaved such that each buffer is processed with one set of parameter values, smoothed or not. That interleaving is a key part of these plugin architectures that transcends any individual API. The API-specific chapters that follow will elaborate with the details as they relate to those plugin specifications. Before we move on, we need to discuss the labels in Figure 2.4 that involve the threads and safe copy mechanisms.

2.5.5 Multi-Threaded Software

If you've studied microprocessor design and programming, you've come across something called the *instruction pointer* (IP). At its lowest level, all software runs in a loop of three stages: *fetch, decode,* and *execute.* The program fetches an instruction from memory, decodes it, and then executes the instruction. The IP points to the current instruction that is being fetched. After the instruction executes—and assuming no branching occurred—the IP will be updated to point to the next instruction in the sequence and the three-step cycle repeats. Branching (*if/then* statements) will alter the IP to jump to a new location in code. Function calls do the same thing and alter the IP accordingly. We say that the IP "moves through instructions" in the program. There was a time when you could write a piece of software with only one IP; most of us wrote our first "Hello World" program like that. An instruction pointer that is moving through a program from one instruction to the next is called a *thread* or a *thread of execution.* This is shown in Figure 2.7a. Most of us spent our formative programming classes writing single-threaded programs that did very important things like calculating interest on a savings account or listing prime numbers.

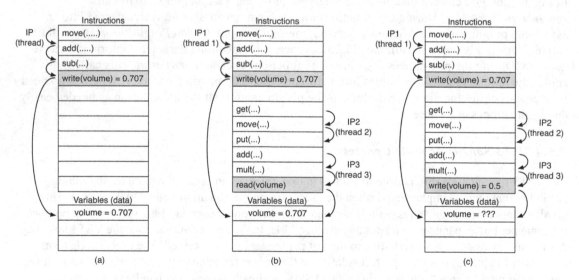

Figure 2.7: (a) A single-threaded application has only one instruction pointer (IP) moving through the instructions and data. (b) A multi-threaded program features multiple IPs accessing the same instructions and data. Here, IP1 tries to write the volume variable while IP3 reads it. (c) In this diagram, a race-condition exists between thread 1 and thread 3 as they compete to write the *volume* variable with different values.

Now think about a computer game that simulates a jet fighter. This is a different kind of program. The graphics are showing us flying through the air: as we launch our missiles and hit targets there is accompanying audio in sync, and when we move the joystick we watch the sky dance around in response. All the while, there is a small radar indicator spinning around to show that the enemy is behind us. Even if we release the controls and do nothing, the airplane still flies, at least for a while.

Now a bunch of things need to appear to happen simultaneously, and the programmers need to write the code accordingly. In this kind of software, there are multiple IPs, each moving through different parts of the code: one IP is running the code that paints the images on the screen, while another IP is looping over the code that plays the music soundtrack and yet another is tracking the motion of the joystick. This is an example of a *multi-threaded* application, shown in Figure 2.7b. Each different instance of an IP moving though code is one of the threads, and this software has multiple IPs. The OS is involved in creating the illusion that all of this is happening at once. On a single-CPU system, each thread of execution is allowed a sliver of processing time; the OS doles out the CPU time-slices in a round-robin fashion, moving from one thread to the next. It certainly looks like everything is happening at once, but in reality it is all chopped up into processing slices so short in duration that your brain can't discern one from the other—it all merges into the beautiful illusion that you really are flying that jet. Even with multi-CPU systems, most programs still execute like this, with one slice at a time per thread; adding more CPUs just adds more time-slices.

Now think about our audio plugin. The two things that need to appear to happen at once—the audio processing and the user interaction with the GUI—are actually happening on two different threads that are being time-shared with the CPU. This means that our plugin is necessarily a multi-threaded piece of software. It has at least two threads of execution attacking it. Each thread can declare its priority in attempt to try to get more (or less) CPU time. The highest priority threads get the most CPU time, and the lowest priority threads get the least. This also requires that the threads run asynchronously. The OS is allowed to adjust the CPU thread-sliver time as it needs to: for example, if the system becomes bogged down with your movie download, it may decide to down-prioritize the already low-priority GUI thread messages so that your plugin GUI becomes sluggish. But the high-priority audio is not affected and audio arrives glitch-free at the driver.

The priority of each thread is really not the issue we need to deal with, though we do need to keep it in mind. The larger problem we need to address is the fact that the two threads need to access some of the same data, in particular the plugin parameters. The GUI needs to alter the parameters and the audio processing thread needs to access them as well to update its processing accordingly. The fact that the two threads running asynchronously need to access the same information is at the core of the multi-threaded software problem. Now think about the threads in that flight simulator: how much data do they need to share to give the user the proper experience? Figure 2.7b shows this issue in which thread 1 and thread 3 both attempt to access the volume variable in the data section of the program.

The problem with sharing the data doesn't seem that bad on first glance: so what if the two threads share data? The CPU lets only one thread access the data at a time, so isn't it already handing the thread synchronization process for us? Isn't it keeping the shared data valid? The answer is no. The problem is that the data themselves are made of multiple bytes. For example, a *float* or *int* is actually 32 bits in length, which is four bytes. A *double* is 64 bits in length or eight bytes. Threads read and write data in bytes, one at a time. Suppose a GUI thread is writing data to a *double* variable, and it has written the first three bytes of the eight that are required when the OS gives it the signal that it must halt and relinquish control to the next thread. For ordinary variables, the thread will stop immediately and remember where it was, then pass execution to the next thread. When our GUI thread gets a new

CPU time-slice, it finishes writing the remaining five bytes of data to the variable and moves on. But what if our audio processing thread that was next in line for CPU time tried to *read* the partially written *double* variable? It would be accessing corrupt data, which is one type of multi-threading issue: inconsistent data between threads. Now suppose that the audio processing thread tried to *write* data to that *double* variable and succeeded in overwriting the three new bytes and five old existing bytes. Later, the GUI thread regains control and finishes writing the rest of the data with the remaining five bytes it believes need to be written. Once again we are left with a corrupt value in the variable. When two threads compete to write data to the same variable, they are said to be in a race condition: each is racing against the other to write the data and in the end the data may still be corrupted. Race conditions and inconsistent data are our fundamental problems here.

The good news is that each of the APIs has its own internal mechanism to ensure thread safety in the plugin. The reason we spend time discussing this is that these mechanisms not only affect how each API is designed but also trickle down into how our plugin code must be written. We'll discuss each API's multi-threading tactics in more detail in the following chapters to give you a deeper understanding and some more insight into the APIs. However, rest assured that the ASPiK is 100% thread-safe in its operation within the various APIs—you will not need to concern yourself with these issues directly when programming.

2.6 Monolithic Plugin Objects

The plugin itself is compiled as a DLL but it implements the plugin as a C++ object or set of objects. The chores of these objects include the following:

- describing the plugin for the host
- initializing the parameter list
- retrieving parameter updates from the user/DAW
- cooking parameter updates into meaningful variables for the DSP algorithm
- processing audio information with the DSP algorithm
- rendering the GUI and processing input from the user, then transferring that information safely into the parameter list

In all APIs, the very last item in the list—rendering the GUI—is preferably done in a separate and independent C++ object that is specifically designed to implement the GUI and nothing else. Trying to combine that GUI and platform-specific code into the same object that handles the other plugin duties is going to be problematic at best and fraught with basic object-oriented design errors at worst. So we will set that C++ object aside for now and look at the other items.

You can roughly divide the remaining duties into two parts, one that handles parameters and another that handles audio processing. The VST3 and AAX specifications both allow two distinct paradigms for creating the underlying C++ objects that encapsulate the plugin. In one version, the duties are split into these two components: processing audio and handling parameters. In the other paradigm, a single C++ object handles both of the chores. In the AAX SDK, this pattern is called the "monolithic plugin object" paradigm. For consistency, I will adhere to this monolithic object archetype throughout our plugin projects and within ASPiK. If you want to later split the code up along the parameter/processing boundary lines, feel free to do so—it is not overly difficult. For example, the monolithic plugin object in VST3 is actually just a C++ object that is derived from both the parameter and audio processing objects, inheriting both sets of functions. However, in no case will our GUI rendering object ever be merged with the monolithic processing object: it will always remain as an independent component and our monolithic processing object will neither require nor know of the GUI's existence. Likewise, our GUI rendering object has no knowledge that there is an associated processing object that is using its information. The API-specific thread-safe mechanisms that ASPiK implements ensure a proper and legal relationship between the GUI and plugin object itself—these will be revealed in Chapter 6.

2.7 Bibliography

Apple Computers, Inc. 2018. *The Audio Unit Programming Guide*, https://developer.apple.com/library/archive/documentation/MusicAudio/Conceptual/AudioUnitProgrammingGuide/Introduction/Introduction.html, Accessed August 1, 2018.

Avid Inc. *AAX SDK Documentation*, file://AAX_SDK/docs.html, Available only as part of the AAX SDK.

Bargen, B. and Donnelly, P. 1998. *Inside DirectX*, Chap. 1. Redmond: Microsoft Press.

Coulter, D. 2000. *Digital Audio Processing*, Chap. 7–8. Lawrence: R&D Books.

Petzold, C. 1999. *Programming Windows*, Chap. 21. Redmond: Microsoft Press.

Richter, J. 1995. *Advanced Windows*, Chap. 2, 11. Redmond: Microsoft Press.

Rogerson, D. 1997. *Inside COM*, Chap. 1–2. Redmond: Microsoft Press.

Steinberg.net. *The Steinberg VST API*, www.steinberg.net/en/company/developers.html, Accessed August 1, 2018.

CHAPTER 3

VST3 Programming Guide

VST3 is the current version of Steinberg's mature VST plugin specification. The venerable VST2 API, which has arguably produced more commercial and free plugins than any other API in the history of audio plugins, was officially made obsolete in October 2018 with its removal from the VST SDK and the inability for new plugin authors to legally write for it, though there is a grandfather clause for older plugin developers who have already signed license agreements with Steinberg. VST3 was redesigned from the ground up and has almost nothing in common with VST2 except some naming conventions and other very minor details. VST3 specifically addressed some of the shortcomings of the VST2 API, including a somewhat vague side-chain specification and a messaging system. VST3 is also designed with upgrading and product extension in mind, and the developers add new features with every release. Finally, due to the fact that the VST2 specification has been gracefully laid to rest, I will use the terms "VST3" and "VST" interchangeably in this text. You may download the VST3 SDK for free from Steinberg (www.steinberg.net/en/company/developers.html) and you are encouraged to join the developer's forum, where you may post questions.

3.1 Setting Up the VST3 SDK

The VST SDK is simply a collection of subfolders in a specific hierarchy. You can download the SDK from Steinberg and unzip it to your hard drive. The VST SDK is already set up to work with CMake (more on that later), so you will notice that the folder names have no whitespaces. The root folder is named *VST_SDK*. As of SDK 3.6.10, the root folder contains two subfolders, one for VST2 and another for VST3. Steinberg has made the VST2 plugin API obsolete and is no longer developing or supporting it. The VST2 folder contains the old VST2 API files; these are used in a VST3-to-VST2 wrapper object that is currently included in the SDK, and is fairly simple to add to projects. Unless you have a previous license to sell VST2 plugins, you are not allowed to sell new VST2 plugins. I'm not sure if future SDKs will contain the VST2 portion or not, or if the VST2 wrapper will be around much longer.

3.1.1 VST3 Sample Projects

With the SDK planted on your hard drive, you need to run CMake to extract the sample projects. Instructions are included in the SDK documentation and a video tutorial at www.willpirkle.com. The VST3 SDK comes with a rich set of sample plugin projects that will generate more than a dozen example plugins in all shapes and sizes. Install the SDK on your system, then use the CMake software to generate a single Xcode or Visual Studio compiler project that contains every one of the sample plugins inside of it. You can build all of the plugins at one time with one build operation. If your compiler and SDK are set up correctly, you will get an error-free build and have a bunch of VST3 plugins to play with. You should verify that your compiler will properly build the SDK sample projects before trying to write your own plugins.

3.1.2 VST3 Documentation

The VST3 API documentation is contained within the SDK and is generated using Doxygen®, which creates HTML-based documents that you view with a web browser. You can open the documentation by simply double-clicking on the *index.html* file, which is the Doxygen homepage. This file is located close to the SDK root at *VST_SDK/VST3_SDK/index.html*; you should definitely bookmark it and refer to it frequently. The VST3 SDK documentation improves greatly with each new SDK release. It contains a wealth of information as well as documentation on all structures, classes, functions, and VST3 programming paradigms.

3.2 VST3 Architecture and Anatomy

VST plugins may be written for Windows, MacOS, iOS, and Linux. This means you must use the proper compiler and OS to target these products—you cannot use Xcode to write a Windows VST, for example. This book targets Windows and MacOS specifically, although the VST API is written to be platform independent. If you are interested in writing for iOS or Linux, I suggest you get the Windows and MacOS versions working first: there is much more information and help available for these. The latest version of the VST3 SDK packages both Windows and MacOS plugins in a *bundle*, which you may think of as simply a directory of subfolders and files. The bundle is set up to appear as a file for a typical user, though in reality it is a folder—in MacOS you may right-click on the plugin "file" and choose "Show Package Contents" to get inside the bundle. The VST3 SDK also includes VSTGUI4 for generating the custom GUIs, though you may use whatever GUI development library you like; VSTGUI4 was designed to interface directly with VST3, and ASPiK uses VSTGUI4 for its GUI library operations.

At its core, VST3's Module Architecture (VST-MA) is based on (but not identical to) Microsoft's Common Object Model (COM) programming paradigm. This allows the VST3 API to be easily updated and modified, without breaking older versions. COM is really just a "way" or approach to writing software and is technically independent of any programming language, though VST-MA supports only C++ as of this writing.

COM programming revolves around the concept of an interface, which we also use in our FX objects that accompany this book (*IAudioSignalProcessor* and *IAudioSignalGenerator*). For the VST-MA, an *interface* is a C++ object that defines only virtual functions so that it represents an abstract base class. There are no member variables, and the classes that inherit an interface must override any pure abstract function it defines. Some, but not all, of the VST-MA interfaces are pure abstract, so that all inherited functions are implemented on the derived class. There is some resemblance to the *protocol* in Objective C programming where a class "conforms to a protocol" by implementing its required methods. The interface concept is powerful as it allows objects to hold pointers to each other in a safe manner: if object A holds an interface pointer to object B (i.e. a pointer to B's *interface* base class) object A may call only those functions that interface defines. Being abstract, the interface has no constructor or destructor, so object A cannot call the *delete* operator on the interface pointer, and trying to do so at compile time produces an error. With multiple inheritance, an object may implement many interfaces (or conform to their protocols) so that it may safely be accessed via one of these base class pointers.

In addition, the COM paradigm provides a device that allows for backwards compatibility when interfacing with older objects that were compiled with earlier SDKs. It consists of a querying mechanism so that one object may safely query another object to ask it if it supports an interface of some kind. If that object replies that it does, the first object may then ask for an interface pointer and use its methods. This means that when new features are added to the VST3 API, they may be implemented as new interface class definitions. A host may query the plugin to see if it has one of these new interfaces. If it does not, then the host does not attempt to use the new features. Or, the host may work its way "backwards" querying the object to see if has an older interface that supports at least part of the features it needs. If it does and the host acquires an interface pointer for it, then the host may call only those older functions that are guaranteed to exist, preventing crashes and providing a robust mechanism for safely updating the API.

3.2.1 Single vs. Dual Component Architectures

There are two paradigms for writing VST3 plugins, which involve how the two primary plugin duties are distributed: audio signal processing and GUI implementation. In the *dual component* pattern, you implement the plugin in two C++ objects: one for the audio processing (called the *Processor* object) and the other for the GUI interface (called the *Controller* object). The VST3 architects went to great lengths to isolate these two objects from one another, making robust communication between the two nearly impossible. This is not meant to be a criticism, as it closely follows the C++ object-oriented design paradigm and it allows the audio processing and GUI connection/implementation to run on separate CPUs. For multi-CPU platforms, which use farm cards full of CPUs, this makes sense. However, if you are trying to work with monolithic programming objects for implementing plugins, then your VST3 versions may be very different than your AU and AAX versions. In addition, there are some plugin architectures whose design patterns prohibit them from being distributed in two objects running on two separate CPUs, or whose implementations become overly complicated with this paradigm.

To address this, the VST3 API also allows for a single component version in which one monolithic C++ object handles both the processing and GUI interfacing chores. This is called the *single component* paradigm. The architects simply created a single C++ object with multiple inheritance from the processor and controller objects. Figure 3.1 shows the two different design patterns. Please note

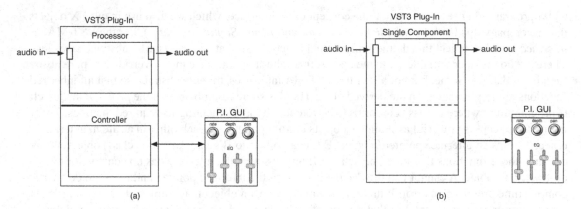

Figure 3.1: (a) The dual component VST3 plugin uses two C++ objects; the thick black bar across the center represents the hard communication barrier between the two components. (b) In contrast, the single component version uses one C++ object; the thin dotted line denotes a less firm barrier (which is really up to the designer).

that, as with AAX and AU, the controller portion deals with communication with the GUI, but does not handle the implementation details—the part that actually renders the knobs, buttons, switches, and views, which is written in a separate GUI object.

3.2.2 VST3 Base Classes

In the dual component version, you create two C++ objects that are derived from two separate base classes: *AudioEffect* (for the processor) and *EditController* (for the controller). In the single component version, which we use exclusively in this book and ASPiK, your plugin is derived from *SingleComponentEffect*, which essentially inherits from the dual components. Notice that unlike AU, the same base classes (listed here) are used regardless of the type of plugin, FX or synth.

- *AudioEffect*: the dual component processor base class
- *EditController*: the dual component controller base class
- *SingleComponentEffect*: the combined single object base class

3.2.3 MacOS Bundle ID

Each bundle in MacOS is identified with a unique string called a *bundle ID*. Apple has recommendations on naming this string, as well as registering it with Apple when you are ready to sell your product commercially. Typically, you embed your company and product names into the bundle ID. The only major restriction is that you are not supposed to use a hyphen ("-") to connect string components, instead you should use a period. For example, *mycompany.vst.golden-flanger* is not legal, whereas *mycompany.vst.golden.flanger* is OK. This is set in the Xcode project's *Info* panel, which really just displays the information inside of the underlying *info.plist* file (see Figure 3.2). Consult the Apple documentation if you have any questions. If you use ASPiK to generate your VST projects, then the bundle ID value is created for you and set in the compiler project in the *Info.plist Preprocessor Definitions*; you can view or change this value in the project's *Build Settings* panel.

3.2.4 VST3 Programming Notes

The VST3 API is fairly straightforward as far as C++ implementation goes. However, the VST3 designers tend to stay close to the forefront of programming technology, and the current version of the SDK will require a compiler that is C++11 compliant. VSTGUI4 also requires C++11, and for MacOS/Xcode, it also requires C++14. So be sure to check the release notes of your SDK to ensure that your compiler will work properly. If you compile the sample projects and immediately get a zillion errors, check to make sure you are compliant. Note that the MacOS versions require that several compiler frameworks are linked with the AU project, including *CoreFoundation, QuartzCore, Accellerate, OpenGL,* and *Cocoa*®; the latter two frameworks are for VSTGUI4. However, the compiler projects are actually very complex—more so than AU and AAX—due in part to the fact that the plugins must link to several static libraries that are compiled along with the plugin and are part of the complier solution. In earlier versions of the VST3 API you precompiled your base library (the same as the current version of AAX), but later versions merged these libraries into the compiler project. Figure 3.2 shows what a stock VST3 plugin project looks like in Xcode. This plugin uses VSTGUI for rendering the GUI and has additional libraries for that support, but the *base* and *sdk* libraries are required. The Visual Studio solution view is similar, showing independent sub-projects that represent the libraries and validator.

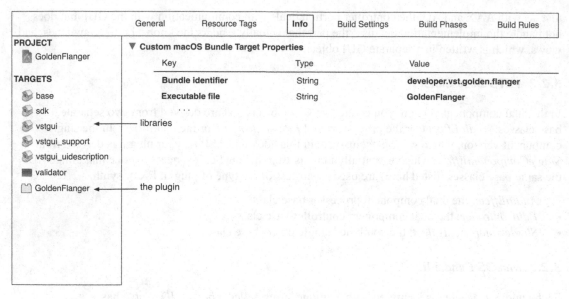

Figure 3.2: A standard VST3 project in Xcode includes multiple libraries, a validator executable, and the plugin target.

As with AAX and AU, there are numerous C++ *typedefs* that rename standard data types, so you should always right-click and ask the compiler to take you to definitions that you don't recognize. Most of the definitions are simple and are supposed to make the code easier to read, though sometimes they can have the opposite effect. In particular, you will see a typedef for *PLUGIN_API* decorating many VST3 functions, which simply renames the *__stdcall* function specifier. This specifier has to do with how the stack is cleaned up after function calls. You can safely ignore it if you don't know what it means (it simply keeps consistency across platforms).

Most of the VST3 plugin functions return a success/failure code that is a 32-bit unsigned integer typedef'd as *tresult*. The success value is *kResultTrue* or *kResultOK* (they are identical) while failure codes include *kResultFalse*, *kInvalidArgument*, *kNotImplemented* and a few more defined in *funknown.h*.

3.2.5 VST3 and the GUID

VST3 uses GUIDs as a method of uniquely identifying the plugin (for the DAW) as well as uniquely identifying each of the processor and editor halves when using the dual component paradigm (perhaps as a way to distribute the components across CPUs). The single component version uses just one GUID. If you use ASPiK, then CMake will generate the GUID for you as part of the project creation; if you do not use ASPiK, then consult your framework documentation. If you are making your own plugin from scratch, then you need to generate this value yourself. On Windows, you may use the built-in GUID generator that is installed with Visual Studio called *guidgen.exe*; on MacOS, you may download one of numerous GUID generator apps for free. You can see the GUID in the class definition codes provided in Section 3.3.

3.2.6 VST3 Plugin Class Factory

VST3 uses the class factory approach for generating multiple instances of a plugin during a typical DAW session. There are a couple of ramifications to this approach. First, you need to be careful about using singletons or other anti-patterns, as they probably won't work properly with the class factory. We don't use them with our ASPiK plugin shells and neither should you (please, no vicious emails). Second, the construction and destruction of your VST3 plugin's dynamically allocated resources do not happen in the C++ object's constructor and destructor as they do with the other APIs. Instead, there are two extra functions that you use for these purposes—*initialize( )* and *terminate( )*—whose roles should be obvious by name. So, you just leave the constructor and destructor of the *SingleComponentEffect* alone. For the dual component paradigm, these functions are part of the processor object.

3.3 Description: Plugin Description Strings

Your plugin description strings, along with the GUID, are part of this class factory definition. There are two macros used to define the class factory; your description information is embedded in them. The class factory macro named *DEF_CLASS2* (for one of the SDK sample projects, *AGainSimple*) follows. It contains the following description strings, which are outlined in Section 2.2 except the FUID, which is unique to VST3: vendor name, vendor URL, vendor email, GUID, plugin name, and plugin type code. These are shown in bold in that order.

```
BEGIN_FACTORY_DEF ("Steinberg Media Technologies",
                   "http://www.steinberg.net",
                   "mailto:info@steinberg.de")

DEF_CLASS2 (INLINE_UID (0xB9F9ADE1, 0xCD9C4B6D, 0xA57E61E3,
                        0x123535FD),
            PClassInfo::kManyInstances,
            kVstAudioEffectClass,
            "AGainSimple VST3",
            0,                      // 0 = single component effect
            Vst::PlugType::kFx,
            1.0.0,                  // Plug-in version
            kVstVersionString,
            Steinberg::Vst::AGainSimple::createInstance)
END_FACTORY
```

Notice that the creation function is the last argument—this function serves up new instances of the plugin in a single line of code using the *new* operator. Also notice the single component effect designator (the 0 which has no predefined constant declaration); for the dual component version it is re-defined as *Vst::kDistributable*. The FX plugin type code follows it in the macro's argument sequence as *Vst::PlugType::kFx*; for completeness, the synth version is *Vst::PlugType::kInstrumentSynth*. VST3 also supports one sub-category layer: for example, if your plugin is for reverb, you could alternatively assign its category as *Vst::PlugType::kFxReverb*. In that case the DAW may group your plugin with the other reverb plugins the user has installed.

3.4 Description: Plugin Options/Features

The plugin options include audio I/O, MIDI, side chaining, latency and tail time. For VST plugins, these are all done programmatically either in the plugin initializer, or in other base class functions that you override.

3.4.1 Side Chain Input

VST3 plugins work on the concept of audio busses. For a FX plugin there is one input buss and one output buss, which may have one or more channels each. Note that there does not necessarily need to be the same number of channels for input as output. The side chain input is declared as a separate input buss and it is typically declared as a stereo buss. We will work with the audio channel I/O declarations shortly, but since implementing robust side chaining was an important part of the VST3 specification, we can look at the code now. You add an audio buss using the *addAudioInput* function and you pass it a name string, a channel I/O setup (called a "speaker arrangement"), and an optional buss type and buss info specifier. That buss type is key for identifying the side chain and is specified as *kAux*. To set up a stereo side chain input, you would write:

```
addAudioInput(STR16("AuxInput"), SpeakerArr::kStereo, kAux);
```

We will discuss the speaker arrangements shortly, but the main concept here is the *kAux* designator. The default value (and if none is supplied in the arguments) is *kMain*. There are no other designations at the time of this writing.

3.4.2 Latency

If your plugin includes a latency that adds a necessary and fixed delay to the audio signal, such as the *PhaseVocoder* in Chapter 20 or the look-ahead compressor in Chapter 18, you may signify this for the VST3 plugin host. The host will call the function *getLatencySamples( )* on your plugin object to query it about this feature. If your plugin includes a fixed latency, you would override this function and return the latency value. If your plugin introduces 1024 samples of latency, you would override the function and write:

```
virtual uint32 PLUGIN_API getLatencySamples() override { return 1024;}
```

3.4.3 Tail Time

Reverb and delay plugins often require a tail time, which the host implements by pumping zero-valued audio samples through the plugin after audio playback has stopped, allowing for the reverb tail or delays to repeat and fade away. The host will query your plugin when it is loaded to request information about the tail time. For VST3 you report the tail time in samples rather than seconds. For algorithms that are sample-rate dependent, this means that your tail time in samples will likely change. The function you use to report the time is named *getTailSamples*; it will be called numerous times during the initialization as well as when the user modifies the session in a way that changes the sample rate. To report a tail time of 44,100 samples, or 1 second at a sample rate of 44.1 kHz, you would write this:

```
virtual uint32 PLUGIN_API getTailSamples() override {return 44100;}
```

You may also specify an infinite tail time, which forces the DAW into sending your plugin a stream of 0s forever after the user stops playback by returning the predefined *kInfiniteTail*.

3.4.4 Custom GUI

The controller portion of the VST3 plugin, whether using the dual or single component versions, inherits from an interface that implements the *createView* function. Your derived class should override this function if you are planning on generating a custom GUI. The host will call the function when the user requests to see the GUI: your plugin must create the GUI and return a specific type of interface pointer called *IPlugView*. Regardless of how the GUI is created and implemented, you must return an *IPlugView* pointer back to the host. This means that your GUI must conform to the *IPlugView* protocol and implement the pure virtual functions that it specifies. The ASPiK VST3 plugin shell defines the *IPlugView* object that implements these functions and handles the lifecycle of the GUI, including resizing it for hosts that allow this. It uses the VSTGUI4 *PluginGUI* object defined in Chapter 6.

There are a couple of items to note: first, your plugin object can always implement the *createView* function and simply return 0 or *nullptr*, which signifies to the host that your plugin does not support a custom GUI—this can be valuable when debugging a new plugin if you think your GUI may have some internal fault or you are trying to narrow down your problems. Second, you should never cache (save) the *IPlugView* pointer you return, nor that of the internal GUI object, in an attempt to set up some kind of communication with the GUI—which is a bad idea to begin with. Make sure you use the VST3-approved mechanisms for sending parameter updates to and from the GUI.

3.4.5 Factory Presets and State Save/Load

Factory presets are part of the *IUnitInfo* interface that the *SingleComponentEffect* object inherits. For the dual component architecture, it is part of the controller object. According to the VST3 documentation, *IUnitInfo* describes the internal structure of the plugin:

- The root unit is the component itself.
- The root unit ID has to be 0 (*kRootUnitId*).
- Each unit can reference one program list, and this reference must not change.
- Each unit using a program list references one program of the list.

A program list is referenced with a program list ID value. The root unit (the plugin) does not have a program list. However, you may add sub-units that contain these program lists, which represent your factory presets. In order to support factory presets, you declare at least two units: one for the root, and another for the preset program list. You then set up a string-list parameter—the VST3 object that is normally used to store string-list parameters for the plugin. Since that object holds a list of strings to display for the user, it may be used to store lists of factory preset name strings. When you set up the special string-list parameter object, you supply optional flags—one of which is *ParameterInfo::kIsProgramChange,* which signifies that the string list is holding preset names. So, the order of operations is as follows:

1. Add a root unit that represents the plugin.
2. Add another unit that represents the preset list.
3. Set up a string-list parameter that holds the names of the factory presets.
4. Make sure to supply the *kIsProgramChange* flag for this special parameter.
5. Add the parameter to the plugins list of other parameters (i.e. GUI controls).

In addition to exposing the preset names, there are several other functions you need to override and support to conform to the *IUnitInfo* interface that you need for presets. None of these functions are difficult or tedious, but there are too many to include here.

VST3 is unique in that it requires you to write your own serialization code for loading and storing the plugin state that occurs when the user saves a DAW session or creates their own presets. You must write the code that reads and writes data to a file. You serialize the data to the file on a parameter-by-parameter basis, one parameter after another. This means you need to take care to ensure that you read data from the file in the exact same order as you wrote the data to the file. The fact that you must supply this code has advantages in that you are in charge of saving and retrieving the exact state for the plugin, and you may write other information into that file that is not directly tied to any plugin parameter, allowing robust customization of the data you save.

There are three functions that your plugin must override to support the state serialization and two of them are for reading state information, while the third is for writing it. Two functions are necessary to support the dual component architecture. The two state-read functions, which you typically code identically when using the *SingleComponentEffect*, are:

```
tresult setState(IBStream* fileStream)
```

```
tresult setComponentState(IBStream* fileStream)
```

The state-writing function is:

```
tresult getState(IBStream* state)
```

You are provided with an *IBStream* interface that you use to read and write data to the file. A helper class *IBStreamer* is defined that provides functions for reading and writing the standard data types, such as *float, double, int, unsigned int*, etc. In order to support Linux and pre-Intel Macs, both big-endian and little-endian byte ordering may be specified. For the book projects, which designed are for Windows and Intel-based Macs, we use little-endian byte ordering; if you want to support Linux, see the VST3 sample code for identifying big-endianness. You use the supplied *IBStream* pointer to create the *IBStreamer* object, and then use the necessary helper function to read and write data. Remember that you usually bind plugin member variables to GUI parameters (Chapter 2). Those bound variables are the ones that you are reading and writing. For example, to write a *double* variable named *pluginVolume* that is bound to the volume parameter, you would use the *IBStreamer::writeDouble* function:

```
tresult PLUGIN_API getState(IBStream* fileStream)
{
    // --- get a stream I/F
    IBStreamer strIF(fileStream, kLittleEndian);
    // --- write the data

    if(!strIF.writeDouble(pluginVolume))
        return kResultFalse;

    return kResultTrue;
}
```

To read that variable in either of the two read-functions, you would use the *readDouble* function, which returns the data as the argument:

```
tresult PLUGIN_API setState(IBStream* fileStream)
{
    // --- get a stream I/F
    IBStreamer strIF(fileStream, kLittleEndian);

    // --- read the data
    if(!strIF.readDouble(pluginVolume))
        return kResultFalse;

    return kResultTrue;
}
```

Have a look at the *fstreamer.h* file and you will find dozens of functions for writing all sorts of data types, including arrays of values as well as individual ones. You can also find functions that allow you to read and write raw blocks of data to serialize custom information that is not part of the plugin parameter system. While being required to write your own serialization functions for even the basic parameter states may be a drag, you can also enjoy the robust functions provided for serializing custom information as well.

3.4.6 VST3 Support for 64-bit Audio

VST3 plugins allow you to support both 32- and 64-bit audio data. To keep consistency between all of the APIs, the ASPiK book projects support only 32-bit audio data. If you want to add your own support for 64-bit audio, then it is fairly straightforward: you tell the host that you support these data, then during the audio processing you write code that may operate on either 32- or 64-bit data. This is accomplished most easily using function or class templates. The host will query your plugin to ask for the supported audio data types in a function called *canProcessSampleSize* and it will pass in a flag that specified the data type of the inquiry, either *kSample32* or *kSample64*. For our ASPiK book projects that support 32-bit audio, we write:

```
tresult PLUGIN_API canProcessSampleSize(int32 symbolicSampleSize)
{
    // -- we support 32 bit audio
    if (symbolicSampleSize == kSample32)
        return kResultTrue;

    return kResultFalse;
}
```

During the buffer processing function that I will address shortly, you will be passed a structure that includes the current "symbolic sample size" that you decode, then you can branch your code accordingly.

3.5 Initialization: Defining Plugin Parameters

In VST3, parameters are moved to and from the GUI as *normalized* values on the range of [0.0, 1.0] regardless of the *actual* value. This is in contrast to AU and AAX, which operate on the actual data value. In VST3 lingo, the actual value is called the *plain* value. For example, if you have a volume control that has a range of −60 to +12 dB, and the user adjusts it to −3 dB, the parameter will store the normalized value of 0.7917 and transmit this to your plugin. If you want to write data out to the GUI parameter (e.g. user loads a preset), then you pass it the normalized value of 0.7917 as well. If you are using ASPiK, then all of this is handled for you, so you never need to bother with the normalized or plain values. However, if you intend to create custom parameters, you will need to work with both versions. The normalized values are always implemented as *double* data types, which VST3 typedefs as *ParamValue,* and the control ID is an unsigned 32-bit integer, which VST3 typedefs as *ParamID.*

VST3 supplies you with a base class named *Parameter* along with two sub-classed versions named *RangeParameter* and *StringListParameter*, which are used to encapsulate linear numerical parameter controls and string-list controls respectively. For our plugins, we define additional parameter types that are based on the control taper (see Chapter 6), including log, anti-log and volt/octave versions. In this case we need to sub-class our own *Parameter* objects that handle the chores of not only converting linear control values to their nonlinear versions, but also for converting from *plain* to *normalized* and back. These are built into the book project VST3 shell code and are independent of ASPiK, so feel free to use them as you like. We've sub-classed the following custom *Parameter* objects.

- *PeakParameter*: for parameters that are already normalized such that both the plain and normalized values are the same (this was borrowed directly from the VST3 SDK sample projects)
- *LogParameter*: for log-controls
- *AntiLogParameter*: for inverse log-controls
- *VoltOctaveParameter*: for controls that move linearly in octaves of 2^N

This gives you six total parameters to use, and you may always define your own hybrid versions—just see the file *customparameters.h* in the VST3 shell code for our versions, which will guide you into making your own.

As with all the other APIs, you declare these parameter objects to expose your plugin parameters to the host (in VST3 lingo, this is called *exporting parameters*). For VST3, you must declare one parameter object for each of your plugin's internal controls, plus an additional parameter to operate as the soft bypass mechanism. For early versions of the VST3 API, this soft bypass was optional, but it is now required. Once you have exposed your parameters to the host, you should not alter their ordering, nor should you add or remove parameters. This makes sense: if you altered the parameter ordering, it would massively screw up the automation system and confuse the host, and the user as well. You expose your parameters in the *initialize* function that we discussed in Section 3.2.6, and this function will be called only once per lifecycle.

The *SingleComponentEffect* base class includes a container for these parameters. You simply call a member function (*addParameter)* to add parameters to the list. Like ASPiK, VST3 parameters are ultimately stored in both a *std::vector* and *std::map* for fast iteration and hash-table accesses. As you learned in Chapter 2, all the APIs are fundamentally the same, and the VST3 parameter objects are going to encode the same types of information: an index value for accessing; the parameter name; its units; and the minimum, maximum and default control values. You just need to use the appropriate sub-classed parameter object that depends on the control type and taper. For example, suppose a plugin has two parameters with the following attributes.

Parameter 1

- type: linear, numeric
- control ID = 0
- name = "Volume"
- units = "dB"
- min/max/default = −60/+12/−3
- sig digits (precision past the decimal point): 4

Parameter 2

- type: string list
- control ID = 42
- name = "Channel"
- string list: "stereo, left, right"

You would declare them as:

```
// -- parameter 1
Parameter* param = new RangeParameter(USTRING("Volume"), 0, // name, ID
                                       USTRING("dB"),       // units
                                       -60, +12, -3); // min,max,default

param->setPrecision(4); // fractional sig digits
parameters.addParameter(param);

// -- parameter 2
StringListParameter* slParam = new StringListParameter(
                                       USTRING("Channel"), 42; // name, ID

slParam ->appendString(USTRING("stereo"));
slParam ->appendString(USTRING("left"));
slParam ->appendString(USTRING("right"));
```

To add the required soft bypass parameter, you first need to assign it a control ID value. Of course this will depend on your own numbering scheme; here we define the ID as SOFT_BYPASS, then use the pre-supplied *RangeParameter* object to create the parameter; we name it "Bypass." There are no units (" "); the minimum value is 0 (off) and the maximum value is 1 (on), with the default value set to off (0). The next (0) value is the parameter's step count which is ignored. Note the use of the *flags* argument, which is used to wire-or the automate and bypass flags together. As an exercise, you can search for the *ParameterFlags* structure to reveal other flags that are used with the parameters, such as the *kIsList* flag, which is automatically enabled for the string-list parameter object. After the declaration, you add the soft bypass parameter to your list with the same *addParameter* function as before:

```
const unsigned int PLUGIN_SIDE_BYPASS = 131072;
int32 flags = ParameterInfo::kCanAutomate|ParameterInfo::kIsBypass;

// -- one and only bypass parameter
Parameter* param = new RangeParameter(USTRING("Bypass"),
                                       PLUGIN_SIDE_BYPASS, USTRING(""),
                                       0, 1, 0, 0, flags);

parameters.addParameter(param);
```

3.5.1 Thread-Safe Parameter Access

The controller component is used for accessing the parameters as normalized values. The *SingleComponentEffect* object inherits the controller component, so your plugin object simply calls the functions directly—their names and parameters are easily identified. The *get* function accesses the parameter while the *set* function writes it all in a thread-safe manner.

```
ParamValue PLUGIN_API getParamNormalized(ParamID id);

tresult PLUGIN_API setParamNormalized(ParamID id, ParamValue value);
```

3.5.2 Initialization: Defining Plugin Channel I/O Support

As of this writing, VST3 has the broadest audio channel support of all of the APIs: it includes 33 traditional formats (mono, stereo, 5.1, 7.1, etc.), three Ambisonics® formats, and a whopping 27 different 3-D formats for a total of 63 supported channel I/O types. These input/output channel formats are named *speaker arrangements* and you can find their definitions in *vstspeaker.h*. When the user sets up a session track with a specific channel format and loads your plugin, the host will query it to find out if the format is supported. The system is quite robust and (as with the other APIs) it allows for different input and output formats. There is some dependence on the host's own capabilities as well, but the specification allows for maximum flexibility. For example, if the user loads a 5.1 audio file, the host will query you to see if your plugin supports 5.1. But what if you have some kind of specialized plugin that operates on the left and right surround channels only? The host will query that as well: the speaker arrangement named *kStereoSurround* is a two-channel format just for this purpose.

For most DAWs, the host will call the function *setBusArrangements* repeatedly to see if your plugin can handle the audio file format, or any of its sub-formats as discussed earlier. Your plugin must not only reply with a true/false answer, but also must set up the audio busses on the plugin component in response. Two functions named *addAudioInput* and *addAudioOutput* are used for this purpose. However, you need to clear out any previously declared busses to adapt to the new scheme. Some very limited or very specialized VST3 hosts may support only one audio channel I/O scheme, so you should actually set up your audio I/O twice: once in the initialize function, and then again in response to the *setBusArrangements* queries. For example, suppose your plugin supports mono in/mono out and stereo in/stereo out operation, and your preferred default arrangement is stereo in/stereo out. First, you would set up the busses inside the initialize function:

```
addAudioInput(STR16 ("Stereo In"), SpeakerArr::kStereo);
addAudioOutput(STR16 ("Stereo Out"), SpeakerArr::kStereo);
```

Next, you would override the *setBusArrangements* function to handle other combinations and schemes. The arguments for *setBusArrangements* will pass you an array of input and output speaker arrangements and channel counts. The following snippet is based on the VST3 sample project *AGainWithSideChain* (you can find more variations in the numerous other sample projects, so definitely check out that code). If you are supporting many formats and sub-variants, this code can become tricky, as you might have a bunch of logic to sort through. There are a few things to observe here. Note how the audio busses are cleared out with *removeAudioBusses:* if you support a side chain input, you must re-create it any time you remove all busses. Here, the side chain is mono (some code has been removed for brevity):

```
tresult PLUGIN_API setBusArrangements(SpeakerArrangement* inputs,
                                      int32 numIns,
                                      SpeakerArrangement* outputs,
                                      int32 numOuts) {
    // first input is the Main Input and the 2nd is the SideChain
    if (numIns == 2 && numOuts == 1)
    {
```

```
                    // the host wants Mono => Mono (or 1 channel -> 1 channel)
                    if (SpeakerArr::getChannelCount (inputs[0]) == 1 &&
                        SpeakerArr::getChannelCount (outputs[0]) == 1)
                    {
                        removeAudioBusses();
                        addAudioInput(STR16("Mono In"), inputs[0]);
                        addAudioOutput(STR16("Mono Out"), inputs[0]);

                        // recreate the Mono SideChain input bus
                        addAudioInput(STR16("Mono Aux In"),
                                        SpeakerArr::kMono, kAux, 0);

                        return kResultOk;
                    }
            etc...
}
```

3.5.3 Initialization: Channel Counts and Sample Rate Information

For VST3, the channel count initialization is part of the channel I/O declaration and is all contained in the *setBusArrangements* function, so there is nothing else to do (note this is somewhat different from AU). That said, the channel configuration in VST3 is very advanced and may be set up in more complex arrangements than what we use for the FX plugins here, as well as the VST3 sample projects. The channel I/O buss configurations are dynamic, and you may set up multiple input and output busses and then activate, deactivate, or reactivate them as required.

To inform the plugin of the current sample rate (and other information), the host will call the function *setupProcessing* and pass it a *ProcessSetup* structure with the relevant information. Extracting the pertinent information is straightforward—see the documentation for the *ProcessSetup* structure. To grab the sample rate and bit depth, you would override the function and implement it like this:

```
tresult PLUGIN_API setupProcessing(ProcessSetup& newSetup)
{
    double sampleRate = processSetup.sampleRate;
    int32 bitDepth = processSetup.symbolicSampleSize;

    // --- base class
    return SingleComponentEffect::setupProcessing(newSetup);
}
```

Note: this function is called when the *ProcessSetup* has changed. The base class object will then save a copy of this structure as a protected member variable; you may access it in other functions to get the sample rate or other information on the fly.

3.6 The Buffer Process Cycle

The VST3 buffer processing cycle follows the same basic paradigm as AAX and AU, as shown in Figure 3.3. The entire operation happens inside of one audio processing function called *process*. This function has one argument called *ProcessData* that contains all of the information needed to complete the procedure.

```
tresult PLUGIN_API process(ProcessData& data)
```

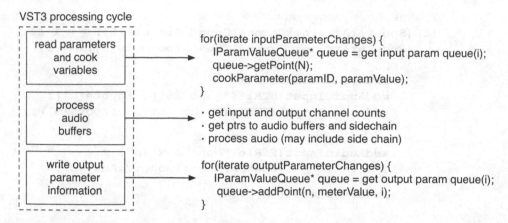

Figure 3.3: The VST3 buffer process cycle follows the standard audio plugin pattern; the pseudo-code for each functional block is shown on the right.

The *ProcessData* members are shown here with the most important ones for our FX plugins in bold:

```
struct ProcessData
{
        int32 processMode;              // realtime, offline, prefetch
        int32 symbolicSampleSize;       // 32 or 64 bit
        int32 numSamples;               // number of samples to process
        int32 numInputs;                // number of audio input busses
        int32 numOutputs;               // number of audio output busses
        AudioBusBuffers* inputs;        // buffers of input busses
        AudioBusBuffers* outputs;       // buffers of output busses

        IParameterChanges* inputParameterChanges;
        IParameterChanges* outputParameterChanges;
        IEventList* inputEvents;
        IEventList* outputEvents;
        ProcessContext* processContext;
};
```

The *processMode* variable indicates real-time or offline processing in addition to a flag for plugins that require process spacing that is nonlinear: for example, time stretching or shrinking. For all book projects, we use the real-time mode. The bit depth variable, block size in samples, and the channels counts are self-explanatory. You access the audio buffers using the *AudioBusBuffers* pointers; these also include the side chain input buffers when a side chain has been activated. The parameter changes that occurred while the audio data was captured are transmitted into the function in the *inputParameterChanges* list; this is where you grab the parameter updates and apply them to your signal processing algorithm before working on the audio data. You write data out to the GUI parameters (mainly for metering) with the *outputParameterChanges* list. We will discuss the *processContext* member shortly. The input and output *events* consist mainly of MIDI messages, which I cover in detail in *Designing Software Synthesizers in C++*. For the most part, the procedure for accessing the audio buffers for processing is fundamentally identical to that of AAX and AU. One

VST3 exclusive is how the GUI parameter changes are sent into this function, which includes sample accurate automation (SAA) that allows the host to interpret recorded automation data to send smoother control change information when automation is running—this has no effect during normal operation with the user manually adjusting controls.

We can break the buffer processing into three parts, exactly following Figure 3.3 where we have the succession:

1. Update parameters from GUI control changes, cooking variables as needed.
2. Process the audio data using the updated information.
3. Write outbound parameter information (meters, signal graphs, etc.).

A nice feature of VST3 (and AAX and RackAFX) that is not available in AU is that the input parameter change list includes only those parameters that actually changed during the audio block capture or playback. If no GUI controls were adjusted and automation is not running, there is no need to check parameters and update internal variables—saving CPU cycles and simplifying the operation.

3.6.1 Processing: Updating Plugin Parameters From GUI Controls

The parameter changes arrive in the *process* function in a list of *queues* that is passed into the function via the *processData* structure. There is one queue for each parameter that has changed. The queue contains one or more timestamped parameter change values that were transmitted during the audio buffer capture/playback. The timestamp is really just an integer value that denotes the sample offset from the top of the audio buffer in which the parameter change event occurred. The information about the parameter change is provided with a function called *getPoint*. Figure 3.4a is a conceptual diagram of how the parameter change events line up with various samples in the audio buffer. Notice that the buffers are rotated vertically so that sample time and buffer index values move from left to right. Each normalized parameter value change event is shown as a diamond.

Under non-automation conditions, with the user manually adjusting the GUI controls, there is usually only one data point in any queue (I've only ever observed one point but the spec does not explicitly limit

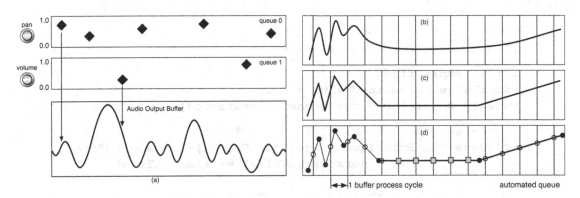

Figure 3.4: (a) In this conceptual diagram of how the parameter change events line up, there is one control change queue for each parameter that needs updating, each queue containing one or more data points. (b) An example automation curve for a single parameter that changes over many buffer process cycles. (c) The DAW-linearized version. (d) This graph (adapted from the VST3 SDK documentation) shows the data points actually transmitted to the plugin (filled circles) and the data points that are implied (hollow circles).

it to one point). However, if the user is running automation with a non-constant automation curve, then there will be multiple points in the queue. The simplest way to handle both of these conditions is to take the first or last parameter change value in the queue and apply that to your plugin processing. If your framework supports parameter smoothing, then it will at least provide smooth ramping over those points.

If the user is running automation and the queue contains multiple points, Steinberg specifies that these points are to be connected using linear interpolation. Consider a single, automated parameter that evolves slowly over many buffer process cycles, with the recorded curve shown in Figure 3.4b. The VST3 API specifies that the host DAW must linearize this curve to produce something like Figure 3.4c. The exact method of linearization, including how many segments used to encode a curve, is up to the host manufacturer. The parameter change points that are actually transferred into the *process* function are the endpoints of the piecewise linear approximation of the curve, shown as filled dots in Figure 3.4d and the hollow dots at the end of each buffer/queue. These hollow endpoints are not re-transmitted at the start of the next buffer/queue. If the automation value remains constant across more than one buffer cycle, then no value is transmitted: these points are implied and are shown as filled gray squares. If the plugin supports SAA, then it is required to first make sense of the queue's data points and understand the concept of non-repeated or implied data points, then provide its own mechanism for slowly changing the parameter value across each sample interval, providing the smoothest possible parameter changes.

The parameter changes arrive in a normalized format on the range of [0.0, 1.0]. Your plugin must decode the control ID and pick up this normalized value, then convert it to the actual value that the plugin will use. The code for grabbing the last data point in the first queue in the list is shown here; notice how the control ID is transmitted as the queue's "parameter ID" value, then the normalized value and sample offset are obtained with the *getPoint* function. The rest of the function (labeled 1, 2, 3) will depend on our plugin's implementation. With your plugin parameters updated, you proceed to the next step and process the audio.

```
// get the FIRST queue
IParamValueQueue* queue =
                data.inputParameterChanges->getParameterData(0);

if(queue)
{
        // --- check for control points
        if(queue->getPointCount() <= 0) return false;

        int32 sampleOffset = 0.0;
        ParamValue normalizedValue = 0.0;
        ParamID controlID = queue->getParameterId();

        // --- get the last point in queue #0
        if(queue->getPoint(queue->getPointCount()-1, sampleOffset
                        normalizedValue) == kResultTrue)
        {
                // 1) un-normalize the normalizedValue
                // 2) cook the actual value to update the plugin algorithm
                // 3) here we ignore the sample offset because we are only
                //    using one control point
        }
}
```

3.6.2 Processing: Resetting the Algorithm and Preparing for Streaming

VST3 plugins may be turned on or off to enable or disable them during operation. The function that is called is named *setActive* and has a Boolean argument that identifies the state: *true* is enabled and *false* is disabled. You may then perform reset operations to flush buffers and prepare the plugin for the next audio streaming operation.

```
tresult PLUGIN_API setActive(TBool state)
{
        if(state){
                // --- do ON stuff
        }
        else {
                // --- do OFF stuff
        }

        return SingleComponentEffect::setActive(state);
}
```

3.6.3 Processing: Accessing the Audio Buffers

In VST3, the audio input buffers arrive in the input buffer list passed into the *process* function in its *processData* argument. The buffer list is named *inputs* and it contains pointers to each input buss. Buss number 0 is the main audio buss, and buss number 1 is the side chain. Each buss may contain one or more audio channels; each audio channel arrives in its own buffer (array) exactly as with the other APIs. To get the array of input buffer pointers for a 32-bit main audio input buss, you would access *data.inputs[0],* where 0 indicates the main buss, and then get the array of pointers in the first slot [0] of the *channelBuffers32* member.

```
float** inputBufferPointers = &data.inputs[0].channelBuffers32[0];
```

You then access the individual channel buffers:

```
float* inputL = inputBufferPointers[0];
float* inputR = inputBufferPointers[1];
```

The output buffers are accessed identically except that you use the *data.outputs* member instead of the *data.inputs* variable, as follows.

```
float** outputBufferPointers = &data.outputs[0].channelBuffers32[0];

float* outputL = outputBufferPointers[0];
float* outputR = outputBufferPointers[1];
```

To get the side chain buffers, you would access buss 1 of the *data.inputs* array. Since the side chain is an input-only buffer, there is no corresponding output buffer. To access buss 1, you should first make sure that the buss actually exists, and then grab the channel count and buffer pointers:

```
BusList* busList = getBusList(kAudio, kInput);

Bus* bus = busList? (Bus*)busList->at(1): 0;
if (bus && bus->isActive())
```

```
{
    float** sidechainBufferPointers =
            &data.inputs[1].channelBuffers32[0];
}
```

The *AudioBusBuffers* variable is actually a structure that contains additional information about the buss, which you may use to find the exact channel count as well as access the silence flag. To get the channel counts, you would write:

```
int32 numInputChannels = data.inputs[0].numChannels
```

```
int32 numOutputChannels = data.outputs[0].numChannels
```

```
int32 numSidechainChannels = data.inputs[1].numChannels
```

If the input buffer is full of zero-valued samples (e.g. the user has muted the track with the plugin), then the silence flag will be set. You may access this and quickly bail out of the function if there is nothing to do.

```
bool silence = (bool)data.inputs[0].silenceFlags;
```

With the buffer pointers acquired, you may then do your audio signal processing: you create a loop that processes the block of samples and you find the sample count in the *data* structure's *numSamples* member. Here we are processing stereo in/stereo out:

```
// --- loop over frames and process
for(int i = 0; i < data.numSamples; i++){
    // -- apply gain control to each sample
    outputL[i] = volume * inputL[i];
    outputR[i] = volume * inputR[i];
}
```

To use information in the side chain buffer list, you treat it the same way as the others:

```
float* sidechainL = sidechainBufferPointers[0];
float* sidechainR = sidechainBufferPointers[1];
```

3.6.4 Processing: Writing Output Parameters

If your plugin writes data to output meters or some kind of custom view (see Chapter 21) then you would do that here as well. With all of the processing done, you would then write the output information to the appropriate GUI parameter that needs to receive it. To do this, you need to write data points into the outbound parameter queues—but you must first find those queues to write into; you are now on the opposite side of the queue paradigm. To create/find the first outbound queue for a meter parameter with a control ID named *L_OUTMTR_ID,* you would write the following:

```
int32 queueIndex = 0; // return variable
```

```
IParamValueQueue* queue =
data.outputParameterChanges->addParameterData(L_OUTMTR_ID, queueIndex);
```

The *addParameterData* function finds the queue for the parameter with the associated control ID and returns that value in the *queueIndex* variable, which you typically won't need to use. With the queue found, you may then add data points to it. For metering, we will add one data point that represents the metering value across the entire audio buffer. The meter is displayed on a low-priority GUI thread that is animated once every 50 milliseconds or so (depending on your framework's GUI code). Adding more points to the queue won't make the meter appear more accurate or "fast," as it is limited primarily by the GUI repaint interval. To add a new point to the queue, you use the *addPoint* function, which accepts the timestamp as a sample offset (we will use 0) and the normalized meter value. The index of the data point in the queue is returned as an argument (*dataPointIndex* here), and you will usually ignore it. Notice that for this meter we transmit the absolute value of the first sample in the left output buffer:

```
int32 dataPointIndex = 0; // return variable
queue->addPoint(0, fabs(outputL[0], dataPointIndex);
```

3.6.5 Processing: VST3 Soft Bypass

All VST3 FX plugins must support the soft bypass function. Remember that we declared that soft bypass parameter along with the other plugin parameters, so it will arrive in its own queue inside of the *inputParameterChanges* list along with those same parameters. You set your own internal plugin flag (*plugInSideBypass* here) when bypass is enabled, then you use it in your processing block—you just pass the input to the output, like this:

```
if(plugInSideBypass)
{
    for(int32 sample = 0; sample < data.numSamples; sample++)
    {
        // -- output = input
        for (unsigned int i = 0; i<info.numOutputChannels; i++)
        {
            (data.outputs[0].channelBuffers32[i])[sample] =
                (data.inputs[0].channelBuffers32[i])[sample];
        }
    }
}
```

3.7 Destruction/Termination

Because of the class factory approach, you destroy any dynamically allocated resources and perform any other cleanup operations in the *terminate( )* function. You may treat the *terminate* method as if it were a typical object destructor except that you should call the base class method at the head of the function and check its status:

```
tresult PLUGIN_API terminate()
{
    tresult result = SingleComponentEffect::terminate ();
    if (result == kResultOk)
    {
        // -- do your cleanup
```

```
    }

        return result;
}
```

3.8 Retrieving VST3 Host Information

VST3 provides a robust amount of host information that may be provided to your plugin. This information arrives in the process method's *processData* argument in the *processContext* member and includes data about the current audio playback such as the absolute sample position of the audio file (the index of the first sample in the input audio buffer for the current buffer processing cycle), the DAW session's BPM and time-signature settings, and information about the audio transport (whether it is playing or stopped, or if looping is occurring). Not all VST3 hosts support every possible piece of host information. However, the absolute sample location and time—along with the host BPM and time signature information—is very well defined across all VST3 clients we've tested. You can find all of the possible information in the *ivstprocesscontext.h* file, which defines the *processContext* structure. Note that this includes the current sample rate as well, so you have a lot of information available on each buffer process cycle. As an example, here is how to grab the BPM and time signature numerator and denominator:

```
if (data.processContext)
{
    double BPM = data.processContext->tempo;

    int32 tsNum = data.processContext->timeSigNumerator;
    int32 tsDen = data.processContext->timeSigDenominator;
}
```

3.9 Validating Your Plugin

Steinberg provides validation software in the VST3 SDK called the *validator*. If you go back and look at Figure 3.2, you can see the validator executable is part of the VST3 plugin project. The VST3 sample projects are packaged this way, as are the ASPiK VST3 projects. The compiler projects are set up to compile the libraries, your plugin, and the validator together. The validator is built before your plugin in the sequence of operations; after your plugin builds without errors, it is run through the validator. If the validator passes, then your plugin binary is written to disk. If validation fails, the output file will not be written. If this occurs, you need to go back through the compiler's output log and find the location where the validator faulted. The validator's text output is verbose, so it is fairly easy to find the spot where it fails. You then need to fix the problem. There are some tactics you may use to do this that you can ask for help at www.willpirkle.com/forum/, but you still have a lot of control. Unlike the AU validator, which is part of the OS, the VST3 validator is part of your plugin project and there are ways to debug through it. You should also use the *printf* statement to pipe text messages from the various plugin functions, as they will be printed to the output along with the validator. This allows you to watch the sequence of operations as the validator runs.

3.10 Using ASPiK to Create VST3 Plugins

There is no substitute for working code when designing these plugins; however, the book real estate that the code requires is prohibitive. Even if you don't want to use the ASPiK *PluginKernel* for your development, you can still use the code in the projects outlined in Chapter 8. If you use ASPiK, its

VST3 plugin shell will handle all of the description and initialization chores, and your *PluginCore* object will do the audio processing—you will only work within two files: *plugincore.h* and *plugincore.cpp*. You may also enable VST3 SAA when you create the new project with CMake. If the parameter you are supporting has a CPU-intensive cooking function, then you may not want to update the parameter value on each sample interval. In this case, you may set the sample accurate granularity in the constructor of your *PluginCore* object by altering one of the plugin descriptors named *apiSpecificInfo*. You alter the member variable set the granularity in samples. For example, to set the accuracy to perform an update after every 32-sample interval, you would write:

```
apiSpecificInfo.vst3SampleAccurateGranularity = 32;
```

3.11 Bibliography

Pirkle, W. 2015. *Designing Software Synthesizers in C++*, Chap. 2. New York: Focal Press.
Steinberg.net. *The Steinberg VST API*, www.steinberg.net/en/company/developers.html, Accessed August 1, 2018.

CHAPTER 4

AU Programming Guide

There are currently two different AU plugin specifications: v2 and v3. The AUv3 spec is fundamentally aimed at iOS programming, and AUv2 is still the API of choice for non-iOS applications. There is also an AU bridge that can help wrap a v2 version as a v3 plugin. For this book we will be using the V2 plugin API for non-iOS applications, as this is the plugin type that all AU DAWs will load. Both versions follow the same audio plugin paradigms that we studied in Chapter 2. You may download the AU SDK for free from Apple.

4.1 Setting Up the AU SDK

The AU SDK consists of just two folders: *AUPublic* and *PublicUtility*. These folders contain all of the AU SDK files. Unlike the other SDKs, they contain no sample code. You may download just these two folders, or you may download sample projects (which contain the same two folders) from https://developer.apple.com/library/content/samplecode/sc2195/Introduction/Intro.html. You can also get just the SDK files from www.willpirkle.com/Downloads/AU_SDK.zip. There are no libraries to build; after installing the two folders, you are done.

4.1.1 AU Sample Projects

The Apple Documentation Archive that contains the SDK folders also contains a set of sample projects. In those projects, the AU SDK folders (*AUPublic* and *PublicUtility*) are embedded inside of the sample project outer folder; it is best to leave them there and simply copy the two SDK folders to your final plugin development folder. There are sample projects for FX, synth and generator (oscillator) plugins, and others. Just open the Xcode project files inside of each one and build them; note that there is a tight dependency on the location of the SDK folders that were packaged with the sample code. Make sure you can build the sample code properly before trying to write your own plugins.

4.1.2 AU Documentation

The AU API documentation is not contained within the SDK but is instead fully implemented on the Internet as web pages and sites. You may wish to download and cache the full documentation in case you need to work in an Internet-limited area. The AU documentation is generally very good; however, you may sometimes run across outdated information or non-working links. The AU programming guide is especially nice and includes details about sample code, examples, and links to some of the OS-specific portions that involve *CoreAudio*, *CoreFoundation*, and other frameworks. The main page of the documentation may be found at https://developer.apple.com/library/archive/documentation/MusicAudio/Conceptual/AudioUnitProgrammingGuide/Introduction/Introduction.html.

4.2 AU Architecture and Anatomy

AU plugins are written for MacOS only. This means you must use Xcode and MacOS to compile AU plugins irrespective of your plugin framework (ASPiK, JUCE, etc.). AU plugins are *components* packaged inside of MacOS *bundles*. A bundle is a special name for a folder that appears to the user as a file. If you right-click on a bundle (file) and choose "Show Package Contents" then you can see the bundle's files, folders, and subfolders. Both AAX and VST have moved to the bundle concept for both MacOS and Windows (the bundle in Windows is also a folder disguised as a file).

From the very start of development you need to decide if the plugin will support a custom GUI or not. AU plugins are unique in that the signal processing and GUI objects must be compiled independently of one another as separate products; in the other APIs, the objects are compiled together into one product. An AU Xcode project must be set up to specifically compile the two components independently, producing two output products: the AU plugin with the *.component* suffix and the *Cocoa* GUI view factory with the *.bundle* suffix, as shown in Figure 4.1a. After the build is completed you then move the GUI component *Into* the AU bundle. This is handled in the compiler as a post-build operation and depicted in Figure 4.1b, though you may also do this move manually. Finally, the plugin file (e.g. *GoldenFlanger.component*) is copied or moved into the final destination folder, as shown in Figure 4.1c. For development, you may do this final move manually, or as a post-build operation. If you are using ASPiK, then all of this copying and moving is handled automatically for you and the compiler projects are set up accordingly. If you are not, consult your plugin framework documentation for these details. If you are designing your plugin from scratch, then you must handle these details yourself.

The AU signal processing portion is written in C++ as an object that derives from one of the AU base classes, in the same manner as AAX and VST plugins and as described in Chapter 2. However, the GUI product must be designed as a *Cocoa* object using the Objective-C programming language. In addition, the GUI object must be fashioned as a class factory, serving up instances of the GUI each time the user opens it. That class factory must likewise be written in Objective C. If you are *not* going to implement a custom GUI, then your plugin can be written as a single C++ object without the GUI component. Now, here is the good news: if you use ASPiK to develop your plugin, you will use the

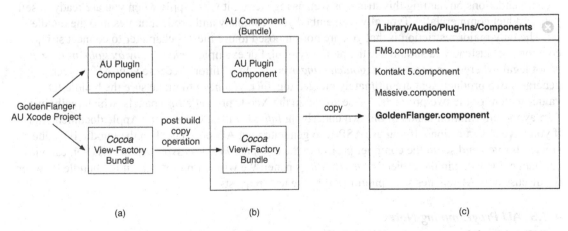

Figure 4.1: (a) The AU plugin project compiles two output files, the AU component and the *Cocoa* GUI bundle. (b) The *Cocoa* bundle is moved into the AU component. (c) Finally, the single bundle is moved into the AU library destination folder.

VSTGUI4 library for the GUI object, which is written in native C++, including any custom view objects you decide to implement (see *plugingui.h* and *plugingui.cpp,* which implement the VSTGUI4 objects). You won't need to bother with a single line of Objective-C code! This is possible due to very clever programming on the part of the VSTGUI4 architects. They use the Objective-C runtime programming API to register and wrap the C++ GUI objects as Objective-C. The Objective-C runtime API is specifically designed to allow other object-oriented programming components to be registered and wrapped as *Cocoa* objects. If you are designing an AU plugin from scratch, then you will also need to write the GUI factory and view objects directly in Objective C.

4.2.1 AU Base Classes

Your AU processing object is written in C++ and is derived from one of several AU base classes, depending on your final product goals. This is consistent with all other plugin APIs. The choices are as follows.

- *AUEffectBase*: for FX plugins with one audio input and one audio output stream (each stream may contain multiple channels) without MIDI support
- *AUMIDIEffectBase*: for FX plugins with one audio input and one audio output stream that would also like to receive or generate MIDI messages
- *MusicDeviceBase*: for synth plugins that have no audio input stream, one audio output stream that renders the synth notes, and the ability to receive or generate MIDI messages

All ASPiK AU plugin projects are created using the *AUMIDIEffectBase* class so that the FX plugins may send and receive MIDI if they wish. There is an implication for side chaining that we will discuss later, since all ASPiK plugins have the option of exposing a side chain for the user. All three of these base classes are derived from the mother class named *AUBase*. As such, they all share many common functions and programming paradigms. For this FX book, you will want to use either the *AUEffectBase* or *AUMIDIEffectBase* classes if you decide to create plugins from scratch.

4.2.2 MacOS Bundle ID

Each bundle in MacOS is identified with a unique string called a *bundle ID.* Apple has recommendations on naming this string, as well as registering it with Apple when you are ready to sell your product commercially. Typically, you embed your company and product names into the bundle ID. The only major restriction is that you are not supposed to use the "-" character to connect string components, instead you should use the period symbol. For example, *mycompany.au.golden-flanger* is not legal, whereas *mycompany.au.golden.flanger* is OK. In addition, because your AU project generates two products that are ultimately merged together, you need to make sure the bundle ID values match for the two products. These are set in the Xcode project's *Info* panels, which really just display the information inside of the two underlying *info.plist* files. Consult the Apple documentation if you have any questions. If you use ASPiK to generate your AU projects, then the bundle ID value is created for you and set in the compiler project in the *Info.plist Preprocessor Definitions*. You can view or change this value in the project's *Build Settings* panel. We will learn more about the bundle ID when we discuss how AU creates your plugin GUI as the user requests it.

4.2.3 AU Programming Notes

AU plugins are written in C++ using Xcode. However, unlike AAX and VST, AU plugins require OS-dependent frameworks and programming paradigms. If you are not familiar with *CoreFoundation* objects such as *CFArrayRef, CFStringRef, CFMutableArrayRef* and the like, then some of the AU plugin code is going to look foreign. If you want to write AU plugins from scratch or you plan on spending time and effort in AU programming that goes beyond what your third party framework implements, then you might want to invest some time in the Apple *CoreFoundation* objects. The frameworks required for an AU plugin with custom GUI include *CoreAudio, CoreMIDI, CoreFoundation, QuartzCore, AudioUnit, AudioToolbox, Accellerate,* and *OpenGL*. The latter two are optional and are for fast DSP algorithms (*Accellerate*) and VSTGUI4 (*OpenGL*). You do not need to be an expert at any of these to create plugins with ASPiK or other plugin frameworks; however, at some point you may want to do some customization that will require at least a bit of poking around in the frameworks.

As with AAX and VST, there are numerous C++ typedefs that rename standard data types, so you should always right-click and ask the compiler to take you to definitions that you don't recognize. Most of the definitions are simple and are supposed to make the code easier to read, though sometimes they can have the opposite effect. The large majority of AU base class functions return an error or success code, which are really just integers and are typedef'd as *ComponentResult* and *OSStatus*. The success code is *noErr* and common failure codes are *kAudioUnitErr_InvalidParameter* and *kAudioUnitErr_InvalidProperty*.

The plugin is a DLL and requires entry points (function names that the host will call) that are defined in an exported symbols file with the suffix *.exp*. One of these is the AU factory function that matches the *info.plist* entry. In general, your plugin framework will define these; you won't need to spend time dealing with them unless you want to write the plugin from scratch.

During instantiation the plugin will be queried repeatedly to ask for information about its attributes and capabilities. There is a big list of standard queries; you may set breakpoints and watch the action if you like. In some cases this is a direct result of your plugin answering *true* to other queries or your plugin exposing string-list parameters. The host will also query the plugin about its factory presets, which the plugin creates in a contiguous block of memory. When the user selects one of the presets, the host will notify the plugin so that it may set the parameters accordingly. We will discuss presets shortly.

The C++ name of the AU plugin object has no restrictions other than it must be a legal C++ class name—you may reuse the same AU plugin object name over and over. However, as noted in Section 4.7.1, the naming of the *Cocoa* view factory and view objects does require your attention and must be unique.

In AU programming, you will also see many references to *scopes* and *elements*, which are passed into many of the AU plugin functions. A scope references the context in which something is being used or accessed in an AU plugin. The AU plugin architecture defines eight scopes, the first three of which we use for FX plugins:

- *global*: having to do with aspects of the plugin as a whole entity
- *input*: having to do with data coming into the AU
- *output*: having to do with data leaving the AU
- *group*: having to do with rendering musical notes (synthesis)
- *part*: having to do with multi-timbral synths
- *note*: having to do with musical notes
- *layer* and *layerItem*: having to do with synth layers in groups

Elements are sub-components of scopes that are zero indexed with integers. For *input* and *output* scopes, these correspond to hardware busses. A *global* scope has one and only one element, which is element 0.

4.3 Description: Plugin Description Strings

Your plugin description strings are contained in the Xcode "Info" tab in its main view. This reflects the information in the associated *info.plist* file; you may edit either that file or the compiler project itself, which I recommend. There is a second *info.plist* file for the *Cocoa* view factory, but it does not contain any information about the audio plugin; you generally only need to make sure that the AU plugin and *Cocoa* view bundle ID values match since the *Cocoa* bundle is embedded inside of the AU bundle. A typical Xcode AU plugin project will have information panels that resemble that in Figure 4.2. The project is named *GoldenFlanger* and it includes two targets, one for the AU plugin and the other for the *Cocoa* GUI. You may name the targets as you wish.

The plugin packaging information in Figure 4.2 consists of the following.

- **bundle identifier**: the unique identifier string for the plugin's bundle (package)
- **executable name**: the name of the final plugin file (bundle/folder); here the plugin is packaged as *GoldenFlanger.component*

More important is the dictionary object, which contains more vital plugin description strings:

- **manufacturer**: a four-character code that defines the manufacturer; here our fictitious company is PluggCo with the four-character code PLCO
- **subtype**: a four-character code that identifies the plugin, within your company's set of plugins; you will likely adjust the value of this string during testing (for our initial build, we will use FLNG, for flanger)

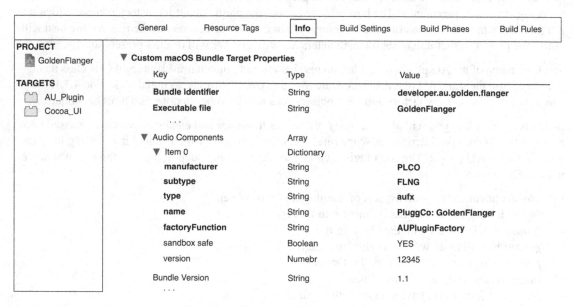

Figure 4.2: The Xcode *Info* panel for an AU plugin project. The critical strings are shown in bold; some information is omitted for brevity.

- **type**: a four-character code that identifies the type of AU plugin, which may be one of the following:
  - *aufx*: an audio effect plugin derived from *AUEffectBase*
  - *aumf*: an audio effect with MIDI capability, derived from *AUMIDIEffectBase*
  - *aumu*: a software synth with MIDI capability, derived from *MusicDeviceBase*
- **name**: this field is crucial because it includes the name of the plugin and manufacturer as you wish them to appear in the DAW's plugin selector menu; a colon separates the plugin manufacturer and DAW plugin name string such as "PlugCo: Golden Flanger"
- **factory function**: the name of the entry point for the plugin, set in the exported symbol file with suffix *.exp*

Notice in particular the *name* field: this allows you to choose the exact name string as you would like for the DAW to expose for the user; it means that this name string may be completely independent from the plugin project or package name.

4.4 Description: Plugin Options/Features

The plugin options include audio I/O, MIDI, side chaining, latency, and tail time. For AU plugins, these are all done programmatically either in the plugin constructor, or in other base class functions that you override.

4.4.1 Side Chain Input

When you derive the AU plugin object from the appropriate base class, it will automatically set up the audio input and output buss counts for you. The first function that the plugin object's constructor calls is *AUBase::CreateElements( )*, which will set up the basic audio I/O busses. You may then modify the audio I/O details afterwards. If you wish to expose a secondary input as a side chain, you must explicitly add it and name it. This is simple, taking only three lines of code:

```
SetBusCount(kAudioUnitScope_Input, 2);
SafeGetElement(kAudioUnitScope_Input, 0)->SetName(CFSTR("Main Input"));
SafeGetElement(kAudioUnitScope_Input, 1)->SetName(CFSTR("Sidechain"));
```

The first line adjusts the input buss count to two while the following lines name each buss. Audio buss 0 is the default audio input, and buss 1 is the side chain input; you may not change the designations, though you may change the name strings. If your plugin does not require a side chain, you may omit this code altogether as the *CreateElements* function call will handle the basic assembly. There is one little caveat we should discuss here that involves using the *AUMIDIEffectBase* class. While this base class appears to be the same as the normal *AUEffectBase* but with MIDI added, it does not allow exposing a side chain for the user. Owing to the fact that many side-chaining plugins may also want to receive MIDI messages, you can fix this in a slightly less than ideal manner by setting the *type* in the AU dictionary to *aufx*, even though the plugin is derived from the MIDI effect base class. This will allow the DAW to expose a side chain for this AU plugin type (don't tell anyone). If your plugin wants MIDI, but does not need a side chain input, then you may specify the more legal *aumf* as the plugin type.

4.4.2 Latency

If your plugin includes a latency that adds a necessary and fixed delay to the audio signal such as the *PhaseVocoder* in Chapter 20 or the look-ahead compressor in Section 18.7, you may signify this for the AU plugin host. The framework will call the function *GetLatency( )* on your plugin object to query it about this feature. If your plugin includes a fixed latency, you override this function and return the latency value. Unlike AAX and VST, the AU latency is reported in seconds rather than samples. This means that most likely the AU plugin's latency value will change with the sample rate. Therefore, you will need to adjust the latency value based on the sample rate. The function is simple to implement for a fixed (not sample rate dependent) latency; for example, if the latency is 1 millisecond, you might simply write:

```
virtual Float64 GetLatency() {return 0.001;}
```

For a latency based on samples, which varies with the sample rate, you may return a variable instead:

```
virtual Float64 GetLatency() {return latencyInSeconds;}
```

The variable *latencyInSeconds* is a member variable for the plugin that we declare (of course you may rename it as you like). We will recalculate this value in two different functions where the sample rate may change, which we will describe in later sections.

4.4.3 Tail Time

Reverb and delay plugins often require a tail time that the host implements by pumping zero-valued audio samples through the plugin after audio playback has stopped, allowing for the reverb tail or delays to repeat and fade away. The host will query your plugin when it is loaded to request information about the tail time. If your plugin does not require a tail time then you may ignore these functions. There are two *AUBase* functions that work together: the first queries the plugin to see if it wants an audio tail; if the plugin returns *true*, then the second function is called to ask for the tail time in seconds. For example, if your plugin would like a five-second audio tail time, you would implement the two functions as follows:

```
virtual bool SupportsTail() {return true;}
```

```
virtual Float64 GetTailTime() {return 5.0;}
```

4.4.4 Custom GUI

During instantiation, the AU host will query the plugin repeatedly for a variety of information, collectively called *Audio Unit properties*. There are many properties that your plugin may wish to support: for example, the host may ask the plugin about its "most important" parameters to supply as an overview. If the plugin implements a custom *Cocoa* GUI, then it needs to reply with *noErr* to the query for the Audio Unit property *kAudioUnitProperty_CocoaUI,* which we'll get to shortly.

4.4.5 Factory Presets and State Save/Load

AU and VST support factory presets that are encoded when the plugin is compiled, so they are "built in" to the C++ object and can never be changed. There are two functions for preset support that you will need to override and complete if not using a third party plugin framework. The first function is a query for a list of the presets, packaged as *AUPreset* structures, which encode the preset name string and the preset index number:

```
typedef struct AUPreset {
     SInt32      presetNumber;
     CFStringRef presetName;
} AUPreset;
```

The *AUPresets* are returned in a statically declared array, meaning that the array of strings must be persistent for the life of the plugin. The array is returned as a function parameter and the function itself returns the success/error code:

ComponentResult GetPresets(CFArrayRef *outData) const

You may write the array directly as a global variable such as:

```
static AUPreset kPresets[3] = {
{0, CFSTR("Factory Preset")},
{1, CFSTR("Really Loud Flanger")},
{2, CFSTR("Crazy Resonance")}};
```

Another option is to dynamically allocate memory to store the array at runtime if the preset names and quantity are not known at compile time. When the user selects a factory preset, the plugin is notified via the function, where it is passed the *AUPreset* structure for the selected factory preset:

OSStatus NewFactoryPresetSet(const AUPreset& inNewFactoryPreset)

The function decodes the preset information, then sets the AU plugin parameters based on the preset data that it has previously stored.

When the user saves a DAW session or creates their own presets, the AU host will handle reading and writing the data to disk (also called *serialization*), so there is nothing that the plugin needs to do directly. However, only AU parameters will be saved with the standard mechanism. The AU base class features two functions for loading and saving state information called *SaveState* and *RestoreState*: if you have some kind of custom information that needs to be saved or loaded with presets that goes beyond the plugin parameter states, you need to override these functions and implement your own saving and loading functionality using the *CFMutableDictionaryRef* object.

4.5 Initialization: Defining Plugin Parameters

In AU, a plugin parameter always encodes and represents the *actual* value of the parameter. For example, if you have a volume control that has a range of −60 to +12 dB, and the user adjusts it to −3 dB, the parameter will store the value of −3 and transmit this to your plugin. If you want to write data out to the GUI parameter (e.g. user loads a preset), then you pass it the actual value of −3 as well. This is in contrast to VST3, which uses *normalized* values on the range of [0.0, 1.0] when moving data between the GUI and the plugin. AU parameters use an integer value as the control ID, so you access the parameters using this numerical value.

As with all the other APIs, AU defines a plugin parameter object that allows parameters to be adjusted in a thread-safe manner. For AU, the object is a *ParameterMapEvent*, which represents the value of the parameter and also allows for ramped parameters to enable parameter smoothing. The parameters are always represented as 32-bit floating point values regardless of the GUI controls they link with. They are accessed using an ID value of type *AudioUnitParameterID,* which is simply an unsigned 32-bit integer which I refer to as the control ID in this book. AU has two internal options for storing your parameters: *std::vector* and *std::map*. You have the option to declare your parameters as *indexed,* meaning that the control ID values are zero indexed with no gaps (0, 1, 2, 3, . . . *N*), in which case the base class will use a vector to store the parameters. For non-indexed parameters, which is the default operation, the map is used instead. For random access, the vector is slightly faster. However, AAX and VST do not have this option. We would like the flexibility of allowing gaps in the ID values (for example to encode some information based on the ID's numerical value), so we will use the non-indexed version.

You get and set the plugin parameters using a helper function named *Globals( ),* which accesses the stored parameter list, then you simply call the *get* or *set* function, passing the control ID value and a variable for the data. To set the parameter with the ID of 42 to a value of 0.707, call the set function:

```
Globals()->SetParameter(42, 0.707);
```

To get the value of the same parameter, you call the *get* function that returns the value into a variable you declare:

```
double auParameter = Globals()->GetParameter(42);
```

The method of defining or exposing the plugin parameters to the AU host during the initialization phase is comprised of three parts. In the plugin constructor, you create a parameter by calling the *SetParameter( )* function and pass it a control ID value. Since the parameter does not exist, then one is created for you, so there is no actual creation function to deal with. For example, suppose a plugin has two parameters with the following attributes:

Parameter 1

- type: linear, numeric
- control ID = 0
- name = "Volume"
- units = "dB"
- min/max/default = −60/+12/−3

Parameter 2

- type: string list
- control ID = 42
- name = "Channel"
- string list: "stereo, left, right"

In the constructor, you would effectively create them with:

```
Globals()->SetParameter(0, -3); // volume = 0, default = -3dB
Globals()->SetParameter(42, 0); // channel = 42, default = item 0
```

After the constructor is finished, the base class will then query your plugin repeatedly, asking for more information about each of the newly created parameters. The query function is:

```
ComponentResult GetParameterInfo(AudioUnitScope inScope,
                            AudioUnitParameterID inParameterID,
                            AudioUnitParameterInfo& outParameterInfo)
```

The scope and parameter ID values are sent into the function (the scope represents whether the parameter is *global* and pertains to the AU as a whole, or if it is specialized—all of our AU parameters are global). The information is returned in an *AudioUnitParameterInfo* structure that encodes the following:

- name string to display for the user
- units (e.g. dB or Hz; may be left blank)
- minimum, maximum, and default parameter values
- type: *indexed* parameters display zero indexed string lists for the user while non-indexed parameters represent typical numerical parameters

For our parameters, you would decode the incoming parameter ID value and then fill in the *outParameterInfo* member. For continuous parameters we apply custom units to fill in our own string, while for the string-list parameter, we notify that this parameter is *kAudioUnitParameterUnit_Indexed*, which signifies the host that it needs to call another function to receive the string list.

```
if(inParameterID == 0)
{
    // --- VOLUME --- //
    CFStringRef name = CFStringCreateWithCString(NULL, "Volume"",
                                        kCFStringEncodingASCII);
    AUBase::FillInParameterName (outParameterInfo, name, false);

    // --- custom, set units
    CFStringRef units = CFStringCreateWithCString(NULL, "dB"",
                                        kCFStringEncodingASCII);
    outParameterInfo.unit = kAudioUnitParameterUnit_CustomUnit;
    outParameterInfo.unitName = units;

    // --- set min and max
    outParameterInfo.minValue = -60.0;
    outParameterInfo.maxValue = +12.0;
    outParameterInfo.defaultValue = -3.0;
}
else if(inParameterID == 42)
{
    // --- CHANNEL --- //
    CFStringRef name = CFStringCreateWithCString(NULL, "Channel"",
                                        kCFStringEncodingASCII);
    AUBase::FillInParameterName (outParameterInfo, name, false);

    // --- set indexed to get query for string list
    outParameterInfo.unit = kAudioUnitParameterUnit_Indexed;

    // --- set min and max
    outParameterInfo.minValue = 0;
```

```
    outParameterInfo.maxValue = 2;
    outParameterInfo.defaultValue = 0;
}
```

Finally, for any parameter that you denoted as indexed in the *GetParameterInfo( )* function, the base class will call another function so that you may supply the list of strings for display, which you pass back in a *CoreFoundation* array:

```
ComponentResult GetParameterValueStrings(AudioUnitScope inScope,
                                AudioUnitParameterID    inParameterID,
                                CFArrayRef*             outStrings)
```

To reply to the query for our parameter 2, we would implement the function as shown next. Notice that we create the string with a comma-separated version, then create the array with the comma-separated list—this is because it is easy to store the string list as comma separated. (Check out the various other *CFStringCreateArray* functions in the *CoreFoundation* documentation.)

```
if(inParameterID == 42)
{
    // --- convert the list into an array
    CFStringRef enumList = CFStringCreateWithCString(NULL,
                    "stereo, left right", kCFStringEncodingASCII);

    CFStringRef comma CFSTR(",");
    CFArrayRef strings = CFStringCreateArrayBySeparatingStrings(NULL,
                                                    enumList, comma);

    // --- create the array COPY
    *outStrings = CFArrayCreateCopy(NULL, strings);
}
```

4.5.1 Thread-Safe Parameter Access

For thread-safe access to the parameters, you use the two functions described above:

```
Globals()->GetParameter(. . .)
Globals()->SetParameter(. . .)
```

4.5.2 Initialization: Defining Plugin Channel I/O Support

An AU plugin declares its audio channel I/O capabilities via the function *SupportedNumChannels( )*, which is called during the initialization phase. The plugin returns its channel I/O support in an array of *AUChannelInfo* pointers, along with the count of possible audio I/O combinations as the return variable.

```
UInt32 SupportedNumChannels(const AUChannelInfo** outInfo)
```

The *AUChannelInfo* structure is simple and contains variables to hold the number of input and output channels:

```
typedef struct AUChannelInfo {
    SInt16 inChannels;
    SInt16 outChannels;
} AUChannelInfo;
```

AU has the least specific audio channel support of all of the APIs: you simply declare all of the channel count combinations the plugin supports. For example, both AAX and VST define two distinct versions of 7.1 surround sound audio: one for Sony and the other for DTS. AU does not discern between the two, and in this case would report only the value eight for the number of input channels and the number of output channels (7 + 1 = 8).

4.5.3 Initialization: Channel Counts and Sample Rate Information

The AU function *Initialize( )* is called when the host wants to place the plugin in an initialized state. The base class version of the function handles the basic audio channel support information—and it calls the *SupportedNumChannels* function as part of this process. It also sets up some low-level audio stream information and performs some basic initialization operations. Your AU plugin should override this method and first call the base class, then do any other work. For our FX plugins, we get the current DAW session sample rate and adjust any rate-dependent variables such as our plugin latency value. Remember that AU requires the latency to be reported in seconds, not samples. You can pick up the current sample rate and file bit depths with a query to the audio stream description, which ultimately leads to a structure that contains audio stream information, aptly named *AudioStreamBasicDescription,* which you acquire with a call to a helper function *GetStreamFormat( )*:

```
struct AudioStreamBasicDescription
{
    Float64          mSampleRate;
    AudioFormatID    mFormatID;
    AudioFormatFlags mFormatFlags;
    UInt32           mBytesPerPacket;
    UInt32           mFramesPerPacket;
    UInt32           mBytesPerFrame;
    UInt32           mChannelsPerFrame;
    UInt32           mBitsPerChannel;
    UInt32           mReserved;
};
```

You can get access to the sample rate like this:

```
Float64 sampleRate = GetOutput(0)->GetStreamFormat().mSampleRate;
```

Because of the various audio stream format attributes (including interleaving or non-interleaving), there are a series of helper functions that translate the stream information into channel counts, bit depths, and other data that you may need. These are defined in *CAStereamBasicDescription.h,* including *SampleWordSize( ),* which you may use to calculate the audio stream bit depth:

```
UInt32 bitDepth = GetOutput(0)->GetStreamFormat().SampleWordSize()*8;
```

If your plugin dynamically changes audio channel counts for some reason, this is where you would handle part of the operation (remember that this function ultimately calls *SupportedNumChannels*). An example implementation is shown here; notice that the base class function is called at the top of the function:

```
ComponentResult Initialize()
{
    // --- call base class FIRST
    ComponentResult result = AUEffectBase::Initialize();
```

```
  if(result == noErr)
  {
    // --- do your initialization here
    Float64 sampleRate = GetOutput(0)->GetStreamFormat().mSampleRate;

    // --- do more stuff . . .
  }
  return result;
}
```

4.6 The Buffer Process Cycle

The AU buffer processing cycle follows the same paradigm as AAX and VST. A stock AU plugin is automatically assumed to process channels independently of one another, and the processing function will get called once for each audio channel. In Chapter 2 we discussed that to be as generic as possible with our plugin design, we would like to process the audio buffers all at once, breaking them into frames so that we may access all channel data at the same time for the same sample period. This is not possible with the stock AU channel-processing paradigm. However, the AU plugin may override the low-level functions that generate those stock monaural channels and access the buffers directly, in exactly the same manner as AAX and VST. This allows you to write processing code that is portable across the different APIs. The two functions we wish to override are named *Render* and *ProcessBufferLists* and are shown in Figure 4.3.

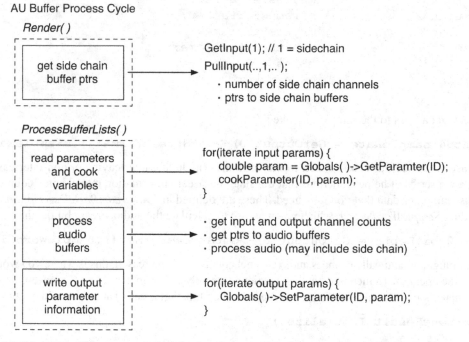

Figure 4.3: The AU buffer process cycle is broken into two functions, *Render* and *ProcessBufferLists*; pseudo-code for each functional block is shown on the right.

You need only override the *Render* function to gain access to additional side chain input buffers; you pick up pointers to these buffers and store them. The *ProcessBufferLists* function is called immediately after *Render*, and you then perform your audio signal processing using pointers to the side chain buffers if your plugin uses the side chain input. Using the buffer list process function makes the processing identical across the APIs. As we saw in Chapter 2, all APIs transmit individual, de-interleaved channel buffers—one buffer per audio channel.

We can break the buffer processing into three parts, exactly following Figure 4.3 where we have the succession:

1. Update parameters from GUI control changes, cooking variables as needed.
2. Process the audio data using the updated information.
3. Write outbound parameter information (meters, signal graphs, etc.).

The paradigm is identical with AU parameters, except for one caveat: there is no way to determine if a GUI parameter has changed since the last buffer process cycle. In other words, there is no Boolean "dirty" flag to signify that the parameter had been altered. Although your plugin may override low-level parameter functions to try to circumvent this (not recommended), thread safety requires that we accept the fact that we will potentially need to update and re-cook every parameter in the plugin at the start of each buffer process cycle. This is not necessarily a bad thing, since the user could set up a DAW session with automation on every single plugin parameter so that they are all in flux on each buffer process cycle. This represents a worst-case scenario and is not unlike the paradigm that hardware designers follow—for example, a hardware synth might assume that every patch parameter may be modulated on every sample interval.

One option is to access each parameter and compare it with a stored value from the last process cycle. You may then subtract the values to try and determine if anything changed—because of rounding, you would usually subtract the two values and compare them to some small rounding delta value, such as 0.0001. One argument here is that you are eating CPU cycles to perform the comparison. However, if a parameter has a very complicated, CPU-intensive cooking function, then we really only want to update the parameter if we are sure it changed. In the sample book projects, we generally write the cooking functions to first check to see if the parameter has changed, then cook the data only if it has changed, offloading that chore to the plugin kernel and out of the plugin shell code. There are also granularity methods in which you only update internal parameters on some granular level that is done by counting sample periods. Ultimately, it is up to you to decide how to handle this operation. For very simple plugins with short and simple cooking functions, it is usually simpler to just blindly update and recalculate all parameters at the start of the buffer process cycle.

4.6.1 *Processing: Updating Plugin Parameters From GUI Controls*

For AU, acquiring the GUI parameter values is simple and straightforward: you just call the parameter's *get* function that we discussed in Section 4.5. You would typically write a for-loop that simply accesses the parameters one at a time and grabs the current parameter value for each. Then you would need to call the cooking functions to process the parameter changes. This is shown in Figure 4.3 inside of the *ProcessBufferLists* function. Your plugin framework should have this built-in to the buffer processing (for ASPiK, see the *preProcessAudioBuffers( )* function). Of course, if you are rolling your own AU plugin from scratch then you will need to handle this yourself.

4.6.2 Processing: Resetting the Algorithm and Preparing for Streaming

All audio plugins need to have a way to prepare for the next audio streaming event, as we saw in Chapter 2. For delay and filtering plugins this usually involves flushing old data out of buffers and resetting linear or circular buffer pointers back to the top. Other plugins may require resetting state variables for each new audio run: for example, some of the *PhaseVocoder* algorithms in Chapter 20 need to keep track of the running phase of different fast Fourier transform (FFT) frequency bins, and these values would likewise need to be reset or cleared at the start of each new audio streaming event. Note that this also includes interactions between the user and the DAW: for example, even with audio streaming, most DAWs allow the user to re-position the audio playback location, which may require resetting internal variables and states. AU provides the *Reset* function for this operation. In ASPiK, I use *both* the *Reset* and *Initialize* functions to check and update the sample rate, as well as to flush buffers and reset variables. The reason is from experience: if you load an AU plugin into Apple Logic, Apple AULab, and Reason®, then set breakpoints to watch how the host and base class interact with the plugin, you will find three different sequences of calls to these functions. Therefore, our *Reset* and *Initialize* functions are fundamentally identical except for the base class call. The *Reset* function is shown here; the *Initialize* function is identical except it calls *AUMIDIEffectBase::Initialize* instead of *AUBase::Reset*.

```
ComponentResult AUFXPlugin::Reset(AudioUnitScope inScope,
                                  AudioUnitElement inElement)
{
    // --- reset the base class
    ComponentResult result = AUBase::Reset(inScope, inElement);

    if(result == noErr)
    {
        // --- do your initialization here
        Float64 sampleRate = GetOutput(0)->GetStreamFormat().mSampleRate;
        // --- do more stuff . . .
    }

    return result;
}
```

4.6.3 Processing: Accessing the Audio Buffers

AU differs from both AAX and VST in that the side chain buffers are accessed in a separate function from the audio signal processing, which is the *Render* function as shown in Figure 4.3. The function arguments include the audio timestamp structure that contains the absolute sample location, other optional timing information like SMPTE and word clock data, and the number of frames that will need to be processed in the upcoming call to *ProcessBufferLists*.

```
OSStatus Render(AudioUnitRenderActionFlags& ioActionFlags,
                const AudioTimeStamp&       inTimeStamp,
                UInt32                      inNumberFrames)
```

If you set up a side chain input during the construction, then you may access it and get its channel count as follows. First you need to check to see if the side chain actually exists by calling the function

HasInput( ) and passing the side chain channel (1)—if none exists then you can bail out and call the base class method:

```
bool scAvailable = false;
try {scAvailable = HasInput(1);}
catch (. . .) {scAvailable = false;}

if(!scAvailable)
      return AUEffectBase::Render(ioActionFlags, inTimeStamp,
                                      inNumberFrames);
```

If the side chain exists, you access it with the function *GetInput( )* and then get the sample count and buffer pointer using the *PullInput( )* function. This requires some of the arguments to *Render*, which are not available to *ProcessBufferLists*, and hence the reason for putting the side chain access in a separate function:

```
// --- get and check side chain input
AUInputElement* SCInput = GetInput(1);
if(!SCInput) return AUEffectBase::Render(. . .);

AudioBufferList* scBufferList = nullptr;

if(SCInput->PullInput(ioActionFlags, inTimeStamp, 1, inNumberFrames) ==
   noErr) {
    int sidechainChannelCount = SCInput->NumberChannels();
    scBufferList = &(SCInput->GetBufferList());
}
```

The audio signal processing is done in the *ProcessBufferLists* function and consists of three parts:

1. Get the input and output channel counts.
2. Get pointers to the input and output buffers.
3. Get the frame count and set up a for-loop to process the from input to output.

The function and its arguments are as follows.

```
OSStatus ProcessBufferLists(AudioUnitRenderActionFlags& ioActionFlags,
                       const AudioBufferList&        inBuffer,
                       AudioBufferList&              outBuffer,
                       UInt32               inFramesToProcess)
```

The input and output buffer pointers are contained in the *AudioBufferList* references, while the *inFramesToProcess* variable contains the frame count. The *AudioBufferList* is a structure that allows access to the individual channel buffers that are indexed the same as the channel I/O interleaving we saw in Chapter 4. For a stereo buffer, index = 0 is the left channel while index = 1 is the right. To pick up the channel counts and buffer pointers, you would write something like this (count and pointer validation is omitted for brevity):

```
SInt16 auNumInputs = GetInput(0)->GetStreamFormat().NumberChannels();
SInt16 auNumOutputs = GetOutput(0)->GetStreamFormat().NumberChannels();
```

To get the left and right buffer pointers out of the *AudioBufferLists*, you need to cast the data block pointer (*mData,* which is declared as a *void**) to a *float**. You might write this:

```
float* inputL = (float*)inBuffer.mBuffers[0].mData;
float* inputR = (float*)inBuffer.mBuffers[1].mData; // 1 = right

float* outputL = (float*)outBuffer.mBuffers[0].mData;
float* outputR = (float*)outBuffer.mBuffers[1].mData; // 1 = right
```

Now that you have the pointers and frame count, you could set up a loop to process the information (perhaps using the side chain as well). A simple volume plugin might be coded like this; here we are processing stereo in/stereo out:

```
// --- loop over frames and process
for(int i=0; i<inFramesToProcess; i++){
    // --- apply gain control to each sample
    outputL[i] = volume * inputL[i];
    outputR[i] = volume * inputR[i];
}
```

To use information in the side chain buffer list, you simply treat it the same way as the others:

```
float* sideaChainInputL = (float*)scBufferList.mBuffers[0].mData;
float* sideaChainInputR = (float*)scBufferList.mBuffers[1].mData;
```

4.6.4 Processing: Writing Output Parameters

If your plugin writes data to output VU meters or some kind of custom view (see Chapter 21) then you would do that here as well. With all of the processing done, you would then write the output information to the appropriate GUI parameter that needs to receive it. The code that follows would take the magnitude (absolute value) of the first output sample from each output buffer and write to a VU meter connected to the appropriate parameter index value:

```
Globals->SetParameter(meterL_ID, fabs(outputL[0]));
Globals->SetParameter(meterR_ID, fabs(outputR[0]));
```

4.7 The AU/GUI Connection

We saw that as part of the plugin description, we may signify that the AU plugin implements a custom *Cocoa* GUI by responding to the *GetPropertyInfo* query. Specifically, we need to return a success code for the property identifier *kAudioUnitProperty_CocoaUI,* but we also need to massage the function parameters slightly to prepare for future calls. The code for doing all of this is here. Notice that if we don't support the property, decoded with the ID value, we call the base class implementation for the return.

```
ComponentResult GetPropertyInfo(AudioUnitPropertyID inID,
                    AudioUnitScope        inScope,
                    AudioUnitElement      inElement,
                    UInt32&               outDataSize,
                    Boolean&              outWritable)
```

```
{
  if (inScope == kAudioUnitScope_Global)
  {
    if(inID == kAudioUnitProperty_CocoaUI)
    {
      outWritable = false;
      outDataSize = sizeof(AudioUnitCocoaViewInfo);
      return noErr;
    }
  }

  return AUEffectBase::GetPropertyInfo(inID, inScope, inElement,
                                       outDataSize, outWritable);

}
```

At this point the host believes we have a GUI to display, so we have several issues to take care of:

- Create and display the GUI when the user opens it.
- Destroy the GUI when the user closes it.
- Set up an AU event listening system to properly connect the *Cocoa* GUI to the underlying AU parameters in a thread-safe manner.

When the user opens the GUI, the host will query the plugin for two pieces of information: the name of the *Cocoa* GUI class factory and the path to the AU bundle, called the *bundle URL*. When you create your AU project, you create and name the *Cocoa* GUI class factory. If you are using ASPiK, the class factory name will be generated for you, including the GUID decoration for uniqueness. If you are using another plugin framework, consult with that documentation. If you're rolling your own, then you make up the name. Getting the bundle path is not as simple and requires several calls to *CoreFoundation* helpers that are used to suss out the final path. The basic code is shown next. You can of course get the complete project code that goes with the book. The return code "*fnfErr*" is a "file not found" designator. Notice that the view factory name and bundle path are placed into a *AudioUnitCocoaViewInfo* structure that is passed back in the *outData* parameter.

```
ComponentResult GetProperty(AudioUnitPropertyID inID,
                            AudioUnitScope        inScope,
                            AudioUnitElement      inElement,
                            void*                 outData)
{
  if (inScope == kAudioUnitScope_Global)
  {
    // --- find the UI associated with this AU
    if(inID == kAudioUnitProperty_CocoaUI)
    {

        // --- Look for a resource in the main bundle
        CFBundleRef bundle = CFBundleGetBundleWithIdentifier(. . .)
        if(bundle == NULL) return fnfErr;
```

```
      // --- get the path
      CFURLRef bundleURL = CFBundleCopyResourceURL(. . .)

      if(bundleURL == NULL) return fnfErr;

      CFStringRef className = "CocoaGUIFactory_234324";

      // --- pack data into AudioUnitCocoaViewInfo structure
      AudioUnitCocoaViewInfo Info = {bundleURL, {className}};
      *((AudioUnitCocoaViewInfo *)outData) = cocoaInfo;

      return noErr;
    }
  }
  return AUEffectBase::GetProperty(inID, inScope, inElement,outData);
}
```

The operations that proceed after this point are going to be highly dependent on your plugin framework, or your own Objective-C code if you are designing from scratch. Because these details are all going to vary, and for ASPiK users are fully documented in both the code and user manual, we will omit the majority of this minutia here. Please refer to your framework documentation.

However, any framework that attempts to implement an AU GUI is going to need to deal with two issues that are exclusive to AU and *Cocoa*:

- *Cocoa's* flat namespace
- the AU Event Listener system

4.7.1 Cocoa's *Flat Namespace*

Cocoa does not use namespaces. This automatically creates an issue with name collisions for GUI objects—meaning that two plugin designers could accidentally name their *Cocoa* view factories, *Cocoa* views, and *Cocoa* GUI objects similarly. For example, the designers might both name their view objects *AUPluginView*. This will cause problems when the user tries to use both plugins in one DAW session, generally manifesting itself as the wrong GUI being opened for one of the plugins (this depends on the order in which the user instantiated the plugins in the DAW). Great care must be taken in naming the GUI view factory, view, and internal objects. Apple recommends decorating the object names with the designer's initials or other symbols. To overcome this problem, VSTGUI4 architects use an object name decoration scheme that is implemented at runtime (when the plugin is instantiated) and is based on the current date and time. This fixes the issue with the GUI object naming, but not the naming of the *Cocoa* view factory or the view object it creates. If you are using ASPiK, when you generate the complier project with CMake, a GUID (see Chapter 2) is created for you and is used (among other things) to decorate the *Cocoa* view factory and object names, guaranteeing by design that you will not have any namespace issues. If you are designing your plugin from scratch, then you need to address the namespace issues on your own, and if using other third party frameworks you should consult with the documentation to understand how they circumvent this problem.

4.7.2 *The AU Event Listener System*

The AU specification goes to great lengths to separate and isolate the GUI from the plugin parameters so that the system operates in a thread-safe manner. Deep inside the AU parameter is an atomic variable access that ensures that data are never left in a half-baked state (see Chapter 2). However, there is a completely separate mechanism for moving parameter values to and from the GUI called the *AU Event Listener system*. All of the code required for implementing the AU Event Listener system is inside of the GUI object. Of course, that will depend on your plugin framework: the implementation details are all going to be different. With ASPiK, we do all of this work in C++ because we are using VSTGUI4. Other frameworks or scratch-built plugins may implement the Event Listener system in Objective-C.

However, the basic idea is consistent across all implementations. It involves information flow in two directions:

1. When the user adjusts a GUI control, we transmit the control change information into the AU parameter using an AU event listener in a *parameter-set* operation.
2. If the user loads a preset, has automation running, or opens a saved DAW session with the AU plugin in a particular state, the GUI needs to be updated with the proper information and this is done with an *event dispatch* operation; with automation running, the GUI is animated (with controls moving on their own) through a series of event dispatch calls.

There are four components to implementing the AU Event Listener system. In order to implement it, the GUI object must maintain two member variables: one is a reference to the AU plugin with which the GUI is connected; the other is an event listener reference, which is used during the parameter-set operation. This is somewhat interesting: in order to isolate the two components, one of them (the GUI) must hold a reference to the other. The GUI object might declare these two variables as follows:

```
AudioUnit          m_AU;
AUEventListenerRef AUEventListener;
```

4.7.2.1 The Event Lister Dispatch Callback Function

When asynchronous parameter changes are being sent to the GUI, such as after a preset is loaded or when automation is running, an old-fashioned C-style callback function is required. This function sits outside of the GUI object's code. You may name it as you wish, but it must have the correct argument set for use. In ASPiK, we name it *EventListenerDispatcher* and it is defined like this:

```
void EventListenerDispatcher(void *inRefCon, void *inObject,
                       const AudioUnitEvent *inEvent,
                       UInt64 inHostTime, Float32 inValue)
```

If you are used to working with callback functions, then the *void** arguments should be familiar. When the GUI object sets up the AU Event Listener system, it passes in a pointer to itself (the *this* pointer). When the callback function is invoked, that pointer is passed into the callback cloaked as a *void** in the *inRefCon* variable. Then, within the function we may uncloak the pointer and use it—for example, to move the location of a control in response to the preset load or automation. The exact type of object that is passed depends on the framework and GUI object design. The *void** *inObject* points to something associated with the event, which may be a control pointer, or the same *this* pointer as we used before. In any case, the *AudioUnitEvent* input value is used to decode the parameter's control ID that is connected to the GUI control and the *inValue* represents the un-normalized, actual AU parameter value that needs to be reflected in the GUI control. To implement the callback function, you may write something like the following, where the class name for the GUI is *PluginGUI* and it contains a member function named *dispatchAUControlChange* to move the location of a control based on the parameter's actual value; of course the details of this function depend on the GUI object implementation. The *this* pointer for the *PluginGUI* object was passed into the AU Event Listener when it was created (see Section 4.8).

```
void EventListenerDispatcher(void *inRefCon, void *inObject,
                            const AudioUnitEvent *inEvent,
                            UInt64 inHostTime, Float32 inValue)
{
    // --- un-cloak the void* to reveal the PluginGUI object
    PluginGUI *pEditor = (PluginGUI*)inRefCon;
    if(!pEditor) return;

    // --- get the control ID from the event's parameter
    UInt32 controlID = inEvent->mArgument.mParameter.mParameterID;

    // --- use the control ID and actual value to move the GUI control
    pEditor->dispatchAUControlChange(controlID, inValue);
}
```

4.7.2.2 Creating the AU Event Listener

During initialization, the GUI object must create the main event listener component, and then it adds an event for each and every parameter that connects to a GUI control. You use the *AUEventListenerCreate* function to set up the listener. The arguments include the name of the event dispatcher callback function, an object-specific pointer to pass into that callback function as the *inRefCon* value, some in *CoreFoundation* information about the message loop, a pointer to the GUI object's *AUEventListenerRef* declared above, and finally some information on the granularity of the event system. You have a bit of leeway in just how accurate you want/need the system to be. An example of this creation function follows, with the important arguments in bold. You can see the name of the callback function and the *this* pointer are the first two arguments, while the member variable for the *AUEventListenerRef* is last.

```
const Float32 AUEVENT_GRANULARITY = 0.0005;
const Float32 AUEVENT_INTERVAL = 0.0005;
enum {kBeginEdit, kEndEdit, kValueChanged, kNumEventTypes};

__Verify_noErr(AUEventListenerCreate(EventListenerDispatcher,
                                     this,
                                     CFRunLoopGetCurrent(),
                                     kCFRunLoopDefaultMode,
                                     AUEVENT_GRANULARITY,
                                     AUEVENT_INTERVAL,
                                     &AUEventListener));
```

With the AU Event Listener created, you then need to add the events for which your GUI would like notifications. There are three primary event types that you need to support.

1. **Begin parameter change:** the user clicks on a control to begin moving it.
2. **Parameter value change:** the user is changing the value of the control.
3. **End parameter change:** the user releases the control and is finished with the change.

The begin and end parameter change events are important for ensuring that automation will work properly as it needs to know when the control is being adjusted (think about your favorite DAW's automation recording system—you touch the control, move it, then release it). The value change event will fire off as the user moves the control, or as the automation systems plays back information into the control, or if a preset is loaded. This means that we need to set up three specific event types for *each* plugin parameter. Adding the events is fairly straightforward; this uses the *AudioUnitParameter* structure, which contains variables for the AU event reference, the parameter's control ID, the scope, and the element. The scope is always global, and it has only one element with index = 0. Each event type is added with the *AUEventListenerAddEventType* function, which also takes an object-specific pointer to be passed into the dispatcher callback. We use the *this* pointer again.

Suppose we have declared an AU parameter for a simple volume control knob. The control ID for this parameter has been predefined as VOLUME_CTRL. To set up the three event types for this control, you would write:

```
// --- to hold the event
AudioUnitEvent auEvent;

// --- parameter information
AudioUnitParameter parameter = {m_AU, VOLUME_CTRL,
                                kAudioUnitScope_Global,
                                0}; // 0 = global element

// --- attach parameter
auEvent.mArgument.mParameter = parameter;

// --- add the begin change gesture (aka "touch")
auEvent.mEventType = kAudioUnitEvent_BeginParameterChangeGesture;

__Verify_noErr(AUEventListenerAddEventType(AUEventListener,
                                           this,
                                           &auEvent));

// --- add the end change gesture (aka "release")
auEvent.mEventType = kAudioUnitEvent_EndParameterChangeGesture;

__Verify_noErr(AUEventListenerAddEventType(. . .);

// --- add the value change event
auEvent.mEventType = kAudioUnitEvent_ParameterValueChange;

__Verify_noErr(AUEventListenerAddEventType(. . .);
```

4.7.2.3 Destroying the Event Listener

When the GUI object is destroyed, it must first destroy the Event Listener that it created. This is done with a single call to *AUListenerDispose,* which requires the saved *AUEventListenerRef* that we used during creation.

```
__Verify_noErr(AUListenerDispose(AUEventListener));
```

4.8 Destruction/Termination

AU plugins are easily destroyed. All of the low-level cleanup work is done for you in the base classes as the destructor for your AU object is called. Your AU plugin object needs only to free up any dynamically declared resources, which you do in the object destructor. The AU plugin object is not a class factory, and once the object is destroyed it is gone forever. You don't need to worry about state-hood or other class factory details.

4.9 Retrieving AU Host Information

AU provides a mechanism for your plugin to query the host for information. This is unique among the three plugin APIs in that host information is delivered to the plugin on each buffer processing cycle. The host information includes data about the current audio playback such as the absolute sample position of the audio file (the index of the first sample in the input audio buffer for the current buffer processing cycle), the DAW session's beats-per-minute (BPM) and time-signature settings, and information about the audio transport (whether it is playing or stopped, or if looping is occurring). Not all AU hosts support every possible piece of host information, and some of them are notoriously incorrect: for example, after the user starts audio streaming for the first time, Apple Logic will report that the transport is in play mode whether the audio is actually playing or not; it has a "sticky-on" Boolean flag. So, you need to be careful about interpreting some of this information. However, the absolute sample location and time, along with the host BPM and time-signature information, is very well defined across all AU clients. Your plugin may call three functions on the AU host at any time it wishes to obtain this information. For ASPiK, we make these calls at the beginning of each buffer processing cycle so that the plugin core object has the updated information at the start of its internal processing. The three functions are as follows.

- *CallHostBeatAndTempo*: get the current BPM and beat value
- *CallHostMusicalTimeLocation*: get the time signature, number of samples to the next beat, etc.
- *CallHostTransportState*: get the transport state (playing or stopped, notoriously wrong in Apple Logic) and looping information

The arguments for each function are all implemented as pass-by-pointer, so you must first declare local variables to hold the information. For example, to get the BPM and current beat value you use *CallHostBeatAndTempo*:

```
// --- return variables
Float64 outCurrentBeat = 0.0;
Float64 outCurrentTempo = 0.0;

OSStatus status = CallHostBeatAndTempo(&outCurrentBeat,
                                       &outCurrentTempo);
if(status == noErr)
{
    hostInfo->dBPM = outCurrentTempo;
    hostInfo->dCurrentBeat = outCurrentBeat;
}
```

You can check the Apple AU documentation for the details of the other functions. If you are using ASPiK, all of the information from all three function calls will be available to your plugin core and will be updated on each buffer processing cycle. Otherwise, consult your plugin framework to see what information is available to you.

4.10 Validating Your Plugin

Once your plugin has compiled, you need to validate it before attempting to load it into a DAW for testing. For Apple Logic, validation is crucial—if you try to load your plugin and it crashes, or does not pass Logic's AU validation, then it will be "black-listed" and the DAW will never load it again. The MacOS contains an AU validation application called *auval* that you invoke from Terminal. Both *auval* and Apple Logic use the two four-character codes for manufacturer and subtype that we discussed in Section 4.3 in your plugin description to select the component for testing. If Apple Logic black-lists your plugin (because you neglected to run *auval* first) then you will need to change at least one of those four-character codes, typically the subtype, to fool Apple Logic into thinking it is a different plugin. The main rule is this: take the time to run *auval*—in the end it will save you pain and headaches. To run *auval*, you need to know the AU type code that your plugin uses, which depends on the base class from which it is derived. This is identical to the type code you set up in the plugin description: *aufx*, *aumf*, or *aumu* (see Section 4.3). The syntax for validating an *aufx* plugin is:

> *auval -v aufx <PLUG> <COMP>*

<PLUG> is your four-character plugin code and <COMP> is your four-character company name. For the plugin description in Section 4.3 you would type:

> *auval -v aufx FLNG PLCO*

The validation will succeed or fail; it will show a very detailed printout of all of the tests that were run and which test(s) passed and failed. It will also give you some hints about some of the failures, but if the plugin fails validation, in many cases you will need to go back and do some sleuthing. The problem is that if your constructor is malformed, you may never get past instantiation, which will make debugging difficult—and of course Apple Logic will prevalidate your plugin. You can get help and more information about figuring out *auval* problems at www.willpirkle.com/forum/. Browse the previous user questions or post your own—all are welcome at the Forum. There are no stupid questions!

4.11 Using ASPiK to Create AU Plugins

There is no substitute for working code when designing these plugins; however, the book real estate that the code requires is prohibitive. Even if you don't want to use the ASPiK *PluginKernel* for your development, you can still use the code in the projects outlined in Chapter 8.

4.12 Bibliography

Apple Computers, Inc. 2018. *The Audio Unit Programming Guide*, https://developer.apple.com/library/archive/documentation/MusicAudio/Conceptual/AudioUnitProgrammingGuide/Introduction/Introduction.html, Accessed August 1, 2018.
Pirkle, W. 2015. *Designing Software Synthesizers in C++*, Chap. 2. New York: Focal Press.

CHAPTER 5

AAX Native Programming Guide

The AAX plugin format is a third generation plugin API built on top of a prestigious pedigree of ultra-high quality plugin APIs including RTAS (Real Time Audio Suite, now obsolete for any version of Pro Tools past v10) and TDM (Time Division Multiplexing, the original DSP-based plugin API requiring DSP farm cards). AAX comes in two flavors: DSP and Native, the former requiring external DSP Integragted Circuits (ICs) along with DSP code and the latter running natively on the user's CPU and written in C++. This chapter and book projects are for AAX Native plugins, which we refer to as simply "AAX" from this point forward. As a third generation product, its architects leveraged on lessons learned from the previous versions to produce a clean and tight plugin API that is simultaneously both advanced and relatively easy to code. The AAX SDK is one of Avid's public APIs. You may download and code for it free of charge—you need only set up a developer's account with Avid to get started, which is also free of charge. Go to http://developer.avid.com to begin the process.

AAX is unique in that it is designed specifically for Pro Tools and other AAX hosts, and Avid hardware that supports it—you won't find AAX supported in DAWs outside of Pro Tools. As an AAX developer, you can also leverage on Avid's desire to publish a variety of high-quality plugins—not only will they will test your plugin for you (!) but they will optionally place it on their "Plugin Store" so that users may download/buy it directly from within Pro Tools, called the Alliance Partner Program (www.avid. com/alliance-partner-program/about-marketplace).

In addition to the SDK you will also need an iLok2® copy protection device so that you may run the developer version of Pro Tools (Pro Tools Dev) that allows you to debug and test your plugin—you can't use the Pro Tools commercial release for this. When you sign up for a developer account and you have your iLok key, you may then contact Avid and request an iLok activation code for your Pro Tools Dev; that version is free to designers with an Avid account but of course it is useless without the iLok. When testing and debugging your plugin in Pro Tools Dev, you do not need to license your own plugin through iLok, you need only it for Pro Tools. When releasing an AAX plugin for commercial sale, you will then need to address the copy protection mechanism. Thanks to Avid for providing the following information.

> Pro Tools uses digital signature checking to prevent the loading of AAX Plug-Ins that are other than those created by the original developer, including attempted unauthorized copies, or 'cracks.' Avid requires that all AAX Plugins bear a digital signature without which it will not load in Pro Tools. Pace Anti-Piracy, Avid's long-time license management partner, licenses and operates the digital signature service that makes this possible. AAX developers who use PACE® Eden for license management do not need to take any additional action related to digital signatures. AAX developers who use other methods for license management can write to partners@avid.com with 'Pace signing Tools Request' in the subject line to request the needed tools, providing company and contact information. Avid will confirm that the requester is a valid AAX developer and request that Pace provide the developer with the license and tools for the PACE® Eden Digital Signature

Service. In an easy process, the developer uses a command line tool to add the signature during the build process for the Plug-In.

There is also an AAX developer's forum that you may access with your AAX developer's account where you may find the latest news, bug reports, and feature requests.

5.1 Setting Up the AAX SDK

You will need to create a free developer account with Avid and download the AAX SDK to create AAX plugins. The SDK is simply a folder full of subfolders in a hierarchy that Avid defines. There is no installer to run. You can create your account and get the SDK by visiting https://my.avid.com/account/orientation. The root folder of the AAX SDK is named according to a convention that encodes the SDK version as letters and numbers. For example, the *AAX_SDK_2p2p1* folder name would be interpreted as "AAX SDK version 2.2.1."

The next step is to compile the AAX base library. Fortunately, you need only compile the AAX library once for each OS and for each compiler build configuration (one for Debug and another for Release). To build the library, navigate into the

..\AAX_SDK\Libs\AAXLibrary\MacBuild

or

..\AAX_SDK\Libs\AAXLibrary\WinBuild

folders, depending on your target OS. In these folders, you will find Xcode and Visual Studio projects for the library. Open the projects and build the libraries for both Debug and Release configurations on the appropriate platforms. If you are going to make both MacOS and Windows AAX plugins, then you need to build the libraries for both compiler/OS targets. After building, the resulting static library files will be located in the ..\AAX_SDK\Libs\Debug and ..\AAX_SDK\Libs\Release subfolders. There is nothing else to do: you do not need to copy or move the library files. Note that if you update your compiler version, you will need to recompile these libraries, in part because the C++ standard library (*std*::) has a different binary for different compilers as well as for different configurations (Debug and Release).

5.1.1 AAX Sample Projects

The AAX SDK comes with a set of sample plugin projects that will generate more than a dozen example plugins in all shapes and sizes. The sample projects are located in the */ExamplePlugins* folder. As of this writing, they do not use CMake, so there are pre-made Visual Studio solution and Xcode project files for each sample plugin—just double-click on the appropriate file to launch the compiler. These include FX and synth plugins as well as multiple sub-variations that demonstrate the different design patterns and the many options for GUI library interfacing. The projects in this book are based on the monolithic example *DemoSynth,* which includes a big chunk of complex code that implements the thread-safe parameter queue system. This was lifted verbatim from the *DemoSynth* project (thanks Avid!).

5.1.2 AAX Documentation

The AAX API documentation is contained within the SDK and is generated using Doxygen, which creates HTML-based documents that you view with a web browser. You can open the documentation by simply double-clicking on the index.html file, which is the Doxygen homepage. This file is located in the SDK root as *docs.html;* you should definitely bookmark it and refer to it frequently. The AAX SDK documentation is generally excellent, improves with each new SDK release, and contains a wealth of information as well as documentation on all structures, classes, functions, and AAX programming paradigms. This includes information not included here, such as the Texas Instruments (TI) DSP documentation for the AAX DSP variant.

5.2 AAX Architecture and Anatomy

AAX plugins may be written for Windows and MacOS. This means you must use the proper compiler and OS to target these products; you cannot use Xcode to write a Windows AAX, for example. AAX packages both Windows and MacOS plugins in a *bundle*, which you may think of as simply a directory of subfolders and files. The bundle is set up to appear as a file for a typical user, though in reality it is a folder; in MacOS you can right-click on the plugin "file" and choose "Show Package Contents" to get inside the bundle.

Within the AAX API, there are two distinct design patterns: one is for monolithic plugin objects and the other is for non-monolithic designs. A monolithic plugin uses an external, non-AAX object as the plugin implementation, handing the audio signal processing as well as dealing with parameter updates from the GUI, MIDI messaging, and other chores. As with the VST3 *SingleComponentEffect* paradigm, this does not necessarily include the code to render the GUI, but rather to deal with parameter changes that result from GUI manipulation and provide a thread-safe, two-way path to transfer parameter changes to and from the GUI. The term "monolithic" we use in this book comes from the AAX SDK: "Plug-ins that share data directly between their data model and algorithm are referred to as monolithic . . . AAX Native plug-ins that are designed to work with a cross-format framework may use a similar design." In this text, I refer to the non-monolithic design pattern as the *standard* AAX design.

This monolithic paradigm follows a similar design pattern as our AU and VST3 (*SingleComponentEffect*) plugins. Therefore, it is attractive for us to use, as it allows a unified method of coding for all three APIs at the same time. The AAX sample code includes examples of both design patterns (monolithic and standard) as well as AAX DSP variants. You are encouraged to study those projects as part of your AAX education, especially if you plan on designing plugins exclusively for AAX. Figure 5.1a shows the top-level conceptual architecture of an AAX plugin that is no different from that of AU or VST—part of the plugin named the *algorithm* does the audio signal processing while another part called the *data model* handles the plugin parameters. The basic structure of an AAX plugin is shown in Figure 5.1b and consists of the algorithm and data model along with a GUI interfacing component and a plugin description. In this non-monolithic case only the GUI interface and data model are implemented as C++ objects, shown as a thick outline around the blocks. For this version, the algorithm sits outside of any C++ object and is implemented as a pure C-style callback function. As such, it uses a *void** mechanism as a data path from the other functional components. This expanded view also includes a *description* block that is implemented as a set of *static* C-style functions that, like the algorithms, are also not part of any C++ object.

The monolithic version is shown in Figure 5.1c, where the algorithm and data model are combined. This means that the signal processing occurs in a C++ member function rather than an external, heavily isolated C-style callback function. This follows the same design pattern as we discussed in Chapter 2, and therefore is the same as AU and VST3. Now you can see why this is the choice for writing cross-API-compatible plugin objects. This processing function that is part of the C++ object is prototyped identically to the standard version (it has the same arguments) and it must be declared as a *static* member function, which prohibits it from accessing non-static C++ member variables and functions.

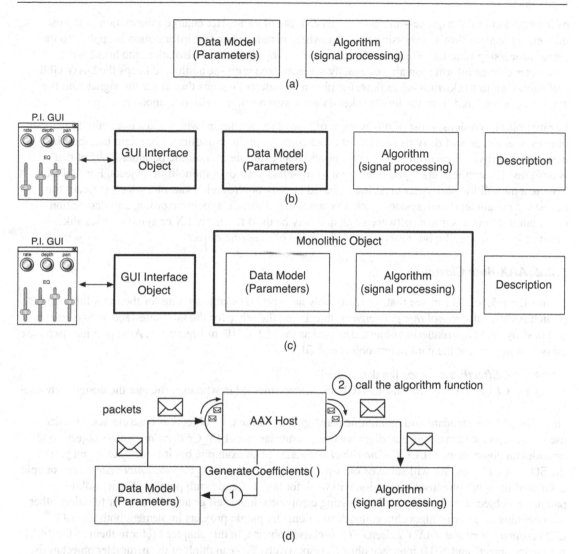

Figure 5.1: (a) The AAX standard plugin model consists of a signal processing component and a component that handles plugin parameters. (b) The non-monolithic AAX plugin model uses C++ objects for the GUI interface and the data model (parameters), while the algorithm and description are static C-functions. (c) The monolithic model combines the data model and the signal processing into one object; the thick block outline denotes a C++ object. (d) The data model communicates with the algorithm via a host-provided message system with a two-step process (note: this is a conceptual figure).

5.2.1 AAX Model-Algorithm Synchronization

AAX uses a message system to allow the data model and algorithm to communicate; this system applies to both monolithic and standard plugins. The host provides the buffered messaging system, as shown in Figure 5.1d (note that this is a conceptual diagram only). At the top of the buffer process cycle, the host will call the *GenerateCoefficients* function on the plugin, which will post a new packet of information into the queue—in general, this consists of parameter change information. The host then calls the algorithm's processing function, where parameter update information is applied to the signal processing function. For a standard design with the algorithm in isolation, the host ensures that parameter change information arrives exactly synchronized with the audio and keeps the DAW GUI that shows audio tracks moving in time, the plugin parameter changes that affect the signal, and the audio that is rendered from the loudspeakers, tightly synchronized with one another.

For monolithic versions, most of this work is offloaded to the plugin, which must maintain thread-safe queues, make copies of dirty parameter lists, and conform to the monolithic paradigm that the SDK describes to provide synchronization. Avid provides an example monolithic parameter object that is designed for synthesizers—you may use it to sub-class your own monolithic objects. For ASPiK, I wrote a monolithic parameter class that is based heavily on the Avid example, which includes the necessary parameter queue system, packet posting, initialization, synchronization, and description code, but does not assume a software synth and may be used for either FX or synth plugins alike. It is essentially a less specific but nearly identical version of the same object.

5.2.2 AAX Base Classes

From Figure 5.1c you can see that we really only need two C++ objects: one for the monolithic object (which is called the *monolithic parameters* object) and the other for the GUI interface, which creates and destroys the GUI rendering object, depicted as the "P.I. GUI" in Figure 5.1. Avid provides two base classes to implement the parameters object and GUI:

- ***AAX_CEffectParameters***: the data model
- ***AAX_CEffectGUI***: maintains the GUI object, implemented in whatever manner the designer chooses

Note that both the standard and monolithic paradigms use these C++ objects as base classes. In order to use the monolithic parameters paradigm we need to subclass the *AAX_CEffectParameters* object. Avid provides an object named *AAX_CMonolithicParameters* as an example, but it is technically not part of the SDK base classes. We will subclass our own version since the *AAX_CMonolithicParameters* example is tailored for synth plugins. Our version will work for both FX and synth plugins. The monolithic parameters object absorbs the signal processing component into itself as a static member function rather than defining an external algorithm callback function. Our plugin projects implement both the *AAX_CEffectParameters* and *AAX_CEffectGUI* sub-classed objects. In this chapter I refer to them as the "AAX parameter object" and "GUI interface object", respectively. You can think of the parameter object as the data model object if you like, but it has a direct interface to the plugin parameters as well.

5.2.3 MacOS Bundle ID

Each bundle in MacOS is identified with a unique string called a *bundle ID*. Apple has recommendations on naming this string, as well as registering it with Apple when you are ready to sell your product commercially. Typically, you embed your company and product names into the bundle ID. The only major restriction is that you are not supposed to use the "-" character to connect string components; instead, you should use the period symbol. For example, *mycompany.aax.golden-flanger* is not legal, whereas *mycompany.aax.golden.flanger* is OK. Consult the Apple documentation if you have any questions. If you use ASPiK to generate your AAX projects, then the bundle ID value is created for you and set in the compiler project in the *Info.plist Preprocessor Definitions*. You can view or change this value in the project's *Build Settings* panel.

5.2.4 AAX Programming Notes

Like VST3, AAX uses a Microsoft COM type of foundation, though it is less exposed in the outer objects, meaning that you can write AAX plugins without thinking too much about COM. Also like VST3, it is not identical to Microsoft COM but shares some of the same design features. This means that it also makes extensive use of C++ interfaces.

The AAX API is fairly straightforward as far as C++ implementation goes, making use of the standard library (*std::*) as needed. The supported compilers for Windows and MacOS are documented in the AAX SDK and tend to stay behind the most recent releases by about a year or so. This is because Avid invests heavily in testing and qualifying the various compliers to ensure that plugins are properly built. Be sure to check the release notes for your SDK to ensure that your compiler will work properly. Note that the MacOS versions do require several compiler frameworks, including *Foundation, CoreData, AppKit, OpenGL,* and *Cocoa*; the latter two frameworks are for VSTGUI4 and ASPiK.

Launch the compiler and build the libraries for both the Debug and Release configurations. If you have a qualified compiler installed, then these libraries will build without errors. You can then find the compiled libraries in */Libs/AAXLibrary/Debug* and */Libs/AAXLibrary/Release* for the two configurations. The sample code that comes with the SDK expects that these files will be located in the above folders. If you use ASPiK, then those plugin projects will also expect the libraries to be in their native locations. With the libraries compiled, your Visual Studio and Xcode projects become extremely simple and lightweight, smaller and lighter than their AU or VST3 counterparts—there is no need to show a figure here because of the simplicity of the resulting projects.

As with VST3 and AU, there are numerous C++ typedefs that rename standard data types. You should always right-click and ask the compiler to take you to definitions that you don't recognize. Most of the definitions are simple and are supposed to make the code easier to read, though sometimes they can have the opposite effect. You can find these definitions in the *AAX.h* file. Most of the AAX plugin functions return a success/failure code that is a 32-bit unsigned integer typedef'd as *AAX_Result*; the success value is *AAX_SUCCESS*. There are more than 50 different error codes, which you can find in the same file.

5.2.5 AAX Class Factory

AAX uses the class factory approach for serving up new instances of your plugin. However, all of those details are buried in the AAX library that you precompile, so they don't really show up in the plugin projects. As with VST3, you most likely want to void singletons and other anti-patterns in your plugin projects.

5.2.6 AAX Effect Categories

When the AAX host adds your plugin to its list for the user, it would prefer to group similar plugins together by category. Part of the effect description is on whether your plugin is for FX or synth, but AAX offers sub-categories. You can find these definitions in *AAX_Enums.h*, including the following (among many others).

- *AAX_ePlugInCategory_EQ*
- *AAX_ePlugInCategory_Dynamics*
- *AAX_ePlugInCategory_PitchShift*
- *AAX_ePlugInCategory_Modulation*

You may also choose the generic *AAX_ePlugInCategory_None* if you desire. If you specify *AAX_ePlugInCategory_SWGenerators,* then you are specifying a synthesizer plugin. The others are all of the FX type, with the exception of *AAX_ePlugInCategory_HWGenerators,* which is reserved for specialized hardware with a plugin interface such as *SampleCell®*.

5.2.7 AAX Algorithms: Channel-Processing Functions

For AAX, an *algorithm* represents an audio processing function; your plugin names and implements this function. Unlike AU and VST, which have only one process function for every possible kind of channel I/O combination, AAX allows you to define a different algorithm function for each channel I/O possibility that your plugin supports. For example, if your plugin supports mono-in/mono-out, mono-in/stereo-out, and stereo-in/stereo-out, then you may define three separate algorithm processing functions, one for each possible scenario. You may also define just one algorithm function and then let it decode the incoming audio channel information and process accordingly—the choice is up to you. This means that part of the description process involves your plugin iterating through every possible channel I/O combination it supports and adding that combination, along with the algorithm's process function name, to a special descriptor called the *component description*.

5.2.8 AAX Algorithm Data

For both monolithic and standard designs, the algorithm's audio processing function is designed in the same manner as a C-style callback function. The host does not necessarily know whether it is a stand-alone callback, or a static C++ member function. With C-style callback functions, you usually use some kind of *void** cloaking system to prototype the function argument. If you've worked with callbacks before, then you understand why this kind of system is used. The specific mechanisms depend on the architecture but the overall idea is basically the same (the AU Event Listener Dispatch method in Section 4.7.2.1 is also an example that uses *void* pointers as a cloaking system). The fundamental idea is shown in Figure 5.2 and consists of several steps:

1. Create a data structure that contains the information that the callback function needs to do its job.
2. Allocate the data structure so that it is persistent for the lifetime of the application and store a pointer to this structure.
3. Give this pointer to the entity that will call the callback function—now it is a shared resource.
4. The pointer is passed into the callback function cloaked as a *void** (i.e. casted as a *void**).
5. Inside the callback function, you uncloak the pointer by re-casting it to the original type, then you access its data.
6. Externally, changes made to the data in the structure may then be propagated into he callback function.

While Figure 5.2 shows a generic version of the *void**/callback mechanism, it is not the literal way that the AAX framework handles this communication. However, the basic concept is the same. In AAX you typically declare a data structure to hold the necessary algorithm information, which consists of pointers to audio buffers, buffer sample count, and pointers to MIDI nodes, as well as a pointer to *private data*. *Private data* is another structure that contains variables that reflect the current state of the plugin parameters that link to the GUI controls. For a monolithic plugin, that private data may simply be a pointer back to the monolithic parameter object.

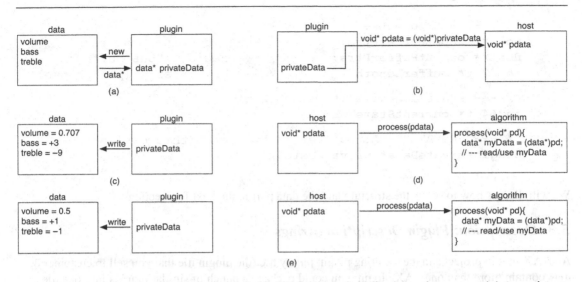

Figure 5.2: In the generic callback system using *void** cloaking, the following actions occur. (a) A persistent private data structure is created. (b) The plugin passes a copy of the pointer cloaked as a *void** to the host which stores it. (c) During the buffer process cycle, the plugin alters the data in the structure. (d) In addition, the host calls the callback function, passing the stored pointer to it. The process function un-cloaks the pointer, casting it back to its original data type and uses the newly changed information. (e) The process continues for each buffer cycle.

5.2.9 Algorithm Data Contents

The algorithm data structure for both monolithic and standard AAX plugins contains some of the same information, including:

- pointers to the audio buffers, including a separate side chain buffer pointer if required
- pointers to MIDI nodes to handle MIDI messaging
- a count variable to hold the sample count for the audio buffers

For the standard design, this structure also holds variables that correspond to each of the plugin parameters required for audio processing. The host will synchronize the information so that these variables are updated in sync with the audio that is streaming. For the monolithic version we include a pointer to the monolithic parameter object along with the buffer pointers and other data. The dirty parameter queues may then be accessed with that pointer. Notice the inclusion of the state index variable: it identifies the current state packet that was delivered for the process cycle. Here is an example of a monolithic algorithm data structure that you might declare; our monolithic parameter object is of type *AAXPluginParameters* and we store a pointer to it as private data.

```
struct pluginPrivateData
{
        AAXPluginParameters* aaxPluginParameters;
};

struct AAXAlgorithmData
{
```

```
    // --- inits ommited for brevity
    float** inputBufferPtrs;           // --- audio input buffers
    float** outputBufferPtrs;          // --- audio output buffers
    int32_t* bufferLength;             // --- buffer size (per channel)

    // --- state index (state timestamp)
    int64_t* currentStateNum;

    // --- PRIVATE DATA: pointer to the monolithic parameters
    pluginPrivateData* privateData;
};
```

We will discuss how to set up the structure and the data port in the next few sections.

5.3 Description: Plugin Description Strings

An AAX plugin project and the resulting plugin binary file (the plugin file that you sell to customers) may contain more than one AAX plugin: you could package a bunch of similar plugins into one file, or even your entire library of plugins. AAX calls this a "collection" and the AAX description process reflects this added complexity. When studying the AAX description mechanism, remember that it is set up to include the added flexibility of multiple plugins, each with multiple algorithms (potentially one algorithm per channel I/O combination). Figure 5.3 shows the concept as it applies to a collection

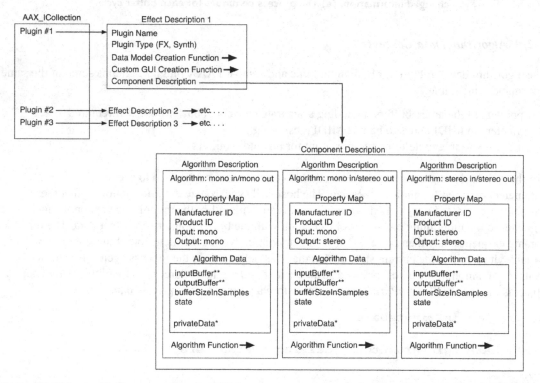

Figure 5.3: This graphic illustrates that an AAX binary may contain more than one plugin effect, each having its own effect and component descriptions. Each component description consists of a set of algorithm descriptions.

of plugins. The effect description holds the high-level attributes of plugin name and type, as well as pointers to the data model and custom GUI creation functions. It also holds component descriptions for the plugin. A component description consists of a set of algorithm descriptions, each of which contains information about one algorithm that is used for a given channel I/O combination.

The AAX library that you precompile contains the code that creates and registers your plugin when the user requests it. This results in a call to a function named *GetEffectDescriptions*. Each AAX plugin project must include a statically declared description component:

```
AAX_Result GetEffectDescriptions(AAX_ICollection* outCollection)
```

The *AAX_ICollection** parameter handles the rest of the description mechanism. Using *AAX_ICollection* involves several steps:

1. Obtain a new *AAX_IEffectDescriptor* from the collection interface using the *NewDescriptor* method.
2. Call methods on the plugin descriptor interface; describe the plugin as a component (name, type).
3. Obtain a new *AAX_IComponentDescriptor* from the effect descriptor interface using the *NewComponentDescriptor* method.

For each channel I/O combination your plugin supports, do the following:

- Add a new property map to the component descriptor with manufacturer ID and product ID as well as the supported input/output channel combination.
- Register a pointer to the algorithm's process function for that channel combination.
- Allocate the algorithm data structure, including pointers to audio buffers, sample count, and pointers to its private data members.
- Add any additional properties for this algorithm.

With the component description complete, then do the following:

- Add the plugin to the collection interface, passing it a unique plugin ID value along with the plugin descriptor interface using the *AddEffect* function.
- Set the manufacturer name, plugin package name (we use the same name as the plugin itself), and version on the collection interface.

Then, repeat the process for each plugin in your binary (note that for book projects, there is only one plugin inside of each binary file). At this point, the AAX host has everything it needs to expose your plugin to the user. The basic code to complete these steps begins in the static description file and *GetEffectDescriptions*:

```
AAX_Result GetEffectDescriptions(AAX_ICollection* outCollection)
{
    AAX_Result result = AAX_SUCCESS;

    // --- obtain descriptor
    AAX_IEffectDescriptor* descriptor = outCollection->NewDescriptor();
    if(descriptor)
        // --- call descriptor functions here . . .
            <SEE BELOW>

    // --- add the plugin to the collection
    outCollection->AddEffect("aaxDeveloper.goldenflanger", descriptor);
```

```
    // --- set some global strings
    outCollection->SetManufacturerName("Plugg Co");
    outCollection->AddPackageName("GoldenFlanger");
    outCollection->AddPackageName("GldnFlng");
    outCollection->SetPackageVersion(1);

    return result;
);
```

Now let's look at the effect descriptor portion, where we define the plugin name and category. Note that we define two names here, one long version and a shortened version (we did the same earlier with the package names). AAX requires a short plugin name that it uses in various locations within Pro Tools or other AAX hosts where long strings are prohibitive. We will revisit the plugin descriptor shortly when we discuss the channel I/O support description component.

```
// --- Effect identifiers
descriptor.AddName("GoldenFlanger");
descriptor.AddName("GldnFlng");

// --- setup category
descriptor.AddCategory(AAX_ePlugInCategory_Modulation);
```

5.3.1 Description: Defining AAX Algorithms

The component description is created with the *NewComponentDescription* method on the effect descriptor:

```
AAX_IComponentDescriptor* cmpDes = descriptor.NewComponentDescriptor();
```

Now we can iterate through our algorithms (one for each channel I/O combination we support), set up a property map, and register the algorithm function pointer. Here is the code to set up one algorithm for stereo in/stereo out. The input/output channel combination is set in the property map using the AAX constants *AAX_e . . .*, and then we register the native processing algorithm *AAX_eProperty_PlugInID_ Native* with an integer constant that we define (*kPlugInID_Native_Stereo*).

```
const AAX_CTypeID kPlugInID_Native_Stereo = 'PCST';

AAX_IComponentDescriptor* cmpDes = cmpDes.NewComponentDescriptor();
cmpDes->Clear();

// --- setup first algorithm: stereo in/stereo out
AAX_IPropertyMap* properties = cmpDes.NewPropertyMap();

// --- IDs (four character codes)
properties->AddProperty(AAX_eProperty_ManufacturerID, 'PLCO');
properties->AddProperty(AAX_eProperty_ProductID, 'GFLG');

// --- channel I/O combination: stereo/stereo
properties->AddProperty(AAX_eProperty_InputStemFormat,
                                  AAX_eStemFormat_Stereo);
properties->AddProperty(AAX_eProperty_OutputStemFormat,
                                  AAX_eStemFormat_Stereo);
```

```
// --- register algorithm with ID: kPlugInID_Native_Stereo
properties->AddProperty(AAX_eProperty_PlugInID_Native,
                        kPlugInID_Native_Stereo);
```

```
// --- add pointer to audio processing function for this property map
cmpDes.AddProcessProc_Native(AAXPluginParameters::StaticRenderAudio,
                        properties);
```

Notice the last line of code here: we have named the algorithm processing function *StaticRenderAudio* and we added a pointer to that function, linked to the properties we set up in the property map.

The last thing left to do is set up the algorithm's data structure, which includes the buffer pointers and private data. We'll use the structure in Section 5.2.9, which is named *AAXAlgorithmData*. To add its members, you use a macro called *AAX_FIELD_INDEX* that generates an integer that may be used as a constant expression when identifying the different algorithm data members (fields). Notice that there are functions for adding each of the different algorithm data components, including *AddPrivateData*.

```
// --- AUDIO stuff
cmpDes.AddAudioIn(AAX_FIELD_INDEX (AAXAlgorithmData, inputBufferPtrs));
cmpDes.AddAudioOut(AAX_FIELD_INDEX (AAXAlgorithmData,
                                    outputBufferPtrs));
cmpDes.AddAudioBufferLength(AAX_FIELD_INDEX (AAXAlgorithmData,
                                    bufferLength));
```

```
// --- private data
cmpDes.AddPrivateData(AAX_FIELD_INDEX(AAXAlgorithmData, privateData),
    sizeof(privateData), AAX_ePrivateDataOptions_DefaultOptions);
```

The final part of the description consists of a few more definitions. First, in Figure 5.1d you can see there is a packet messaging system. You add a data input port that will accept packets from the AAX *privateData*. The *currentStateNum* member of the algorithm structure is used to identify this port. Once it is added, the plugin can send information into the port using the *PostPacket* function. This effectively sets up the messaging system from the plugin side.

```
cmpDes.AddDataInPort(AAX_FIELD_INDEX(AAXAlgorithmData,
                    currentStateNum), sizeof(uint64_t));
```

After the port is added, the algorithm description is complete. You then repeat the process for each channel I/O combination you support. This is true even if all of your algorithms call the same process function: you keep registering the same function pointer and algorithm data structure for each channel I/O combination. Once this is complete, add the component descriptor to the effect:

```
descriptor.AddComponent(cmpDes);
```

Now there are only two steps left. We need to register the creation functions with the descriptor. The first step is to implement our own creation plugin mechanism as a C-style callback function that is external to the monolithic parameter object. Our AAX parameter object is named *AAXPluginParameters*; it inherits from the *AAX_CEffectParameters* base class, so the creation function simply uses the new operator to return a base class pointer. This is standard fare in the C++ plugin paradigm.

```
// --- creation function
AAX_CEffectParameters * AAX_CALLBACK AAXPluginParameters::Create()
{
        return new AAXPluginParameters();
}
```

In the second step, we register that function on the effect descriptor with the *kAAX_ProcPtrID_Create_ EffectParameters* flag, as follows.

```
// --- Data model
descriptor.AddProcPtr((void*) AAXPluginParameters::Create,
                      kAAX_ProcPtrID_Create_EffectParameters);
```

With the AAX algorithms defined, we can move on to the plugin options and features that are common to the other APIs.

5.4 Description: Plugin Options/Features

The plugin options include audio I/O, MIDI, side chaining, latency, and tail time. For AAX plugins, these are all implemented programmatically either in the plugin description, or in other base class functions that you override.

5.4.1 Side Chain Input

Exposing a side chain input requires three steps:

1. Declare a side chain input *index* pointer in the algorithm data structure.
2. Add a description input for the side chain index to the effect descriptor (similar to adding the audio input and output buffers, except you only add the side chain index pointer, which you use later to acquire the side chain buffer).
3. In the description process (Section 5.3.1), create a new property map and add the property *AAX_ eProperty_SupportsSideChainInput*.

The code to do all of this is here, beginning with the side chain index declaration:

```
struct AAXAlgorithm
{
      // --- inits ommitted for brevity
      float** inputBufferPtrs;    // --- inputs buffer

      <SNIP SNIP SNIP>

      int32_t* sidechainChannel; // --- sidechain channel pointer
}
```

Next is the effect descriptor declaration, where you add that side chain index for the input during the algorithm description phase:

```
AAX_IPropertyMap* scProps = descriptor.NewPropertyMap();

descriptor.AddSideChainIn(AAX_FIELD_INDEX(AAXAlgorithm,
                          sidechainChannel));
scProps ->AddProperty(AAX_eProperty_SupportsSideChainInput, true);
```

5.4.2 Latency

The AAX host implements an interface called *AAX_IController*, which allows the plugin to query the host for information, such as the current sample rate. The plugin may get a pointer to the controller at any time by calling the *Controller( )* method. The majority of its member functions are for the plugin-to-host queries. Setting the latency is one of two plugin-to-host *set* functions, the other is for DSP plugins and sets the expected DSP CPU cycle count. To set the latency time in samples, you get the controller interface and call the *SetSignalLatency( )* function. Here we set the value to 1024 samples:

```
Controller()->SetSignalLatency(1024);
```

You may call this function when the algorithm latency changes; however, you must wait for a notification from the host that the new latency value has been applied. You do that with the notification system described in Section 5.4.6 and listen for the *AAX_eNotificationEvent_SignalLatencyChanged* message. At that point, your plugin may apply the new latency to the processing. Avid states that you should not automate any component that will change the latency time as it will result in audio clicks and glitches.

5.4.3 Tail Time

AAX does not have a tail-time setting. The AAX host will stream zeros through the plugin, even when the transport is stopped, so there is an effective infinite tail time and no need for the plugin to establish one.

5.4.4 Custom GUI

A custom GUI is implemented with the AAX GUI interface class named *AAX_CEffectGUI*. You create a derived class that implements the GUI portion of the plugin work. In AAX, this object is relatively simple: its methods are called when the user opens and closes the GUI. You create and destroy your GUI view object from those methods. The GUI interface object also plays a role in the thread-safe read access to plugin parameters, which are transmitted from the data model (parameter) object. The GUI rendering object that receives value-changed messages from its GUI controls must either provide the thread-safe write part of the operation, or provide a way for the GUI interface object to do it. The GUI object must implement its own creation function as a C-style callback that is outside of the C++ object definition or implementation. It simply calls the *new* operator and returns the *AAX_CEffectGUI* base class pointer. Our AAX GUI interface object is named *AAXPluginGUI*. The creation function is:

```
AAX_IEffectGUI* AAX_CALLBACK AAXPluginGUI::Create()
{
    return new AAXPluginGUI;
}
```

You register this creation function during the description phase by adding the creation function pointer with the flag *kAAX_ProcPtrID_Create_EffectGUI* on the effect descriptor:

```
descriptor.AddProcPtr((void*)AAXPluginGUI::Create,
                    kAAX_ProcPtrID_Create_EffectGUI);
```

When the user opens the GUI, the interface object's *CreateViewContainer( )* method is called; when the GUI is closed, the *DeleteViewContainer( )* method is invoked. When plugin parameters are updated externally and the GUI needs to be adjusted to reflect the changes, the *ParameterUpdated( )* method is called, passing the AAX control ID value into the function. The GUI object may then use one of the thread-safe functions (Section 5.5.1) to get the parameter value and update the GUI.

The sample code includes a project that shows how to use several different GUI libraries with AAX. If you already have a GUI library of choice, check the *DemoGainGUIExtensions* sample project for an example implementation. If you are using ASPiK, note that the plugin GUI interface object was designed with VSTGUI4 and already has everything needed to connect with the AAX GUI interface, including specific AAX keyboard/mouse click combinations and other AAX-specific GUI interactions to fully support automation.

5.4.5 Factory Presets and State Save/Load

AAX factory presets are not hard-coded into your plugin as they are with AU and VST. Instead, you set up the factory presets within the Pro Tools Dev host using your plugin under development, then you save the preset files in their native *.tfx* format with the mechanism that Pro Tools provides. You then add the presets to the final package, which you will need to have code-signed as part of the final release of the plugin (see the introduction text of this chapter). If your plugin is like most, which save the parameter (GUI control) values only, then you don't need to do much else because the host will take care of serializing those values. If your plugin needs to save additional information, there is a data chunk load/save mechanism available (see *CChunkDataParser* in the SDK documentation). You may also receive notifications when presets are loaded using the notification system discussed in Section 5.6 and listening for the *AAX_eNotificationEvent_PresetOpened* event; preset information is then loaded from the *.tfx* file.

5.4.6 AAX Notification System

AAX features a robust notification system that allows the data model (AAX parameter object) to send and receive messages. Your plugin may receive notifications from the AAX host—there are a bunch of them, listed in *AAXEnums.h*—by overriding the *NotificationReceived( )* method. To listen for the notification that the latency value changed, you would implement the function like this:

```
AAX_Result NotificationReceived(AAX_CTypeID inMessageType,
                                const void* inNotificationData,
                                uint32_t inNotificationDataSize)
{
    if(inMessageType == AAX_eNotificationEvent_SignalLatencyChanged)
    {
    // --- do latency change stuff
    }

    return AAX_SUCCESS;
}
```

Some notifications are simply messages that have no "payload data" associated with them, such as the latency update notification. For notifications that arrive with extra data, you use the *inNotificationData* and *inNotificationDataSize* parameters to handle the incoming data. See the definitions for each notification message for the presence of incoming payload data.

On the transmit side, your plugin may send notifications to its GUI interface object, if that object implements the *NotificationReceived* function. There are two outbound notification message functions for messages with and without payload data. The two functions are:

```
virtual AAX_Result SendNotification(AAX_CTypeID inNotificationType);

virtual AAX_Result SendNotification(AAX_CTypeID inNotificationType,
                              const void * inNotificationData,
                              uint32_t inNotificationDataSize);
```

Both functions will result in a *NotificationReceived* function call on the receiver, but the first will pass a null pointer for the data and a size of zero while the second function may be used to pass data as well.

5.4.7 AAX Custom Data

There is a second method of moving data to the GUI interface object called *custom data*. You may define as many types of custom data as you like. You must set up the data block you wish to move (typically a C-structure) along with a custom index value to identify the type of custom data. Your AAX parameter object may then override the two functions *GetCustomData( )* and *SetCustomData( )* as needed. If you override *GetCustomData*, then an external caller may query the AAX parameter object and get the data. If you override *SetCustomData*, then an external caller may send custom data to it. In ASPiK, we use this method to give read-only access to the *PluginKernel* for the GUI to use for initialization prior to being displayed—it is never used again afterwards.

5.4.8 AAX EQ and Dynamics Curves

AAX hosts (which may also include hardware like the Avid S6) allow for a couple of extra custom views as a convenience. These include EQ curves, dynamics processor transfer functions (the knee-shaped curve with a ball that tracks the audio signal level), and a gain reduction curve for dynamics processors. Your AAX parameter object may optionally override the *GetCurveData( )* function to allow an AAX host to query for an EQ curve, dynamics knee curve, or gain reduction curve.

```
virtual AAX_Result GetCurveData(AAX_CTypeID iCurveType,
                              const float* iValues,
                              uint32_t iNumValues,
                              float* oValues
```

The curve type is used to decode the request and may be *AAX_eCurveType_EQ, AAX_eCurveType_Dynamics,* or *AAX_eCurveType_Reduction.* A plugin's custom GUI may also call this function to draw these curves as well. You may define your own custom curve types for use with your plugin GUI. The SDK contains additional documentation if you would like to use this feature.

5.4.9 AAX Gain Reduction Meter

The AAX architecture allows for a variety of metering options that are external to your own plugin GUI metering and take on several different forms. The AAX host will automatically provide the plugin input/output metering for you, or you may send plugin metering data to it instead: for example, if you would rather meter the point after some input gain stage.

AAX hosts may also implement one or more gain reduction meters, which are inverted meters that show gain reduction in a dynamics processor. In theory, there may be more than one of these meters for a plugin, but I have only ever observed one meter in Pro Tools as well as in the S6 hardware. In Pro Tools, it manifests itself as a narrow, single channel meter next to the mixer fader. Setting up your plugin to send information to these meters is fairly straightforward. In ASPiK, we implement the gain reduction metering as an option for plugins that request it during the CMake project generation. You can easily modify the code to allow for other metering options by using the gain reduction metering code as an example.

The first thing to do is to add a member to the algorithm data structure to access the meters from your algorithm processing function. In this example, we will set up a gain reduction meter for a dynamics processor plugin.

```
struct AAXAlgorithm
{
    // --- inits omitted for brevity
    float** inputBufferPtrs;        // --- inputs buffer
    float** outputBufferPtrs;       // --- outputs buffer
    float** grMeterPtrs;            // --- gr meter(s)
    etc . . .
```

The next step is to add the meters to the effect descriptor during the algorithm description phase (see Section 5.3.1)—note that the *GR_MeterID* is a value that your plugin defines for meter identification purposes:

```
descriptor.AddMeters(AAX_FIELD_INDEX(AAXAlgorithm, grMeterPtrs),
                     &GR_MeterID, 1); // --- one meter
```

The last step is to complete the effect description by adding a property map that defines the meter type (input, output, or gain reduction) and other properties. This is very similar to the process for setting up the side chain input. Check the SDK for the metering options; here we define the meter type as *AAX_eMeterType_CLGain*, which stands for "Compressor-Limiter Gain reduction." The orientation is set to default, which is inverted for a gain reduction meter. This happens after the algorithm descriptions are completed.

```
AAX_IPropertyMap* meterProps = descriptor.NewPropertyMap();

meterProps->AddProperty(AAX_eProperty_Meter_Type,
                        AAX_eMeterType_CLGain);

meterProps->AddProperty(AAX_eProperty_Meter_Orientation,
                        AAX_eMeterOrientation_Default);

descriptor.AddMeterDescription(GR_MeterID, "GRMeter", meterProperties);
```

Your algorithm may then access the gain reduction meter for writing.

5.5 Initialization: Defining Plugin Parameters

The AAX parameter object exposes the plugin parameters in the *EffectInit( )* function and uses the same list-of-parameters design paradigm along with the other APIs. For AAX, each parameter is encapsulated in an *AAX_CParameter* object. This object stores the same kind of information that defines a plugin parameter in a very similar manner as the other APIs. The object stores the typical parameter name; units; and minimum, maximum, and default values; along with the control tapering attributes.

AAX uses a parameter ID value as the parameter identifier—but unlike AU and VST, which use an integer value, the AAX ID is a string value. You can keep consistency across the other APIs by simply encoding the control ID integer value as its string equivalent. In addition, AAX FX plugins need to support a soft bypass function, and like VST3, the bypass switch is a plugin parameter. Unlike VST3, this must always be the first parameter in the list of parameters, or parameter 0. There are two constructors for the *AAX_CParameter* object, differing only in the way the control ID string is defined, either as a *char** or as an *AAX_IString* (a very simple, ultra light string container).

The control's minimum, maximum, and tapering information is contained in one control taper delegate, which inherits from the *AAX_ITaperDelegate* interface and implements methods for generating control taper curves. There are linear and log taper delegates built into the AAX base classes; you may add your own to extend that capability. ASPiK plugins define their own taper delegate objects for linear, log, anti-log, and volt/octave, and you are free to use them in your AAX plugins, ASPiK or not. There are two more delegate objects: a string display and a units decorator delegate. You use the string display delegate to set up string-list parameters and the units decorator delegate to add a units string to your numerical data value. When you initialize the *AAX_CParameter* object, you pass in the appropriate taper delegate, string display delegate, and/or units decorator as required. You also specify the type of variable the parameter represents through several template class definitions.

For example, suppose a plugin has two parameters with the following attributes.

Parameter 1

- type: linear, numeric
- control ID = 0
- name = "Volume"
- units = "dB"
- min/max/default = −60/+12/−3
- bound variable type: double

Parameter 2

- type: string list
- control ID = 42
- name = "Channel"
- string list: "stereo, left, right"
- bound variable type: 32-bit integer

To keep the control ID consistent as bypass is parameter 0, we will use the "add 1" method of just incrementing the established control ID values by 1. Notice the use of a member object to add the parameter to our list (*mParameterManager*) and the last line of code for adding a "synchronized" parameter, which is part of the monolithic design paradigm that we will discuss later.

The first volume parameter is defined as follows.

```
// --- taper (min, max)
AAX_CLinearTaperDelegate<double> taperDelegate =
                    AAX_CLinearTaperDelegate<double>(-60.0, +12.0);
```

```
// --- we want a units decorator
AAX_CUnitDisplayDelegateDecorator<double> displayDelegate =
AAX_CUnitDisplayDelegateDecorator<double>
                       (AAX_CNumberDisplayDelegate<double>(), "dB");

// --- the parameter: -3 = default value
AAX_CParameter<double>* param = new AAX_CParameter<double>(
                       "1",        /* ID + 1 */
                       AAX_CString("Volume"),
                       -3.0, /* -3 = default value */
                       taperDelegate, displayDelegate,
                       true); /* automatable */

param->SetType(AAX_eParameterType_Continuous);
mParameterManager.AddParameter(param);
AddSynchronizedParameter(*param); // < --- MONOLITHIC PLUGIN CHORE
```

The second string-list parameter is defined a bit differently; we need to use a string display delegate for the list of strings which are arranged in a *std::map*. Notice also the function to set the exact number of steps (clicks) for the control:

```
std::map<int32_t, AAX_CString> enumStrings;

enumStrings.insert(std::pair<int32_t, AAX_CString>
                (0, AAX_CString("stereo")));

enumStrings.insert(std::pair<int32_t, AAX_CString>
                (1, AAX_CString("left")));

enumStrings.insert(std::pair<int32_t, AAX_CString>
                (2, AAX_CString("right")));

AAX_CLinearTaperDelegate<int32_t,1> switchTaper =
AAX_CLinearTaperDelegate<int32_t,1>(0, 2); /* 3 strings (indexed 0, 1
and 2) */

AAX_CParameter<int32_t>* swParam = new AAX_CParameter<int32_t>(
                       "43",    /* ID + 1 */
                       AAX_CString("Channels"),
                       0,      /* default value */
                       switchTaper,
                AAX_CStringDisplayDelegate<int32_t>(enumStrings),
                       true);   /* automatable */

swParam->SetNumberOfSteps(3);
swParam->SetType(AAX_eParameterType_Discrete);
mParameterManager.AddParameter(swParam);
AddSynchronizedParameter(*swParam);
```

The *SetNumberOfSteps( )* method is also be used for continuous parameters (the typical value is 128 or 256).

To add the required soft bypass parameter, you create a binary (on/off) parameter with a special ID value (*cDefaultMasterBypassID*):

```
// --- bypass
AAX_CString bypassID = cDefaultMasterBypassID;
AAX_IParameter* masterBypass = new AAX_CParameter<bool>(
                            bypassID.CString(),
                            AAX_CString("Master Bypass"),
                            false, /* default */
                            AAX_CBinaryTaperDelegate<bool>(),
                        AAX_CBinaryDisplayDelegate<bool>
                                ("bypass", "on"), true);
masterBypass->SetNumberOfSteps(2);
masterBypass->SetType(AAX_eParameterType_Discrete);
mParameterManager.AddParameter(masterBypass);
AddSynchronizedParameter(*masterBypass);
```

5.5.1 Thread-Safe Parameter Access

The AAX parameter base class defines a thread-safe set of function designed specifically to ensure that the GUI and other external objects have safe access to the parameters. The GUI object may obtain a safe pointer to the data model (AAX parameter object) by calling *GetEffectParameters( )*, then it may access parameters with any of several safe access functions. (Note that these functions work with normalized values.)

```
GetParameterNormalizedValue(AAX_CParamID iParameterID,
                            double* oValuePtr);

SetParameterNormalizedValue(AAX_CParamID iParameterID,
                            double iValue);
```

There are additional functions that allow you to get the parameter value as a string, and methods to convert values to strings and back. These are all found in *AAX_CEffectParameters.h.*

5.5.2 Initialization: Defining Plugin Channel I/O Support

For AAX plugins, channel I/O support is done as part of the plugin description that involves setting up the algorithms for each of the channel input/output combinations your plugin supports. This is very different from either AU or VST, in which channel support information is accomplished via host querying—so this has already been taken care of during the algorithm description process in Section 5.3.1.

5.5.3 Initialization: Channel Counts and Sample Rate Information

For AAX, the plugin channel count capability is built in to the algorithm description process and needs no other definition or support. The AAX host sample rate is simple to query using the same controller interface as for setting channel latency—except that it is a query (get) rather than a set operation, as follows:

```
AAX_CSampleRate sampleRate = 44100; // AAX_CSampleRate = float
Controller()->GetSampleRate(&sampleRate);
```

Your parameter object can get the current input and output channel configuration, which AAX names "stems" using similar functions:

```
AAX_EStemFormat inputStemFormat = AAX_eStemFormat_None;
AAX_EStemFormat outputStemFormat = AAX_eStemFormat_None;

Controller()->GetInputStemFormat(&inputStemFormat);
Controller()->GetOutputStemFormat(&outputStemFormat);
```

AAX defines more than a dozen mono- and multi-channel audio formats for plugins, plus several ambisonic formats. You can find the enumeration list in *AAXEnums.h*; these are the same constant enumeration values you use when describing the plugin algorithms. You may also use these functions to decode the audio channel formats if your plugin uses a single process function for all algorithms.

5.6 The Buffer Process Cycle

The AAX buffer processing cycle follows the same basic paradigm as VST3 and AU, as shown in Figure 5.4—except that it includes a callback to the host to deliver updated plugin data information. The cycle proceeds in three steps:

1. AAX host calls *GenerateCoefficients()* on the plugin.
2. Plugin calls *PostPacket()* on the host (controller) with updated plugin information in a timestamped packet.
3. AAX host calls the algorithm processing function that you specified when you set up the algorithm descriptions and passes in the plugin packet for that buffer; the plugin performs the audio signal processing using the private plugin data along with the buffer pointers, MIDI node pointers, and other connections that you set up during the describe phase.

Both the monolithic and standard designs follow the process in Figure 5.4, but monolithic plugins are required to play a role in the parameter synchronization that keeps the audio, video, automation, and music in sync for the user. This has to do with the fact that the monolithic object owns the processing function. In the standard design, the host adds packets into a queue and synchronizes them to line up correctly with the audio being processed. This is especially crucial for automation to work properly and identically for each audio playback run. To increase timing accuracy, the host will divide the audio buffers into progressively smaller sub-buffers, down to a minimum length of 32 samples, to get the parameter change information to line up as closely as possible with the streaming audio.

When using a monolithic object with the standard design, parameter control changes would be applied immediately to the audio processing, causing the audio and parameter controls to slip out of sync. The monolithic design requires that the monolithic object does most of the synchronization work by effectively delaying the delivery of control information so that it lines up with the proper audio events. By using the existing AAX packet delivery system, the monolithic plugin emulates the same synchronization process as the host.

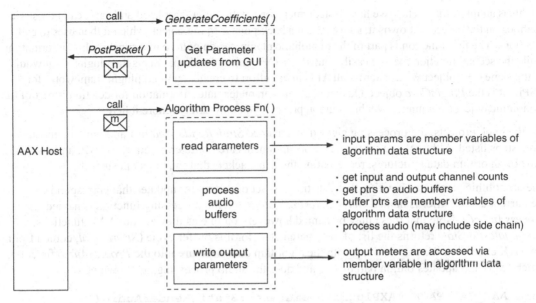

Figure 5.4: The AAX buffer process cycle follows the standard audio plugin pattern but involves a series of calls from and to the AXX host; the parameters, buffer pointers, and any other information needed for the algorithm are delivered in the algorithm data structure.

You can probably guess that the code for implementing this synchronization chore is not trivial. It involves atomic queues, a dirty parameter list, and a layer of list copying along with timestamping in order to post packets to the messaging system so that they synchronize correctly. The good news is that Avid supplies the AAX base class extension for monolithic parameters that you may use, almost right out of the box. The ASPiK monolithic AAX parameter object uses all of the synchronization code from the example, but is modified to be both more generalized, and more easily interfaced to the *PluginKernel's* core. Due to the complexity of the code, I won't repeat it here (but it is well documented).

Part of the parameter update process is an AAX host call to the *AAX_CEffectParameters* member function named *UpdateParameterNormalizedValue( )*. This is the method that alters the plugin parameter value. A monolithic parameter object must override this function and queue up the parameter changes in its dirty parameter list, which will ultimately be applied on the correct buffer process cycle later on.

For the monolithic algorithm data structure, we include a pointer back to the AAX parameter object so that the static process function can perform three tasks:

1. Call a function to update the parameter states for the current buffer process cycle; this updates the plugin's internal variable from the GUI and/or automation control changes that occurred during the buffer capture.
2. Call a processing function that actually does the DSP work—this function may freely access its newly updated plugin variables to affect the audio processing for that buffer cycle.
3. Queue up the parameter state information for deletion.

So this is an interesting setup: we have a static member function that is isolated and can't call non-static methods on the object that owns it, so we pass it a back-pointer to the object, which it then uses to call functions. The first function is part of the monolithic plugin design pattern that synchronizes information, while the second function may be easily routed to a C++ object that implements the plugin—that would be the same C++ object we use across all APIs in our effort to create a unified plugin framework; for ASPiK it is the *PluginCore* object. Queueing up the parameter state information for deletion is part of the monolithic object's parameter synchronization process; you may safely ignore it here.

For the book projects, the processing algorithm is named *StaticRenderAudio( )* and implements the three steps listed. Although we set the name, its arguments are predetermined by the API and consist of a list of algorithm data structures, representing the data packets that the object posted.

The monolithic parameter function that gets the next set of parameter updates that correspond to the current audio buffer is named *GetUpdatesForState( )* and the processing function is named *ProcessAudio( )*. Note that these may be named however you like as they are not AAX functions. *GetUpdatesForState* returns the list of dirty parameters for this buffer cycle (*SParamVal*, copied from the AAX example). We pass it along with the algorithm data structure into the *ProcessAudio* function, which reads and updates plugin parameters and does the audio processing, as shown here.

```
void AAX_CALLBACK AAXPluginParameters::StaticRenderAudio(
              AAXAlgorithm* const inInstancesBegin[],
              const void* inInstancesEnd)
{
  // --- process "packets"
  for (AAXAlgorithm* const* instanceRenderInfoPtr =
      inInstancesBegin;
      instanceRenderInfoPtr != inInstancesEnd;
      ++instanceRenderInfoPtr)
    {
        // --- get our private data from algorithm
        pluginPrivateData* privateData =
                            (*instanceRenderInfoPtr)->privateData;

        if(!privateData) continue:

        // --- get AAX parameter object pointer from the algo struct
        AAXPluginParameters* parameters =
              privateData->aaxPluginParameters;

        if(!parameters) continue:

        // --- get parameter changes for timestamp (currentStateNum)
        SParamValList paramValList =
        parameters->GetUpdatesForState(
                        *(*instanceRenderInfoPtr)->currentStateNum);

        // --- call the process function on the parameter object
        parameters->ProcessAudio(*instanceRenderInfoPtr,
                  (const TParamValPair**)paramValList.mElem,
                  paramValList.mSize);
```

```
            // --- Queue the parameter value pairs for later deletion
            for (int32_t i = 0; i < paramValList.mSize; ++i)
            {
                    <SNIP SNIP SNIP> // sync stuff
            }
        }
}
```

The static algorithm processing function really just sets up the parameters and then calls another function to do the audio work.

5.6.1 Processing: Updating Plugin Parameters From GUI Controls

At the start of the static buffer processing function, we call the function *GetUpdatesForState,* which generates a list of dirty parameters for that buffer cycle. This list is sent to the audio processing function. You iterate through this map and grab the ID and value pairs; you can then update the plugin's internal variables accordingly, depending on your plugin framework.

```
SParamValList paramValList = parameters->GetUpdatesForState(. . .);
for(int32_t i=0; i<paramValList.mSize; ++i)
{
    AAX_CParamID const ID = inSynchronizedParamValues[i]->first;
    const AAX_IParameterValue* v = inSynchronizedParamValues[i]->second;
}
```

5.6.2 Processing: Resetting the Algorithm and Preparing for Streaming

For AAX, there is one main reset function that is called just after the plugin is instantiated; this is called *ResetFieldData( )*. In this function you would do any one-time resets to prepare for audio streaming. Once the AAX host begins streaming audio into the plugins, it will not cease. For monolithic parameter objects, the *ResetFieldData* function is where we set the private data pointer (the back-pointer to our parameter object, or the *this* pointer) in the algorithm's data structure. For our AAX parameter object named *AAXPluginParameters*, this is fairly straightforward:

```
AAX_Result ResetFieldData (AAX_CFieldIndex inFieldIndex,
                           void* oData, uint32_t iDataSize) const
{
    // --- is this our private data field?
    if (inFieldIndex == AAX_FIELD_INDEX(AAXAlgorithm, privateData))
    {
        // --- check and clear pointer
        AAX_ASSERT(iDataSize == sizeof(pluginPrivateData));
        memset(oData, 0, iDataSize);

        // --- get private data pointer we allocated in describe
        pluginPrivateData* privateData =
                static_cast <pluginPrivateData*> (oData);
```

```
            // --- store our THIS pointer here!
            privateData->aaxPluginParameters =
                                    (AAXPluginParameters*) this;
            return AAX_SUCCESS;
    }

    // --- base class does its thing
    return AAX_CEffectParameters::ResetFieldData(inFieldIndex,
                                        oData, iDataSize);
}
```

5.6.3 Processing: Accessing the Audio Buffers

In AAX, the audio input buffers arrive in the algorithm data structure that we pass into the monolithic plugin's process function (you may set up your processing function however you wish). For our projects it is prototyped as:

```
void ProcessAudio(AAXAlgorithm* ioRenderInfo,
                const TParamValPair* inSynchronizedParamValues[],
                int32_t inNumSynchronizedParamValues);
```

We pass it the algorithm data structure (*ioRenderInfo*) along with the dirty parameter list and a count of items in that list. At the top of this function, we process the incoming parameters and update the plugin variables. Next, we grab the audio buffers and channel count information to set up the processing loop. Note that the main audio inputs are stored in slot 0 of the buffer pointer list (this is identical to both AU and VST3).

```
// --- get number of samples to process
const int32_t buffersize = *(ioRenderInfo->bufferLength);

// --- get input and output channel formats
AAX_EStemFormat inStemFormat = AAX_eStemFormat_None;
AAX_EStemFormat outStemFormat = AAX_eStemFormat_None;

Controller()->GetInputStemFormat(&inStemFormat);
Controller()->GetOutputStemFormat(&outStemFormat);

// --- get channel counts for AAX stem types:
const int32_t inChans = AAX_STEM_FORMAT_CHANNEL_COUNT(inStemFormat);
const int32_t outChans = AAX_STEM_FORMAT_CHANNEL_COUNT(outStemFormat);

// --- get buffer pointers
float** inputBufferPointers = &ioRenderInfo->inputBufferPtrs[0];
float** outputBufferPointers = &ioRenderInfo->outputBufferPtrs[0];
```

You then access the individual channel buffers as:

```
float* inputL = inputBufferPointers[0];
```

```
float* inputR = inputBufferPointers[1];
```

The output buffers are accessed identically:

```
float* outputL = outputBufferPointers[0];
float* outputR = outputBufferPointers[1];
```

With the buffer pointers acquired, you may then do your audio signal processing; here we are processing stereo in/stereo out:

```
// --- loop over frames and process
for(int i = 0; i < buffersize; i++){
    // --- apply gain control to each sample
    outputL[i] = volume * inputL[i];
    outputR[i] = volume * inputR[i];
}
```

To access the side chain, we first find the index in the input buffer pointer list. The AAX host sets this value.

```
// --- get index
const int32_t scChanIndex = *ioRenderInfo->sidechainChannel;

// --- get buffer of channel pointers
float** scBuffers = &ioRenderInfo->inputBufferPtrs[scChanIndex];
```

At the time of this writing, AAX side chains are limited to mono sources only, so you access the side chain in the first side chain buffer array slot:

```
float* sidechain = scBuffers[0];
```

5.6.4 Processing: Writing Output Parameters

If your plugin writes data to output VU meters or some kind of custom view (see Chapter 21) then you do that here as well. With all of the processing done, you would then write the output information to the appropriate GUI parameter that needs to receive it. To do this, you call the thread-safe asynchronous set parameter function *SetParameterNormalized( )* and pass it the control ID value (as a string) for the meter and the meter's new control value. If the meter control ID is the integer 20, then you'd set parameter number 21 (because the bypass parameter is always the first parameter) as a string, and send the value. Here it is 0.8:

```
SetParameterNormalizedValue("21", 0.8);
```

To write data to the gain reduction meter for a dynamics processor, you just write the information into the meter array that you added to the algorithm data structure (Section 5.2.8).

```
// --- get gain reduction meter pointer
float* const AAX_RESTRICT grMeters = *ioRenderInfo->grMeterPtrs;

// --- set the meter value
grMeters[0] = 0.8.
```

5.6.5 Processing: AAX Soft Bypass

All AAX FX plugins must support the soft bypass function. The bypass parameter is used to signal the bypass condition. Since it is packaged with the other parameters it will arrive in the dirty parameter list only if it has been changed. You can identify the bypass parameter from the AAX control ID and then set your bypass flag according to the value:

```
AAX_CParamID const ID = inSynchronizedParamValues[i]->first;
const AAX_IParameterValue* v = inSynchronizedParamValues[i]->second;

if(AAX::IsParameterIDEqual(cDefaultMasterBypassID, ID))
// --- if v = 0, bypass = FALSE, if v != 0, bypass = TRUE
```

Implementing the soft bypass is easy: just copy the input samples into the output buffer. This code copies all of the input channels to all of the output channels:

```
for (int ch = 0; ch < inChans; ch++)
{
    const float* const AAX_RESTRICT pInputBuffer =
                        ioRenderInfo->inputBufferPtrs[ch];

    float* const AAX_RESTRICT pOutputBuffer =
                        ioRenderInfo->outputBufferPtrs[ch];

    for (int t = 0; t < buffersize; t ++)
    {
        pOutputBuffer[t] = pInputBuffer[t];
    }
}
```

5.7 Destruction/Termination

The AAX plugin is destroyed via its destructor—there are no termination functions or other processes to call. Your parameter and GUI interface objects should include destructors that clean up any dynamic allocations or other de-initializing operations.

5.8 Retrieving AAX Host Information

AAX provides a robust amount of host information that may be provided to your plugin; however, you need to create a "transport node" in your algorithm data structure. This is really part of the MIDI sub-system in AAX, so you need to describe it as such. First, add the transport pointer to your algorithm data structure:

```
struct AAXAlgorithm
{
```

```
        // --- inits ommitted for brevity
        float** inputBufferPtrs;        // --- inputs buffer
        float** outputBufferPtrs;       // --- outputs buffer
        <SNIP SNIP SNIP>

        // --- transport
        AAX_IMIDINode* midiTransportNode; // --- host BPM, and info
        etc . . .
}
```

During the algorithm description phase, you add the transport node with the *AAX_eMIDINodeType_ Transport* flag and a MIDI channel mask (here we want all channels):

```
description.AddMIDINode(AAX_FIELD_INDEX(AAXAlgorithm,
                                    midiTransportNode),
                AAX_eMIDINodeType_Transport,
                "Transport", 0xffff); // 0xffff = mask
```

The transport node will then arrive along with the other algorithm data. You can pick up the node and access its member functions. The code to get the host BPM and time signature follows:

```
AAX_IMIDINode* const transportNode = ioRenderInfo->midiTransportNode;
AAX_ITransport* const midiTransport = transportNode->GetTransport();

// --- get info
int32_t nTSNumerator = 0;
int32_t nTSDenominator = 0;
bool playing = false;
double BPM = 0.0;

midiTransport->IsTransportPlaying(&playing);
midiTransport->GetCurrentTempo(&BPM);
midiTransport->GetCurrentMeter(&nTSNumerator, &nTSDenominator);
```

See the documentation for *AAX_ITransport* for a list of functions you may use to get much more information about the current playback state and DAW session.

5.9 Validating Your Plugin

At the time of this writing, there is no AAX validation software provided in the SDK (this could change). However, once you are ready to market your plugin, you may send it to Avid for testing. In addition, you may find more information the AAX users forum that you can access with your Avid developers account.

5.10 Using ASPiK to Create AAX Plugins

There is no substitute for working code when designing these plugins; however, the book real estate that the code requires is prohibitive. Even if you don't want to use the ASPiK *PluginKernel* for your development, you can still use the code in the projects outlined in Chapter 8. If you use ASPiK, its AAX plugin shell will handle all of the description and initialization chores and your *PluginCore* object will do the audio processing. You will work within two files only: *plugincore.h* and *plugincore.cpp*.

5.11 Bibliography

Avid Inc. *AAX SDK Documentation*, file://AAX_SDK/docs.html, Available only as part of the AAX SDK.

CHAPTER 6

ASPiK Programming Guide

In order to simplify writing plugins for all APIs simultaneously, we will need some objects to handle the chores of both audio signal processing and GUI design and implementation. This is accomplished with the ASPiK *PluginKernel*. ASPiK stands for *Audio Specific Plugin Kernel*. The kernel is actually a set of three main objects: the *PluginCore*, *PluginGUI,* and *PluginParameter*. The *PluginCore* (or simply "core") object implements the audio signal processing component and contains a list of *PluginParameter* objects, each of which represents a GUI control or other parameter that may be stored or loaded with DAW projects. The *PluginGUI* object creates the GUI as well as the drag-and-drop GUI designer that each plugin project carries with it. In addition, we will need a system that can generate new (blank) projects for both MacOS and Windows, with the compiler project files already set up to contain all of the relevant files we need to start coding the *PluginCore* object immediately. ASPiK is the name of the plugin system that creates compiler projects and installs the *PluginKernel* files. The *PluginKernel* is the independent C++ component, and where you will spend most of your time. When using ASPiK, you may also set flags that will include all of the book C++ objects at once, and/or enable linking to the FFTW library, which is required for the advanced FFT-based plugins in Chapters 20 and 22.

PluginKernel

- is a set of C++ objects designed to handle the main plugin chores common to all APIs
- is fully portable and may be copied to and from any project in any API and on any platform (Windows or MacOS) with little or no modification
- is platform independent
- is complete yet extendible
- is 100% free to use with no licensing for your commercial or noncommercial plugin development

PluginParameter

- stores plugin parameters as atomic variables for thread-safe operation
- encapsulates each parameter specific to the plugin as a C++ object
- can store all types of input parameters; *ints, floats, doubles*, string lists, and custom user types
- can implement audio meters, complete with meter ballistics (attack and release times) and various envelope detection schemes: peak, MS, RMS, and in linear or log format
- implements optional automatic variable binding to connect GUI parameter changes to plugin variables in a completely thread-safe manner across all APIs
- has an optional auxiliary storage system to maintain other information along with any of the plugin parameters allowing you to easily customize and extend the object
- implements optional parameter smoothing for glitch-free GUI controls with two types of smoothing available: linear and exponential

- automatically implements four types of control tapering: linear, log, anti-log, and volt/octave
- implements optional SAA for VST3 plugins (VST3 is the only API that has a specification for SAA)

PluginCore

- handles the audio signal processing and implements the DSP functionality
- is straight C++ and does not contain any API-specific code or require any API-specific files or SDK components
- does not link to any precompiled libraries
- defines and maintains a set of *PluginParameter* objects, each of which corresponds to a GUI control or other plugin parameter that may be stored and loaded with DAW sessions and presets
- defines factory presets for AU, VST, and RAFX2 plugins (AAX factory presets are done in an entirely different manner and are discussed in Chapter 5)
- exists independently from the *PluginGUI* and does not know, or need to know, of *PluginGUI*'s existence
- does not create the *PluginGUI* object
- does not hold a pointer to the *PluginGUI* object or share any resources with it

PluginGUI

- handles all of the GUI functionality including the GUI designer
- is built using the VSTGUI4 library
- is platform independent
- does not link to any precompiled libraries
- encodes the *entire* GUI in a single XML file, including the graphics file data; this one file may be moved or copied from one project to another, allowing the whole GUI to be easily moved or reused in other projects; advanced users may define multiple GUIs in multiple XML files that may be swapped in and out
- contains API-specific code in the few places where it is absolutely needed (namely for the AU event listener system)
- supports the VTSGUI4 *Custom View* and *Sub-Controller* paradigms to extend its functionality
- exists independently from the *PluginCore* and does not know, or need to know, of the *PluginCore's* existence
- does not hold a pointer to the *PluginCore* object or share any resources with it

6.1 Plugin Kernel Portability and Native Plugin Shells

All plugin frameworks require that you obtain the SDK for each plugin API that you wish to support, and that you compile plugins with the correct compiler for each target OS. For example, an AAX plugin must be compiled as an AAX plugin and linked to the AAX library that is part of the AAX SDK. The plugin kernel was designed to tackle the problem of supporting multiple APIs and SDKs on Windows and MacOS. The kernel files may be easily moved or copied from project to project with little or no modification.

The kernel was designed for maximum portability between plugin projects and APIs. Figure 6.1a shows this concept. Each plugin product (AAX, AU, VST) consists of a native plugin shell that includes a plugin kernel inside of it. The native plugin shell (or simply "shell") is the API-specific plugin project code that includes the kernel objects as part of its design. We have made it very easy

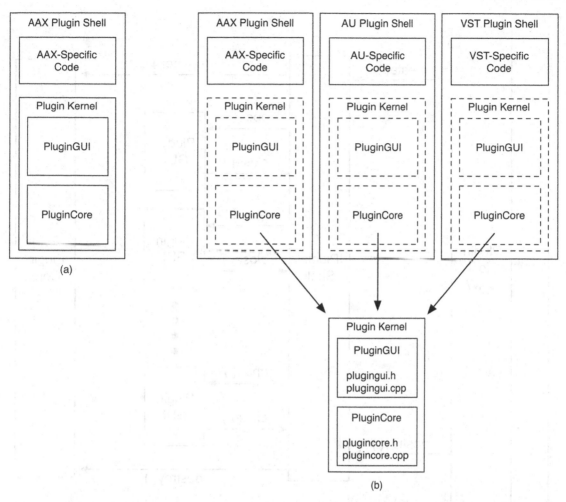

Figure 6.1: (a) As this graphic illustrates, each native plugin shell contains API-specific code as well as a plugin kernel. **(b)** In combined API projects, multiple plugin shells share the same plugin kernel code—so changes to the core are easily propagated to the other shells. The kernel objects may be moved freely across all products and projects.

to support all of the APIs and also to combine plugin projects on similar compilers. For example, you may easily create a plugin project for Xcode that creates the MacOS AAX, AU, and VST plugins in one project and using one set of plugin kernel files, shown in Figure 6.1b. You build the project once, and it produces all three plugins in succession. If you modify the plugin kernel code, you need only rebuild the project again to propagate your changes into all three of the target plugin products. You may just as easily target single plugin APIs: for example, you might be interested only in AU plugins on MacOS. If you decide later that you want to support another API, it is simple to generate the project for that API and simply move your kernel into the new project without needing to modify any code.

The native plugin shell creates the *PluginCore* object upon instantiation. The core object exists for the duration of the lifetime of the shell and is only ever created once and destroyed once. The user may open

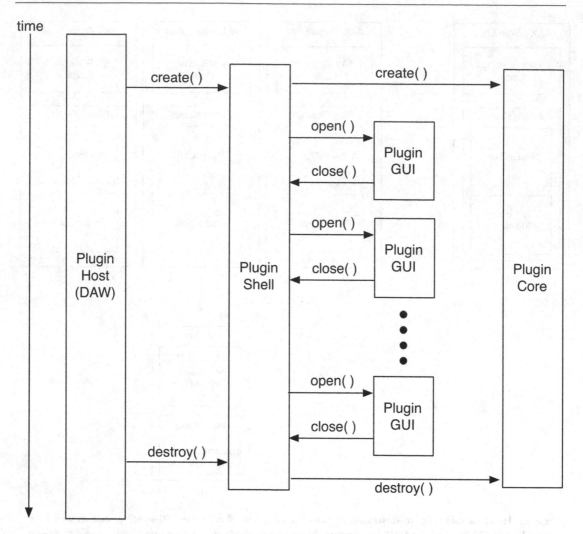

Figure 6.2: The *PluginCore* and *PluginGUI* lifecyles; the GUI object is created and destroyed each time the user opens and closes it.

and close the plugin's GUI at will. The shell object will likewise create and destroy the *PluginGUI* object each time. The plugin shell does not coordinate the two objects nor forms any kind of relationship between them; the two objects never trade pointers with one another. The object lifecycles are shown in Figure 6.2.

You have several options to use as your plugin design pattern, based on your preferred OS and API. One option that is especially nice for beginner plugin programmers is to use RackAFX; however, this is not required and you do not need the RackAFX software to use this book or its project code. If you use this design paradigm, then you use the RackAFX software to create a RAFX2 plugin project, add plugin parameters, and test your code. When you are happy with the results, you can then use the *Export* function to generate AAX, AU, and VST individual or combined projects, which you then compile against the AAX, AU, and VST SDKs. These exported projects do not contain any RackAFX code whatsoever. During the exporting process, only your *PluginKernel* objects (and any objects that they include and use for processing) are copied into the exported projects. After this, you run CMake, exactly as you do for the AAX, AU, and VST projects here.

The other option is to develop your plugin straight away from the plugin template project code, which is very simple. You may also easily copy and reuse old plugin projects without needing to change any code. You may use any combination of APIs and platforms you choose for development. Your plugin kernel code is not API-specific, so you will not be wasting any time developing for any given platform.

6.2 Organizing the SDKs: AAX, AU, and VST

There are two options for developing your AAX, AU, and VST plugin projects, and you may easily flip and flop back and forth between two design paradigms at any time. In the first method, you target a single plugin API and develop your project *within* the SDK for that particular plugin. In the second method, you target one or more APIs at once and your plugin project exists *outside* of any of the SDKs. In both paradigms, your plugin project code will not interfere with nor change your SDK files whatsoever. When a new SDK is available, you simply install it and drop your project folder into the appropriate location. If you want to change the relative locations of the SDKs, then you may do that as well by modifying the CMake files accordingly (more on CMake later). We have designed the system to be flexible and mutable. For your first projects, you should stick to the same default SDK arrangement. After that, you may experiment with different SDK folder locations and other options.

6.2.1 Your C++ Compiler

You will need a proper compiler for MacOS and Windows. For Xcode, you will need at least Xcode 8; for Visual Studio, you will need at least VS2015 Community. Both of these compilers are free to download from Apple and Microsoft. The VSTGUI library requires C++11 extensions for both MacOS and Windows, and C++14 extensions for MacOS. In addition, you should check the compiler certification list for AAX plugins from Avid. At the time of this writing, Visual Studio 2017 was not yet officially supported. You will need to download and install the compiler of choice for your target OS(es) prior to doing any plugin development. As an acid test, you might want to create a simple C++ project and build it to make sure everything is ready.

6.2.2 Setting Up the AAX SDK

Follow the instructions in Chapter 5 to install the SDK and compile the base class libraries. The SDK outer folder is decorated to tell you the compiler version (e.g. 2p3p0 refers to version 2.3.0). However, these numbers may be problematic when you use CMake to generate your compiler project files, which has to do with browsing to the SDK location in order to run CMake. To overcome this issue and to standardize our SDK naming, rename the root folder to simply *AAX_SDK* (remember that you can always change your naming convention later). Once that is done, you will have a SDK folder hierarchy that resembles Figure 6.3a. Notice the folder named *myprojects* that is located in the top-level SDK folder alongside the rest of the subfolders. This is the folder you need to create that will contain all of your AAX-specific projects. You may name this folder whatever you wish, though you should keep the name simple and CMake-friendly with no whitespaces, numbers, or special characters. We will use the name *myprojects* exclusively for all plugin project folders. Go ahead and add that empty folder to your SDK to get started.

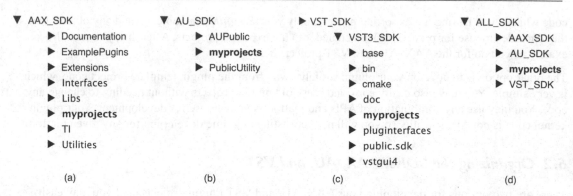

Figure 6.3: Folder hierarchies for the following SDKs: (a) AAX, (b) AU, (c) VST, and (d) universal (combined).

6.2.3 Setting Up the AU SDK

Follow the instructions in Chapter 4 to download the AU SDK sub-folders. Next, combine these two folders together in one outer folder named *AU_SDK,* as shown in Figure 6.3b. You need to create the *myprojects* subfolder inside the *AU_SDK* folder the same way you did for AAX, noting again that you may name the folder whatever you like as long as it is a legal CMake folder name. There are no libraries to build—so after adding that folder, you are done.

6.2.4 Setting Up the VST SDK

Follow the instructions in Chapter 3 to install the VST3 SDK and build the sample projects. With that done, you need only add the *myprojects* subfolder as you did with AAX and AU (following the same rules). You should then have a folder hierarchy like that in Figure 6.3c.

6.2.5 Creating the Universal SDK Folder Hierarchy

You can combine all of the SDKs into one outer folder and create universal projects that target any combination of APIs that you wish. In this case you will need to create an outer container folder named *ALL_SDK* and then populate it with the three SDK folders we've just discussed. In parallel with these three subfolders, create the *myprojects* subfolder to contain your universal projects. Since you may target any single API or any combination of APIs, this is probably the simplest choice for your design paradigm. Because I am packaging projects with CMake, you may re-target the universal projects to different combinations of APIs any time you wish, so nothing is set in stone here. If you want to use this option, you should wind up with the *ALL_SDK* folder hierarchy shown in Figure 6.3d. You should also remember that you may freely move projects around and re-target them for specific APIs or universal combinations. You don't need to make any commitments during this process!

6.2.6 Adding the VSTGUI4 Library

Our *PluginGUI* object is built around the VSTGUI4 library. This package is free and open source; you may download it from the Sternberg VST web page here: https://github.com/steinbergmedia/

vstgui Simply use the folder that comes with the VST3 SDK, or install it from RackAFX. New VSTGUI4 versions are timed to be released with new VST3 SDKs so if you upgrade your VST3 SDK, you will automatically get the latest *VSTGUI4* library. The VSTGUI4 engineers are constantly updating and improving this awesome library. However, in the update process they may change objects and functions in a way that sometimes requires tweaking on our end. In order to keep our plugin project folder completely isolated and separate from the SDKs, we will make a copy of the *vstgui4* folder and place it inside each of the *myprojects* subfolders. The library does not contain precompiled code and is fairly small in size (less than 40 MB) so it is not a burden to copy this folder to our various *myprojects* subfolder(s). Notice that we will follow this rule in all cases, including the VST3 SDK. The reason has to do with updating: if a new VST3 SDK comes out that makes a significant change to the *VSTGUI4* library, we are not affected because we will use our own, local individual folder. Of course we will make every attempt to keep the *PluginGUI* object updated for the latest version of VSTGUI4, however we will always have our own fully functioning version at hand if needed. The *VSTGUI4* library comes packaged like the rest of the SDKs with a root folder full of subfolders. The root folder is named *vstgui4* and we will not modify this name for our use. But as with everything else, you may change the name if you wish by modifying the CMake files slightly.

6.2.7 CMake

Simply put, CMake is a utility that generates Visual Studio and Xcode projects from one or more script files, each named *CMakeLists.txt*. You can download it for free from the CMake website (http:// cmake.org) and install it on MacOS and/or Windows. Seasoned programmers are likely already familiar with this product. All of the CMake script files are written for you, but you will need to alter a few lines of text to set your project's name as well as the options for building the various plugins in any combinations you like. Making these changes is super easy and involves changing only a few lines of text. If you want to rename your SDK root folders or change their relationship or locations relative to one another you will need to do some more editing. Although this is also very easy, it requires a bit of deeper CMake knowledge. CMake is most often run directly from the command prompt (Windows) or Terminal (MacOS), and this is how we will be using it. You will need to navigate to your project folder and then execute a single CMake command. This will then generate your compiler project file (*.sln* for Visual Studio and *.xcodeproj* for Xcode) complete with all the files needed for all APIs as well as your *PluginKernel*. You may then open the compiler project and rebuild it straight away to begin working on it. Typically, navigating to the project folder is the most difficult part of this process. The command prompt and Terminal can be finicky about your project folder naming, so be sure to refrain from names that use whitespaces, numbers, or other special characters—and use the underscore (_) symbol as needed in place of whitespaces.

One thing is very important when installing CMake: you need to make sure that the path to the CMake executable is added to your operating system so that you can run it with a simple "cmake" command and without needing to type in the entire path to the executable. For Windows, this is done during the CMake installation process when you check an option to add the path to CMake to your system path variable. For MacOS, it is a little more complex, requiring that you create a *SymLink* (shortcut) to CMake in your *usr/local/bin* folder. If you don't know how to do this, you can find numerous tutorial websites that contain the instructions: it is a common task, especially for programmers. We will discuss the details of using CMake in Section 6.3.2.

6.3 *Creating a Plugin Project With ASPiKreator:* IIRFilters

The ASPiK system is heavily documented and available through a variety of outlets including www.
willpirkle.com as well as GitHub. Rather than hash through the minutiae of every detail, I will present
an example project named *IIRFilters*—the first book plugin project, in Chapter 11. You may want
to refer to that chapter, or bookmark this one and come back. If you are using RackAFX, you will
create your RAFX2 project using the instructions in Chapter 7, then use the *Export ASPiK* function to
generate the ASPiK project. RackAFX will automatically alter the *CMakeLists.txt* file with your project
details. You are also free to manually alter the file at a later date and re-run CMake to enable other
options or target other APIs.

You can also use the built-in ASPiKreator software that is included in the ASPiK SDK and is found
in the root folder *ASPIK_SDK/ASPiKreator*. There are precompiled versions for both Windows and
MacOS—just launch the appropriate application for your OS. ASPiKreator consists of two panels:
the *ASPiK Project Creator* is for generating new projects while the *GUI Code Creator* is for adding
GUI control code quickly and easily to your project. While all of this can be done manually—from
creating the project to writing the GUI code—this application will save you a lot of time as it handles
the tedious chores for you. If you need to create your project manually, see the documentation in the
ASPiK SDK for step-by-step instructions on editing the *CMakeLists.txt* file and writing the GUI code.
The first step is to set your default project folder, which is named *myprojects* and may be located in
the SDK of your choice or in the outer *ALL_SDK* folder. Next set your company name, four-character
company code, URL, and email address. ASPiKreator will remember these settings for future software
runs. Then, set the project name to *IIRFilters*. The plugin may be named anything you like, and
you may use the same name as the project itself—here we will use *Filterizer*. Remember to set the
AAX-specific names as well as selecting the AAX category if you are targeting AAX in your project.
Figure 6.4 shows the *ASPiK Project Creator* panel with the selected options for this example.

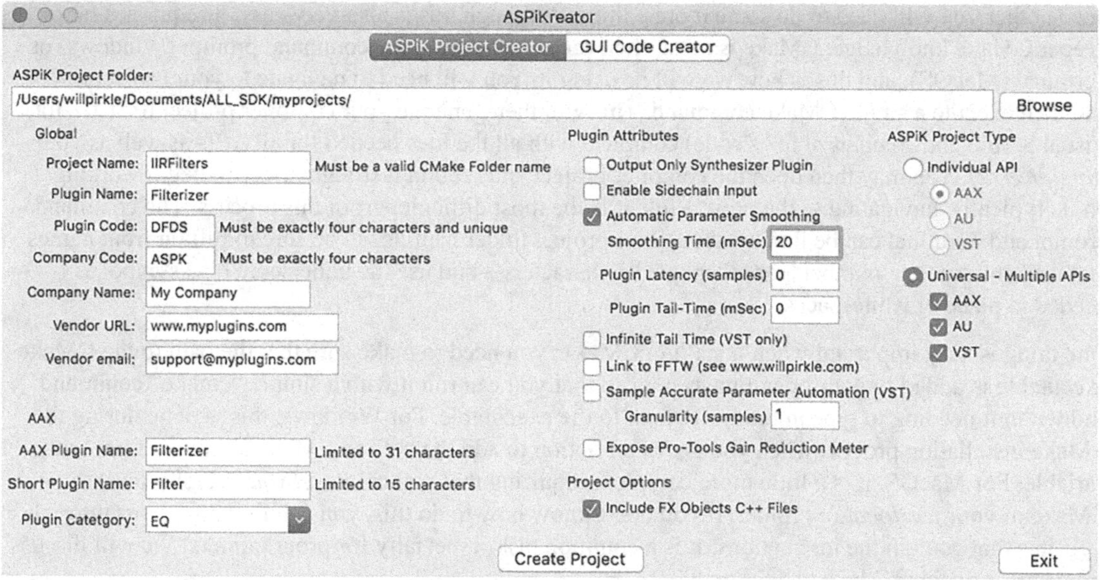

Figure 6.4: The *ASPiK Project Creator* tab allows you to quickly create new projects; here is our example plugin.

On the right side of the *ASPiK Project Creator* panel are the plugin attributes and the targeted APIs. For the plugin options, select and edit the fields as needed. The attributes and their meanings are as follows.

- **Output Only Synthesizer Plugin:** check this box for synth plugins (all the book projects are FX types).
- **Enable Sidechain Input:** check this box if you would like the DAW to expose side chain inputs to your plugin.
- **Automatic Parameter Smoothing:** check this box to enable parameter smoothing, which removes glitches from GUI control movements (see Section 6.4.3)
- **Smoothing Time:** if you are using parameter smoothing, set this value to the smoothing time in milliseconds.
- **Plugin Latency:** this is a special setting for look-ahead compressors or FFT processing plugins that necessarily introduce a latency or time delay to the audio signal. Set this parameter in samples.
- **Plugin Tail and Infinite Tail Times:** the tail time may be set so that you can hear your reverb or delay tails; for VST3, there is an option for an infinite tail. If this variable is set to true, the DAW will run a continuous stream of zeros through your plugin forever after the user stops playback for the very first time.
- **Link to FFTW:** check if you are linking with the FFTW library for the advanced plugins in Chapters 20 and 22; note that you must install FFTW on your system prior to setting the link flag.
- **SAA and Granularity:** check the box to enable the VST3 SAA function (see Chapter 3); you may also adjust the granularity of the updates (higher values will spread out the update interval) to reduce CPU loads.
- **Expose Pro Tools Gain Reduction Meter:** this enables the Pro Tools gain reduction meter on the Pro Tools mixer surface (see Chapter 5).
- **ASPiK Project Type:** you may target individual APIs, or all of them at once, using these controls; make sure you are writing to the proper target folder for your API of choice, or to the universal folder.
- **Include FX Objects C++ Files:** check this box to include all of the C++ objects in this book into your project.

6.3.1 ASPiK Project Folders

Your *myprojects* folder will now look like Figure 6.5a with the newly copied *IIRFilters* folder sitting along side of the *vstgui4* folder. As you add more projects, you will place each one in the *myprojects* folder in parallel with the rest of its folders.

If you open this new project folder (*IIRFilters*), you will see the hierarchy in Figure 6.5b. The *mac_build* folder will be the target location for running CMake on MacOS, and your Xcode compiler project will reside inside it after CMake runs. For Windows, you will target the *win_build* folder for CMake instead. Opening the *project_source* folder will reveal more subfolders, including the important *source* folder, shown in Figure 6.5c. The code you write and modify will be inside this folder. Here is a rundown of the contents of each *source* subfolder:

> *aax_source*: contains the native plugin shell code for AAX
> *au_source*: contains the native plugin shell code for AU
> *vst_source*: contains the native plugin shell code for VST
> *CustomControls*: contains a bunch of pre-built custom VSTGUI controls we will use in Chapter 21.

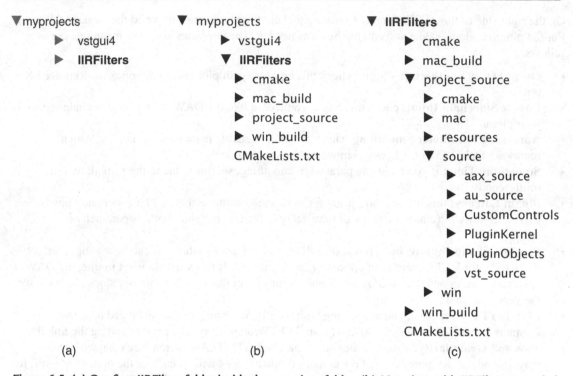

Figure 6.5: (a) Our first *IIRFilters* folder inside the *myprojects* folder. (b) *Myprojects* with *IIRFilters* expanded to show the top-level folders. (c) *IIRFilters* expanded to show the contents of the *source* folder the *IIRFilters* plugin GUI and block diagram.

PluginKernel: contains all of the plugin kernel object code; this is where you will spend most of your time coding

PluginObjects: for storing any and all objects that we will use in the plugin core code. It comes pre-loaded with *fxobjects.h* and *fxobjects.cpp,* which contain all of the helper objects we design and use throughout this book. If you add more of your own C++ objects or other code, place it in this folder.

The *xxx_source* folders contain the native plugin shell code for each API. You will generally not need to modify this code at all, but the code is there if you would like to examine it. The *PluginKernel* folder contains the *plugincore.h* and *plugincore.cpp* files. **These are the only two files you need to edit to create your plugin** (unless you make changes to the *fxobjects* files as part of your homework). After running CMake, your compiler project will contain a set of similarly named project filters (also called *folders*) where you may find your plugin files.

NOTE: If you combine APIs into a single compiler project, you will see the same *PluginKernel* files listed for each project in the compiler's solution browser (i.e. the VST, AU, and AAX targets appear to have their own *PluginKernel* files). For multiple-API projects, there is only one set of files that are shared across the compiler targets. This means that you may edit *any* of the *plugincore.h* or *plugincore.cpp* files and changes will affect *all* APIs in the combined project simultaneously.

6.3.2 Running CMake

In Windows, you will need to run the command prompt "with administrator privilege," which is called *Command Prompt (Admin)* in the Windows start menu. For MacOS, you will use Terminal. Navigate into the new project and then into the proper build folder for your OS (*mac_build* or *win_build*). You will notice that these folders are empty and that the project's *CMakeLists.txt* file exists one folder above them in the hierarchy. This is proper and correct for running CMake. With the command prompt or Terminal navigated to this folder, it is time to issue the CMake command. This will vary with OS and compiler, as listed here.

MacOS and Xcode:
```
cmake —GXcode ../
```

Windows and Visual Studio 2015:
```
cmake —G"Visual Studio 14 2015 Win64" ../
```

Windows and Visual Studio 2017:
```
cmake —G"Visual Studio 15 2017 Win64" ../
```

You can find the CMake command for your compiler on the CMake generators web page (https://cmake.org/cmake/help/v3.11/manual/cmake-generators.7.html).

You will see the output from CMake as it checks your compiler and creates your project. I've placed some CMake messages in the *CMakeLists.txt* files to show the progress of your CMake build. After CMake finishes, your *mac_build* or *win_build* folder will be populated with more subfolders as well as the compiler project file (*.xcodeproj* or *.sln*). Double-click on the compiler to launch it and open the project.

6.4 Adding Effect Objects to the PluginCore

Figure 6.6 shows the GUI and block diagram for the *IIRFilters* project. You can see that there are two *AudioFilter* effect objects that do the signal processing along with a simple GUI. The GUI contains controls that link to plugin parameters. The plugin parameters dump their values into plugin variables that we bind with them. The variable binding is a feature of ASPiK projects, and you can opt to perform the binding operation yourself if you like; variable binding is provided as a convenience.

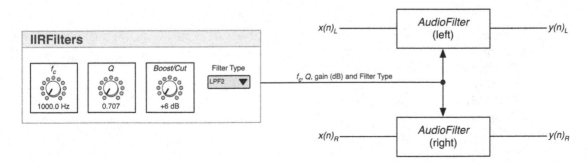

Figure 6.6: The *IIRFilters* plugin GUI and block diagram.

Each of the chapter plugin projects is presented in the exact same fashion, starting with the GUI and block diagram. From that you automatically know that we need two *AudioFilter* objects and four parameters: f_c, *Q*, *boost/cut*, and *filter type*.

6.4.1 The **PluginCore** *Constructor*

When you generate the project, a default set of information will already be written in the constructor. For many projects, you will not need to alter this information. Perhaps the most important information your plugin must describe has to do with the channel I/O that your plugin will support. If you want to write exotic multi-channel plugins, then this is something you will need to study.

All of the plugins in this book support three different combinations of audio input and output channels that are the three most common formats. When you create a new ASPiK project, it will automatically declare three channel I/O combinations for you in the *PluginCore* constructor:

1. mono input to mono output ($1 \rightarrow 1$)
2. mono input to stereo output ($1 \rightarrow 2$)
3. stereo input to stereo output ($2 \rightarrow 2$)

AU, AAX, and VST all support different types of multi-channel operation that are beyond simple stereo. However, there are variations within each API, and there are some formats that are not supported in every API. So, if you decide you want to write a plugin for the multi-channel Ambisonics formats, you will likely not be able to write it for all APIs. For example, AU does not support channel I/O greater than 10 inputs and 10 outputs. In comparison, VST3 supports up to 22.1 channel plugins, including variations for both Sony and DTS surround sound, while AAX supports Sony, DTS, and most of the obscure Ambisonics formats. The first thing you need to decide is how many formats to support. Then, you need to describe each supported combination. If you support a side-chain input, you will need to describe the channel combinations your plugin will support (though there are some limitations on side chains that are API specific). This is best explained with the sample code, so let's continue looking at the *PluginCore* constructor. There are two base class functions that will make setting up your supported channels a simple operation. You just call these functions to add your supported combinations: *addSupportedIOCombination( )* and *addSupportedAuxIOCombination( )*. The channel configurations consist of a simple array that stores an input and output channel configuration pair. Enumerations for each type of channel configuration are found in the *pluginstructures.h* file.

```
enum channelFormat

{
        kCFNone,
        kCFMono,
        kCFStereo,
        kCFLCR,
        kCFLCRS,
        kCFQuad,
        kCF5p0,
        kCF5p1,
etc. . . .
```

You define a channel I/O configuration as an array pair using those enumerations, then pass this value to the helper functions. For ASPiK projects, this is pre-coded for you as follows.

```
// --- for FX plugins
if(getPluginType() == kFXPlugin)
{
    addSupportedIOCombination({kCFMono, kCFMono});
    addSupportedIOCombination({kCFMono, kCFStereo});
    addSupportedIOCombination({kCFStereo, kCFStereo});
}
```

In addition to mono and stereo, the most popular multi-channel format might be 5.1, which is encoded as *kCF5p1* in the previous sample. You can also see that there are both Sony and DTS versions of higher channel count formats. After describing your plugin's normal input/output channel configurations, you do the same for the side chain using the aux IO variables. If you enable the side chain during the ASPiK CMake setup, as follows, then this code is written for you.

```
// --- for sidechaining, we support mono and stereo inputs;
addSupportedAuxIOCombination({kCFMono, kCFNone});
addSupportedAuxIOCombination({kCFStereo, kCFNone});
```

The rest of the constructor needs no modification and simply calls the helper functions we will implement next:

```
    // --- create the parameters
    initPluginParameters();

    // --- create the presets
    initPluginPresets();
}
```

6.4.2 IIRFilters: *GUI Parameter Lists*

Each plugin parameter object requires an attribute that uniquely identifies the parameter for both the DAW and the plugin itself. The AU, VST, and RAFX2 APIs specify these unique attributes as integer values. AAX specifies unique strings instead: in this case, it is simple enough to convert our unique integer values into unique string values; this is done automatically for you in the AAXplugin shell. Each time you create a plugin parameter, you must specify this unique value—and it is important that the value never changes. There are several ways to do this. We are going to use an un-scoped enumeration value that we specify at the top of the *plugincore.h* file. We will use this whenever we need to access plugin parameters. We will name this unique integer identifier the *controlID* value. A parameter's *controlID* value is used to look up and access the parameter's current value, which is transferred to the underlying plugin variable. If you are using RackAFX, then these unique values will automatically be generated and written for you as you add and remove controls from the RackAFX user interface. If you are not using RackAFX, then you will need to write this enumeration yourself. You will also notice that the plugin parameter tables that follow do not list the *controlID* values, though I present suggested values in code for you. We will use a naming convention so that you may easily remember the *controlID* without having to memorize numerical values. At the top of the *plugincore.h* file, you add the enumeration; you can see one enumeration for each parameter. In the book projects, we name these identically as the parameter's linked variable. This makes them both unique and easy to remember. Instead of memorizing an integer for the filter's Q control, you remember the variable name, then append it as *controlID::filterQ*, which evaluates to the value 1 in the enumeration here.

```
enum controlID {
        filterFc_Hz = 0,
        filterQ = 1,
        boostCut_dB = 2,
        filterType = 3
};
```

Each plugin project will list a GUI control table that explains how to set up each plugin parameter as a *PluginParameter* object defining its attributes: name, minimum, maximum, default, units, and in the case of a string-list parameter, the comma-separated list. Table 6.1 shows the GUI parameter table for the *IIRFilters* plugin. Each row shows the parameter attributes, including the control taper and the name of the linked variable that the parameter will place its data into. In this case, the comma-separated string is too long to fit in the table, so it is listed below. Each row in the table will become a *PluginParameter* object.

Table 6.1: The *IIRFilter* plugin's GUI parameter and linked-variable list

Control Name	Units	Min/max/default or string-list	Taper	Linked Variable	Linked Variable type
fc	Hz	20 / 20480 / 1000	1V/oct	*filterFc_Hz*	*double*
Q	–	0.707 / 20 / 0.707	linear	*filterQ*	*double*
Boost/Cut	dB	–20 / +20 / 0	linear	*boostCut_dB*	*double*
Filter Type		string-list follows this table	–	*filterType*	*int*

Here is the filter type string list:
"LPF1P, LPF1, HPF1, LPF2, HPF2, BPF2, BSF2, ButterLPF2, ButterHPF2, ButterBPF2,ButterBSF2, MMALPF2, MMALPF2B, LowShelf, HiShelf, NCQParaEQ, CQParaEQ, LWRLPF2, LWRHPF2, APF1, APF2, ResonA, ResonB, MatchLP2A, MatchLP2B, MatchBP2A, MatchBP2B, ImpInvLP1, ImpInvLP2"

6.4.3 Parameter Smoothing

Recall from Section 2.5 that the GUI control values will be transferred into the plugin core at the start of each buffer processing cycle. With variable binding, these values will then be transferred into our predefined plugin variables. For continuous controls, this may result in audible clicks or glitches if the user moves the GUI control rapidly and its value changes drastically between buffer processes. This is especially true of volume controls, where small changes in dB values may result in large changes in the final volume value calculation. To overcome this problem, we may employ parameter smoothing to effectively slow down the parameter change over the course of the buffer processing. Usually, we will smooth the plugin parameter slightly on each sample interval to result in a glitch-free user interface. The *PluginParameter* object already contains the code for parameter smoothing. We define two types of smoothing: *linear* (ramp) and *exponential* (LPF). The LPF smoothing uses a simulated Resistor-Capacitor (RC) time constant in a 1st order LPF to smooth the transients in the GUI controls. This is the default smoothing method. If you want to make your own smoother, see the *ParamSmoother* object in the *guiconstants.h* file. The linear and LPF smoother have approximately the same CPU overhead, but care should be taken with parameter smoothing. If your plugin needs 100 continuous parameters, then smoothing each of these on every sample interval is going to eat up a lot of CPU power. Notice also that parameter smoothing is only used on continuous controls. There is no need to smooth integer values. You enable smoothing and set the smoothing time after the plugin parameter has been created.

6.4.4 Handling the String-List Parameters

The string-list parameter is a special case with regards to how we handle it in our C++ code. The user is presented with a list of strings in the GUI. Selection of a string will result in the transfer of an *unsigned integer* value to the plugin. The strings are zero indexed, so the first string in the list is 0, the second string is 1, and so on. We need a way to identify the selection in code in a way that makes sense to us as programmers. Suppose our string list consists of: "LPF, HPF, BPF." We might try putting the strings into arrays that are indexed with [0], [1], [2], etc. Then we would need to make string comparisons in code, accessing the array with the control selection value. The problem there is that "LPF" and "lpf" are considered to be two different strings, so we would need to be careful about the cases of the characters in the strings—a mistyped string in C++ code like "HpF" would not be flagged as a compiler error, and the user's selection would never be translated properly. Although string comparisons might be easy to read in code, there are too many issues with them when making comparisons—of course this is only exacerbated if your plugin is *localized*, meaning that you provide different sets of strings for different languages.

C++11 provides us with a solution to this problem by defining a *scoped enumeration*. In this case, the enumeration is named *name::enum*—and we use this syntax to uniquely identify a particular item in the enumeration. We could then declare the two lists separately, while re-using one or more enumeration strings like this:

```
enum class inputFilterEnum {LPF, HPF, BPF};
enum class outputFilterEnum {LPF, BPF, BSF};
```

The compiler does not produce an error because *inputFilter::LPF* and *outputFilter::LPF* are distinct and different. In addition to this benefit, the underlying data type for the scoped enumeration is a signed integer (*int*) and we can use the integer that the GUI control delivers to identify the selected string. The difficulty is that we cannot use the simple syntax of an un-scoped enumeration to make the comparison:

```
if(inputFilter == inputFilterEnum::LPF) << this will not compile!
    // do LPF stuff
```

We must first statically cast the enumeration to make the comparison, like this:

```
if(inputFilter == static_cast<int>(inputFilterEnum::LPF))
    // do LPF stuff
```

Now, this is slightly less readable because of the casting operation, but it overcomes our problem with repeated enumeration values. To simplify coding, we've declared some C++ macros that you may use in place of the casting:

```
#define enumToInt(ENUM) static_cast<int>(ENUM)
#define compareEnumToInt(ENUM,INT) (static_cast<int>(ENUM) == (INT))
#define compareIntToEnum(INT,ENUM) ((INT) == static_cast<int>(ENUM))
#define convertIntToEnum(INT,ENUM) static_cast<ENUM>(INT)
```

We will be using these macros in our control statements when accessing the scoped enumeration values. The macros are simple—and you might even write your own version that makes more literal sense if you like. Notice that these macros not only allow comparisons of *int* and *scoped enum* values, but also provide a conversion macro for passing an integer value into a function that expects a scoped enumeration.

Now we can write code like this:

```
if(compareIntToEnum(inputFilter, inputFilterEnum::LPF))
      // do LPF stuff
```

```
int someFilter = enumToInt(inputFilterEnum::LPF);
```

Consider a function we've written that specifically accepts a scoped enumeration as a function parameter:

```
void doInputFilter(inputFilter filterType);
```

We can use one of the macros to call the function like this:

```
doInputFilter(convertIntToEnum(inputFilter));
```

The upshot of all of this is that each time we declare a plugin variable that will accept the GUI control change for a string-list parameter, we will also declare the scoped enumeration we will use for comparisons and function parameters. These string-list variables will always occur in pairs (*int* and *scoped enum*), and we will use a naming convention that will always give us unique names by appending "Enum" to the integer variable name. So, we will declare our *inputFilter* and *outputFilter* string-list variables like this:

```
int inputFilter = 0;
enum class inputFilterEnum {LPF, HPF, BPF};
```

```
int outputFilter = 0;
enum class outputFilterEnum {LPF, BPF, BSF};
```

Because we adhere to this naming convention, we will not list the name of the scoped enumerator in our plugin parameter tables. We will always encode the name using our naming convention of *<variable_name>Enum*.

For two-state switches, we know that the GUI control will always transmit either 0 or 1 for *off* and *on* respectively, so we really don't need the enumeration and we can easily make comparisons as if the integer values were booleans. For example:

```
int enableMute = 0;
enum class enableMuteEnum {SWITCH_OFF, SWITCH_ON}; <<< not really needed
```

Then, later in code:

```
if(enableMute)
      // do mute stuff
else
      // do un-mute stuff
```

6.4.5 IIRFilters: *Declaring Plugin Variables*

Ultimately, we know that each GUI control will affect an underlying plugin variable. With the variable enumeration defined, we can now declare these variables in the *plugincore.h* file. You have two options: enter the variables manually or use the *ASPiKreator*. The variable names and data types are listed in Table 6.1. If you want to enter the variables manually, edit the class declaration with a *private* access-specifier and add them directly. The object will now look like the code that follows. Notice how the string-list parameters use a *pair* of variables (*int* and *enum class*) for their declarations and that the (massive) filter type enumeration matches the string list in Table 6.1, but without the quotation marks.

```
class PluginCore: public PluginBase

{
public:
    PluginCore();
    virtual ~PluginCore(){}

<SNIP SNIP SNIP>

private:
    // --- Continuous Plugin Variables
    double filterFc_Hz = 0.0;
    double filterQ = 0.0;
    double boostCut_dB = 0.0;

    // --- Discrete Plugin Variables
    int filterType = 0;
    enum class filterTypeEnum {
    LPF1P,LPF1,HPF1,LPF2,HPF2,BPF2,BSF2,ButterLPF2,ButterHPF2,
    ButterBPF2,ButterBSF2,MMALPF2,MMALPF2B,LowShelf,HiShelf,
    NCQParaEQ,CQParaEQ,LWRLPF2,LWRHPF2,APF1,APF2,ResonA,ResonB,
    MatchLP2A,MatchLP2B,MatchBP2A,MatchBP2B,ImpInvLP1,ImpInvLP2 };
etc. . . .
```

However, the simpler option is to use the *ASPiKreator* and generate the plugin variables at the same time that you generate the GUI control code. You do this using buttons that will copy the code to your clipboard, then you simply paste the code into the files as needed.

6.4.6 *Parameter Object Enumerations for Attributes*

The *PluginParameter* object was designed to handle all of the duties of the various parameter types. It manages the parameter value and can be designed as any of the parameter types your plugin may need for GUI controls and audio meters. It performs numerical conversion routines, control tapering, parameter smoothing, conversion to string lists, meter ballistics, and many other chores that we need. In general you won't need to modify this object unless you would like to add other tapering options. The *GUI Code Creator* panel automatically writes C++ object declarations using the various

PluginParameter constructors, and then it adds code that binds the plugin variables. If you look at the *pluginparameter.h* and *pluginparameter.cpp* files, you can see the different constructors that are customized for each parameter type. You will create the objects in the *initPluginParameter( )* function in the *PluginCore* object either manually or using the *GUI Code Creator*. You can find all of the details on manual GUI object creation in the SDK; we will use the *GUI Code Creator* since it is simple and quick. But first, we need to discuss some scoped enumeration classes that will be used to define our parameters, as shown here.

```
enum class controlVariableType {kFloat,
                                kDouble,
                                kInt,
                                kTypedEnumStringList,
                                kMeter,
                                kNonVariableBoundControl};
```

The *controlVariableType* defines the kind of parameter we are creating. The *kFloat* and *kDouble* types specify a continuous floating point parameter or a meter variable (*kFloat* only). The *kInt* defines a discrete integer control, while *kTypedEnumStringList* defines a string-list parameter. We use *kMeter* for meters. We will reserve *kNonVariableBoundControl* for parameters that are not associated with GUI input controls and not tied to any underlying plugin parameters—you may use them to store non GUI information if you like.

The continuous controls may also define a control taper. This makes the knobs or sliders behave in a way that is consistent with user's experiences with other plugins or the hardware products they often emulate. There are four built-in tapers to choose from: *linear*, *log*, *anti-log* and *volt/octave*.

```
enum class taper { kLinearTaper, kLogTaper, kAntiLogTaper,
                   kVoltOctaveTaper };
```

The meters may be calibrated for linear or logarithmic displays; these are encoded in another *enum class* delaration:

```
enum class meterCal {kLinearMeter, kLogMeter };
```

We define data types for our variable binding option for the four most common control variables—*float*, *double*, *int* and *unsigned int*—as shown here.

```
enum class boundVariableType { kFloat, kDouble, kInt, kUInt };
```

Finally, we define two types of parameter smoothing: linear and exponential (LPF).

```
enum class smoothingMethod { kLinearSmoother, kLPFSmoother };
```

If you are using RackAFX for your plugin development, then RackAFX will write this *PluginParameter* instantiation code for you as you add, remove, and modify controls and meters on the RackAFX user interface. Also, it will automatically add variable binding to the variables you describe when you set up the controls. These will all get exported into your ASPiK project, so you may safely skip these next few sections if you like. If you are not using RackAFX, have no fear—we can easily translate Table 6.1 into code using the *GUI Code Creator*.

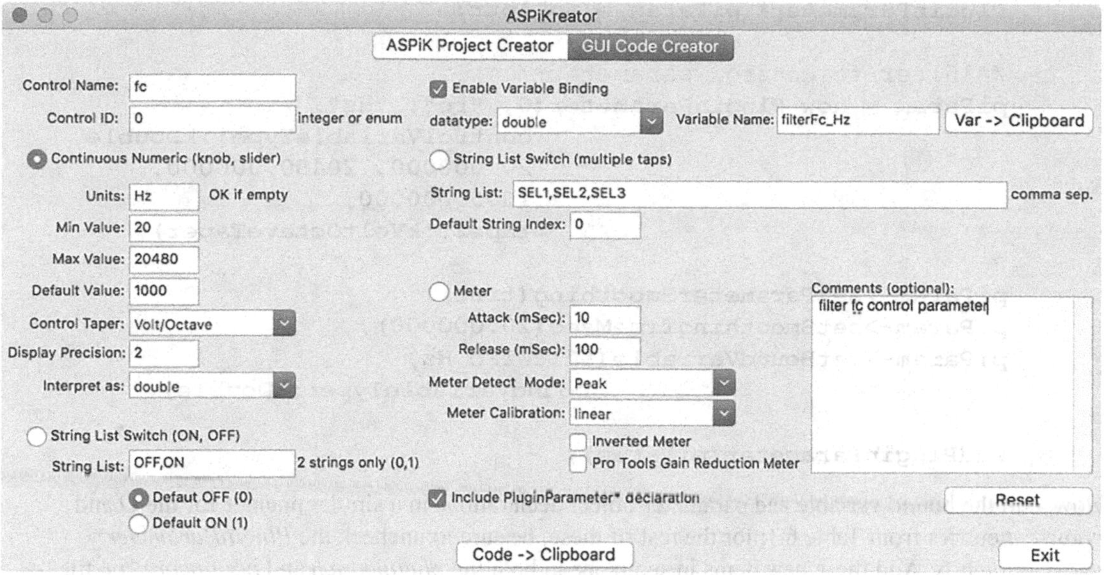

Figure 6.7: Data entry for the f_c parameter using the *GUI Code Creator* tab, note that a comment has been entered in the optional Comments section.

6.4.6.1 Continuous Floating Point Parameters and Discrete Integer Parameters

Let's step through the process of adding the filter f_c parameter first, then you can add the Q and *boost/cut* from Table 6.1 as an exercise. Refer to the entry in Table 6.1 and remember that we enumerated the control ID as 0 in Section 6.4.2, then compare this with Figure 6.7, which shows the data entry into the GUI Code Creator—note that the *Include PluginParameter* declaration* box is checked. You want to do this for the very first parameter you add, then uncheck it for future parameters.

First, click on the *Var → Clipboard* button to copy the variable declaration to your clipboard. In the *plugincore.h* file, add this new variable to the private section; just paste it into the class declaration.

```
double filterFc_Hz = 1000.000000;
```

Next, copy the parameter declaration to the clipboard by clicking the *Code → Clipboard* button. Open the *plugincore.cpp* file and add the new declaration after the *if (pluginParameterMap.size() > 0)* statement. The *GUI Code Creator* generated the bold code here—note the *PluginParameter* piParam = nullptr* declaration, which we will reuse for the rest of the parameters. For automatic parameter binding, the first argument to *setBoundVariable( )* is a pointer to the bound variable using the address-of operator (&) while the second parameter indicates the data type of the bound variable (*double*). The last statement—*addPluginParameter(piParam)*—adds the new object to the *PluginBase* parameter lists.

```
bool PluginCore::initPluginParameters()
{
        if (pluginParameterMap.size() > 0)
             return false;
```

```
PluginParameter* piParam = nullptr;

/* filter fc control parameter*/
piParam = new PluginParameter(0, "fc", "Hz",
                              controlVariableType::kDouble,
                              20.000000, 20480.000000,
                              1000.000000,
                              taper::kVoltOctaveTaper);

piParam->setParameterSmoothing(true);
piParam->setSmoothingTimeMsec(20.000000);
piParam->setBoundVariable(&filterFc_Hz,
                          boundVariableType::kDouble);

addPluginParameter(piParam);
```

Now, add the bound variable and parameter object declarations in a similar manner for the *Q* and *boost/cut* entries from Table 6.1; for the rest of these, be sure to uncheck the *PluginParameter* declaration* box. Add these new items in sequence in both the *plugincore.h* and *plugincore.cpp* files—be sure to include the correct *Control ID* value when generating the object declarations.

6.4.6.2 String-List Parameters

There are two variations on the string-list parameter. The first is a specific kind of list for an on/off switch; you can find this on the lower left side of the *GUI Code Creator*. The string list consists of the "off" string that will map to the number 0, and the "on" string that maps to the number 1. You may name these as you like, but make sure that the "off" string is first. You then have the option of setting the default state to on or off. The second type is for a generic string list with any number of arguments. You list the comma-separated string, and then choose the index of the default value that you wish to display for the user. Our *Filter Type* variable will use the second option, as shown in Figure 6.8. Notice that the bound variable data type is set to *int*: we will use the scoped enumerator method described in Section 6.4.4 to access and decode the parameter information. Clicking the *Var → Clipboard* button produces the pair of entries for the bound variable declarations in the *plugincore.h* file:

```
int filterType = 0;
enum class filterTypeEnum {LPF1P, LPF1, HPF1, LPF2,HPF2, BPF2, BSF2,
ButterLPF2, ButterHPF2, ButterBPF2,ButterBSF2, MMALPF2, MMALPF2B,
LowShelf, HiShelf, NCQParaEQ, CQParaEQ, LWRLPF2, LWRHPF2, APF1, APF2,
ResonA, ResonB, MatchLP2A, MatchLP2B, MatchBP2A, MatchBP2B, ImpInvLP1,
ImpInvLP2};
```

The parameter declaration is different from the continuous controls because there are multiple constructors for the *PluginParameter* object, tailored specifically for each plugin type:

```
piParam = new PluginParameter(3, "Filter Type", "LPF1P, LPF1, HPF1,
                              LPF2, HPF2, BPF2, BSF2, ButterLPF2,
                              ButterHPF2, ButterBPF2,ButterBSF2,
                              etc . . . ");
```

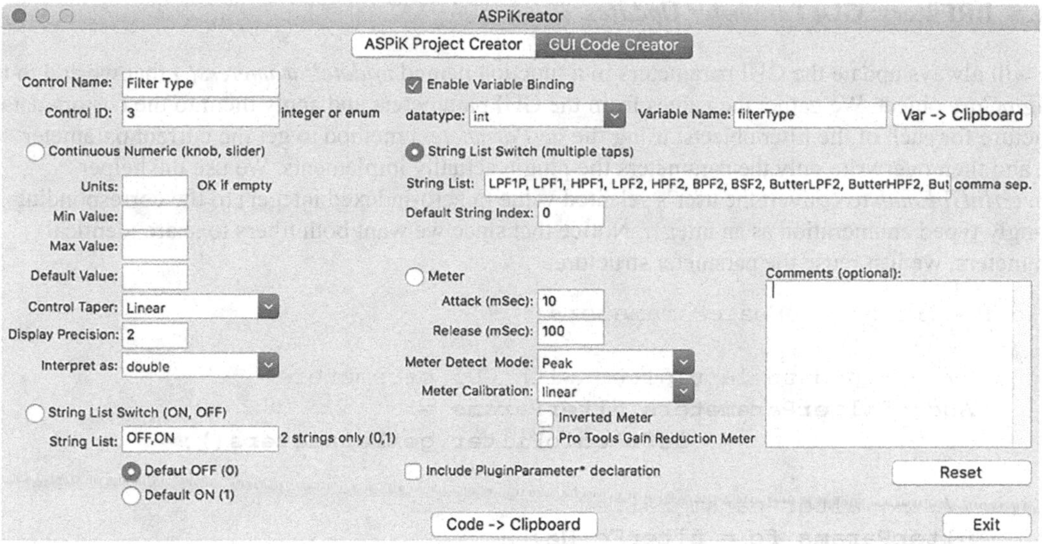

Figure 6.8: Data entry for the *Filter Type* parameter using the *GUI Code Creator*.

```
piParam->setBoundVariable(&filterType, boundVariableType::kInt);

addPluginParameter(piParam);
```

6.4.7 IIRFilters: *Object Declarations and Reset*

Note: this section is repeated in more detail in Section 11.7.2.

You declare the *AudioFilter* objects in the *plugincore.h* file as protected member variables of the *PluginCore* object. In the first plugin, we declare separate filter objects for each channel (we will use arrays of objects in later chapters). We always declare a function named *updateParameters* that will transfer data from the GUI-linked variables into the *AudioFilter* objects.

```
// --- in your processor object's .h file
#include "fxobjects.h"

// --- in the class definition
protected:
    AudioFilter leftAudioFilter;
    AudioFilter rightAudioFilter;
    void updateParameters();
```

The objects require the current DAW sample rate. To apply the sample rate and reset the objects for audio streaming, you call the *reset* function from the *PluginCore::reset( )* method. The *PluginCore's* reset method is passed a data structure with information about the sample rate and bit depth. We use the sample rate member to reset the object's effects:

```
// --- reset filters
leftAudioFilter.reset(resetInfo.sampleRate);
rightAudioFilter.reset(resetInfo.sampleRate);
```

6.4.8 IIRFilters: *GUI Parameter Updates*

We will always update the GUI parameters in a function named *updateParameters( )* that we add to the *PluginCore* object. We gather the values from the GUI parameters and apply them to the custom data structure for each of the filter objects, using the *getParameters* method to get the current parameter set, and then overwrite only the parameters the plugin actually implements. We use the helper *convertIntToEnum* to convert the user's selected value (a zero-indexed integer) to the corresponding strongly typed enumeration as an integer. Notice that since we want both filters to share identical parameters, we just reuse the parameter structure.

```
void PluginCore::updateParameters()
{
        // --- update left filter with GUI parameters
        AudioFilterParameters filterParams =
                            leftAudioFilter.getParameters();

        // --- alter param values
        filterParams.fc = filterFc_Hz;
        filterParams.Q = filterQ;
        filterParams.boostCut_dB = boostCut_dB;
        filterParams.algorithm = convertIntToEnum(filterType,
                                                    filterAlgorithm);

        // -- set on objects
        leftAudioFilter.setParameters(filterParams);
        rightAudioFilter.setParameters(filterParams);
}
```

6.4.9 IIRFilters: *Processing Audio Data*

In Chapter 2 we saw how the DAW de-interleaves the incoming audio and sends individual channel buffers of data into the plugin on the audio signal processing thread. During that time, the user might have also been moving the GUI controls around and thus created parameter change values that were stored locally in the native plugin shell's API-specific parameter list. Or alternatively, the automation systems may have made changes to the local parameter list. During the buffer processing cycle, the audio data are delivered to the plugin and the parameter changes are made available. Amazingly, every one of our supported APIs delivers the audio buffers to the plugin in the same manner: one buffer per audio channel. However, the exact mechanism for getting parameter changes into the plugin depends on the API. Each one has its own method for performing this duty in a thread-safe manner, but each method involves making a safe copy of the parameter values that are transferred to the plugin—in other words, your plugin is never able to access the native shell's actual updated parameters; instead, you receive a copied value.

The *PluginCore* and native plugin shells coordinate all of this for you so that the parameter updates are applied to your plugin variables prior to the audio buffer processing. After processing occurs, we then send parameter data back to the GUI. For our plugins, those parameter data are for meters. In Chapter 21 we will send audio signal information out to a custom view for plotting a waveform histogram. This is also done after the buffer is processed. Figure 6.9 shows a diagram of the buffer

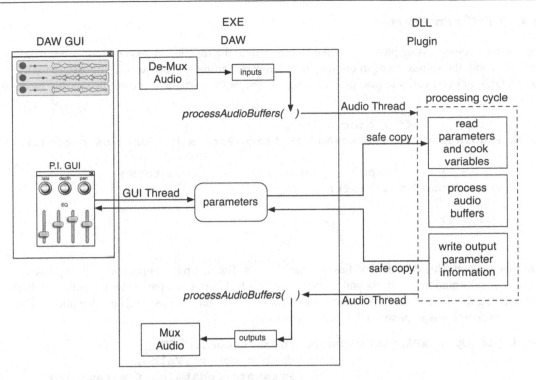

Figure 6.9: As shown here, the DAW gathers incoming audio data and GUI control changes, then delivers them to the plugin in a thread-safe manner.

processing cycle. Notice the block labeled "parameters" inside of the DAW block—this is the native parameter list that the plugin shell created during instantiation. The GUI ultimately connects with and exchanges information with this list. Our local parameters are updated only during the buffer processing cycle in a safe manner.

You can see from Figure 6.9 that the buffer processing cycle consists of three components: reading incoming parameter updates from the GUI controls, processing the audio, then writing output parameter information to the meters or other output-only controls. The *PluginBase* defines three functions that correspond to this three-step architecture.

```
// --- preProcess: sync GUI parameters here
virtual bool preProcessAudioBuffers(ProcessBufferInfo& processInfo);


// --- process buffers—the DSP
virtual bool processAudioBuffers(ProcessBufferInfo& processBufferInfo);


// --- postProcess: do update meters, plots, graphs
virtual bool postProcessAudioBuffers(ProcessBufferInfo& processInfo);
```

These three functions will always be called in this exact order during each and every buffer processing cycle.

6.4.10 Buffer Pre-Processing

During the pre-processing phase, we make a single call to the function that synchronizes our plugin variables with the values stored in our plugin parameters. During this phase, it is impossible for the DAW, GUI, or other entity to manipulate the parameters or the variables. The pre-processing function follows.

```
// --- do pre buffer processing
bool PluginCore::preProcessAudioBuffers(ProcessBufferInfo& processInfo)
{
    // --- sync internal variables to GUI parameters
    syncInBoundVariables();

    return true;
}
```

The *syncInBoundVariables* method has two components. The first part simply copies the updated parameter information into the bound variables we supplied during the parameter instantiation. In the second part, this function calls the function that allows you to do any post-update calculations. This base class function is implemented in the *PluginCore*:

```
bool postUpdatePluginParameter(int32_t controlID,
                               double controlValue,
                               ParameterUpdateInfo& paramInfo)
```

This function will be called for every parameter that is updated and will give you the chance to "cook" any variables that may need alteration as a result. The incoming *ParameterUpdateInfo* structure contains boolean flags that will indicate why this function was called. Once the pre-processing phase is complete, your plugin needs to be fully set up to process audio.

6.4.11 Buffer Post-Processing

The *postProcessAudioBuffers* method is called after all incoming buffers have been processed. The default implementation for this function is the following.

```
bool PluginCore::postProcessAudioBuffers(ProcessBufferInfo&
                                                    processInfo)

{
    // --- update outbound variables
    updateOutBoundVariables();

    return true;
}
```

This function simply calls the sub-function *updateOutboundVariables,* which copies output parameter values into a special list that will be delivered back to the native parameter list, and then back to the GUI in a thread-safe manner. You won't need to modify that function, but this is the opportunity for your plugin to do any other processing after the buffer processing is completed.

6.4.12 Buffer vs. Frame Processing

We have finally arrived at the function that will do all of the cool audio processing stuff our plugin was designed for. At this point, you need to make a fairly critical decision about how that processing will occur—by buffer or by frame. All APIs deliver audio data to the buffer processing function in separated buffers, one for each channel. If we assemble one audio sample from each channel for each sample period, we create a *frame*. So a set of channel buffers containing *M* samples each would be broken into *M* frames. A stereo input frame would consist of an array that contains two audio samples, one for the left channel and one for the right. We might indicate that as {left, right}. A frame of surround sound 5.1 audio data is an array of six audio samples, one from each channel, organized in the following manner: {left, right, center, LFE, left surround, right surround} (where LFE stands for "low frequency effects" or the sub-woofer channel).

In the *PluginCore*, you may choose whether you want to process complete buffers or frames of data. There may be good reasons for choosing one or the other. For example, if your plugin implemented a simple filter operation, you might process the channel buffers, since the audio data in them are unrelated as far as your plugin is concerned. But if your plugin required information from each channel to affect processing on the other, you would need to process frames. An example of this might be a stereo ping-pong delay plugin where left and right channel data are both needed on a sample-by-sample basis. Another example is a stereo-linked compressor, whose side-chain information is derived from both left and right inputs in a way such that the left channel cannot be processed without the right channel, and vice-versa. All projects in this book will use frame-based processing. The reasons are as follows:

*   Frame processing is simpler to understand—you do not need to keep track of buffer sample counters or pointers.
*   Frame processing is universal—all plugin algorithms may be viewed as frame processing systems, even if the channels do not interact; the opposite is not true for buffer processing.
*   Frame processing follows more closely how our algorithms are encoded, both in DSP difference equations as well as flowcharts and signal flow diagrams.

If you want to process buffers instead of frames, the method is simple: override the *processAudioBuffers* function and implement it. If you look in your *plugincore.h* file, you will see that the buffer processing function is commented out. To use it, un-comment it and implement the function in the *plugincore.cpp* file.

```
// --- reset: called when plugin is loaded
virtual bool reset(ResetInfo& resetInfo);

// --- Process audio by frames (default)
virtual bool processAudioFrame(ProcessFrameInfo& processFrameInfo);

// --- uncomment and override this for buffer processing
//virtual bool processAudioBuffers(ProcessBufferInfo&
                                   processBufferInfo);
```

You can use the *processAudioBuffers* implementation in the *pluginbase.cpp* file to help you understand the arguments in the *ProcessBufferInfo* structure (it is actually very straightforward). If your algorithm processes blocks of data, then you will be responsible for filling your own local buffers with data.

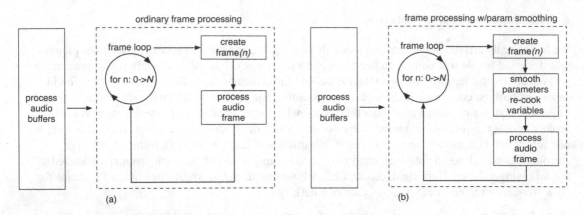

Figure 6.10: (a) As shown here, frame processing requires breaking the buffers into sample interval frames and then processing all channels for that frame together. (b) When parameter smoothing is enabled, parameters are smoothed, updated and cooked on every sample interval.

You should expect that you might receive larger or smaller buffers than your processing calls for. You should also expect to receive partially filled or incomplete buffers from time to time, based on system overhead or the end-of-audio-file situation. Another issue with buffer-based processing involves parameter smoothing. In order to make GUI control changes less glitchy, we can smooth and update their parameters on each sample interval. If you are processing buffers, then this operation is going to break your buffer processing into sample interval chunks. So, if you want to use parameter smoothing, you might consider always processing frames rather than buffers. Figure 6.10 shows ordinary frame processing and frame processing with parameter smoothing.

6.4.13 processAudioFrame: *Information About the Frame*

The *processAudioFrame* function might be the most important function you implement since it is where you will do your DSP magic for the plugin. If you want to process buffers, then you should still make note of this section, as some of the paradigms are similar. The frame processing function is shown here with its function parameter passed as a *ProcessFrameInfo* structure.

```
bool PluginCore::processAudioFrame(ProcessFrameInfo& processFrameInfo)
```

Let's examine that function argument's members (the structure's built-in constructor that simply sets all pointers to null and all integers to 0 is not shown, for brevity):

```
struct ProcessFrameInfo
{
        ProcessFrameInfo(){ }
        float* audioInputFrame = nullptr;
        float* audioOutputFrame = nullptr;
        float* auxAudioInputFrame = nullptr;
        float* auxAudioOutputFrame = nullptr; // --- for future use

        uint32_t numAudioInChannels = 0;
        uint32_t numAudioOutChannels = 0;
```

```
        uint32_t numAuxAudioInChannels = 0;
        uint32_t numAuxAudioOutChannels = 0;

        ChannelIOConfig channelIOConfig;
        ChannelIOConfig auxChannelIOConfig;
        uint32_t currentFrame = 0;

        // --- for future use
        float* controlSignalInputs = nullptr;
        float* controlSignalOutputs = nullptr;
        uint32_t numControlSignalInputs = 0;
        uint32_t numControlSignalOutputs = 0;

        HostInfo* hostInfo = nullptr;
        IMidiEventQueue* midiEventQueue = nullptr;
};
```

At the top of the structure are the members that are the audio input and output pointers. Each frame of data consists of one sample from each channel packed into an array of *float* values. Float values are chosen because they are common across all APIs, though VST3 does allow for processing *doubles* instead. The arrays are passed-by-pointer into the frame processing function. Just below these array pointer declarations are the unsigned integers (*uint32_t*) that tell you how many channels there are in a given frame array. These are easy to interpret: *numAudioInputChannels* is the array size of the *audioInputFrame* array while *numAudioOutputChannels* is the array size of the *audioOutputFrame* array. The audio samples are packed into the arrays according to the standardized audio file formats. These are published on the Internet, and you can also find a very concise description of the channel packing formats in the *vstspeaker.h* file that is inside the VST3 SDK.

The *audioInputFrame* and *audioOutputFrame* arrays hold the frames for the audio input and output busses. The *auxAudioInputFrame* array holds audio data that is coming from a side chain. We will use this information in some of our side-chain-able plugins in later chapters. The *auxAudioOutputFrame* is reserved for future use as plugin APIs begin to implement auxiliary data channels.

The *ChannelIOConfig* variables describe more information about the audio I/O configurations than we can with the channel counts alone. The *ChannelIOConfig* structure holds two member variables, one that describes the input and the other that describes the output channel configurations encoded as *channelFormat* enumerations. You can find the *channelFormat* declaration in the *pluginstructures.h* file. These are especially important for multi-channel plugins. For example, you will see that there are multiple channel formats for a given channel count; *kCF7p1Sony*, *kCF7p1DTS*, and *kCF7p1Proximity* all specify eight total channels (known as "7.1") yet there are three different channel format specifications to deal with. In addition, your plugin may feature very different configurations from input to output; a fold-down plugin may accept a 7.1 input and deliver a plain stereo output.

6.4.14 processAudioFrame: *Input and Output Samples*

We will need to decode the *ChannelIOConfig* variable for the audio data and pick up the audio input data samples, process them, and then write our audio outputs into the *audioOutputFrame* array according to the output channel IO configuration. You may adapt this code as your plugin requires it. The stock *processAudioFrame* function your plugin is created with looks something like this at the head of the function. We will discuss the MIDI event queue and VST3 sample accurate updating later, so you can safely ignore them now. The first thing we do is check our plugin type to differentiate FX and synth plugins.

```
// --- fire any MIDI events for this sample interval
processFrameInfo.midiEventQueue->fireMidiEvents
                                (processFrameInfo.currentFrame);

// --- do per-frame updates; VST automation and parameter smoothing
doSampleAccurateParameterUpdates();

// --- Synth or FX Plugin?
if (getPluginType() == kFXPlugin)
{
        // do FX stuff
}
```

From this point on, we will assume that we are processing for a FX plugin and we'll omit those enclosing curly brackets for brevity. Since all of our plugins will by default accommodate mono-mono, mono-stereo, and stereo-stereo formats, this section will look the same for each plugin in the book. You can add more channel formats as you like. You will then modify the code to do some meaningful processing. The audio data samples arrive in slots in the *audioInputFrame* and *audioOutputFrame* arrays; we use standard C++ array notation to access them. We might write something like this to pick up the input samples, process them through some interesting audio processing function, and then pass the resulting samples to the outputs:

```
// --- get left and right input samples
double xn_L = processFrameInfo.audioInputFrame[0];
double xn_R = processFrameInfo.audioInputFrame[1];

// --- process them to create outputs
double yn_L = leftAudioFilter.processAudioSample(xn_L);
double yn_R = rightAudio.Filter processAudioSample(xn_R);

// --- send audio out
processFrameInfo.audioOutputFrame[0] = yn_L;
processFrameInfo.audioOutputFrame[1] = yn_R;
```

From this example, it should be obvious how to pick up input samples and write output samples. If we were processing a 5.1 input, we might write something like this instead:

```
double leftIn =         processFrameInfo.audioInputFrame[0];
double rightIn =        processFrameInfo.audioInputFrame[1];
double lfeIn =          processFrameInfo.audioInputFrame[2];
double leftSurroundIn = processFrameInfo.audioInputFrame[3];
double rightSurroundIn = processFrameInfo.audioInputFrame[4];
```

However, all of our plugins will have at most two input or output channels, so we will simply use the array index values 0 and 1 to represent the left and right channels exclusively. Of course we will need to check the audio cannel configurations before simply grabbing audio samples to make sure that the values do exist in the input and output arrays. We do know for a fact that the left (mono) input and output channels will always exist, so we can compact the code a bit later. To decode our three I/O combinations, we use the following code:

```
// --- Mono-In/Mono-Out
if(processFrameInfo.channelIOConfig.inputChannelFormat == kCFMono &&
    processFrameInfo.channelIOConfig.outputChannelFormat == kCFMono)
        // --- DO mono-in -> mono out processing

// --- Mono-In/Stereo-Out
if(processFrameInfo.channelIOConfig.inputChannelFormat == kCFMono &&
    processFrameInfo.channelIOConfig.outputChannelFormat == kCFStereo)
        // --- DO mono-in -> stereo out processing

// --- Stereo-In/Stereo-Out
if(processFrameInfo.channelIOConfig.inputChannelFormat == kCFStereo &&
    processFrameInfo.channelIOConfig.outputChannelFormat == kCFStereo)
        // --- DO stereo-in -> stereo out processing
```

You can add your own channel decoding logic as well. One thing to note is that this function (like all of the *PluginBase* functions) returns a boolean variable to signify whether we actually processed data or not. We can return *true* from inside each of those curly bracket sets and return *false* if none of our channel combinations are supported.

6.5 Defining Factory Presets

A preset is really just a defined state of the plugin parameters. You defined your presets inside of the *PluginCore::initPluginPresets( )* function. A *PresetInfo* structure is used for each factory preset you wish to encode. The structure contains members to hold the preset name and its index, which determines its order within your set. The preset contains some number of parameter values stored in a *std::vector*. You may save non-variable bound parameters and other information. To create a preset you first create a *PresetInfo* structure.

```
struct PresetInfo
{
        uint32_t presetIndex; // --- zero indexed
        std::string presetName;
        std::vector<PresetParameter> presetParameters;
};
```

Then, use three *PluginBase* helper functions: *initPresetParameters* initializes a preset structure, *setPresetParameter* sets the value of an individual parameter (you call this as many times as needed), and finally *AddPreset* to add it to the preset list. The plugin shell code then accesses that list to build your factory presets for a given API. For example, you could declare the following presets for the *IIRFilters* plugin:

Preset Name: *DCBlocker*
Parameters:

- filter f_c: 2 Hz
- filter Q: 0.707
- boost/cut: 0 dB
- filter type: HPF2

```
PresetInfo* preset = new PresetInfo(0, "DCBlocker");
initPresetParameters(preset->presetParameters);
```

```
setPresetParameter(preset->presetParameters,
                   controlID::filterFc_Hz, 2.0);
```

```
setPresetParameter(preset->presetParameters,
                   controlID::filterQ, 0.707000);
```

```
setPresetParameter(preset->presetParameters,
                   controlID::boostCut_dB, 0.000000);
```

```
setPresetParameter(preset->presetParameters,
                   controlID::filterType,
                   enumToInt(filterTypeEnum::HPF2));
addPreset(preset);
```

For plugins that have many parameters (e.g. some synth plugins easily have 50+ parameters), it can be tedious to write the code. You can use a helper GUI object that allows you to add a button to your GUI. You adjust the controls to a desired preset, then use the button to write the control information into a text file. That file contains the exact C++ code as shown in this section. You may then simply cut and paste the code into the *initPluginPresets* function. When the development period is over, you can then remove the button from the GUI. You can get more information about this feature in the ASPiK SDK documentation.

6.6 Basic Plugin GUI Design With ASPiK's PluginGUI

The *PluginGUI* object handles all of the GUI chores for the plugin kernel. This includes the following:

- generating the GUI graphically
- initializing the GUI controls
- transmitting control changes from the GUI to the native plugin shell in order to affect parameter changes
- transmitting automation gesture messages to the native plugin shell and host DAW
- receiving control updates from project saved-states
- receiving control updates from running automation
- creating the drag-and-drop GUI designer for you to use in the graphical creation of your plugin GUI

It is important to remember that the *PluginGUI* and *PluginCore* objects are completely independent of one another and do not need each other to operate—nor do they know about each other's existence. The native plugin shell creates the GUI and core objects. An interface named *IGUIPluginConnector* allows for thread-safe communication between the *PluginGUI* object and the native plugin shell. Each AAX, AU, VST, and RAFX2 native plugin shell object owns an *IGUIPluginConnector* interface object and implements the interface functions using the API-specific, thread-safe mechanisms for parameter manipulation. The *PluginGUI* object uses this interface to request information about the plugin parameter values when it is opened so that it can place the GUI controls in their proper positions. It also uses the interface to issue GUI parameter changes to the native plugin shell object (thread-safe) including parameter change gestures for automation. The details of the interfaces and *PluginGUI* object's low-level functionality are not covered here, but you are encouraged to examine the code, set breakpoints, and trace the code as it executes if you are interested in fully understanding the GUI object. There are no static libraries here—every line of code is available for your perusal.

Figure 6.2 shows the lifecycles of the plugin host DAW, the native plugin shell, and the *PluginCore* and *PluginGUI* objects. The user requests a plugin instance from the host, which creates the API-specific native plugin shell object. The shell creates the *PluginCore* object as part of its construction. The *PluginCore* exists as long as the shell object does and is destroyed in the shell's destructor. However, the user may open and close the GUI as many times as they wish during the plugin's lifetime. If the GUI is open at the time the user destroys the plugin, it will be closed first, then destroyed.

6.7 GUI Design With VSTGUI4

There are several GUI design packages available for use in C++ software. Some of these packages are generic and not designed specifically for audio plugins. Steinberg developed the VSTGUI GUI generation package around the same time as their first plugin API (VST) and implemented platform-independent GUI controls packaged in C++ objects. This software has been refined and updated to its current fourth-generation state known as *VSTGUI4*. We will refer to it simply as VSTGUI. This GUI design software was developed with audio plugins in mind right from the beginning. It has matured into a quite remarkable open-source library of GUI objects that may be used for any plugin API. The VST3 API was designed to integrate tightly with VSTGUI, though it is not a hard requirement. However, nothing in the VSTGUI library specifically connects it to the VST3 API, and the library does not require VST3 SDK files for its operation. VSTGUI is ideal for our plugin kernel because:

- It is not plugin API–specific (as is the plugin kernel).
- It is platform independent (like the plugin kernel).
- It is a mature, fourth-generation product.
- It is well maintained and updated with new features and bug fixes.
- Its designers stay on top of the latest C++ developments and extensions, incorporating them into each new major revision.
- It is designed to be extendible with *Custom View* and *Sub-Controller* objects and APIs.
- It is deep: you can really go far down the rabbit hole with VSTGUI and it can be used to develop UIs for all kinds of software, not just audio plugins; for example, the main UI in RackAFX is done entirely with VSTGUI.
- The complete GUI description, including the graphics file data, may be written into a single XML file, making it ultra portable.
- It comes prepackaged with a platform independent drag-and-drop GUI designer, greatly simplifying the process of GUI design.

That last point is one of the supreme advantages of VSTGUI, enabling you to design a fully functional and professional GUI without writing a single line of code. The plugin kernel makes use of this functionality so that you may design your GUI in the plugin API and platform of your choice. Since your entire GUI is described in one XML file, you can easily move your GUI from one platform and API to another, without needing to import graphics files, change a single line of XML code, or do any additional programming whatsoever.

6.7.1 Modifier Keys

The *PluginGUI* object is already set up to implement the various modifier key and mouse click combinations used to fine tune the GUI and reset its controls to their default values. Modifier keys include the command <CMD> (MacOS), control <CTRL>, and alt (or option) <ALT> or <OPTION> keys. Pro Tools defines its own API-specific set of modifiers. This code is already written for you so there is nothing to implement—however, you will need to document these modifier combinations for your users. The modifier keys are listed next. Each involves clicking on a control with the left mouse button while holding a modifier key down.

Fine Adjustment: hold modifier key while clicking/dragging mouse over knob or slider

Pro Tools MacOS:	<CMD>
Pro Tools Windows:	<CTRL>
All others (MacOS and Win):	<SHIFT>

Go to Default Value: hold modifier key while clicking mouse over knob, slider, button

Pro Tools MacOS:	<OPTION>
Pro Tools Windows:	<ALT>
AU:	<OPTION>
VST MacOS:	<CMD>
VST Windows:	<CTRL>
RackAFX:	<CTRL>

Pro Tools: Enable Automation <CTRL><ALT><CMD> while clicking on a GUI control

6.7.2 Zooming (Scaling the GUI)

VSTGUI has a built-in GUI scaling mechanism that allows you to easily expand or shrink the entire GUI window. Sometimes this is required for high-resolution displays (shrinking) or for visually impaired users (expanding). There is a more accurate method for dealing with the Hi-DPI displays that you may find in the Advanced GUI section of the ASPiK SDK. However, we can use the built-in scaling to made moderate adjustments to the GUI size. There is already a preset GUI control reserved for this operation. Your *PluginGUI* object implements the GUI scaling, which VSTGUI calls "zooming." The *PluginGUI* object can scale down to 65% or up to 150% of the normal GUI size. Values outside this range generally do not give acceptable results. However, as with everything else, all the code is there for you to see and modify if you wish.

6.7.3 Reserved Control Tags

TRACKPAD	131073
VECTOR_JOYSTICK	131074
PRESET_NAME	131075
WRITE_PRESET_FILE	131076
SCALE_GUI_SIZE	131077

6.7.4 VSTGUI4 Objects

The VSTGUI library is chock full of GUI control objects. These include the most common plugin GUI controls:

- knobs
- sliders
- buttons and switches
- text labels
- text edit controls
- menus and drop-down displays
- VU meters
- track pads (aka XY-pads)
- background images

In addition, there are many other types of controls and views, including:

- view containers (for aggregating sets of controls or other views)
- splitter views with drag-able pane-splitters
- table views (rows and columns)
- animation views (for embedding animated content)
- a piano keyboard control
- custom views for displaying plugin information (e.g. waveforms, graphs, etc.)

All of the VSTGUI library control objects that may be visible on the GUI are derived from the mother-of-all views called *CView*. This object primarily defines functions for drawing to the screen and handling mouse click events. Any library object that is capable of transmitting or receiving a plugin parameter value of some kind is derived from *CControl,* which defines functions for getting and setting values as well as establishing a link to a specific parameter via a control tag, or what I call a *controlID*.

CControl is derived from *CView* so it inherits the view's drawing and mouse event functions. One interesting example is the *CTextLabel* object. Ordinarily, you use these labels as static text controls: for example, you might place a "Filter Type" text label above a control that allows the user to select a filter, and a "Filter Q" text label below the knob that adjusts the filter Q. In these cases, there are no links to any underlying plugin parameters: there is only static text to display. However, you might decide to place text on the UI that is constantly updated and changing, such as a counter or other timing display. In this case, you would need to link the text control to an underlying plugin parameter's *controlID*. Therefore, the *CTextLabel* is derived from *CControl* rather than just *CView*.

6.7.5 Creating a GUI With VSTGUI

VSTGUI defines two fundamental mechanisms for creating a GUI: you may write your GUI as pure C++ code or describe it in a simple XML file instead. In the first method, your GUI is designed completely in code: you create VSTGUI C++ object instances and set their lengths and widths using pixel values. Then, you place the objects on the GUI according to their destination screen coordinates. You implement functions to handle GUI parameter initialization and control changes. When your GUI is closed, you destroy your C++ objects, and when it is re-opened, you re-create everything and start anew. The main UI for RackAFX is written this way as pure C++ code.

The fourth generation of VSTGUI introduced the concept of packaging the GUI description inside of an XML file. This is loosely aligned with the way Visual Studio and Xcode package their GUI descriptions in the *.RC* and *.NIB* (or *.XIB*) files respectively. The VSTGUI4 library also comes prepackaged with an XML parser object and a UI-description object. The XML parser reads the XML file, and the description object implements view factories to create the various GUI views, controls, and elements. When the user closes the GUI, the description object handles the chores of destroying the objects and cleaning up. This means that you really need only write a relatively simple XML file to create a GUI.

The VSTGUI4 library also includes a built-in XML file generator that uses a drag-and-drop interface to generate the XML description file. RackAFX also includes its own drag-and-drop GUI designer that likewise reads and writes the final XML file as part of its operation. The GUI file may be named anything we wish; in the plugin kernel, it is named *PluginGUI.uidesc*, which maintains the VSTGUI paradigm of suffixing the filename with *.uidesc,* which stands for "UI description." The name of the C++ object that creates the GUI from an XML file is *UIDescription*. So you will see the terms "description" and "UI Description" a lot when using the XML file mechanism. In addition, there is an option we will use to embed the graphic file data inside this XML file. This makes the GUI completely portable as a single file; there is no need to add graphics files to your compiler project's resources when using this method. This allows you to copy and paste the *.uidesc* file from one plugin project to another without modification of any kind while automatically pulling in all of the GUI graphics data. We will be using the built-in GUI designer to handle all of our GUI design tasks, with the exception of *Custom View* objects, where we will need to write some custom code for displaying graphical data for the user.

6.7.6 Important GUI Designer Terms

There are a few terms that will show up all the time while working with your GUI designer software. Before moving on, let's establish these.

Platform Independent

Platform independence means that C++ GUI objects will automatically look and feel the same across the different operating systems. You do not need to know anything about the different GUI systems in Windows and MacOS, and the platform independent code is buried deep inside of VSTGUI. For example, in the advanced GUI design chapter, we will create a view that displays an audio waveform. The drawing code we write is platform neutral. The low-level rendering code handles the translation to the different platforms and we never need to be concerned with any of those details.

Template

A template is a collection of view controls or GUI objects. You can assemble your GUI from pre-made templates that contain clusters of related controls, or you can drag and drop each control, one at a time, into the main GUI designer window. You can make copies of templates to generate sets of similarly grouped controls for a repeated look and feel.

Editor

The top-level template that houses the GUI's main view is named the *Editor*. We will reserve this name for that top-level template only.

View Container

A view container is a C++ object that holds some number of GUI view or control objects called *children*. View containers may be nested and you will often aggregate containers together so that you have a container full of other containers, each of which contains a set of grouped controls.

Views

The GUI view objects that VSTGUI implements are simply named *Views*.

Attributes

Attributes are used to describe each GUI object. The attributes may be numerical or text strings. You must manually type in some attributes while others are set with drop-list or check box controls. The attributes are stored as XML string-data pairs. You can view the XML file in your compiler or any text editor.

6.8 VSTGUI C++ Objects

Although there is absolutely zero C++ code for you to write in order to generate your plugin GUI, there are some concepts and details about the various VSTGUI objects that you need to know in order to make the GUI design process as simple as possible. There are some common design aspects to all of the VSTGUI controls that you should understand.

1. The control's size is specified in screen pixels.
2. The control's location on the GUI is specified as an *x–y* coordinate pair that designates the top and left coordinates of the rectangle that defines the object's dimension; these coordinates are automatically set for you when you use the drag-and-drop GUI designer.
3. Most controls are linked to an underlying plugin parameter via the parameter's *controlID* value, which is called a *control tag* or simply *tag* in VSTGUI parlance.
4. Most controls require a graphics file (always packaged as a Portable Network Graphics, or PNG, file) to render the final control object; if you forget to assign a graphics file, then the control will appear to be invisible, though clicking on the invisible control will usually still work properly. A few controls require two graphics files, for example the VU meter object requires a graphic for the on state and another for the off state.

Here is a breakdown of the GUI controls we will be using, along with their requirements or other information related to their creation. We will discuss some techniques to simplify the GUI design even further a bit later, allowing you to create GUIs very rapidly. These include *auto-sizing* and GUI *templates*. The VSTGUI objects presented here can be divided into two groups: controls that specifically require one or more graphics (PNG) files and those for which graphics are optional or not used. Figure 6.11 shows the controls and their PNG file graphics.

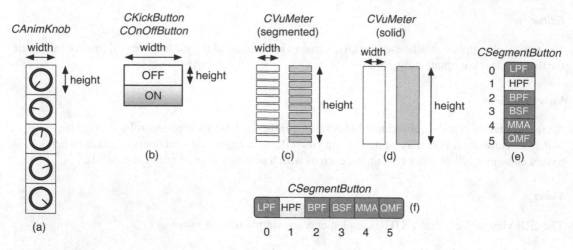

Figure 6.11: The VSTGUI4 controls that require graphics are shown here (a) *CAnimKnob* **(strip animation); (b)** *CKickButton* **and** *COnOffButton* **(on and off states arranged vertically); (c)** *CVuMeter* **(segmented) and (d)** *CVuMeter* **(solid), with off and on states in two separate files; and (e)** *CSegmentButton* **(vertical) and (f)** *CSegmentButton* **(horizontal), which may be used with or without individual graphics files for each of its buttons in each of its states.**

Objects that require graphics files:

- *CAnimKnob*
- *CKickButton*
- *COnOffButton*
- *CVuMeter*

Objects that do not require graphics files:

- *COptionMenu*
- *CSegmentButton*
- *CTextButton*
- *CTextLabel*
- *CTextEdit*

6.8.1 Basic GUI Design

The GUI designer will open from the top and left corner of the blank GUI and has the functional sections shown in Figure 6.12. The first thing you need to do is drag the GUI designer window into the middle of the screen and resize it to your preferred size. A few DAWs (notably Reaper) will provide you with a resizing grip on the lower right corner of the window, which you can use to resize it. Most DAWs do not allow resizable GUIs. If this is the case then you can use the menu items I've placed in the GUI designer's File menu shown in Figure 6.13a. There are four selections you can make: increase or decrease width and increase or decrease height. Each time you select these items, the GUI designer window will grow or shrink by 10% in the dimension you select. Once the window size is to your liking, you may grab the splitter window edges to adjust the relative sizes of the four windowpanes shown in Figure 6.4a–d.

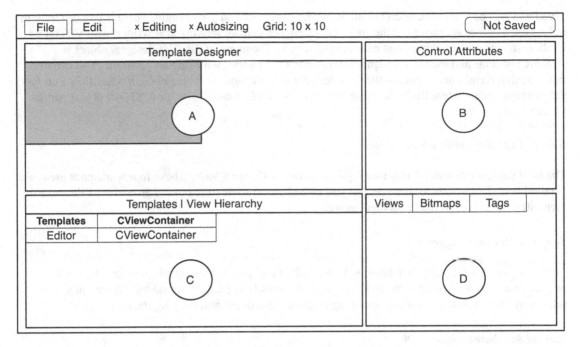

Figure 6.12: The VSTGUI4 GUI designer interface.

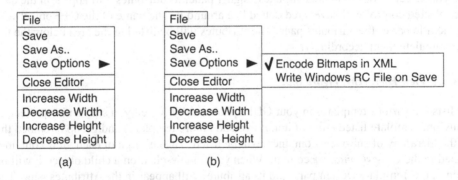

Figure 6.13: (a) The GUI designer *File* menu. (b) The *Save Options* menu.

After adjusting the GUI designer's workspace size, you should save the resulting XML file. This is the XML file that packages the description of your GUI as well as the PNG file data. Make sure this is set correctly by examining the File menu's Save Options as shown in Figure 6.13b. Make sure the option to "Encode Bitmaps in XML" is checked. You do not need to check the Windows RC file option—we won't need to alter this file since we are saving all our graphics data directly in the XML file. A red warning label will show at the upper right that says "Not Saved" to remind you when your GUI design needs saving. Choose "Save As" from the File menu and browse to your project's *resources* folder. This folder is located in *<Project Name>* → *project_source* → *resources*.

Save the file as *PluginGUI.uidesc* and answer "yes" when prompted to overwrite the existing file. From this point on, you need only choose "Save" from the File menu. The path to your XML file is saved and will be used to re-open the file later when you work on your plugin. There is one last item to reflect upon before we move on: the entire GUI designer you are looking at is written with VSTGUI4 itself, and your project carries that code around with it. The name of the GUI designer object is *UIEditController* and the GUI designer is written in C++ rather than using an XML file. You can find the code that creates and launches the GUI designer in the *PluginGUI::createGUI( )* function. Fun fact: the platform independent file Save dialog box you see that is created with the VSTGUI object named *CFileSelector*.

6.8.2 The GUI Designer Workspace

The GUI designer is divided into four regions, shown in Figure 6.9a–d. These four workspace areas are detailed here. In addition to these details, you should spend some time poking around the File and Edit menu items and get familiar with them as well.

Template Designer, Figure 6.9a

This is the canvas for your GUI designs. You will drag and drop GUI controls onto templates you create. Your GUI will come with the Editor template already in place. You use the Edit menu's selections to manipulate templates including creating, destroying and copying them.

Control Attributes, Figure 6.9b

Each time you select a control in the template designer pane, its attributes will appear in the attribute pane. The selected control will have a red dotted line around it. The name of the C++ object will appear in the title bar of the attributes pane. The attributes will be listed in the area below the title and you will manipulate them accordingly.

Templates and View Hierarchy, Figure 6.9c

This area lists the various templates in your GUI alphabetically. When you start a new project, there will be only one template listed—the Editor. As you add more templates and group or embed them together, the hierarchy of sub-view containers will be listed. The children of each view container will be listed in the *CViewContainer* column. When you double-click on a child object, it will be highlighted in the Template Design pane and its attributes will appear in the Attributes pane. This goes for all children objects including views, sub-views, and controls.

Views, Tags, Colors, Gradients, Bitmaps, and Fonts, Figure 6.9d

This pane contains a tab controller that allows you to select a GUI design element. The tab controller that you see that lists "Views, Tags . . ." with icons next to each one is actually a *CSegmentButton* control that has been customized with icon images.

Views

This tab contains a list of all the view objects in VSTGUI4. You will select a row in the list with the control you wish to place by clicking on the row. While holding down the mouse button, drag the row up into the Template Designer pane. When the mouse enters the template designer, you will see the control appear or you will see an empty rectangle that will house a control that requires an external graphics file. Once you've indicated the proper graphics file to use in the Attributes pane, the control will become visible. You cannot edit the Views list.

Tags

This is where you will define your tag names and enumeration index values. These may be any integer value—negative or positive—and the tags do not need to be zero indexed. We will add the tags now for each control, but we will not connect them up to the plugin core until the next chapters. To add a tag, click on the "+" button at the bottom of the pane. This adds a new row to the list. You then name the tag in the first column and then assign a unique index to it in the second column. The "−" button will delete a selected tag. You may name the tags as you wish; however, their numerical index numbers must match the *controlID* values you define for your plugin parameters in order for the GUI controls to affect the plugin parameters.

Colors

This control lets you create colors. Two colors have been defined for you in the empty project template: light gray and dark gray. These colors are used in the basic *CSegmentButton* design. You can blend and save all the colors you wish for your GUI here. Click the "+" button to add a new color and the "−" button to delete one.

Gradients

A gradient is a color that morphs between two or more colors or shades. The colors morph from left to right in the designer window. You can create or edit gradients using the buttons at the bottom of the pane. You add color stops to the Editor and drag them to define the color morphing start and end points. Gradients are used mainly for *CTextButton* and *CSegmentButton* controls. Your empty plugin project has two gradients defined that use two colors, and two more that are a single solid color only. You may use either set in your button designs, or you may create your own. When used properly, gradients can give the GUI a modern and polished look and feel.

Bitmaps

In VSTGUI terminology, a bitmap is a graphics file. VSTGUI uses PNG files, so the term "bitmap" is not to be taken literally as the Windows *.bmp* file type. We will use PNG files exclusively. You add and remove PNG files using this control pane. When you add files, the image data will be stored in the XML files and when you remove them, that data will be deleted from your XML file. Usually one of the first steps in GUI design is to import all the graphics files you will be using, then assemble the GUI. Of course you may add or remove graphics files at any time during the GUI design process.

Fonts

You can add fonts to your GUI design using this control pane. However, you need to be careful with this. If you add exotic fonts that are not normally installed in Windows or MacOS, then you will need to package these fonts along with your plugin when you deliver it to your customers. Typically, you make an installer to install the fonts along with your plugin. We will be using stock fonts for all of our plugin GUIs. This will simplify our designs and free us from needing to ship font installers with the plugin.

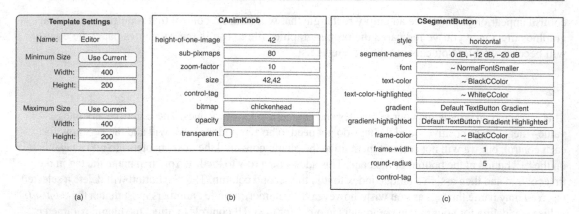

Figure 6.14: (a) The template settings panel. (b) Selected attributes for the *CAnimKnob* object. (c) The *CSegmentButton* object.

6.8.3 Changing Your GUI Canvas Size

The GUI in your template project code is pre-wired to have the dimensions 400 × 200 pixels (width × height) and to use a solid gray color as the background. During development, you may need to grow the size of the GUI canvas. First, select the Editor template in the Templates | View Hierarchy pane. Then use the Edit → Template Settings function to open the Template Settings dialog box. Your GUI requires both maximum and minimum size settings.

We will use the same dimensions for each setting (min and max), and we will use our own internal zoom mechanism to allow the user to grow or shrink the GUI. The Template Settings dialog box is shown in Figure 6.14a. To change the sizing, type in the values that you prefer. One strategy is to make the main GUI Editor template really large (e.g. 1024 × 768) and then add your GUI controls as you wish. Then, you may use the Edit → Size To Fit menu item to automatically shrink the GUI down to contain only your controls.

6.8.4 Setting Up the Control Tags

You assign the control tags using the *Tags* tab in section D of the GUI designer. There are buttons to add and remove the control tags. We name the control tags identically to the control ID enumerations such as:

```
controlID::filterFc_Hz
```

This would correspond to a control tag of 0 from the enumeration. If you want to change the naming of the tags, you can do that here as well.

6.8.5 Importing the Graphics Files

You will need to import the graphics files for the various GUI controls. The files for the book projects were made with two pieces of software named *KnobMan* and *SkinMan*. See the references for links. These products allow you to create all sorts of GUI controls as well as background skins for your plugin GUIs. Each ASPiK project comes with a small set of graphics files for you to use located in the */project_source/resources* folder. You can download hundreds more from the KnobMan Gallery [see references]. It is important that you know the dimensions to enter into the attributes fields for each GUI control. For knobs, you need to know the dimensions of one cell, and the number of sub-pixmaps. For the rest of the controls, you only need to know the pertinent dimensions that are specified in the ASPiK documentation. When you create new graphics files, you need to be careful with the dimensions so that your GUI components are properly sized relative to one another. VSTGUI also allows special filename encoding of high-resolution files for "Hi-DPI" displays (also called "Retina Display"); that information is likewise available in the ASPiK documents.

The graphics files are located in your project's *resources* subfolder; however, we will be importing the data into the GUI XML file, so the file location is unimportant. Once the file has been imported, you can move or delete it as you like. But you need to remember that if you change the graphics file in any way, you will need to remove and then re-import it in order to transfer the PNG data into the XML file.

In the GUI designer, select the "Bitmaps" tab in the lower right pane. Click on the "+" button to add a graphics file. Browse to the file you want to import and select it. You will need to do this for each graphics file you want to use. Once the file has been loaded, click the *More* button to view the file. You will see the entire file contents. The size value you see will be for the entire file, and not necessarily what you will enter for the graphics attributes. Remember that the XML file can be easily copied from project to project, so once you've spent time importing the graphics, you can reuse the same file (which will save time later on). With all of the graphics files imported and their pertinent dimensions noted, you can move on to assemble the GUI.

6.8.6 Assembling the GUI

Many consider this to be the fun part of plugin development; it is a way for you to personalize your product. We will assemble our simple volume plugin GUI by dragging and dropping each control on to the canvas. You are encouraged to play with the templates as well; the sample projects that ship with this book use templates for the knob clusters so you may use them as a basis for creating your own.

6.8.7 Setting the Background

We will use the *halfrackspace.png* file for many of the book project backgrounds. It has the dimensions of 450 × 100 pixels. The first thing to do is to change the minimum and maximum size of the final GUI using the *Template Settings* function from Section 6.8.3. Change both the minimum and maximum sizes to be 450 × 100. Next, select the Editor in the Templates | View Hierarchy pane and the view container attributes appear in the upper right pane. Click on the bitmap attribute and select *halfrackspace* from the list of graphic files that appears. The background will instantaneously change to reflect this new choice. If you don't like the background files, you can set the bitmap attribute to "None" and then use a solid color instead with the background-color attribute. With the background set, you can move on and add the GUI elements.

6.8.8 Adding the GUI Elements

For each GUI element we will follow the same process:

- Drag an instance of the object on to the canvas.
- Set the object's view attributes: bitmap files, dimensions, and other properties.
- Connect the GUI control to a control tag that will ultimately link it to a plugin parameter.

6.8.9 Saving and Re-Building

You can now use the File → Save menu item to save your GUI changes. Unload your plugin, go back to your compiler, and rebuild the project. Now when you load the project, your new GUI will appear and GUI controls will alter the plugin parameter accordingly. If your controls do not work as anticipated, go back and check that your control tags are assigned correctly. If you would like to trace the activity that occurs when you move a control, place a breakpoint in the *PluginGUI::valueChanged* function, which is called any time the user interacts with any GUI object. You can follow the chain of events all the way to the API-specific function calls that safely alter plugin parameters.

6.8.10 Scaling the GUI

VSTGUI allows you to zoom in or zoom out of the entire GUI, making all of the controls grow or shrink. This is useful for allowing the user to scale the GUI to make it look correct for the type of screen being used. Different video screen resolutions may make the GUIs appear too large or small. There is a reserved *controlID* called SCALE_GUI_SIZE that has been set aside for this task. You add this functionality in the *PluginCore::initPluginParameters* function; typically you will add this bonus parameter after all of the normal plugin parameters. The GUI control will need to be a *COptionMenu* or *CSegmentButton* that can transmit an unsigned integer value. You declare the bonus parameter like this:

```
// --- BONUS Parameters
// --- SCALE_GUI_SIZE
piParam = new PluginParameter(SCALE_GUI_SIZE, "Scale GUI",
                              "tiny,small,medium,normal,large,giant",
                              "normal");
addPluginParameter(piParam);
```

The GUI scaling parameter expands or shrinks the GUI according to Table 6.2.

With the parameter in place in your *PluginCore* object, you simply add a *COptionMenu* to the GUI and connect it with the *controlID* value SCALE_GUI_SIZE. It will automatically display the strings in the list above "tiny,small,medium,normal,large,giant" and will default to the normal size. When the user chooses a scaling amount, the *PluginGUI* object will handle the scaling automatically.

Table 6.2: The scaling factors for scaling the GUI

Size	Scaling Factor
tiny	65%
small	75%
medium	85%
normal	100%
large	125%
giant	150%

6.8.11 More ASPiK Features

This chapter only scratches the surface of what you can do with ASPiK, including the GUI designer, which allows very complex structures like tabbed GUIs and custom views. You can download more example projects that detail the many other ASPiK features from www.willpirkle.com. Everything is documented in the ASPiK documentation files.

6.9 Bibliography

ASPiK Documentation, www.willpirkle.com/aspik/, Accessed August 1, 2018.

CHAPTER 7

Using RackAFX to Create ASPiK Projects

RackAFX is designed to rapidly assemble, debug, test, and verify real-time audio signal processing algorithms in C++. The first version was designed for algorithms that would be burned into a hardwired DSP chip. Over the years it became a teaching tool for demonstrating a simple plugin API and was later modified to export AAX, AU, and VST plugin projects along with GUIs designed with VSTGUI4. The modifications were made on top of an existing platform (now called the RAFX1 API). Each exported project was very different and had to be compiled in a strict manner within the different SDK folder hierarchies. New releases of SDKs tended to cause problems as the exported project files needed constant modification.

ASPiK was developed independently of RackAFX as a means of unifying the plugin project APIs (AAX, AU, and VST) with a simple software kernel that acted as a monolithic plugin object. It was designed to be less susceptible to manufacturer API and SDK updates and changes, and fully open source for easy modification. Once ASPiK was finalized, I returned to RackAFX and built a new API that acted as a thin container for the ASPiK kernel. This allows you to build and test the kernel directly without any RackAFX-specific code getting in the way. This new API is known as RAFX2. However it is important to understand that neither RAFX1 nor RAFX2 is built off of any preexisting API, so neither attempts to be a polymorphic cross-platform API or any kind of "plugin substitute." Likewise, there is no desire to push any new API into the existing set of commercially available products. Therefore, the RackAFX software does not export RAFX1 or RAFX2 plugins or code. Instead, it exports only ASPiK projects, which you may then compile into AAX, AU, or VST plugins using the proper SDKs that you download and install on your system. There is no attempt to circumvent any API, and there are absolutely no SDK-specific files within any ASPiK project. Likewise, ASPiK projects contain no RackAFX code. If you don't care about exporting projects then that is fine: the exported ASPiK project is the bonus you get when you develop algorithms with RackAFX.

You can use RackAFX:

- to create FX and synth plugins
- as an academic tool to experiment with newly acquired algorithms in real time
- to digitally test and verify application specific algorithms using high performance analysis tools
- to export ASPiK projects to compile your project for commercial use as AAX, AU, and VST devices
- to capture impulse responses for export as *.wav* files
- to create custom audio views for waveforms, FFTs, etc.
- as a real-time live audio effects or synthesis engine with MIDI control

It is extremely simple and quick to assemble a prototype plugin and add the processing code. There are added tools for testing your plugin:

- high performance sinusoid oscillator with sweep (chirp) capabilities
- white and pink noise oscillators and ideal waveform oscillators for testing filter resonance

- digital oscilloscope with traditional scope controls (triggers, offsets, etc.) including Lissajous X/Y feature
- FFT spectrum analyzer with resolution up to 131,072 points in stereo
- unique impulse, step, frequency, and phase response tester that fires digital impulses into your plugin and captures the results for analysis; new analyses are made each time you nudge a control
- built-in Optimal and Frequency Sampling Method FIR designer plugin
- built-in 1024 point *.wav* file convolver
- side-chain input from *.wav* file or from live audio line-in
- full MIDI capability for receiving MIDI messages and mapping CCs to your plugin controls
- present management: stored presets are exported as factory presets in ASPiK projects
- built-in MIDI piano and CC generator
- ability to process live audio from line input through plugins for real-time testing
- save impulse responses as *.wav* files
- use offline processing into *.wav* files
- drag-and-drop GUI designer

New features are added all the time.

7.1 Installing RackAFX

If you wish to use RackAFX for your plugin development, you may download the latest version at www.willpirkle.com/rackafx/downloads/. Be sure to read any important installation notes or other details provided at the website. You can find numerous tutorial videos and sample projects there as well. RackAFX runs in the Windows operating system and requires a Visual Studio compiler that is C++11 compliant. Once you've installed the software, open the Preferences (View → Preferences) to set up your complier of choice and your RackAFX projects folder. Once you have chosen the projects folder, you should install VSTGUI4 into it. RackAFX comes packaged with the latest version of VSTGUI4; you can install it to any location using the Utilities → Install VSTGUI4 function. The *vstgui4* folder is required for your plugin projects to work properly.

7.2 Getting Started With RackAFX

RackAFX is designed for the Windows operating system. It will run on all current versions of Windows, from Windows Vista and forward (though Windows 8 or 10 are preferred). It is designed to run in tandem with Microsoft Visual Studio, which is used to compile the plugin projects. As of this writing, Visual Studio 2015 or later is required. You must first install your Visual Studio compiler that includes the C++ option, then you may use RackAFX, which you can download at www.willpirkle. com/rackafx/downloads/ and install. At this site you will find a user forum for posting questions and dozens of videos that show exactly how to install and use the software including plugin algorithm development, GUI design, and ASPiK exporting/compiling as well as use of the built-in an analysis and testing tools. This chapter serves as an introduction and fast-track guide to using RackAFX to create your plugin gems.

When you open RackAFX, you will be looking at the main window, which is the project prototyping area. This view contains 1024 blank control slots into which you may place a variety of controls that you will link with underlying plugin variables. You set up the controls with a simple form, and RackAFX writes the parameter declaration code for you. Figure 7.1 shows the main prototyping view in RackAFX.

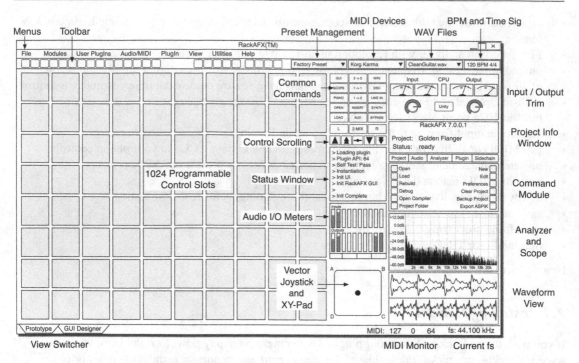

Figure 7.1: The main prototyping panel in RackAFX.

The majority of the view consists of 80 visible control slots that are scrollable to reveal a total of 1024 possible controls. The banks of controls may be scrolled one row at a time, or an entire panel of 80 at a time, allowing you to quickly move around between banks of controls. This allows you to easily set up a basic GUI for testing your plugin algorithm. You do not need to commit to anything during the prototyping phase, and you can easily remove or copy controls as needed. This main panel is also where you launch the test controls such as the oscillator, oscilloscope, FFT analyzer, and other tools. Once your plugin algorithm is tested, you may then optionally move on to a professional GUI design. RackAFX features a built-in drag-and-drop GUI designer that you access with the tab at the lower left of the main view. You may also use the ASPiK GUI designer that is documented in this book, or move back and forth between the two GUI designers. Because you can use the ASPiK GUI designer and because the built-in GUI designer is already very well documented in video tutorials, we won't get into the GUI designer here.

On the right of the main view is the master control module that contains a bunch of embedded menus and controls. You use this area to manage your plugin projects, work with the side-chain audio, configure the plugin channel I/O and routing, and perform other global functions. The small strip to the left contains a bank of buttons with the most common functions so that you don't waste time scrolling through menus or tabbed pages looking for them. In addition there is a status window that updates you on internal operations involving your plugin, and you may also send text information back to that window from your plugin as debug and testing information. Below that are the audio I/O meters and an assignable vector joystick and XY-pad controls.

7.3 Setting Up Your Project Preferences and Audio Hardware

The first thing to do is set your global preferences from View → Preferences or using the *Preferences* button in the command module. On this panel, you need to set up your default folder locations.

Next, set up your audio hardware using the Audio/MIDI → Setup Audio Devices menu item. Here you may choose your input and output device drivers in stereo pairs. After changing your configuration, RackAFX will reset and initialize the audio driver for operation. You may now test the system and play some audio files. You do not need to have a plugin loaded to play audio through RackAFX—and that includes the built-in oscillator.

7.4 Installing VSTGUI4

Your RackAFX project uses VSTGUI4 to render the custom GUI that is part of the final design phase. If you are using RackAFX only to design and test real-time audio algorithms, and you don't care about custom GUIs or ASPiK exporting, you will still need to have VSTGIU4 installed so that your plugin projects will compile correctly. Then, you may safely ignore it. However, VSTGUI4 is a powerful library: if you are going to design your own GUIs then it is worth studying. Installing VSTGUI4 really just means copying a giant folder hierarchy into a destination. There are no files to alter or CMake scripts to run. Fortunately, this is very easy for two reasons: the authors of VSTGUI4 make the library fully open source, public, and administered through GitHub; and updates are always timed to coincide with VST3 SDK updates, so you can get both at the same time. You may also install VSTGUI4 from RackAFX via the Utilities menu. Make sure you install the VSTGUI4 library into your RackAFX projects folder so that it sits in parallel with your RackAFX projects, exactly as the ASPiK project folders in Chapter 5. The root folder is named *vstgui4*.

7.5 Creating a Project and Adding GUI Controls

You create new project with the File → New Project menu or any of the dedicated buttons within the main view. When you create the project, a dialog box will appear that you fill in with the project properties (i.e. the name, synth, or FX, etc.), including some plugin details like the latency and the reverb tail time. You can go back and edit these properties later. After you create the project, Visual Studio will automatically start. A blank RAFX2 project will be created for you that contains the *PluginKernel* object files. You will spend your time editing code primarily in only two files: *plugincore.h* and *plugincore.cpp*. If you decide to export your project as an ASPiK solution, then it will pull out only the kernel files and the RAFX2 files will be left behind. You will never need to edit or concern yourself with any of the RackAFX-specific files (i.e. the RAFX2 stuff). You may add third party files or other objects, and these will be exported along with the *PluginKernel*.

With the project created, you now want to add prototype GUI controls that will link to your plugin variables. These GUI controls appear on the RackAFX prototyping panel; RackAFX will write the parameter generation code for you. It is important to understand that if you decide to export your project and generate a professional GUI, the design of that GUI will be completely independent of the prototyping GUI. This is by design: often during the design stage, we expose many more controls than we will allow for the user in the finished product. This is to allow us to properly voice the instrument, choosing the proper variables and ranges that we will allow the user to adjust.

Your prototyping GUI and your finished product GUI may be very different. You are allowed to change the string lists and many other visible features it the final GUI that you are not allowed to alter during the design phase. Add all the GUI controls you wish during the prototyping phase, then expose only the most important ones for the user during the final GUI design.

To add a new control, right-click on any of the available control slots and a menu will appear asking you what kind of control you would like to insert. These consist of the following.

- **knob**: may be linked to a continuous variable or a string list, in which case the knob will behave as an N-pole rotary switch
- **switch**: you may place on/off or momentary switches in a variety of graphics including toggle and slide switches as well as on-off push buttons
- **four-bank radio buttons**: for short string-list controls that require four or fewer choices, you may place a radio button bank that you populate with up to four strings for selection
- **drop-down menu/list:** for long string-list controls you may also choose the drop-down list control, which allows the user to open and choose values the same way as with a drop-down menu
- **analog VU meter**: an analog VU meter with a needle; this may also be used in "reverse" to act as a gain reduction meter
- **LED meter:** a digital meter that may be used in an inverted manner as well for gain reduction

The type of control you choose for a given plugin variable will likely depend on your algorithm and the manner in which you've chosen to expose controls for the user to manipulate. Figure 7.2a shows a mock-up of the set of controls that you may place, starting with the continuous knob control on the left and moving to meters on the right. There are several dozen different knob graphics you may choose from, and you may design your prototype GUI as you like—in rows, columns, or other clustering or ordering. There are no rules about controls lying in continuous strips; you may lay the controls out in any fashion you choose. Figure 7.2b shows how you might lay out the prototype GUI for a dynamics processor plugin; notice the use of inverted LED meters to indicate gain reduction (GR). Figure 7.2c shows one of many alternate configurations; notice that you may also effectively use the spaces between controls to segment the GUI as you like; here the meters are isolated on the right side, and common controls are grouped in short columns (threshold and ratio, attack and release, etc.).

There are three fundamental types of GUI controls you may work with:

1. continuous controls for transmitting floating point or integer values (usually knobs or sliders)
2. string-list controls that display a zero-indexed list of strings for the user; these technically also include on/off switches (whose strings are "SWITCH_OFF" and "SWITCH_ON") as well as radio button banks and discrete knobs (N-pole switches) in addition to the familiar drop-down menu/list type of control
3. meters, for output data that usually transmit audio information but may also be used to monitor internal nodes within a plugin, downward gain values, or other non-audio information

In all cases, we would like to link the control to an underlying variable; these will be listed in tables for each plugin project in the book. The string-list controls require two pieces of information: the underlying unsigned integer that will receive the selection change, and the list of strings to display for the user. We will use the variable binding option with the *PluginParameter* object to make the transfer automatic.

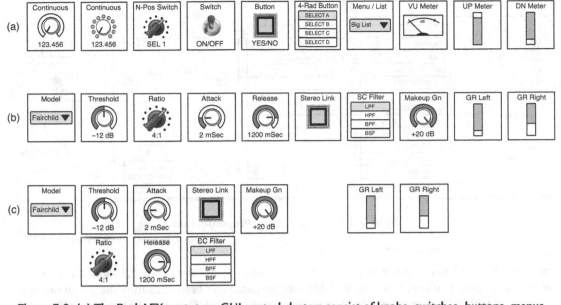

Figure 7.2: (a) The RackAFX prototype GUI control clusters consist of knobs, switches, buttons, menus, and meters. (b) An example of a prototype dynamics processor GUI arranged in a single row. (c) Another way of re-arranging the dynamics processor controls; you may place controls as you like and use blank slots to separate control clusters.

Table 7.1: The volume plugin's input parameter list; notice that the string list is shown in curly brackets and will be listed in place of the min/max/default values for string-list controls

Control Name	Units	Min/max/default or string-list	Taper	Linked Variable	Linked Variable type
Volume	dB	–60.0 / +12.0 / 0.0	linear	*volume_dB*	*double*
Mute	–	"off, on"		*enableMute*	*int*
Channel Select	–	"stereo, left, right"		*channels*	*int*

Table 7.1 shows the GUI parameter table for the example volume plugin that is included as an example in the ASPiK SDK. You can see on the right that the volume control is linked to a *double* data type. This will connect to a knob control in RackAFX. Both the mute and channel controls link to the *int* data type, but they each contain a string list in curly brackets { } so these are string-list variables and are handled differently.

7.5.1 Numerical Continuous Controls

Numerical controls in RackAFX are knobs that transmit values encoded as *float, double,* or *int* data types. We will use either *double* or *int* variables in the book projects. The entry for the volume control in Table 7.1 shows the units, minimum/maximum/default values, and the control taper. When you add a new knob control to the RackAFX prototype panel, you will need to specify all of this information in a simple form that you fill out. It is set up so that you may immediately begin typing and simply hit the <Enter> button to advance through the form. Some items are drop-list menu items that you may

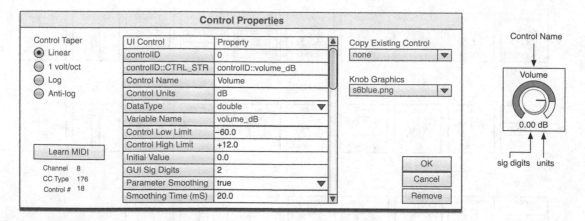

Figure 7.3: The knob Control Properties form.

simply hit the first letter of the selection (e.g. "d" for double) to choose. Figure 7.3 shows the knob control entry form; you simply fill out the table with the parameter properties from the plugin project GUI control table. Here we entered the data for the volume control. Compare the figure with Table 7.1 and it should be obvious how the variables from the table translate to the form.

Some notes on numerical control defintions:

- The *controlID* and *controlID::CTRL_STR* are set for you; using our convention (Chapter 5) the control ID enumeration will always be a concatenation of "controlID::" and your variable name; since the variable name must be unique, the control ID enumeration string will also be unique.
- You enter the name of the variable that you want RackAFX to create for you; it must be a legal C++ variable name.
- You can enable and disable parameter smoothing individually for each knob (this is preferred as parameter smoothing requires a few CPU cycles per control).
- The MIDI learn function on the lower left allows you to connect an external hardware MIDI controller to the RackAFX prototype panel; you can adjust values from your MIDI hardware. Note that this is a RackAFX-only MIDI option and does not attempt to set this in the ASPiK exported projects.
- The DataType and Parameter Smoothing options are set with drop-down boxes in the form.

7.5.2 String-List Controls

String-list controls present the user with a list of strings from which they may only choose one at a time. The strings are usually short, fairly representative characters that are easily decoded. The GUI control transmits integer values, starting with zero for the first string in the list, one for the next string and so on. The bound variable type is an *int*—and the reason for using it over the *unsigned int* is that strongly typed enumerations are automatically bound as integers, not unsigned integers. We use the strongly typed enumeration for making comparisons with string-list control strings (see Chapter 5).

There are four types of string-list controls you may add to the prototyping panel:

- **on/off switch or button**: the strings are { STRING_ON, STRING_OFF }
- **radio button set** of four segments or less: you may set the two to four strings as you like

- **drop-down menu/list**: you enter as many strings as you like and the user chooses them from a drop-down list box (aka combo-box)
- **N-pole switch-knob**: you may also assign string-list parameters to knobs, in which case the knob behaves as an N-pole switch and displays the text string in the edit box below the knob

In each case, you declare a variable name that will become the control's bound *int* data type and you supply a comma-separated string list that the control will use to display for the user. This string list has some restrictions: the strings will be used in a strongly typed enumeration that RackAFX will also write for you to use as a comparison mechanism when evaluating the control's broadcasted value. Therefore, they must not contain any spaces. However, you may use the underscrore (_) character when creating compound strings to connect them. RackAFX will remove the underscore in the GUI control so that "Lowpass_Filter" will be displayed as "Lowpass Filter." With RackAFX, you may set these strings differently in the prototyping panel vs. the GUI designer. For example, you may want to present the user with a list of filters: { LPF, BPF, BSF, BPF } but you later decide that these acronyms may not be obvious to every user. You may wait and set up new strings when you design the GUI—and you don't need to worry about everything matching up. Remember, rapid development is the name of the game here!

The control setup dialog boxes for each of the control types look slightly different. The dialog box for setting up the drop-down list and radio button bank are the same (see Figure 7.4). You assign the control name and linked variable the same way as for a knob. Below the table is a box for entering the string list of comma-separated values. For the radio button bank, you may enter up to four string values. For drop-down lists, you may enter as many values as you like. You may also share and reuse existing lists, which you select by double-clicking the entry in the list at the bottom of the form. On the right of Figure 7.4 you can see the two types of controls. The drop-down list may very well exceed the height of the control slot. In both cases the control transmits a zero-indexed value that corresponds to the user's selection.

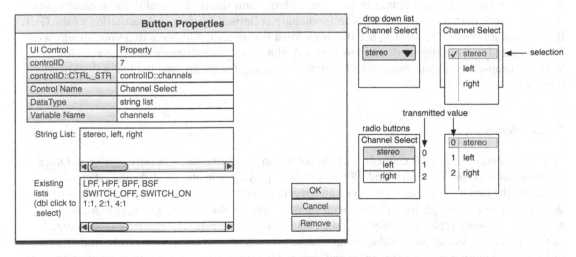

Figure 7.4: The dropdown list and radio button controls use a string-list to initialize and show the user a set of values to choose from.

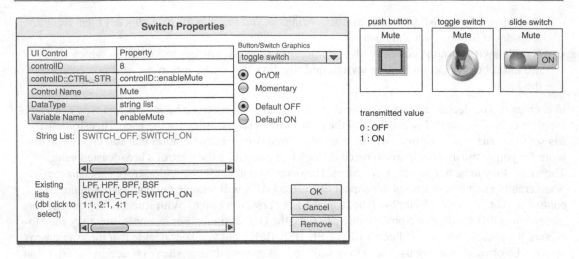

Figure 7.5: The two-state switch setup dialog box.

The dialog box for setting up the on/off buttons is very similar (see Figure 7.5). In this case, you cannot alter the string list as it is disabled and is hardwired to SWITCH_OFF and SWITCH_ON, which correspond to transmitted control values of zero and one respectively. For the two-state switches, you may also choose between momentary and on/off as well as the default state of the switch when the plugin opens: on or off. You may also choose the type of button or switch graphic for the prototyping GUI. More graphics are added all the time, so your version may include many more switch and button graphics options.

To set up a knob as an N-pole switch, you first create a knob control, and then you choose "string list" from the *DataType* field. When you choose this option, the form will change and alter its appearance to display the same string-list and existing string-list boxes you see in Figure 7.4. You enter the comma-separated strings in the same fashion. You should be careful not to enter string lists with more than 32 values as the knob switching is limited by the pixel resolution of the GUI. If you need more than 32 strings, use the drop-down list instead. Also note that your strings will be compacted to fit in the small edit box below the knob, so your strings cannot be very long. With the drop-down menu, there is no restriction on string length and the list box will expand as required.

7.5.3 Meters

Meters are output-only GUI controls: they do not transmit information, but only receive it. Other examples include the waveform and FFT spectrum displays in Chapter 21. Like the other GUI elements, the meters eventually link to an underlying variable. This variable is either a *float* or *double* data type—though we will stick with the *float* type for the book projects, as there is no need for the *double* type's accuracy with such pixel-limited controls. The meters are normalized, so they will show the absolute value (magnitude) of the variable you link them with, from 0.0 (off) to 1.0 (on). Negative values are ignored and clamped to 0.0; values greater than unity are clamped

Table 7.2: The volume plugin's output parameter list; all meter variables are of type *float*

Meter Name	Inverted	attack/release	Detect Mode	Calibration	Variable Name
VU	no	10.0 / 500.0	RMS	logarithmic	*vuMeter*

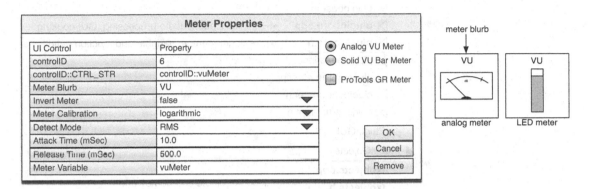

Figure 7.6: The RackAFX meter setup dialog box.

to 1.0. Setting up the meters in RackAFX is just as simple as the other controls. Table 7.2 shows the VU meter for the simple volume plugin. The stock meters in VSTGUI are fairly plain. The RackAFX custom VU meters include embedded envelope detectors that allow you to precisely set the meter ballistics, including attack and release time. In addition, meters may be calibrated as linear or logarithmic (which indicate dB levels) and the detection mode may be varied as peak, mean-squared (MS), and root-mean-squared (RMS). Finally, a meter may be set up to work in an inverted manner so that positive values cause it to move in the opposite direction from normal—these are used to indicate gain reduction in dynamics processing plugins. The meter properties dialog box in RackAFX allows you to set all of these parameters at once; the code will be written for you in the initialization function.

Figure 7.6 shows the meter designer panel. Comparing the form with Table 7.2 is straightforward. The inversion, calibration, and detection modes are selected via drop-down lists. On the right you may choose the meter style. The Pro Tools GR Meter check box maps this meter to the one and only gain reduction meter found on the Pro Tools interface.

7.6 Anatomy of Your RackAFX Project

When you create a new project or open an existing one, Visual Studio will start and will already have your project files loaded into it. A few of the files you see are for the RAFX2 API: the thin container for the *PluginKernel* objects, the *PluginCore,* and *PluginGUI.* You will see all of the *PluginKernel* files grouped together. These are arranged identically to the ASPiK project versions; you will find that it is simple to move between your RackAFX project and your ASPiK exported versions because they share so many common files. The system was designed so that you really

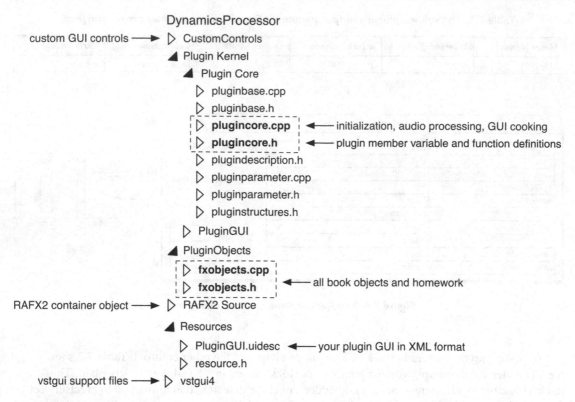

Figure 7.7: A RackAFX project folder contains the familiar ASPiK core files along with a few RAFX2 interface files; the four files in the dotted boxes are the only ones you will alter for all of the book projects except for those in Chapter 20.

only need to modify the two *PluginCore* files to have a working plugin. Your Visual Studio Solution folder will look like the one in Figure 7.7. You only need to concern yourself with the four files in the dotted boxes. The GUI is encoded as XML as described in Section 6.7; you can find that file in the Resources location. Each time you edit your custom GUI, in either the RackAFX or ASPiK GUI designers, this file will be updated, and this is the file you want to point the ASPiK GUI designer towards on your first GUI save (Chapter 6).

If you used RackAFX to create the GUI according to Tables 10.1 and 10.2, your *PluginCore* files would be modified as follows.

plugincore.h

- At the top of the file, an enumeration is added for each of the control ID values.
- The GUI-linked variables are declared and initialized.
- For string-list parameters, a helper strongly typed enum will be written for you to use when evaluating the parameter changes; the name of the enum will always be the variable name + "Enum."

You will find the variables declared in the private area of the class declaration. Notice that the discrete (string-list) variables include the helper enumeration, and also note the hex codes that decorate the variables: RackAFX uses these codes (all top secret), so do not modify or delete them. As you add, remove, or re-locate controls, these definitions will change accordingly.

```
private:
    //**-0x07FD-**

    // --- Continuous Plugin Variables
    double volume_dB = 0.0;

    // --- Discrete Plugin Variables
    int enableMute = 0;
    enum class enableMuteEnum { SWITCH_OFF,SWITCH_ON };

    int channels = 0;
    enum class channelsEnum { stereo,left,right };

    // --- Meter Plugin Variables
    float vuMeter = 0.f;

    //**-0x1A7F-**
    // --- end member variables
```

pluginancore.cpp

- The GUI objects will be written for you using exact ASPiK specifications, one object per GUI control, in the *initPluginParameters* function.
- The variables you describe when you fill out the GUI control forms will be automatically bound to the GUI parameters.
- Any presets you generate will be written in the *initPluginPresets* function.
- It will be up to you to modify the *processAudioFrame* function as well as adding any other functions that the plugin requires.

The plugin parameter initializations are written for you, and they follow the same construction as in Chapter 6. At this point you are ready to add the signal processing code and the FX objects as described in the plugin design sections of Chapters 11 through 22. After you've added the code and compiled it, you use RackAFX to test it.

7.7 Testing Audio Algorithms With RackAFX

RackAFX contains several tools for testing your plugin. First, you may play *.wav* files through it and manipulate the controls as you wish to see if the plugin is functionally operational. Second, you may also use the onboard oscillator to send sinusoids, noise, or other waveforms (including chirp signals) into the plugin. Third, you may use the audio adaptor's line input to pump live audio into the plugin, allowing RackAFX to act as a stand-alone audio FX unit (attach a MIDI controller for more real-time fun). Finally, you may do offline processing with RackAFX and write output data directly to a *.wav* file. Use Audio/MIDI → Process Into WAV File to do your offline processing. You may then open and analyze the audio file with another application or export it elsewhere for analysis.

If no audio is present, or other catastrophic failures occur, then you need to address them—just remember that if you were doing this work in a typical API inside of a typical DAW and your plugin crashed, you would bring down the DAW as well, and in some cases, the DAW might black-list your plugin never to be loaded again. If you do have issues, go to www.willpirkle.com/forum and get some help, including the video tutorials on how to debug a plugin in real time to catch constructor bugs. If your audio flows and the controls appear to work as designed, then you can scrutinize your plugin's performance using the RackAFX Analyzer window. You can run the oscillator into your plugin and monitor it on the oscilloscope and FFT spectrum analyzers. But the really interesting part of the analyzer is its ability to shoot impulses into your plugin and capture the results. It can then take an FFT of the impulse response to show you the frequency and phase responses, as well as the impulse and step responses. In order to do this you open the analyzer and stop the flow of audio. You will see four buttons labeled *Frequency, Phase, Step* and *Impulse.* Simply click one of these buttons to take the impulse response of the plugin. Now it gets really interesting: as you move the GUI controls for your plugin, RackAFX will continually fire and analyze impulse responses in succession, giving you a precise and accurate measurement of your plugin's characteristics—in real time, of course. You can study how the phase response changes as the cutoff frequency of a filter is altered, or how your reverb IR changes as you alter its delay line lengths. You can also take snapshots of the plugin responses (these are used exclusively all over this book to display actual plugin characteristics). Figure 7.8a–d shows the measured frequency, phase, impulse and step responses for a 2nd order low-pass filter plugin and the RackAFX analyzer. In the frequency or phase response plots, you may also click on the plot and move the mouse around: the cursor will snap to the response curve and you may read exact values directly off of the screen. This is shown in Figure 7.8e, where we are measuring the peak gain at resonance for the filter.

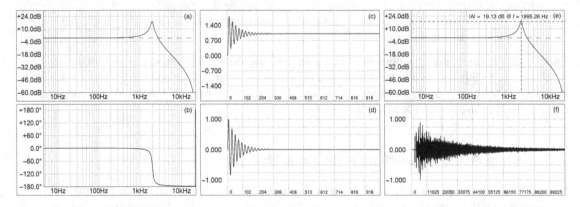

Figure 7.8: The RackAFX analyzer generates the (a) actual plugin frequency, (b) phase, (c) impulse, and (d) step responses. In addition, (e) clicking on the analyzer window allows you to measure exact values on the response curves. (f) An example of a reverb IR that is 2.5 seconds in length.

For reverb and delay plugins that require long impulse response times, you can change to long IR mode on the IR/Graphics Options tab. There you may enter the long impulse time in milliseconds (as long as you wish) and then perform the IR capture with the *Impulse* button—in long IR mode, only the IR will be shown as the phase and frequency responses were likely time varied. An example of a long IR for a reverb is shown in Figure 7.8f.

7.8 RackAFX Impulse Convolver and FIR Design Tools

RackAFX contains numerous other options to help speed up the process of audio algorithm development and testing. This includes the ability to capture and store the impulse response of any loaded plugin as a *.wav* file. You may also import impulse responses into the RackAFX IR library and use the built-in impulse convolver to test them. Figure 7.9a shows the impulse convolver with a previously loaded plugin's captured and stored IR frequency response. In Figure 7.9b, the list on the right contains a library of guitar amp head and cabinet impulse responses (Mesa-Boogie Dual Rectifier through varied microphones), one of which has been selected for frequency response calculation, viewing, and live auditioning—you may play audio through the impulse responses while selecting them from the list. Here are some things you can do with the impulse convolver:

- Play audio through the impulse convolver and click on any IR to slot it into the plugin in real time.
- Select an IR from the list, then hold the down-arrow key to scroll through the IRs and morph between them both visually and aurally—this is useful for testing morphing IR plugins (see the RackAFX videos on YouTube for an example of morphing IR plugins).
- Click on the FFT plot to make super accurate magnitude measurements.
- View the phase response of the IR.
- View the impulse and step responses.
- Export FFT data from the IRs and copy and paste the IR as C++ code directly into your projects (including non-RackAFX stuff).

The FIR Designer plugin allows you to create Optimal (Parks-McClellan) and frequency sampling method FIR filters shown in Figures 7.9c and d, respectively. You can play audio through the filters in real time *while you are designing them* and you can likewise store the impulse responses in the IR library and export them as C++ code to other applications, or convolve them with the impulse convolver later. Please refer to www.willpirkle.com/support/video-tutorials/ for video tutorials on using these bonus audio design tools with RackAFX.

7.9 Designing Your Custom GUI

With RackAFX you have two options: you may use the built-in GUI designer, or you may use the ASPiK GUI designer. The RackAFX GUI designer has some advantages in that you may define blank control templates and easily create complex GUIs including tabbed panels and custom views. However, it implements only about half of the possible VSTGUI4 controls. The ASPiK GUI designer uses the built-in VSTGUI4 editor, which allows access to all of the possible GUI objects. Launching the ASPiK GUI designer is identical to the ASPiK projects: open the GUI and hold the <SHIFT> key while right-clicking, then select "Open UI Editor" to launch the design tool. Make sure to point the GUI designer to your *PluginGUI.uidesc* XML file for saving (see Chapter 6).

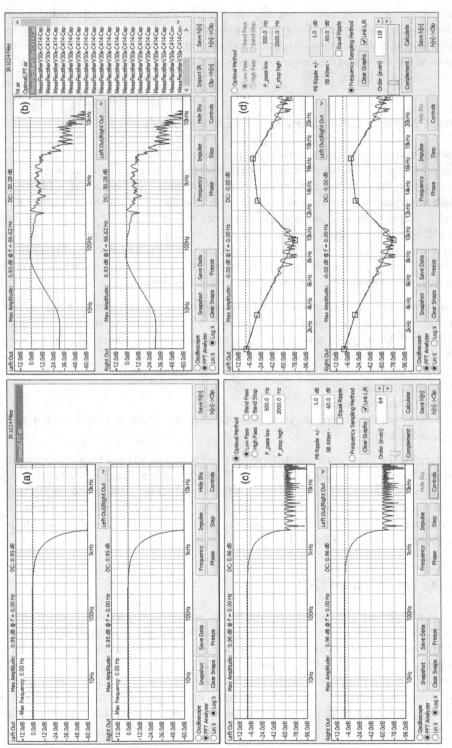

Figure 7.9: The RackAFX impulse convolver can be used to display the frequency response and live-audition the processed audio of both (a) internal, previously captured plugin IRs as well as (b) external IRs that have been loaded from .wav files while the RackAFX FIR designer can help you design both (c) Optimal (Parks-McClellan) and (d) frequency sampling method filters; you may listen to audio processed through the filters as they are being designed, with design updates affecting the audio in real-time.

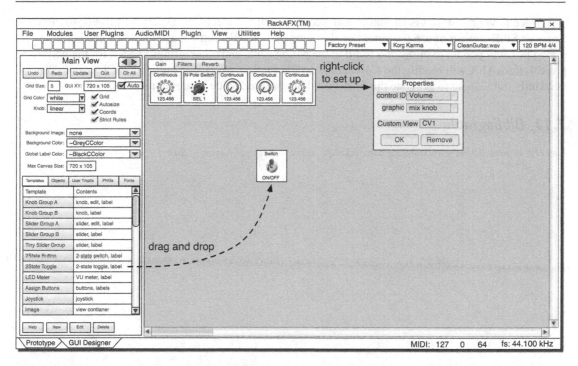

Figure 7.10: The RackAFX GUI designer features an easy to use drag-and-drop interface with zero coding; right-click to set up controls and create endless variations and sub-views on the off-screen views.

To launch the RackAFX GUI designer, click on the tab at the lower left of the main screen. When you use the RackFX GUI designer, the XML file will automatically be selected and updated each time you open and close the designer. Figure 7.10 shows the GUI designer's main view. You import and organize your graphics (PNG) files and drag and drop individual objects or control clusters from the list on the left side. Right-click on any control or control-cluster to edit—you may then link the GUI control to an underlying plugin variable, alter the graphics, save custom controls, define custom views and sub-controllers, and much more. There are an unlimited number of "off screen views" that you may use to assemble pages of templates and control views that may be assembled as tabbed panels on the main GUI view. Please refer to the numerous video tutorials and help at the RackAFX forum for more information on using this simple but powerful GUI design tool.

7.10 Exporting Your ASPiK Project

Because ASPiK was designed independently from RackAFX (to be as compatible as possible with AAX, AU, and VST) exporting projects is simple. Only the ASPiK core is exported, leaving all RackAFX API files behind and generating ASPiK projects for you to build as normal, just like all of the projects in this book. RackAFX did a bit of code writing for you with the GUI parameter and presets, but you did the rest by adding the signal processing code and any parameter specific functions as required. If you want to add third party files to your project, just be sure to add them to the *PluginObjects* folder—all of its contents and subfolders (as many layers as you like) are exported and added to your ASPiK Visual Studio and Xcode compiler projects. To export the project, use File → Export ASPiK Project or use the button on the command module in the master module section. A dialog box will pop up to allow you to fill in the plugin name and other strings needed for the ASPiK project. RackAFX will modify the *CMakeLists.txt* file for you, exactly as outlined in Chapter 6. You need only run *CMake* to generate the final Visual Studio and Xcode compiler projects.

7.11 Bibliography

RackAFX Documentation, www.willpirkle.com/rackafx/downloads/, Accessed August 1, 2018.

CHAPTER 8

C++ Conventions and How to Use This Book

The book consists of DSP processing theory as well as C++ code and object-oriented programming. To use the C++ projects that accompany the book, you are going to need to know at least the basics of C++ as well as how to use your C++ compiler—Visual Studio (Windows) or Xcode (MacOS). For the C++ portion, you should be familiar with the following terms; if you are unsure about what they mean, or never really learned the concepts fully, take some time to get caught up.

- data types: *int, float, double, unsigned int*
- objects: member variables and functions
- pointers, arrays, arrays of pointers
- encapsulation
- inheritance; base classes and derived classes
- virtual functions and abstract base classes

You need to have Visual Studio 2015 or later for Windows and Xcode 8 or later for MacOS. There are numerous resources you can find to help you get started, including www.aspikplugins.com/forum/, where you may ask questions and look for previous issues that have been solved.

You also need to choose a third party plugin framework, and you will likely fall into one of three categories:

1. You have already chosen a plugin framework and you've invested time in understanding its basic operations: define a GUI, get GUI control changes, and process audio samples.
2. You are new to audio plugin programming and need the simplest possible framework that also produces professional results.
3. You've looked at other frameworks and were more confused than anything else, or you don't want to pay licensing fees to a third party for your commercial plugins.

We developed a new professional audio plugin framework from the ground up called *ASPiK* that you may use as well. The plugin projects are packaged as ASPiK projects; however, they were designed in a way that makes them simple to implement in any other plugin framework, or with plugins designed from scratch. None of the projects actually requires ASPiK-specific code to function.

8.1 Three Types of C++ Objects

The book contains more than 55 C++ objects for you to use in your own plugins. These objects fall into three groups:

1. **DSP objects:** these small objects implement fundamental DSP algorithms.
2. **Effect objects:** these larger objects are designed to interface directly with a GUI. They implement specific audio effects and do all of the necessary work—they only need to be connected to a GUI and handed input samples to operate.

3. **WDF Ladder Filter Library objects:** we include an extensive WDF ladder filter library that accompanies Chapter 12. These objects are very different from the others and have their own base classes and connection mechanisms.

The DSP, Effect and Ladder Filter Libarary objects are combined together in the files *fxobjects.h* and *fxobjects.cpp* that are included in the ASPiK SDK. You may easily copy them from the SDK to include in your own projects if you decide not to use ASPiK as your framework. They do not contain any code that binds them to the ASPiK framework and are easily moved between other frameworks.

All of the book project algorithms that accompany the signal processing chapters are packaged as simple and relatively small C++ objects. Each object's audio processing code is derived directly from the signal processing algorithm and usually printed in whole or in part, with those chapters. The signal processing code is done with straight C++ that is generic and platform independent. With the exception of the FFT processing objects that require the FFTW library, none of the other objects require or include third party code. The signal processing code is decoupled from the plugin framework and requires no support code.

8.1.1 Effect Objects Become Framework Object Members

If you are using a plugin framework, there will be one object that is assigned to do the audio signal processing. In ASPiK, it is always the *PluginCore* object, while other frameworks typically specify an audio processing base class from which you derive your plugin object. It doesn't matter what framework you use—the audio algorithm objects are designed to be *member objects* of the main processing core object. You may declare them as static members or dynamically with pointers. For the book projects, I use static members as the objects handle their own dynamic allocations, but you may certainly change this. Almost all of the effect objects process one channel of audio at a time. For multi-channel operation, you will use multiple objects, one for each channel. We did not aggregate the majority of objects into multi-channel objects in order to provide maximum flexibility in your designs.

Your framework object simply declares the various objects that it needs for processing and holds the member objects as shown in Figure 8.1. For example, if your plugin object implements audio filtering,

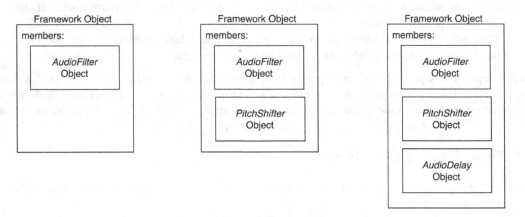

Figure 8.1: As shown here, algorithm objects are members of the outer framework object; you simply declare objects for each kind of processing algorithm and channel count.

you would use the *AudioFilter* object that I describe in Chapter 11. The object's class declaration would look like this:

```
// --- ASPiK uses PluginCore (always same name)
class PluginCore: public PluginBase
{
public:
    PluginCore();
    virtual -PluginCore(){}

    // --- PluginBase Overrides ---
    //
    // --- reset:
    virtual bool reset(ResetInfo& resetInfo);
    // --- etc.. others here

private:
    // --- one filter for the left channel and one for the right
    AudioFilter leftFilter;
    AudioFilter rightFilter;

    // --- etc ... rest of object
};
```

A stereo plugin that incorporates an audio filter and pitch shifter would declare channel objects for each algorithm; we also use arrays as the chapters progress:

```
private:
    // --- one filter for the left channel and one for the right
    AudioFilter filters[2];

    // --- one pitch shifter for each channel
    PitchShifter psmVocoder[2];
```

8.1.2 All Effect Objects and Most DSP Objects Implement Common Interfaces

Every effect object implements one of two interfaces, which makes their use predictable and easy. For these projects, an interface is defined as a pure abstract C++ base class that defines only member functions and no member variables; these objects have no constructor or destructor. The book algorithm objects are sub-classed from these interfaces and implement their required functions. This makes the initial setup and audio processing code fundamentally identical across the objects. The majority of the processing algorithms accept audio input samples and produce audio output samples. In almost all cases, the user interacts with GUI controls that send parameter information to the objects (more on that later). These objects all inherit from and implement the *IAudioSignalProcessor* interface. You are free to modify the interface and add more functionality if you like. The *IAudioSignalProcessor* interface is defined as follows.

```
class IAudioSignalProcessor
{
    // --- reset
    virtual bool reset(double _sampleRate) = 0;

    // --- process one sample in and out
    virtual double processAudioSample(double xn) = 0;

    // --- return true if the derived object can process a frame
    virtual bool canProcessAudioFrame() = 0;

    // --- optional processing function
    virtual bool processAudioFrame(const float* inputFrame,
                                   float* outputFrame,
                                   uint32_t inputChannels,
                                   uint32_t outputChannels)
            {return false;}

    // --- switch to enable.disable the aux input
    virtual void enableAuxInput(bool enableAuxInput) {}

    // --- for processing objects with a sidechain
    // --- the return value will depend on the subclassed object
    virtual double processAuxInputAudioSample(double xn)
    {
        // --- do nothing
        return xn;
    }
};
```

Each derived class must implement the three pure abstract functions; the last three functions
have a default implementation that does nothing. There are really only two main functions
here: *reset( )*, which is called to initialize the object with the current or new sample rate, and
processAudioSample( ), which processes one input sample through the object to provide one output
sample. Frame processing is optional; each object must implement *canProcessAudioFrame( )* and
return *true* or *false* depending on whether it can process frames. The frame processing capability is
required only for objects that require all channels of an input signal to produce the channel outputs,
for example, a stereo ping-pong delay where left and right channel information is swapped back and
forth between channels. All objects will use *reset* for the "prepare for playing" notification and will
accept the current sample rate as the argument. If the object is non-sample-rate dependent, it merely
ignores the sample rate. Our book projects are based on sample-by-sample audio signal processing
and the two processing functions work on either single or multi-channel signals. There are two more
optional virtual functions that deal with objects that accept a side chain, or other auxiliary audio input
or control signal: *enableAuxInput( )* and *processAuxInputAudioSample( )*. The former enables the
input, and the latter processes an audio sample from that signal; we will use these in the *Dynamics*
plugin in Chapter 18.

A few of the algorithms do not accept an audio input, but only render information such as an oscillator. These objects will implement the *IAudioSignalGenerator* interface; note that every function is declared as pure virtual so derived classes must implement them all.

```
class IAudioSignalGenerator
{
        // --- reset()
        virtual bool reset(double _sampleRate) = 0;

        // --- set the frequency
        virtual void setGeneratorFreq(double _oscFrequency) = 0;

        // --- set the waveform
        virtual void setGeneratorWaveform(generatorWaveform _wvfrm) = 0;

        // --- render the generator output
        virtual const SignalGenData renderAudioOutput() = 0;
};
```

The use of the functions should be obvious just by the function name. An oscillator will need to know its frequency and waveform (*setGeneratorFreq( )* and *setGeneratorWaveform( )*). The generator waveform is set with a strongly typed enumeration; the enumerated names indicate the type of waveform:

```
enum class generatorWaveform {kTriangle, kSin, kSaw, kWhiteNoise,
                              kRandomSampleHold};
```

The output is rendered into a structure so that it may expose numerous outputs rather than a single signal, including normal and inverted versions along with the quadrature phase signals (90 degrees out of phase), especially useful in our stereo delay and modulated delay plugins:

```
struct SignalGenData
{
        SignalGenData() {}

        double normalOutput = 0.0;
        double invertedOutput = 0.0;
        double quadPhaseOutput_pos = 0.0;
        double quadPhaseOutput_neg = 0.0;
};
```

The WDF ladder filter library objects implement their own interface named *IComponentAdaptor,* which is used at the lowest levels of operation only, so you won't need to deal with it unless you are making modifications to the internal parts of the library to extend its use.

8.1.3 DSP and Effect Objects Use Custom Data Structures for Parameter Get/Set Operations

In order to simplify and unify the operation of getting and setting object parameters, we implement custom data structures that define member variables for each of the object's parameters (e.g. f_c and Q

for a filter). In the case of our dynamics processor effect object, the structure also includes outbound parameter data (the calculated gain reduction value) that the plugin may display on a gain reduction meter. Each of these custom data structures includes an overridden = operator. This allows us to implement get/set functions that operate on copies of the data. Setting the parameters on these objects always follows the same pattern:

1. Call *getParameter* on the object to retrieve a copy of its current parameter settings.
2. Modify the settings that your plugin supports (some objects have optional parameters that may not be exposed to the end user).
3. Call *setParameter* and return the modified structure to the object; it will examine the incoming parameters for changes. If parameters have changed, the object will update itself accordingly. It will then make a copy of the new parameter list to store for future use and signal processing.

Many of the objects include one or more custom strongly typed enumerations that identify algorithms or modes of operation. These strongly typed enumerations are grouped along with the custom data structure and the object definition. The data structures usually include one of these enumerations as a member. For example, the *AudioFilter* object defines an enumeration for the filter algorithm—it implements every filter in Chapter 11.

```
enum class filterAlgorithm {
      kLPF1P, kHPF1P, kLPF1, kHPF1, kLPF2, kHPF2, kBPF2, kBSF2,
      kButterLPF2, kButterHPF2, . . ., kImpInvLP2
}; // --- you can add more here . . .
```

The *AudioFilter* has an accompanying custom data structure to pass parameter information. All of these custom structures are named alike—the object name plus "Parameters" such as *AudioFilterParameters:*

```
struct AudioFilterParameters
{
      AudioFilterParameters(){}
      AudioFilterParameters& operator=(const AudioFilterParameters&
                                        params)  // need this
      // --- individual parameters
      filterAlgorithm algorithm = filterAlgorithm::kLPF1;
      double fc = 100.0;
      double Q = 0.707;
      double boostCut_dB = 0.0;
};
```

The overloaded = operator code is not shown—you should always remember to add this code if you add more members to these structures. Very often, your objects will use identical parameter updates: for example, the two *AudioFilter* objects you would use for stereo operation would both use identical values for their f_c and Q parameters. In this case, it is simple to set up one channel, then copy its parameter structure to the others. For example, to set parameters on two *AudioFilter* objects declared in a static array you would write the following.

```
AudioFilter filters[2];

// --- get left params
AudioFilterParameters params = filters[0].getParameters();
```

```
// --- update params
params.fc = 100.3; ///< this will usually come from a GUI control
params.Q = 0.707; ///< this will usually come from a GUI control

// --- update objects
filters[0].setParameters(params);
filters[1].setParameters(params);
```

8.1.4 Effect Objects Accept Native Data From GUIs

To make the effect objects as framework-independent and platform neutral as possible, all of the parameters are designed to be set directly from GUI controls. For example, the filter plugin GUI exposes a frequency knob in Hertz (*Hz*) and a *Q* control (unit-less). The underlying audio filtering object accepts parameter changes directly as frequency in Hz and unit-less Q. It then does any post-processing to convert these values into its internal coefficients. This means that there is no glue code between the plugin framework and the C++ algorithm objects, allowing them to be used in any framework with little or no modification. If your GUI library or plugin framework uses normalized parameters, then part of the GUI handler will convert (cook) the parameters into meaningful data for the objects. This conversion is usually baked into the object itself. Figure 8.2 illustrates this concept. At the top of each buffer cycle, your plugin will update these member objects with GUI control information. Then, you will process audio through them with either the single-channel *processAudioSample* or multi-channel *processAudioFrame* function.

8.1.5 Effect Objects Process Audio Samples

All effect objects handle single channel processing by overriding the *IAudioSignalProcessor* function:

```
virtual double processAudioSample(double xn);
```

The input is passed as the argument *xn* and the function returns the output sample *y(n)*.

8.1.5.1 Series Objects

Each output sample of one algorithm object becomes the input sample for the next object when they are connected in series. You may use nested processing functions or use more verbose (but perhaps easier to

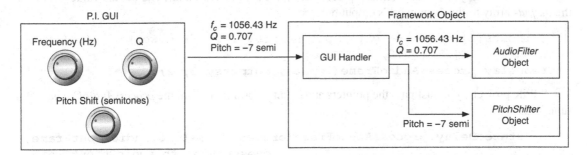

Figure 8.2: The framework object initializes its GUI parameters and receives control change information with the same units and types as the object's handle.

understand and remember) coding. To process the left channel of the example plugin, you need to get the left sample, then process it serially and write it to the output (note that these functions are easily nested).

```
double xn = // --- Get from your plugin framework input buffer

double filterOut = filter[0].processAudioSample(xn);
double pitchShiftOut = psmVocoder[0].processAudioSample(filterOut);

// --- Send pitchShiftOut to your plugin framework output buffer
```

8.1.5.2 Parallel Objects

For two objects in parallel, you send each the same input sample (*xn*), and then mix the outputs of the two objects via simple summation. To avoid clipping, you might attenuate each parallel channel by half, which we usually do by default in these algorithms.

```
double filterOut = filter[0].processAudioSample(xn);
double pitchShiftOut = psmVocoder[0].processAudioSample(xn);

double finalOut = 0.5*filterOut + 0.5*pitchShiftOut;
```

8.1.6 Effect Objects Optionally Process Frames

The *AudioDelay* object must process frames because a few of its algorithms require knowing both left and right channel inputs to calculate each output. In this case, they implement the optional *IAudioSignalProcessor::processAudioFrame* function and change their answer to true for *IAudioSignalProcessor::canProcessAudioFrames*. We define a frame as one set of samples from every audio channel. For a stereo in/stereo out plugin there are two input channels and two output channels in one frame. The multiple channels are passed as *float* data types in arrays—one for the input samples and the other for the output samples. This stems from the fact that all three APIs support processing *floats*. The *processAudioFrame* function takes an array of input samples and an input sample count, and produces output samples for each channel in the output array. If you are using ASPiK, this is the same set of arguments as your plugin core object receives from the plugin shell, so you may simply pass that data directly to these objects. A generic implementation might look like this: you pick up the input samples from your framework and place them in the *inputs* array, and write out the values from the *outputs* array to the framework's output buffers.

```
float inputs[2] = // --- get input samples from framework
float outputs[2] = { 0.0, 0.0 };
stereoDelay.processAudioFrame(inputs, outputs, 2, 2);
```

For ASPiK projects, you just pass the pointers and count information from the *processAudioFrame* function:

```
stereoDelay.processAudioFrame(processFrameInfo.audioInputFrame,
                            processFrameInfo.audioOutputFrame,
                            processFrameInfo.numAudioInChannels,
                            processFrameInfo.numAudioOutChannels);
```

8.2 Book Projects

In an effort to streamline the book code and reduce waste, I cannot show every line of code for every project—and in this day and age we don't need to do so. All of the signal processing code the book describes is contained within the C++ *DSP, effect,* and *WDF* objects in the two files *fxobjects.h* and *fxobjects.cpp* that you can download with the book projects. If you are a seasoned programmer you may simply grab the files, read the chapters, and go. In order to use the projects in this book, you will need to know how to perform the following operations:

- defining plugin *parameters* that are exposed to the host (these are usually shown as the GUI controls)
- receiving *control change* information from the GUI (if the data are normalized, you will need to put it into its un-normalized or "plain" format)
- accessing audio buffers to grab audio samples from the input buffer, and write audio samples to the output buffer (this implements your *processing* code)
- writing information to audio *meters* on the GUI

8.2.1 ASPiK Users

If you want to use ASPiK for your plugin framework, even if only to go through the book examples, then refer to Chapter 6 for all of the pertinent ASPiK documentation and an example, showing how to declare your GUI interface, bind GUI controls to variables you declare, and access the audio data buffers to get samples into and out of your *PluginCore* object. Your plugin algorithm objects are members of the *PluginCore*. You may optionally use RackAFX to export projects in ASPiK format that contain no RackAFX or other support code. You can find information on RackAFX in Chapter 7.

8.2.2 JUCE *and Other Non-ASPiK Users*

We want the book code and projects to be easily accessible. In order to use JUCE, wdl-ol, or some other framework, you will simply need to know how to perform the operations in Section 8.2. In the chapters that contain projects, I generally show the code to process one channel of data, unless the other channel-processing code is absolutely required. Since all objects receive control change information in the same format as you specify on the GUI, there is no parameter update code to show in the text—it is assumed that you know how to get GUI parameter change information from your GUI object and format it properly. Since each algorithm object uses its own data structure, you simply fill in the structure with your GUI control information. One easy strategy you might use is to download the book projects. Inside of each project you will find the *plugincore.h* and *plugincore.cpp* files. These are the audio processing object files and where you will spend most of your time. The ASPiK projects follow the same design pattern, starting with the plugin core member declarations, and include the *reset( )*, *updateParameters( )*, and *processAudioFrame( )* functions as follows:

- In the class definition within the *plugincore.h* file you will see the member declarations for the various effect objects; the methods are all implemented in the *plugincore.cpp* file.
- The *PluginCore::reset( )* function is where we do our per-run object initializations such as setting the sample rate; this corresponds to a prepare-for-audio-streaming function in your framework.
- We define the GUI update function with the same name in all projects as *updateParameters( )*— see its implementation in the *plugincore.cpp* file to watch us load up the data structures from

the GUI controls at the start of each buffer process cycle; you need to do the same within your framework.

- The audio is processed one frame at a time in *PluginCore::processAudioFrame( )*. You can find the code that grabs audio input samples, routes them into the member objects, and writes the output samples; it should be simple to adapt that code to your own framework object.

Figure 8.3 shows how audio data are routed through your processing object; you will need to extract the individual channel samples, send them to the audio algorithm objects, and then write their outputs.

8.2.3 A Sample Plugin Project: GUI Control Definition

As an example, suppose we defined a simple volume control plugin with a GUI shown in Figure 8.4. There are four controls: a volume knob labeled in dB; an on/off mute switch; a channel control that displays the three strings *stereo*, *left*, and *right;* and a VU meter that shows the output level. In plugin project Chapters (11–22), we first specify this GUI in two tables: one for the normal (input) controls and one for the (output) meters. Those are shown in Tables 8.1 and 8.2. Regardless of the plugin framework you choose, you will need to understand how to convert the information from those tables into the code that works in your chosen framework. Notice in particular the columns labeled "Linked Variable" and "Linked Variable Type": each GUI control links to a member variable in your plugin framework object. You will need to know how to connect GUI updates to these variables (for ASPiK, this is done automatically with its variable binding option). At the top of each buffer process cycle, we

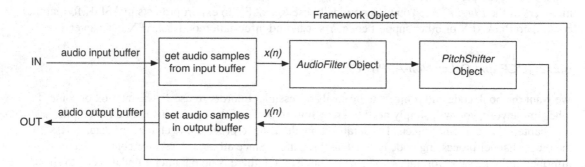

Figure 8.3: A simplified view of processing samples through the audio objects.

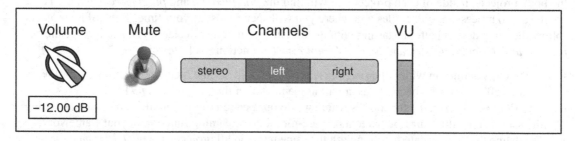

Figure 8.4: A simple plugin GUI.

Table 8.1: The volume plugin's input parameter list; notice that the string list is shown in curly brackets and will be listed in place of the min/max/default values for string-list controls

Control Name	Units	Min/max/default or string-list	Taper	Linked Variable	Linked Variable Type
Volume	dB	−60.0 / +12.0 / 0.0	linear	*volume_dB*	*double*
Mute	-	"OFF, ON"		*enableMute*	*int*
Channel Select	-	"stereo, left, right"		*channels*	*int*

Table 8.2: The volume plugin's output parameter list; all meter variables are of type *float*

Meter Name	Inverted	attack/release	Detect Mode	Calibration	Variable Name
VU	no	10.0 / 500.0	RMS	logarithmic	*vuMeter*

will update the object with fresh GUI control information that is encoded in the linked variables. We always use the *float* data type for meter variables, so that is omitted from the meter tables. For ASPiK users, this is documented in Chapter 6.

All DSP and effect objects are documented in the same manner in each plugin project section, following the same progression each time. Once you understand one of them, the rest will be simple.

CHAPTER 9

How DSP Filters Work (Without Complex Math)

DSP filter theory involves complex algebra, that is, the algebra of complex numbers that involve the $\sqrt{-1}$. Complex numbers contain real and imaginary parts. Rather than simply throw a bunch of equations and algebra at you, I prefer to ease into the theory by starting with a method of examining, stimulating, and testing a DSP filter without complicated math. We may also predict its frequency and phase responses from the data we collect in the process. This method will only require some tabulated bookkeeping. The way you shuffle audio samples through the tables is going to ultimately correspond to C++ code that you will eventually need to write, making that process easier to understand. We've traded complicated math for some tedium in keeping track of modified audio sample values. To remove the monotony, we'll approach the problem from a higher level and apply some math concepts to get the same results but without needing to keep track of a bunch of audio samples.

9.1 Frequency and Phase Response Plots

The frequency response plot shows how a filter modifies the frequency content of a signal. The frequencies are plotted on the x-axis and the relative output amplitudes of the frequencies on the y-axis. The x-axis is often plotted logarithmically and the y-axis is often plotted in dB. Let's consider the most fundamental of all filters, the low-pass filter (LPF), which allows low frequencies to pass through it while attenuating higher frequencies. Figure 9.1a and c show the frequency responses of two LPFs, each with the same cutoff frequency (f_c = 500 Hz). This frequency is also called the *−3 dB frequency* because the filter attenuates this frequency component by −3 dB. The cutoff frequency f_c denotes the pass-band (frequencies below it) and the stop band (frequencies above it). The only real difference between Figure 9.1a and c is the slope of the curve above the cutoff frequency, called the *roll-off,* where the filter in Figure 9.1c displays a steeper roll-off than that in Figure 9.1a. We characterize the roll-off in dB/octave (or sometimes in dB/decade) where an octave represents a doubling of frequency, while a decade represents a 10x increase in frequency. A LPF whose roll-off slope is −6 dB/octave is called a 1st order filter and a LPF with a −12 dB/octave roll-off is called a 2nd order filter. We'll worry about where that naming convention comes from later on.

Whereas the frequency response plot is somewhat intuitive in that it shows us the gain or attenuation of different frequencies that have moved through the filter, the phase response plots shown in Figure 9.1b and d are not as easy to grasp. These show the phase shift through the filter. The only difference in the two plots is the amount of phase shift that the various frequencies encounter when moving through the filter. But what does that really mean? And how do frequencies become amplified, attenuated, or phase-shifted through a digital filter? These are the questions I'll answer in the next few sections (without complex math).

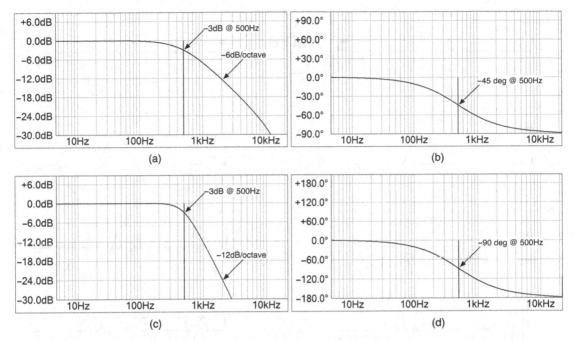

Figure 9.1: (a) The frequency response of a 1st order filter. (b) The phase response of a 1st order filter. (c) The frequency of a 2nd order filter. (d) The phase response of a 2nd order filter. In both of these cases the cutoff frequency f_c = 500 Hz.

9.2 Frequency and Phase Adjustments From Filtering

Consider an audio waveform entering the filter. Fourier showed that a continuous waveform could be decomposed into a set of sinusoids with different frequencies and amplitudes (which is called *Fourier decomposition*). This means that the complicated time domain signal may be considered to be a sum of sinusoidal components. When the sinusoids are added back together (called *Fourier re-synthesis*), the original waveform is obtained. Figure 9.2 shows a filter in action. The input is decomposed into four sinusoids, a fundamental, and three harmonics. The peak-amplitudes of the components are shown as dark bars to the left of the y-axes. For simplicity, we have chosen the input signal as a combination of four sinusoids that have the same amplitude, and that start at 0.0 at time $t = 0$, so that they have identical starting phases of zero degrees. We can observe the following:

- The input waveform is smoothed out and is less "bumpy."
- The lowest frequency component (called the *fundamental frequency*) comes out of the filter with the same amplitude and phase shift as it had when it entered the filter.
- The three harmonics have decreasing amplitudes and more phase shift as they go higher in frequency.

Figure 9.3 shows the same filter with the information plotted differently: here the amplitudes and phases are plotted against frequency rather than a set of time domain sinusoids. You can see by the output frequency response plot that this filter is a kind of low-pass filter. Its curve is similar to Figure 9.1a.

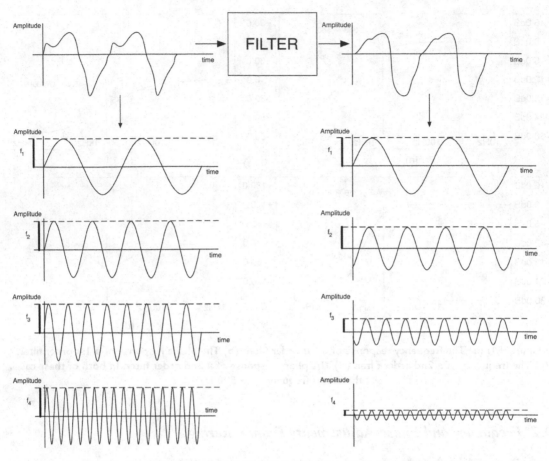

Figure 9.2: An audio waveform is filtered into a smoothed output; the input and output are decomposed into their Fourier components to show the relative amplitudes and phase shifts that each frequency encounters while being processed through the filter.

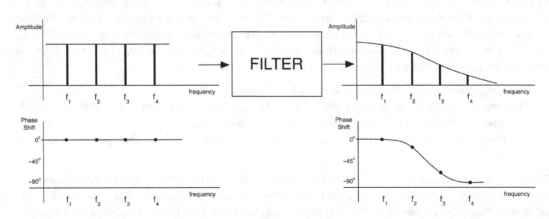

Figure 9.3: The same information from Figure 9.2 is plotted as frequency and phase response curves.

9.3 1st Order Feed-Forward Filter

In the last chapter, I introduced a simple filter to show an example of DSP building blocks consisting of coefficients, summers, and storage registers. We'll be analyzing this filter for most of the chapter, so let's look at how an input signal $x(n)$ would flow through the filter to become the output $y(n)$. Figure 9.4a shows this filter along with the values of the samples at the various nodes of the circuit. The input sample is $x(n)$ and the output of the storage register marked z^{-1} is $x(n-1)$, which was the *previous* input to the filter, one sample period prior to our observation.

If you trace the signal path all the way through, adding the two summer inputs together, you arrive at the *difference equation* of the filter shown in Equation 9.1.

$$y(n) = a_0 x(n) + a_1 x(n-1) \tag{9.1}$$

The difference equation shows the sampled-time domain relationship between the input and output. Don't let the name "difference" complicate things—this is a generic term and there is no requirement that a difference or subtraction has taken place in the equation. You can tell why the structure is called *feed-forward*: the input branches feed-forward into the summer. The signal flows from input to output. There is no feedback from the output back to the input. Now, suppose we let the coefficients a_0 and a_1 coefficients both equal 0.5 as shown in Figure 9.4b. What kind of filter is it? What are its frequency and phase responses?

In order to analyze this filter we can go the easy but tedious route or the difficult but elegant route. Let's start with the easy way. In order to figure out what this does, you apply the five basic digital test signals you learned in Chapter 1 to the filter, then manually push the values through and see what comes out. You need only a pencil and paper or a simple handheld calculator. The five waveforms we want to test are:

- DC (0 Hz)
- Nyquist
- ½ Nyquist
- ¼ Nyquist
- Impulse

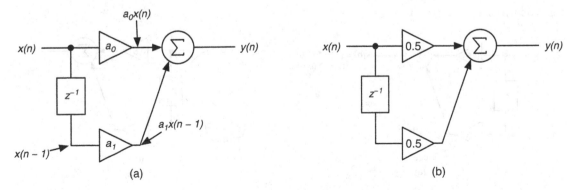

Figure 9.4: (a) The 1st order feed-forward filter showing values at intermediate nodes. (b) The same filter with the a_0 and a_1 coefficients set to 0.5.

For each audio sample that enters the structure, there are two phases to the operation:

- **Process phase**: the sample is read in and the output is formed using the difference equation and the previous sample in the delay register; the input is processed through to the output.
- **State update phase**: the delay element is overwritten with the input value—the older sample stored in the single z^{-1} register is effectively lost.

The value in a filter's z^{-1} register is sometimes called its "state," and the second phase is sometimes called the *state update* phase or more simply *updating the state* of the filter. The z^{-1} register itself is sometimes called the *state register* and we refer to it this way throughout the text. It is during this second phase of operation that we prepare the z^{-1} registers for the *next* processing iteration on the next sample period. It is important that we always update the filter's state *after* the output *y(n)* has been calculated. This will become a rule for our filtering objects later on.

To analyze the structure with the given coefficients, we start with the DC/step input and begin sequentially applying the samples into the filter, performing one iteration through the filter on each sample interval. Figure 9.5a shows the first iteration of the structure where we process the input x(n) into an output y(n) and then update the filter's state by shifting the input value into the delay register, as shown in Figure 9.5b. Notice that we lose the value in the state register, symbolically sending it to the trash can, and overwriting it with the current input value. In complex filter structures with many state registers, we need to take care to ensure that the state registers are updated in the proper sequence so that only the oldest state information is ever lost or thrown away.

Figures 9.6a and b shows the second iteration on the next sample period; we simply perform the exact same operation, but use the next input value in the sequence.

Figures 9.7a and b shows the third and fourth iterations, at which point we observe a pattern and with great confidence can predict the rest of the output sequence:

Input sequence: {...0.0, 1.0, 1.0, 1.0, 1.0...}
Output sequence: {...0.0, 0.5, 1.0, 1.0, 1.0...}

We can plot the input and output sequences as shown in Figure 9.8, where we can make further observations about the filtering operation: the output amplitude eventually settles out to a constant 1.0 or unity gain condition, so at DC or 0 Hz, the output equals the input. However, there is a one-sample

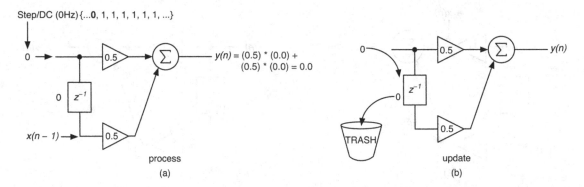

Figure 9.5: (a) The process and (b) state update phases of one iteration through the filter; *x(n)* = 0.0 and the resulting *y(n)* = 0.0.

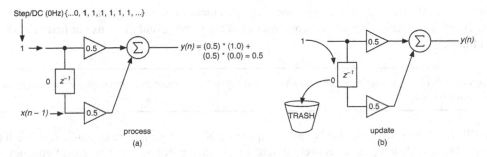

Figure 9.6: (a) The process and (b) state update phases of the second iteration through the filter; $x(n) = 1.0$ and the resulting $y(n) = 0.5$.

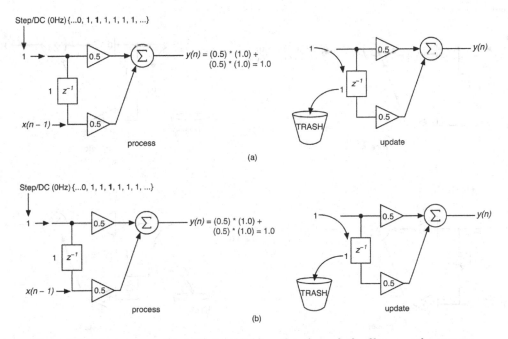

Figure 9.7: (a) The second and (b) the third iteration through the filter reveal a pattern.

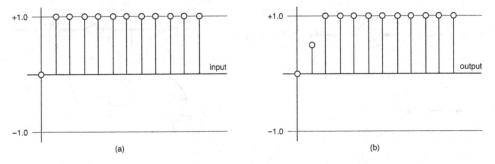

Figure 9.8: The input (a) and output (h) sequences for the filter in Figure 9.4b at DC or 0 Hz.

delay in the response causing the leading edge of the step input to be smeared out by one sample interval. This time smearing is a normal consequence of the filtering and gives us our first rule for feed-forward filters.

> In a feed-forward filter, the amount of time smearing is equal to the maximum delayed path through the feed-forward branches.

Next, we'll repeat the process for the Nyquist frequency (DC and Nyquist are the easiest, so we'll do them first). The filter behaves in an entirely different way when processing the Nyquist frequency—we can spot the pattern after only a few iterations, as shown in Figure 9.9. Now, we make our observations about amplitude and phase by examining the input and output sequences in Figure 9.10. The amplitude

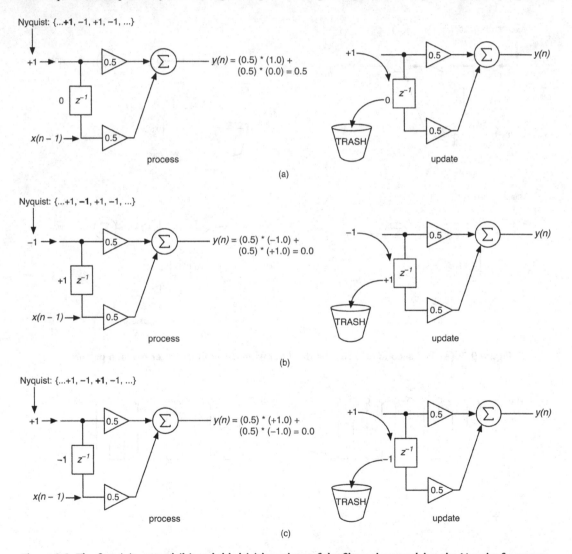

Figure 9.9: The first (a), second (b) and third (c) iterations of the filter when applying the Nyquist frequency.

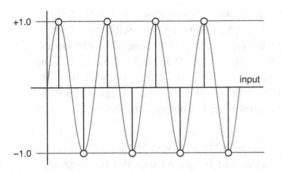 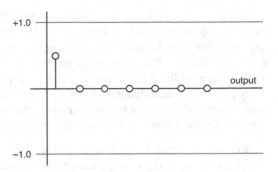

Figure 9.10: The input and output sequences of the filter at Nyquist reveals that the output completely disappears after the first sample period.

Table 9.1: The feed-forward filter input $x(n)$, output of state register $x(n - 1)$ and final filter output $y(n)$ for the input sequence of ½ Nyquist

$x(n)$	$x(n - 1)$	$y(n) = 0.5x(n) + 0.5x(n - 1)$
0	0	0
1	0	0.5
0	1	0.5
−1	0	−0.5
0	−1	−0.5
1	0	0.5
0	1	0.5
−1	0	−0.5
0	−1	−0.5

at Nyquist eventually becomes zero after the one-sample delay time. The phase is hard to characterize because the signal has vanished.

Why did the amplitude drop all the way to zero at Nyquist? The answer is one of the keys to understanding digital filter theory: the one-sample delay introduced exactly 180 degrees of phase shift at the Nyquist frequency and caused it to cancel out when re-combined with the input branch through a_0. This leads us to our next rule.

> Delay elements create phase shifts in the signal. The amount of phase shift depends on the amount of delay as well as the frequency of the input signal.

In the case of Nyquist, the one-sample delay is exactly enough to cancel out the original signal when they are added together in equal ratios. What about other frequencies like ½ and ¼ Nyquist? They are a bit more laborious to work through but worth the effort. By now you can see how the data moves through the filter, so let's use Table 9.1 and move the data through *it* instead.

In Table 9.1, we tabulate the values for the various nodes in the structure. You can see that on each successive row, the previous input $x(n)$ becomes the output of the state register $x(n - 1)$ and each value in the first column appears in the second column shifted down by one row. That shifting shows the delay register's operation. In Figure 9.11a we line up the input sequence (top) with the output sequence (bottom). We observe that the output sequence is periodic, but with reduced amplitude and with a phase offset. The output sequence is $y(n) = \{0, +0.5, +0.5, -0.5, -0.5, +0.5, +0.5, \ldots\}$. At first it might seem difficult to interpret the sequence $\{\ldots -0.5, -0.5, +0.5, +0.5, \ldots\}$, but we know that ½ Nyquist is also encoded with a repeating sequence of four values $(0, 1, 0, -1)$. We can estimate the output amplitude as about 0.7 by overlaying the input sinusoid over the output samples and scaling the sinusoid so that it lines up with the sample values.

Finally, we can continue the process and arrive at the results in Table 9.2 for ¼ Nyquist. The ¼ Nyquist frequency is encoded with eight samples, so it will take some time before we can observe a pattern. Ultimately, we do find a repeating sequence of $y(n) = \{\ldots, +0.354, +0.854, +0.854, +0.354, -0.354, -0.854, -0.854, -0.354, +0.354, \ldots\}$. The input and output sequences are plotted in Figure 9.11b. Here we can estimate that the output amplitude is about 90% of the input size, and that the one sample of time smearing results in a phase shift of 1/16 of a cycle, or 22.5 degrees.

Now, we can combine the frequency and amplitude observations for a frequency response plot and the frequency and phase angles for a phase response plot shown in Figure 9.12. We observe that this digital filter is a low-pass variety with a typical LPF frequency response. However the phase response is quite interesting: it is linear instead of nonlinear like the example at the beginning of the chapter.

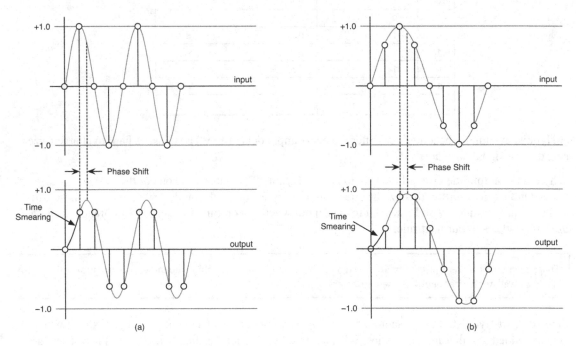

Figure 9.11: (a) The input/output relationship in time at ½ Nyquist, where the ½ Nyquist frequency is attenuated by about 70% and the output is also phase-shifted by 45 degrees or one-eighth of the wavelength. (b) The input/output relationship in time at ¼ Nyquist, where the ¼ Nyquist frequency is attenuated by about 90%; the output is also phase-shifted by 22.5 degrees (or half of that from ½ Nyquist). In both cases the leading edge of the first cycle is smeared out by one sample's worth of time.

Table 9.2: The feed-forward filter input *x(n)*, output of state register *x(n − 1)* and final filter output *y(n)* for the input sequence of ¼ Nyquist

x(n)	x(n − 1)	y(n) = 0.5x(n) + 0.5x(n − 1)
0	0	0
0.707	0	+0.354
1	0.707	+0.854
0.707	1	+0.854
0	0.707	+0.354
−0.707	0	−0.354
−1	−0.707	−0.854
−0.707	−1	−0.854
0	−0.707	−0.354

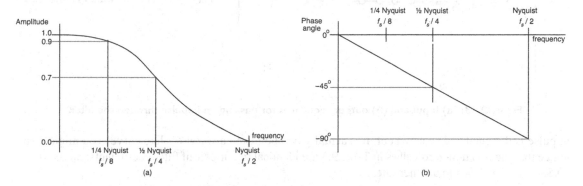

Figure 9.12: The frequency response (a) and the phase response (b) of our simple 1st order feed-forward filter.

In fact, this simple filter is a *linear phase filter*. A feed-forward filter will be a linear phase filter if its coefficients are symmetrical about their center. In this case, (0.5, 0.5) is symmetrical. Another example would be (−0.25, −0.25). Linear phase filters shift the phase of the output signal linearly across the entire frequency range. These are interesting for two reasons: first, this kind of response cannot be obtained from an analog filter; the closest we can get is an analog filter whose phase response is linear through the pass band, but then goes nonlinear in the stop band. Second, this kind of filter has useful applications where phase linearity across the spectrum is important, such as filters for loudspeaker crossovers.

Finally, let's run the impulse signal through the filter structure. When we do this and capture the result, we are finding the *impulse response* of the filter. The impulse response defines the filter in the time domain, like the frequency response defines it in the frequency domain. The basic idea is that if you know how the filter reacts to a single impulse you can predict how it will act to a series of impulses of varying amplitudes. But it can also reveal other information for us as well, including the frequency and phase responses. Table 9.3 shows our analysis of the structure by applying the impulse to the input; Figures 9.13a and b show the input and output sequence plots, respectively. Here you can see that the

Table 9.3: Processing the impulse through the filter yields the impulse response

x(n)	x(n − 1)	y(n) = 0.5x(n) + 0.5x(n − 1)
0	0	0
1	0	0.5
0	1	0.5
0	0	0
0	0	0
0	0	0

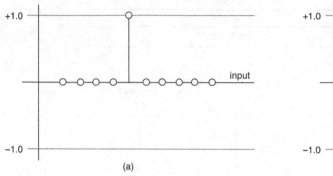

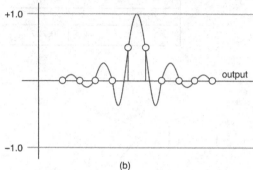

Figure 9.13: (a) Input and (b) output sequences for passing an impulse through the filter.

impulse is flattened and smeared out. It is actually two points on a *sin(x)/(x)*-like curve. You might also notice that the two non-zero values in Table 9.3 are identical to our pair of filter coefficients: a_0 and a_1 = {0.5, 0.5}. This leads to another rule.

> For a pure feed-forward filter, the impulse response is finite and consists a set of values that are identical to the filter coefficients.

What makes this filter a low-pass filter? It is a combination of the coefficients and the filter topology (1st order feed-forward). There are three basic topologies: Feed-forward (FF), feedback (FB), and a combination of FF/FB. So once the topology has been chosen, it's really the *coefficients* that determine what the filter will do. To illustrate this and crystallize it as a concept, we've created a simple plugin with GUI controls that allow the user to vary the a_0 and a_1 coefficients of the filter. Figure 9.14 shows a composite set of results for various values of the two coefficients revealing low pass and shelving types of responses. Many DSP books devote significant numbers of chapters solely to the task of calculating coefficients for filters to obtain various types of frequency responses. In this book Chapters 11, 12, and 16 contain a plethora of methods for finding coefficients for audio-specific filters.

We can now make some important generalizations of our filters and think about what a plugin that implements filters would need to do:

- The topology of the filter determines its difference equation.
- The coefficients (a_N) of a feed-forward filter determine its filter frequency and phase responses and therefore its type (e.g. low-pass filter or LPF) and its sonic qualities.

- A filtering plugin must implement the difference equation to process input samples $x(n)$ into output samples $y(n)$.
- As the user adjusts the plugin controls (filter type, cutoff frequency, etc.) we must recalculate the filter coefficients accordingly so that the filter is adjusted properly.

9.4 1st Order Feedback Filter

The 1st order feedback filter is shown in Figure 9.15 and consists of two coefficients: a_0 in the forward path and b_1 in the feedback path. There is only one state register, but this time it is in the feedback path, and it stores previous filter outputs $y(n-1)$ instead of previous filter inputs $x(n-1)$. For now you can safely ignore the fact that the b_1 coefficient has been negated (this will all make sense in the next chapter). You can also see that there is no b_0 coefficient in this structure—and the fact is that we will never have a b_0 coefficient. The difference equation for this filter is:

$$y(n) = a_0 x(n) - b_1 y(n-1) \qquad (9.2)$$

We can apply the same concept of pushing analytical sequences through this structure and examining the output to determine the frequency and phase responses. But a little thought will show that this is

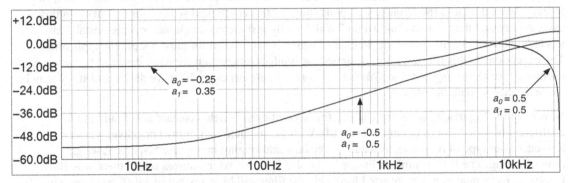

Figure 9.14: The feed-forward filter with a variety of coefficient values yields different types of filter curves.

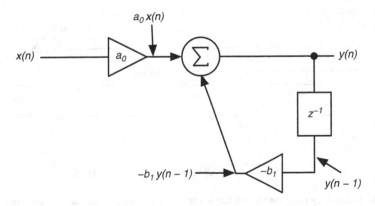

Figure 9.15: A 1st order feedback structure.

going to become exceedingly tedious, especially if the b_1 coefficient has a large value—this will keep recycling the output back into the summer for many sample intervals and it might take hundreds or even thousands of iterations before we notice a pattern. This in itself is good motivation for us to find a better way to analyze and design these filters, even if it requires some heavier mathematical lifting on our parts. You need to be willing to make this trade—more difficult math, but far less tedium—in order to advance your studies of DSP theory. Rather than presenting those mindless tables for a feedback filter, we've coded a simple 1st order feedback filter plugin and have plotted the results using various values for the a_0 and b_1 coefficients. The results are shown in Figure 9.16.

Once again we observe that the various coefficient pairs produce a variety of frequency responses ranging from low pass to high shelf in nature, and many in-between combinations. So, the coefficients (a_N and b_M) of a feedback filter determine its filter frequency and phase responses. Examine Figures 9.17a, b, and c which show pairs of plots for frequency and impulse responses of this 1st order feedback structure. In each case, the a_0 coefficient is held constant at 0.5, and the b_1 coefficient is varied from 0.5 to 0.9 to 1.0. Notice what happens to the impulse response as the value for the b_1 coefficient increases: you can see that it becomes a damped oscillation, which we never observed in the feed-forward filter. This is called *ringing* and has an audible effect that is referred to as *ringing,* or sometimes *pinging*. The oscillations take longer to die out when the b_1 coefficient is increased from 0.5 to 0.9. In the third plot with the b_1 coefficient set to 1.0, we observe that the impulse response has become a pure sinusoidal oscillation—the impulse response for a feedback filter can, in fact, ring forever. More drastic is the frequency response plot for this coefficient setting—it is a mess. In this case, we say that the filter has "blown up," which really just means that it has become unstable or has entered a pure oscillation mode, neither of which is very useful for most FX plugin applications. The exceptions to this rule are filters designed for synthesizers, which we often desire to become self-oscillatory. In *Designing Software Synthesizers in C++*, almost every filter we derive and study may be made to self-oscillate like this.

Now examine Figures 9.18a, b and c where the a_0 coefficient is held constant at 0.5, and the b_1 coefficient is varied from −0.5 to −0.9 to −1.0. With the b_1 coefficient turned negative, we get a low-pass filter type of response. Instead of overshoot and ringing in impulse responses, we get time smearing. The last example in Figure 9.18c also blew up, which makes sense: there was 100% feedback. Even though it was inverted feedback, the filter still blew up, but in a different way. This time, the gain at DC went to infinity and it now "rings" at 0 Hz.

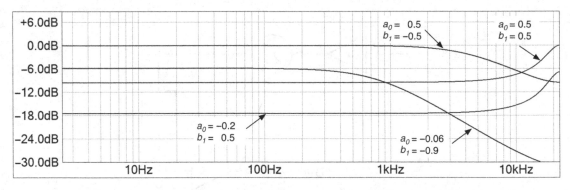

Figure 9.16: Frequency response plots for a 1st order feedback filter with a variety of a_0 and b_1 coefficient values.

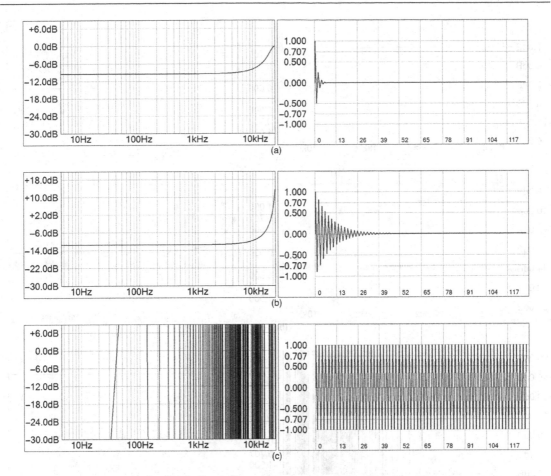

Figure 9.17: The 1st order feedback filter's frequency response (left column) and impulse response (right column) for various coefficient values of (a) $a_0 = 0.5$, $b_1 = 0.5$; (b) $a_0 = 0.5$, $b_1 = 0.9$; and (c) $a_0 = 0.5$, $b_1 = 1.0$ (notice that the impulse response here rings forever, and the frequency response is garbage).

9.5 Final Observations

We can combine our observations about feed-forward and feedback filters together to generate a brief summary of these filter structures and consider the ramifications of using them:

Feed-Forward Filters

- operate by making some frequencies go to zero; in the case of $a_0 = 1.0$ and $a_1 = 1.0$, the Nyquist frequency went to zero; this is called a *zero of transmission* or a *zero frequency* or just a *zero*
- step (DC) and impulse responses show smearing (amount of smearing is exactly equal to the total amount of delay in the feed-forward branches)
- don't blow up
- are called *finite impulse response* (*FIR*) filters because their impulse responses, though they may be smeared, are always finite in length

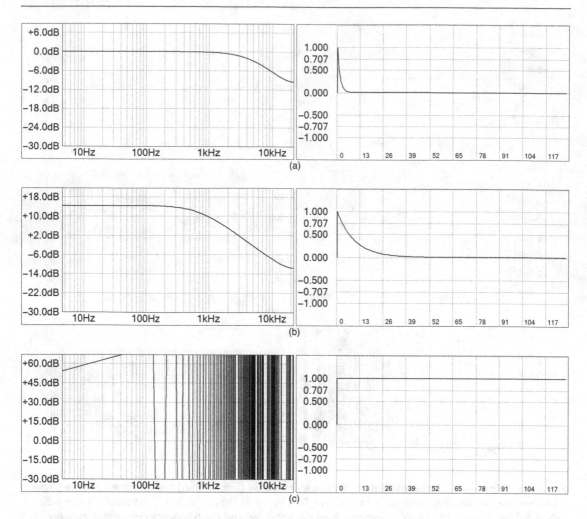

Figure 9.18: The 1st order feedback filter's frequency response (left column) and impulse response (right column) for various coefficient values of (a) $a_0 = 0.5$, $b_1 = -0.5$; (b) $a_0 = 0.5$, $b_1 = -0.9$; and (c) $a_0 = 0.5$, $b_1 = -1.0$ (notice that the impulse response here) "rings forever" at 0 Hz, and the frequency response is garbage.

Feedback Filters

- operate by making some frequencies go to infinity; in the case of $a_0 = 0.5$ and $b_1 = 1.0$, the Nyquist frequency went to infinity but in the case of $a_0 = 0.5$ and $b_1 = -1.0$, the DC or 0 Hz went to infinity; this is called a *pole of transmission* or a *pole frequency* or just a *pole*
- step and impulse responses show overshoot and ringing or smearing depending on the coefficients (amount of ringing or smearing is proportional to the amount of feedback)
- can blow up (or go unstable) under some conditions
- are called *infinite impulse response* (*IIR*) filters because their impulse responses can become infinitely long

The problem now is that we want to be able to specify the filter in a way that makes sense in audio—a low-pass filter with a cutoff of 100 Hz or a band-pass filter with a Q of 10 and center frequency of 1.2 kHz. What we need is a better way to analyze the filtering than just randomly trying coefficients. We need methods that calculate filter coefficients easily, based on how we specify the filter. In the next chapter, we'll work on the basic DSP theory that will make this happen.

9.6 Homework

In the 1st order feed-forward filter, we saw that one sample of delay was enough to phase shift the signal in the feed-forward branch enough to cause it to cancel out the original input signal if the input is the Nyquist frequency. How would you modify this structure so that it cancels out the ½ Nyquist frequency instead? How about the ¼ Nyquist frequency?

9.7 Bibliography

Dodge, C. and Jerse, T. 1997. *Computer Music Synthesis, Composition and Performance*, Chap. 3, 6. New York: Schirmer.
Steiglitz, K. 1996. *A DSP Primer with Applications to Digital Audio and Computer Music*, Chap. 4. Menlo Park: Addison Wesley.

CHAPTER 10

Basic DSP Theory

You want to get a grip on the underlying DSP theory of filters for several reasons: it helps to understand the anatomy of the filter because you want to be able to implement new filter designs as they appear, rather than wait for someone else to code it; a deeper understanding of the theory can only help your coding strategy; and the same DSP analysis and mathematical models can be applied to all sorts of non-EQ effects like delay, chorusing, reverb, and compression. In order to intuitively understand the foundation of DSP theory, you need to review some math and engineering concepts.

10.1 The Complex Sinusoid

The analysis and design of digital filters uses the sinusoid as its basic stimulus function. Since Fourier showed that a signal can be decomposed into sinusoids, if you know how the system reacts to a sinusoidal stimulus at a bunch of different frequencies, you can plot the frequency and phase responses like you did by hand in the last chapter. This is akin to using the impulse response: since the input signal is a train of impulses of varying amplitudes, if you know how the filter responds to a single impulse, you can figure out how it will respond to multiple impulses. You also did this by hand when you took the impulse response of the low-pass filter in the last chapter.

Everyone is familiar with the sine and cosine functions—sine and cosine are related by an offset of 90 degrees and the sine function starts at 0.0 whereas cosine starts at 1.0. In Figure 10.1a you can identify the sine and cosine waveforms by their starting position.

But what do you call a sine-like waveform that starts at an arbitrary time in Figure 10.1b? Is it a sine that has been phase-shifted backwards or a cosine that has been phase-shifted forward? You have to be careful how you answer because sine and cosine have different mathematical properties: their derivatives are not the same and it usually becomes difficult when you try to multiply them or combine them in complex ways. Add a phase offset to the argument of the $sin()$ or $cos()$ function and then it really turns into a mess when you do algebra and calculus with them. What you need is a function that behaves in a sinusoidal manner, encapsulates both sine and cosine functions, and is easy to deal with mathematically. Such a function exists, and it's called the complex sinusoid.

$$\text{Complex Sinusoid} = e^{j\omega t} \tag{10.1}$$

You might recognize this as involving Euler's equation (N.B. Euler is pronounced like "oiler"):

$$e^{j\theta} = \cos(\theta) + j\sin(\theta)$$
$$j = \sqrt{-1} \tag{10.2}$$

You can see that it includes both sine and cosine functions. The j term is the imaginary number, the square root of -1 (mathematicians call it *i* but since we already use that as the symbol for current

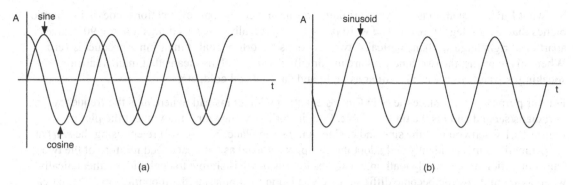

Figure 10.1: (a) The sine and cosine functions. (b) A sinusoid.

in engineering, we rename it *j* instead). The *j* is known as the phase rotation operator; multiplying a function by *j* rotates the phase by 90 degrees.

Suppose you want to shift the phase of a waveform by 180 degrees, thereby inverting it. Mathematically, you can do this by multiplying the waveform by −1, inverting the values of all its points. Suppose you wanted to invert the waveform again (which would bring it back to its original shape): you could do that by multiplying by −1 again. But suppose that you wanted to shift the phase by 90 degrees only? Is there a number η you could use as a multiplier to shift by 90 degrees? In other words,

$$90\text{-degree shifted waveform} = (\text{original waveform})(\eta)$$

You don't know what η is yet, but you can figure it out. Suppose you then wanted to shift the waveform by another 90 degrees, which would be the same as shifting the original waveform by 180 degrees. You would multiply it by η again. You'd then have the following:

$$180\text{-degree shifted waveform} = (\text{original waveform})(\eta)(\eta)$$
$$= (\text{original waveform})(\eta^2)$$
$$= (\text{original waveform})(-1)$$

This leads to Equation 10.3.

$$\eta^2 = -1$$
$$\eta = \sqrt{-1} = j \tag{10.3}$$

So, you can perform a conceptual 90-degree phase shift by multiplying a waveform by *j*. A −90 degree phase shift is accomplished by multiplying by −*j*. Some other useful relationships with *j* are:

$$e^{-j\theta} = \cos(\theta) - j\sin(\theta)$$
$$j^2 = -1 \tag{10.4}$$
$$\frac{1}{j} = -j$$

Euler's equation is complex, containing a real part (*cos*) and imaginary part (*sin*)—and the plus sign (+) in the equation is not a literal addition (you can't add real and imaginary numbers together). In a complex number of the format $a + jb$, the two parts coexist as part of one complex number.

So, what Euler's equation is really describing is a sine and cosine pair of functions, coexisting in two planes that are 90 degrees apart. The word *orthogonal* generally involves things that are 90 degrees apart or at right angles. Mathematically, two functions are orthogonal if their inner product is zero. When this happens, the two functions are maximally dissimilar. This means that mathematically speaking, *sin* and *cos* are as dissimilar as night and day, hot and cold, etc.

For our purposes, we replace the Θ in Euler's equation with ωt instead, where ω is the frequency in radians/second and t is the time variable. Plug in various values for t and you get the plot in Figure 10.1 when you plot the sine and cosine in the same plane. So, we will reject using the *sin()* and *cos()* functions independently and adopt the complex sinusoid as a prepackaged mixture of the two. Our motivation here is partly mathematical—as it turns out, e is simple to deal with mathematically, whereas *sin* and *cos* can become difficult once you begin manipulating them on higher mathematical levels. You only need to learn four rules in Equation 10.5 in addition to Euler's equation (also called *Euler's identity*).

$$\text{Euler's equation:}\quad e^{j\omega t} = \cos(\omega t) + j\sin(\omega t)$$

Polynomial behavior:

$$e^a e^b = e^{(a+b)} \quad \text{or} \quad e^{at} e^{bt} = e^{(a+b)t}$$

$$\frac{e^a}{e^b} = e^{(a-b)} \quad \text{or} \quad \frac{e^{at}}{e^{bt}} = e^{(a-b)t}$$

(10.5)

Calculus:

$$\frac{d(e^{at})}{dt} = ae^{at}$$

$$\int e^{at} dt = \frac{1}{a} e^{at}$$

The first two equations in Equation 10.5 demonstrate that e behaves like a polynomial ($x^2 * x^5 = x^7$) even when the argument is a function of time, t. The next pair of equations demonstrates how simply it behaves in calculus. Before we leave this topic, make sure you remember how to deal with complex numbers—you'll need it to understand where the frequency and phase responses come from.

10.2 Complex Math Review

Because a complex number has both real and imaginary parts, it cannot be plotted on a single axis. In order to plot a complex number, you need two axes—a real and imaginary axis. The two axes are aligned at right angles, just like the *x*- and *y*-axes in two-dimensional geometry. They are orthogonal. The *x*-dimension is the real axis (*Re*), and the *y*-dimension is the imaginary axis (*Im*). So, a complex number is plotted as a point on this *x*–*y* plane, also called the *complex plane*. Complex numbers are usually written in the form $a + jb$ where a is the real part and b is the imaginary part. Notice that the notation always stays in the form $a + jb$ even when common sense might contradict it. For the point in the second quadrant, you still write $-2 + j2$ even though $j2 - 2$ might look nicer. Also notice that I will write $2 - j1$ instead of just $2 - j$ to firmly identify that the imaginary component $b = 1$. The points are plotted the way you would plot (x,y) pairs in a plane as shown in Figure 10.2a.

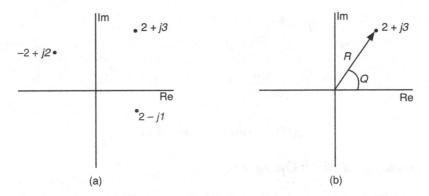

**Figure 10.2: Complex numbers plotted in the complex plane in
(a) Cartesian coordinates and (b) polar form.**

Figure 10.2a plots points as $(x + jy)$ pairs—this is called a Cartesian coordinate system. You can also use polar notation to identify the same complex numbers. In polar notation, you specify the radius (R) and angle (Θ) for the vector that results from drawing a line from the origin of the axes out to the point in question. For example, consider the point $2 + j3$ in Figure 10.2a. You could plot this point in polar form as in Figure 10.2b. We are leaving the Cartesian form $(2 + j3)$ for reference. Normally, all you would see is R and Θ. Most engineers prefer polar notation for dealing with complex numbers. The polar number is often written as $R < \Theta$ Fortunately for us, the conversion from Cartesian to polar notation is simple. Starting with a complex number in the form $a + jb$, you can find the resulting radius and angle with Equation 10.6. You may convert polar to coordinates to Cartesian coordinates with the second half of Equation 10.6. We will perform this type of conversion in the phase vocoder (pitch shifter) object in Chapter 20.

$$
\begin{array}{cc}
\text{(a, jb) to polar} & \text{polar to (a, jb)} \\[4pt]
R = \sqrt{a^2 + b^2} & a = R\cos(\Theta) \\[8pt]
\Theta = \tan^{-1}\left[\dfrac{b}{a}\right] & b = R\sin(\Theta)
\end{array}
\tag{10.6}
$$

The radius (R) is called the *magnitude* and the angle (Θ) is called the *argument, phase, phase angle,* or more simply *angle.* You sometimes only care about the square of the magnitude (called the *magnitude-squared* or $|R|^2$) as shown in Equation 10.7:

$$
|R|^2 = a^2 + b^2
\tag{10.7}
$$

The frequency response plots of filters are actually magnitude responses of a complex function called the *transfer function* of the filter. The phase response plots are actually argument responses of this function. The transfer function is complex because it contains complex numbers; many transfer functions are actually quite simple. We use the letter H to denote a transfer function. The transfer function relates the output of a system to its input (i.e. it describes the transfer of information from the input to the output).

Equation 10.8 shows how to extract the magnitude and phase from a transfer function; we'll use these simple relationships soon.

$$H = \frac{num}{denom}$$

then

$$|H| = \frac{|num|}{|denom|} \tag{10.8}$$

and

$$Arg(H) = Arg(num) - Arg(denom)$$

10.3 Time Delay as a Math Operator

The next piece of DSP theory you want to understand is the concept of time delay as a mathematical operator. This will be fairly easy since we are going to exploit the simple mathematical behavior of the $e^{j\omega t}$ function. First, consider an analog complex sinusoid and a delayed version of the same complex sinusoid as shown in Figure 10.3. Notice that these are analog waveforms and the x-axis is labeled in analog time t.

How does the delay of n seconds change the complex sinusoid equation? Since positive time goes in the positive x direction, a delay of n seconds is a shifting of $-n$ seconds. In the complex sinusoid equation, you would then replace t with $t - n$. In other words, any point on the delayed curve is the same as the non-delayed curve minus n seconds. Therefore, the delayed sinusoid is:

$$\text{delayed sinusoid} = e^{j\omega(t-n)} \tag{10.9}$$

But, by using the polynomial behavior of e and running Equation 10.5 in reverse, you can re-write it as:

$$e^{j\omega(t-n)} = e^{j\omega t} e^{-j\omega n} \tag{10.10}$$

It's a subtle mathematical equation but it says a lot: if you want to delay a complex sinusoid by n seconds, multiply it by $e^{-j\omega n}$—this allows us to express time delay as a mathematical operator.

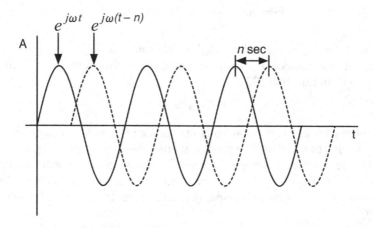

Figure 10.3: A complex sinusoid (solid line) and a delayed version of the same function (dotted line). The delay time is n seconds.

10.4 The Sampled Sinusoid

The sinusoid we've been discussing so far is technically an analog function of time t. The sampled version will be valid only on each sample interval T. If the sample rate $f_s = 44.1$ kHz, then its period T is ~22.676 microseconds. The first few sample intervals of a sinusoid that starts at 0.0 at time $t = 0$ would be {0.0μS, 22.7μS, 45.4μS}, which can be seen in Figure 10.4a as $e^{j\omega 0}$, $e^{j\omega 22.7uS}$ and, $e^{j\omega 45.4uS}$. Halving the sample rate to 22.050 kHz doubles the sampling interval and results in Figure 10.4b as $e^{j\omega 0}$, $e^{j\omega 45.5uS}$ and, $e^{j\omega 90.7uS}$.

The generic way to identify a sampled sinusoid is to replace the analog time variable t with the digitized nT, where n is a counter that starts at zero and counts up by one each simple interval, and T is the sample period. Then, the normal and delayed versions would be related by samples rather than continuous analog time. For M samples of delay, we could write:

$$\text{sampled complex sinusoid} = e^{j\omega nT}$$
$$\text{delayed sampled sinusoid} = e^{j\omega(nT-M)}$$

and

$$e^{j\omega(nT-M)} = e^{j\omega nT}e^{-j\omega M}$$
$$M = \text{samples of delay}$$

(10.11)

So, if you want to delay a digitized complex sinusoid by M samples, multiply it by $e^{-j\omega M}$.

Time delay can be expressed as a mathematical operator by multiplying the analog signal to be delayed n seconds by $e^{j\omega n}$. For the digitized signal to be delayed by M samples you multiply it by $e^{j\omega M}$. This is useful because neither $e^{j\omega n}$ nor $e^{j\omega M}$ is dependent on the time variables t or nT.

In the last two sections you've learned that phase rotation and time delay can both be expressed as mathematical operators.

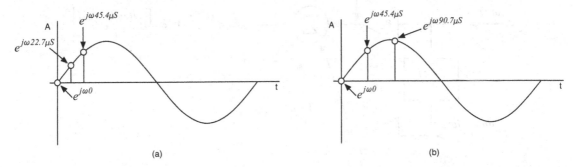

Figure 10.4: The first three samples of a sinusoid with (a) f_s = 44.1 kHz and (b) f_s = 22.050 kHz.

10.5 1st Order Feed-Forward Filter Revisited

Being able to express M samples of delay as the mathematical operation of multiplication by $e^{-j\omega M}$ means you can take the block diagram and difference equation for a DSP filter and apply a sinusoid to the input in the form of $e^{j\omega nT}$ rather than having to plug in sequences of samples as you did in Chapter 2. Then, you can see what comes out of the filter as a mathematical expression and evaluate it for different values of ω (where $\omega = 2\pi f$ with f in Hz) to find the frequency and phase responses directly rather than having to wait and see what comes out, and then try to guesstimate the amplitude and phase offsets. Consider the 1st order feed-forward filter from the last chapter but with $e^{j\omega nT}$ applied as the input signal, shown in Figure 10.5. In this case, the z^{-1} state register can be replaced with multiplication by $e^{-j\omega l}$ to implement the one-sample delay element.

The different equation that results from applying $e^{j\omega nT}$ as the input signal is:

$$y(nT) = a_0 e^{j\omega nT} + a_1 \left[e^{j\omega nT} e^{-j\omega 1} \right] \tag{10.12}$$

Now we can perform an (almost) amazing little mathematical stunt by separating the original input $x(nT)$ and the newly formed output $y(nT)$.

$$y(nT) = a_0 e^{j\omega nT} + a_1 \left[e^{j\omega nT} e^{-j\omega 1} \right]$$
$$= e^{j\omega nT} \left(a_0 + a_1 e^{-j\omega 1} \right)$$

and the input $x(nT) = e^{j\omega nT}$ so

$$y(nT) = x(nT)\left(a_0 + a_1 e^{-j\omega 1} \right) \tag{10.13}$$

The transfer function is defined as the ratio of output to input therefore

$$\frac{y(nT)}{x(nT)} = a_0 + a_1 e^{-j\omega 1}$$

What is so cool about this is that the transfer function is not dependent on time nT, even though the input and output signals are functions of time. In fact, the transfer function is dependent on frequency

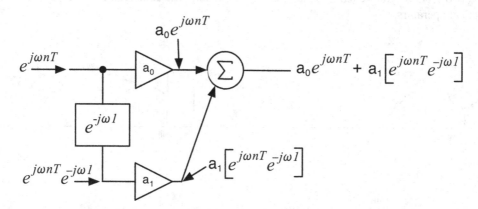

Figure 10.5: The 1st order feed-forward filter with a complex sampled sinusoid applied to the input.

ω only. That means we can re-write the transfer function and call it $H(\omega)$. Some engineers prefer to notate the transfer function as dependent on *complex* frequency, or $j\omega$, and they will write transfer functions as $H(j\omega)$ instead. Here we are going to stick with the simpler terminology and just call it $H(\omega)$.

$$H(\omega) = a_0 + a_1 e^{-j\omega 1} \tag{10.14}$$

The *transfer function* of the filter is the ratio of output to input.

The *frequency response* of the filter is the magnitude of the transfer function.

The *phase response* of the filter is the argument (or angle) of the transfer function.

To produce the frequency and phase response plots, you evaluate the function for various values of ω then find the magnitude and argument at each frequency.

The evaluation always uses Euler's equation to replace the **e** term and produce the real and imaginary components, which may then be converted to magnitude and argument with Equation 10.8.

But what values of ω are to be used in the evaluation? We know that $\omega = 2\pi f$, but do we really care about the frequency in Hz? In Chapter 2 when you analyzed the same filter, you applied DC, Nyquist, ½ Nyquist, and ¼ Nyquist without thinking about the actual sampling frequency. This is called *using the normalized frequency* and is usually the way you want to proceed in analyzing DSP filters. The actual sampling rate determines Nyquist, but the overall frequency range (0 Hz to Nyquist) is what we care about. To normalize the frequency, you let $f = 1$ Hz in $\omega = 2\pi f$, then ω varies from 0 to 2π or across a 2π range. There is also one detail we need to be aware of: negative frequencies.

10.5.1 Negative Frequencies

You may have never thought a frequency could be negative, but it can. When you first learned about the concept of a waveform's frequency, you were taught that the frequency is $1/T$ where T is the period, as shown here in Figure 10.6a. The reason the frequencies came out as positive numbers is because the period is defined as $t_2 - t_1$ that makes it a positive number. But, there's no reason you couldn't define the period to be the other way around: $T = t_1 - t_2$ (except that it implies that time is running backwards). Mathematically, time can run backwards. This means that for every positive frequency that exists, there is also a negative "twin" frequency. When you look at a frequency response plot, you generally look at the positive side only. Figure 10.6b shows a textbook low-pass response up to the highest frequency in the system: Nyquist. Notice that the frequency response shows magnitude of the transfer function, $|H(f)|$.

In reality, the filter also operates on the negative frequencies, just as in a mirror image. In this case, as the negative frequencies get higher and higher, they are attenuated just like their positive counterparts as in Figure 10.7a—and it makes sense, too. If you take an audio file and reverse it in time, then run it through a low-pass filter, the same frequency filtering still occurs. For filter evaluation, ω varies on a 0

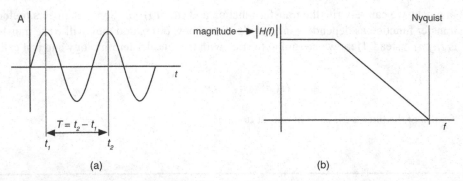

(a)

(b)

Figure 10.6: (a) The classic "textbook" definition of defining the period forces the resulting frequency to always be a positive value, as shown here. (b) The common way to show a frequency response plot it to only show the positive frequencies, or right-hand side.

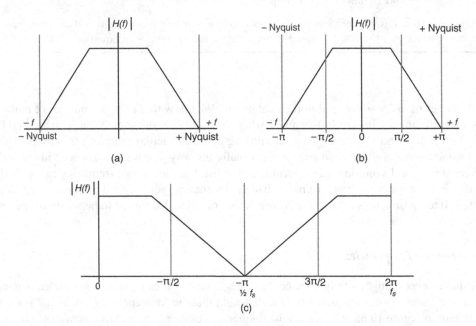

(a)

(b)

(c)

Figure 10.7: (a) The more complete frequency response plot shown here contains both positive and negative frequencies. (b) One way to divide up the 2π of frequencies includes both positive and negative halves. (c) We can also map the 0 to 2π of frequencies across the 0 to f_s range.

to 2π radians/second range; one way to think about this 2π range is to split it up into the range from $-\pi$ to $+\pi$, corresponding to –Nyquist to +Nyquist, as in Figure 10.7b.

The sampling theorem sets up the Nyquist criteria with regards to completely recovering the original, band-limited signal and without aliasing. But, frequencies above Nyquist and all the way up to the sampling frequency f_s are also allowed mathematically—and, in theory, any frequency could enter the system and you could sample it without limiting the maximum frequency component to Nyquist. For a frequency or phase response plot, the frequencies from Nyquist up to the sampling frequency are a mirror image about Nyquist. This is another way to divide up the 2π range by going from 0 Hz to the sampling frequency as shown in Figure 10.8c. Notice that in either method the same information is conveyed as we get both halves of the curves, and in both cases, Nyquist maps to π and 0 Hz to 0 and positive frequencies map to the range 0 to π.

To evaluate the transfer function, let ω vary from 0 to π and get the first half of the response. The other half is a mirror image of the data.

10.6 Evaluating the Transfer Function H(ω)

DSP filter transfer functions will contain $e^{-j\omega n}$ terms that need to be evaluated over the range of 0 to π. The way to do this is by using Euler's equation to decompose the sinusoid into its real (*cos*) and imaginary (*sin*) components. Then, evaluate the *cos* and *sin* terms at the frequency of evaluation. In the Chapter 9 we manually calculated the input/output relationship of a filter by cranking through the filter operation one step at a time: we did this four times, once for each input frequency ω. In this improved method, we need only solve the transfer function equation once, and then evaluate it at four different values for ω. Let's start with the block diagram in Figure 10.6, our old friend the 1st order feed-forward structure and its transfer function:

$$H(\omega) = a_0 + a_1 e^{-j\omega 1} \tag{10.15}$$

We will use the coefficients $a_0 = 0.5$ and $a_1 = 0.5$, and we will evaluate the transfer functions at the same frequencies as in Chapter 2, only this time we will use our π-based equivalents to denote the various fractions of the sample rate. We can pre-calculate the *sin* and *cos* terms, and create Table 10.1 to make operations simpler.

Evaluation is a two-step process for each frequency:

1. Use Euler's equation to convert the *e* terms into real and imaginary components.
2. Find the magnitude and argument of the complex equation at each evaluation frequency.

In each case, we will compare the results we got with the graphical results from Chapter 2, which are condensed into Figure 10.8.

Table 10.1: The various *sin* and *cos* evaluations
at frequencies between 0 and Nyquist

Test Signal	Ω	$\cos(\omega)$	$\sin(\omega)$
DC	0	1.0	0.0
¼ Nyquist	$\pi/4$	0.707	0.707
½ Nyquist	$\pi/2$	0.0	1.0
¾ Nyquist	$3\pi/4$	−0.707	0.707
Nyquist	π	−1.0	0.0

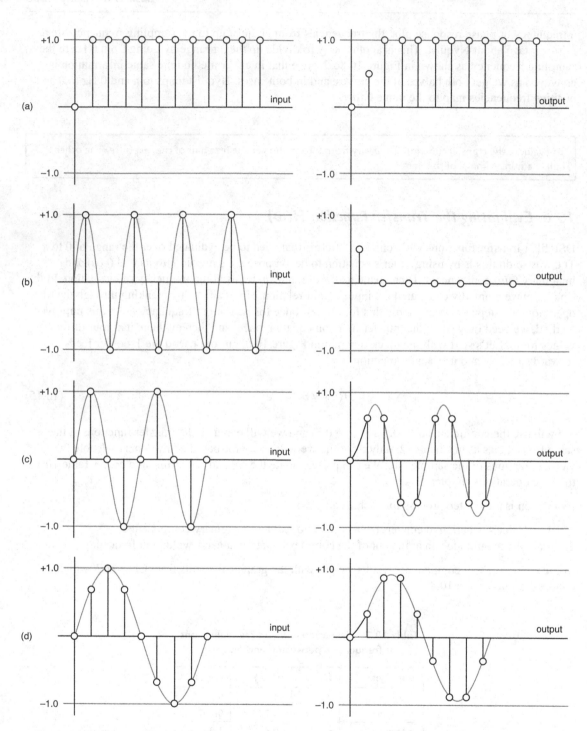

Figure 10.8: The graphical results we obtained from analysis of the 1st order feed-forward structure in Chapter 2 for (a) DC (0 Hz), (b) Nyquist, (c) ½ Nyquist, and (d) ¼ Nyquist.

10.6.1 DC (0 Hz)

First we apply the Euler equation to the e term, then we find the magnitude and phase using Equations 10.6 and 10.8.

$$H(\omega) = 0.5 + 0.5e^{-j\omega 1} \qquad\qquad \omega = 0$$

Apply Euler   Find Magnitude and Phase

$$
\begin{aligned}
H(\omega) &= 0.5 + 0.5(\cos(\omega) - j\sin(\omega)) & |H(\omega)| &= \sqrt{a^2 + b^2} \\
&= 0.5 + 0.5(\cos(0) - j\sin(0)) & &= \sqrt{1^2 + 0^2} \\
&= 0.5 + 0.5(1 - j0) & &= 1.0 \\
&= 0.5 + 0.5 - j0 & Arg(H) &= \tan^{-1}(b/a) \\
&= 1.0 - j0 & &= \tan^{-1}(0/1) = 0.0
\end{aligned}
\tag{10.16}
$$

Now compare these mathematical results with the results from the last chapter in Figure 10.8a. Our observed filter magnitude response at DC was unity gain (output = input) with a phase offset of 0 degrees (this is arguably inconsequential as a DC waveform has no phase). Now, let's repeat the process for the Nyquist frequency and compare that with our observation in Figure 10.8.

10.6.2 Nyquist (π)

Repeat the procedure for evaluation at $\omega = \pi$:

$$H(\omega) = 0.5 + 0.5e^{-j\omega 1} \qquad\qquad \omega = \pi$$

Apply Euler   Find Magnitude and Phase

$$
\begin{aligned}
H(\omega) &= 0.5 + 0.5(\cos(\omega) - j\sin(\omega)) & |H(\omega)| &= \sqrt{a^2 + b^2} \\
&= 0.5 + 0.5(\cos(\pi) - j\sin(\pi)) & &= \sqrt{0^2 + 0^2} \\
&= 0.5 + 0.5(-1 - j0) & &= 0.0 \\
&= 0.5 - 0.5 - j0 & Arg(H) &= \tan^{-1}(b/a) \\
&= 0 - j0 & &= \tan^{-1}(0/0) = 0.0
\end{aligned}
\tag{10.17}
$$

The inverse tangent argument is 0/0 and the phase or *Arg(H)* is defined to be zero under this condition. The C++ function you use is *arctan2(im,re)*, which performs the inverse tangent function; it will also evaluate to 0 in this case. Now, compare our results to the last chapter's graphical results, shown in Figure 10.8b. Once again, we've obtained the same results; the phase shift is equally negligible since there is no frequency component to shift as its magnitude is 0.0.

10.6.3 ½ Nyquist (π/2)

$$H(\omega) = 0.5 + 0.5e^{-j\omega 1} \qquad\qquad \omega = \pi/2$$

Apply Euler   Find Magnitude and Phase

$$
\begin{aligned}
H(\omega) &= 0.5 + 0.5(\cos(\omega) - j\sin(\omega)) & |H(\omega)| &= \sqrt{a^2 + b^2} \\
&= 0.5 + 0.5(\cos(\pi/2) - j\sin(\pi/2)) & &= \sqrt{0.5^2 + 0.5^2} \\
&= 0.5 + 0.5(0 - j1) & &= \sqrt{0.5} = 0.707 \\
&= 0.5 - j0.5 & Arg(H) &= \tan^{-1}(b/a) \\
& & &= \tan^{-1}(-0.5/0.5) \\
& & &= -45°
\end{aligned}
\tag{10.18}
$$

Compare this to the last chapter's graphical results in Figure 10.8c; the magnitude is 0.707 with a phase shift of −45 degrees: once again the results agree. Since 0.707 = −3 dB, it also means we've found the filter's cutoff or −3 dB frequency, f_c = ½ Nyquist.

10.6.4 ¼ Nyquist (π/4)

$$H(\omega) = 0.5 + 0.5e^{-j\omega 1} \qquad\qquad \omega = \pi/4$$

Apply Euler · Find Magnitude and Phase

$$H(\omega) = 0.5 + 0.5(\cos(\omega) - j\sin(\omega)) \qquad |H(\omega)| = \sqrt{a^2 + b^2}$$

$$= 0.5 + 0.5(\cos(\pi/4) - j\sin(\pi/4)) \qquad = \sqrt{0.853^2 + 0.353^2}$$

$$= 0.5 + 0.5(0.707 - j0.707) \qquad\qquad = 0.923$$

$$= 0.5 + 0.353 - j0.353 \qquad\qquad Arg(H) = \tan^{-1}(b/a)$$

$$= 0.853 - j0.353 \qquad\qquad = \tan^{-1}(-0.353/0.853)$$

$$= -22.5° \qquad\qquad (10.19)$$

Compare this to Figure 10.8c; you can see how much more accuracy we get with the mathematical calculation. The magnitude and phase shift look about right when compared to the graphs. Now, you can combine all the evaluations together and sketch out the frequency response of the filter in Figure 10.9—and behold, it is the same as what we got in Chapter 2, but far more accurate. Hopefully,

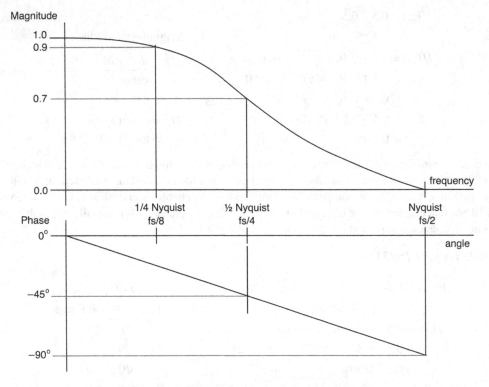

Figure 10.9: Our predicted frequency and phase response plots show the same results as Chapter 2, but with a lot less tedium.

this quick example has convinced you that it is better to do a little complex math than have to analyze and design these filters by brute force analysis of time domain input sequences.

10.7 Evaluating $e^{j\omega}$

In the evaluation of the transfer function, you had to substitute values of ω from 0 to π into the $e^{j\omega}$ terms of the equation. But what would the plot look like if you evaluate a single $e^{j\omega}$ term on its own? You saw that the use of Euler's equation produced the real and imaginary components of the term; now it's time to plot them over the range of 0 to π. If you evaluate $e^{j\omega}$ over more frequencies and plot the resulting values in the complex plane, you get an interesting result: the frequencies in Table 10.2 from 0 to $+\pi$ map to an arc that is the top half of a circle, with radius = 1.0 shown in Figure 10.10a. Remember, the magnitude is the radius and the argument is the angle when using polar notation, which simplifies the analysis. You don't have to keep track of the real and imaginary parts. The evaluation at $\omega = \pi/4$ is plotted on the curve. The circle this arc is laying over would have a radius of 1.0 and is called the *unit circle*. If you evaluate $e^{j\omega}$ over the negative frequencies that correspond to 0 to $-\pi$, you get a similar but inverted table (Table 10.3) that translates to a mapping across the lower half of the same unit circle, shown in Figure 10.10b. The negative frequencies increase as you move clockwise from 0 Hz, while the radius stays 1.0 during the entire arc.

Why bother to evaluate $e^{j\omega}$? It will be useful very soon when we start picking apart the transfer functions in an effort to figure out how to design filters. It also shows the limited "frequency space" of the digital domain. All the frequencies that could exist from −Nyquist to +Nyquist map to a simple unit circle. In contract, the analog domain has an infinitely long frequency axis and an infinite frequency space.

10.8 The z Substitution

It's going to get messy having to write $e^{j\omega}$ so often, and we know $e^{j\omega}$ behaves like a polynomial mathematically. So, we can simplify the equations by making a simple substitution of Equation 10.20.

$$z = e^{j\omega} \tag{10.20}$$

This is just a substitution right now and nothing else. Making the substitution in Equation 10.20 and noting the resulting transfer function is now a function of z, not ω, we can write it like this:

$$H(z) = a_0 + a_1 z^{-1} \tag{10.21}$$

The reason this is useful is that it turns the transfer function into an easily manipulated polynomial in z. In this case, the polynomial is a 1st order polynomial (the highest exponent absolute value is 1) and this is the real reason the filter is called a 1st order filter—it's the polynomial order of the transfer function.

Table 10.2: The magnitude and angle of $e^{j\omega}$ from DC to Nyquist

ω	$E^{j\omega} = cos(\omega) + jsin(\omega)$	$\lvert e^{j\omega} \rvert$	$Arg(e^{j\omega})$
DC (0 Hz)	$1 + j0$	1.0	0
¼ Nyquist	$0.707 + j0.707$	1.0	$\pi/4$
½ Nyquist	$0 + j1$	1.0	$\pi/2$
Nyquist	$-1 + j0$	1.0	π

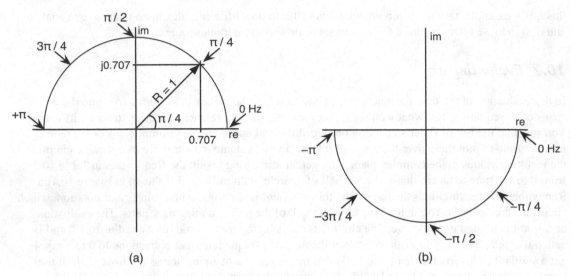

Figure 10.10: (a) The positive frequencies map to the upper half of the unit circle. (b) Negative frequencies map to the lower half.

Table 10.3: The magnitude and angle of $e^{j\omega}$ from DC to −Nyquist

ω	$e^{j\omega}cos(\omega) + jsin(\omega)$	$|e^{j\omega}|$	$Arg(e^{j\omega})$
DC (0 Hz)	1 + j0	1.0	0
−¼ Nyquist	0.707−j0.707	1.0	−π/4
−½ Nyquist	0−j1	1.0	−π/2
−Nyquist	−1 + j0	1.0	−π

> The *order* of a filter is the order of the polynomial in the transfer function that describes it. The order of the polynomial is the maximum absolute exponent value found in the equation.

10.9 The z Transform

The *z* Substitution does a good job at simplifying the underlying polynomial behavior of $e^{j\omega}$ and it lets us use polynomial math to solve DSP problems. But there is an interesting application of $z = e^{j\omega}$ that simplifies the design and analysis of digital filters. In Figure 10.11 you can see how the indexing of the samples determines their place in time. The future samples have positive indices and the past (delayed) samples have negative indices.

We've seen that $e^{j\omega}$ is the delay operator, so $e^{-j\omega 1}$ means one sample of delay, or one sample behind the current one. Likewise $e^{-j\omega 2}$ would be two samples behind and $e^{+j\omega 4}$ indicates four samples ahead or in the future. That means that $e^{j\omega}$ could also be used as a bookkeeping device since it can relate the position in time of something (a sample, for us).

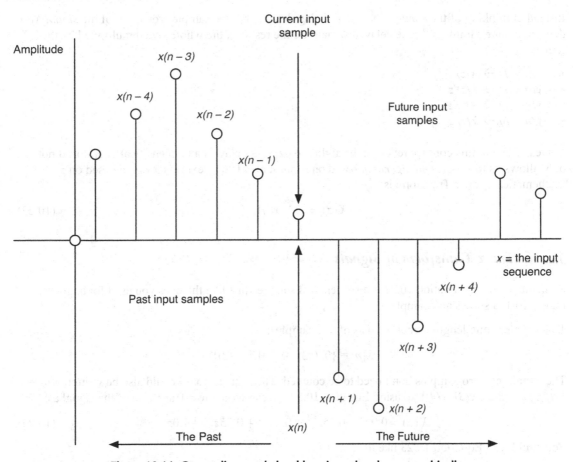

Figure 10.11: Our audio sample bookkeeping rules shown graphically.

The *z Transform* changes a sequence of samples in *n* to a sequence of samples in *z* by replacing the indices . . . n – 1, n, n+1 . . . with . . . z^{-1}, z, z^{+1} . . . This works because multiplication by z = $e^{j\omega}$ represents the operation of delay or time-shift.

The resulting transformed sequence now is a function of the complex frequency $e^{j\omega}$, therefore it transforms things from the *time domain* into the *complex frequency domain*.

The rules for implementing the *z* Transform on a discrete signal or difference equation are simple and can usually be followed by inspection. The current sample *x(n)* or *y(n)* transforms into the *signal X(z)* or *Y(z)*. Instead of thinking of the *sample x(n)* you think of the *signal X(z)* where *X(z)* is the whole signal—past, present, and future

- *x(n)* ➔ *X(z)*
- *y(n)* ➔ *Y(z)*

Instead of thinking of the *sample x(n − 1)* as being delayed by one sample, you think of the *signal X(z)* delayed by one sample, z^{-1}. The delayed signals are the result of the whole signal multiplied by the z^{-N} terms:

- $x(n − 1) \rightarrow X(z)\, z^{-1}$
- $y(n − 1) \rightarrow Y(z)\, z^{-1}$
- $x(n − 2) \rightarrow X(z)\, z^{-2}$
- $y(n + 6) \rightarrow Y(z)\, z^{+6}$

You can see that this concept relies on the ability to express delay as a mathematical operator. It not only allows us to express an *algorithm* based on z, but it also lets us express a *signal* based on z. Mathematically the *z* Transform is:

$$X(z) = \sum_{n=-\infty}^{n=+\infty} x(n)z^{-n} \tag{10.22}$$

10.10 The z Transform of Signals

Remember, *x(n)* in equation 10.22 is a sequence of samples just like the ones you used for analysis. Figure 10.12a shows an example.

This simple, finite length signal consists of five samples:

$$x(n) = \{0, 0.25, 0.5, 0.75, 1.0\}$$

The remaining zero samples don't need to be counted. The sequence *x(n)* could also be written *x(n)* = {x(0), x(1), x(2), x(3), x(4)} so using Equation 10.22 we can write the *z* Transform of the signal as:

$$X(z) = 0.0z^0 + 0.25z^{-1} + 0.5z^{-2} + 0.75z^{-3} + 1.0z^{-4} \tag{10.23}$$

You could read Equation 10.23 like this:

The whole signal, *X(z)* consists of a sample with an amplitude of 0.0 at time 0 followed by a sample with an amplitude of 0.25 one sample later and a sample with an amplitude of 0.5 two samples later and a sample with an amplitude of 0.75 three samples later and a sample with an amplitude of 1.0 four

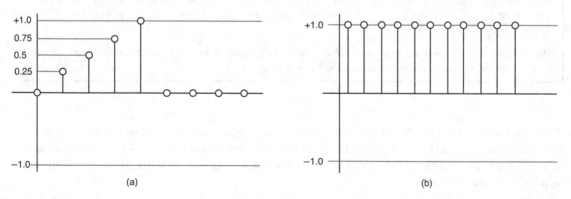

(a)   (b)

Figure 10.12: Two signals for the *z* Transform: (a) a finite ramp-burst of five samples, and (b) an infinite DC sequence of all 1.0 valued samples.

samples later. This should shed light on the fact that the transform really involves the book keeping of sample locations in time and that the result is a polynomial. You can multiply and divide this signal with other signals by using polynomial math. You can mix two signals by linearly combining the polynomials.

Let's do one more example regarding the transformation of an input signal. This time, let's choose the DC signal—it goes on forever. Figure 10.12b shows the DC signal with the first sample at 1.0 and all preceding samples at 0.0 with a sequence of $x(n) = \{1.0, 1.0, 1.0, 1.0, 1.0 \ldots\}$. Using Equation 10.22 (and remembering that $1z^0 = 1$) we can directly write the z Transform as:

$$X(z) = 1 + z^{-1} + z^{-2} + z^{-3} + \ldots + z^{-\infty} \tag{10.24}$$

While that looks like an ugly, infinitely long equation, it can also be represented in a closed form. In fact, a closed form representation exists for this polynomial using a polynomial series expansion:

$$\begin{aligned} X(z) &= 1 + z^{-1} + z^{-2} + z^{-3} + \ldots + z^{-\infty} \\ &= \frac{1}{1 - z^{-1}} \\ &= \frac{z}{z - 1} \end{aligned} \tag{10.25}$$

10.11 The z Transform of Difference Equations

The z Transform of signals is interesting, but something fascinating happens when you take the z Transform of a difference equation, converting it into a transfer function in z all at once—and the same simple rules apply. Let's do that with the basic 1st order feed-forward filter:

$$y(n) = a_0 x(n) + a_1 x(n-1)$$

Taking the z transform:

$$\begin{aligned} Y(z) &= a_0 X(z) + a_1 X(z) z^{-1} \\ Y(z) &= X(z)[a_0 + a_1 z^{-1}] \\ H(z) &= \frac{Y(z)}{X(z)} = a_0 + a_1 z^{-1} \end{aligned} \tag{10.26}$$

This is a really interesting result—you got the final transfer function in just a handful of steps, using the simple z Transform rules. Let's recap what you had to do before you learned how to take the z Transform of a difference equation:

1. Redraw the block diagram with $e^{-j\omega n}$ operators in the n-sample delay elements.
2. Apply the complex sinusoid $e^{-j\omega nT}$ to the input.
3. Find out what comes out, $y(nT)$ and formulate the transfer function $H(\omega)$.
4. Apply the z Substitution to the transfer function.

Taking the z Transform does all these steps at once, and we're left with the same simple polynomial in z. If we evaluate the transfer function for different values of $z = e^{j\omega}$ we can find the frequency and phase plots. You'll get more practice taking the z Transforms of more difference equations soon.

10.12 The z Transform of an Impulse Response

Taking the z Transform of a difference equation results in the transfer function. But what if you don't have the difference equation? Suppose you only have a black box that performs some kind of DSP algorithm and you'd like to figure out the transfer function, evaluate it, and plot the frequency and phase responses. In many cases, this can be done without knowing the algorithm or any details about it by taking the impulse response of the system. You apply the input sequence $x(n) = \{1,0,0,0,0\ldots\}$ and capture what comes out, which we'll call $h(n)$. If you take the z Transform of the impulse response, you get the transfer function expanded out into a series form.

> The **z Transform** of the impulse response $h(n)$ is the transfer function $H(z)$ as a series expansion. Evaluate the transfer function to plot the frequency and phase responses.

In fact, this is exactly what the RackAFX software's audio analyzer does: it takes the impulse response of the filter and then runs a z Transform on it to make the magnitude and phase plots you see on the screen. Mathematically Equation 10.27 is identical to Equation 10.22 except that I've changed the signal from X to H:

$$H(z) = \sum_{n=-\infty}^{n=+\infty} h(n)z^{-n} \tag{10.27}$$

Try this on the 1st order feed-forward filter we've been working on; you already have the impulse response "captured" from the last chapter in Figure 10.13. The impulse response is:

$$h(n) = \{0.5, 0.5\}$$

Applying the z Transform yields:

$$\begin{aligned} H(z) &= 0.5z^0 + 0.5z^{-1} \\ &= 0.5 + 0.5z^{-1} \end{aligned} \tag{10.28}$$

Notice that this is the identical result as taking the z Transform of the difference equation and the filter coefficients $(0.5, 0.5)$ *are* the impulse response $\{0.5, 0.5\}$.

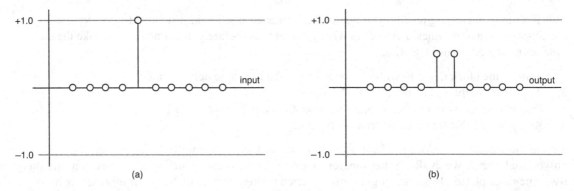

Figure 10.13: The impulse response input (a) and output (b) for the 1st order feed-forward filter of Chapter 2.

10.13 The Zeros of the Transfer Function

When we used the coefficients $a_0 = a_1 = 0.5$ in our 1st order feed-forward structure, we wound up with a filter that completely destroys the Nyquist frequency—you saw how its output became 0 in both the manual and complex sinusoid evaluations. We noted that feed-forward filters have *zeros of transmission* or *zero frequencies* (or just *zeros*) when their output becomes zero. In both the manual and complex sinusoid evaluations, we just got lucky when we stumbled upon this value since Nyquist happened to be one of the signals we were evaluating. There's a way to precisely find these critical frequencies by using the polynomial result of the z Transform. You probably remember factoring polynomials in high school or college. When you did that, you set the polynomial equal to 0 and then you factored to find the roots of the polynomial. What you were really doing was finding the zeros of the polynomial, that is, the values of the independent variable that make the polynomial become zero. You can do the same thing with the transfer function by setting it equal to zero and then factoring the polynomial. Suppose $a_0 = a_1 = 0.5$ and we factor the transfer function in Equation 10.28 to arrive at the following variation:

$$H(z) = 0.5 + 0.5z^{-1}$$
$$= 0.5 + \frac{0.5}{z} \tag{10.29}$$

You can find the zero by inspection: it's the value of z that forces $H(z)$ to be 0. In this case we say that "there is a *zero* at $z = -1.0$." But what it does really mean to have a zero at -1.0? This is where the concept of evaluating $e^{j\omega}$ comes into play. When you did that and plotted the various points, noting they were making a unit circle in the complex plane, you were actually working in the z-plane (i.e. the plane of $e^{j\omega}$). The location of the zero at $z = -1.0$ is really at the location $z = -1.0 + j0$, and is purely on the real axis and at Nyquist. In Figure 10.14 the zero is shown as a small circle sitting at the location $z = -1.0$.

There are several reasons to plot the zero frequencies. First, you can design certain filters directly in the z-plane by deciding where you want to place the zero frequencies first, then figuring out the transfer function that will give you those zeros. Second, plotting the zeros gives you a quick way to sketch the frequency response without having to evaluate the transfer function directly. You can estimate a phase plot too, but it is a bit more involved. So, you have two really good reasons for wanting to plot the zeros: one for design and the other for analysis.

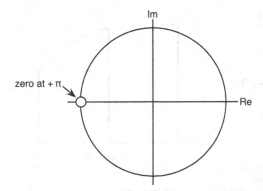

Figure 10.14: Zero is plotted in the z-plane as a small filled circle located at $z = -1 + j0$.

10.14 *Estimating the Frequency Response: Zeros*

An interesting property of the z-plane and z Transform is that you can measure the frequency response graphically on the z-plane. In the simplest case of only one zero, the method is as follows:

1. Locate each evaluation frequency on the outer rim of the unit circle.
2. Draw a line from the point on the circle to the zero and measure the length of this vector. Do it for each evaluation frequency.
3. The *lengths* of the lines will be the *amplitudes* at each frequency in the frequency response.

In Figure 10.15 you can see the complete operation: first drawing and measuring the lines (you can use graph paper and a ruler, if you want) then building the frequency response plot from them. Notice also that the magnitude of a line drawn from Nyquist to the zero at −1 has a length of zero. The lengths of the vectors *are* the mathematical definition of magnitudes at the various evaluation frequencies; you are evaluating the whole filter at once. These z-plane plots are going to be useful for filter design. You can also derive the phase response, but it's a bit more tedious and involves summing the angles of incidence of each vector on the zero. With multiple zeros, it becomes cumbersome—but estimating the frequency response is fairly simple, even for more complex filters.

You might notice that even though this frequency response looks like the one we produced earlier, the gain values are not the same. In this filter, the gain is 2.0 at DC; in ours, it's half of that. In fact, this filter's magnitudes at the evaluation frequencies are all twice what ours are. In the next section, we will see why.

10.15 *Filter Gain Control*

The last thing we need to do is remove the overall gain factor from the transfer function so that overall filter gain (or attenuation) can be controlled by just one variable. This is actually fairly simple to do, but it requires re-working the transfer function a bit. The idea is to pull out the a_0 variable as a multiplier for the whole function. This way, it behaves like a volume knob, gaining

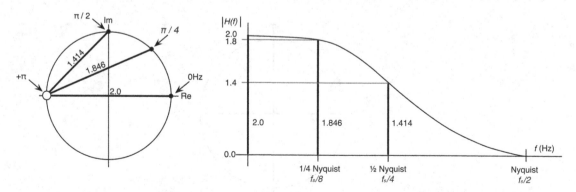

Figure 10.15: The geometric method shows how the length of each vector from the evaluation frequency to the zero is really a magnitude in the final response.

the whole filter up or down. The way you do it is to normalize the filter by dividing out a_0 as follows:

$$H(z) = a_0 + a_1 z^{-1}$$

$$= a_0 \left[1 + \frac{a_1}{a_0} z^{-1} \right]$$

$$\text{Let } \alpha_1 = \frac{a_1}{a_0}$$

$$H(z) = a_0 \left[1 + \alpha_1 z^{-1} \right]$$

(10.30)

By factoring out the a_0 coefficient and using the α_1 terminology we can produce a transfer function that looks basically the same in the polynomial but pulls a_0 out as a scalar multiplier—a gain control.

Where is the zero of this new transfer function?

$$H(z) = a_0 \left[1 + \alpha_1 z^{-1} \right] = a_0 \left[1 + \frac{\alpha_1}{z} \right]$$

(10.31)

Once again, by inspection of Equation 10.31 we can see that if $z = -\alpha_1$ then the function will become 0 regardless of the value of a_0. This transfer function has a zero at $z = -\alpha_1$. If we plug our values of $a_0 = a_1 = 0.5$, we still get the same zero at $z = -1.0$. The difference is in the gain of the filter.

> The *graphical interpretation* method of evaluating a filter in the z-plane assumes the filter is normalized so that $a_0 = 1.0$.

So the last step you need to do after extracting the magnitude response from the z-plane plot is to scale it by your a_0 value. This makes the response in Figure 10.24 match ours because everything gets multiplied by $a_0 = 0.5$. The idea of controlling the gain independently of the magnitude response is useful in audio filters so we will keep it and use it in all the further analyses.

At this point, you have all the DSP theory tools you need to understand the rest of the classical DSP filters (1st order feedback, and 2nd order feed-forward and feedback) as well as many filter design techniques. The rest of the chapter will be devoted to applying these same fundamentals to the other classical DSP filters—but we will move much more quickly, applying each analysis technique to the other algorithms. For example, we will dispense with the evaluation of $e^{j\omega}$ terms and start off directly in the z Transform of the difference equations.

10.16 1st Order Feedback Filter Revisited

Now let's go through the same analysis technique on the 1st order feedback filter from the last chapter. We can move much more quickly now that we have the basic DSP theory down. There will be many similarities but also several key differences when dealing with feedback designs. You already saw that the feedback topology can blow up or ring forever and that the feed-forward design cannot. We will find a way to figure out if this is going to happen and how to prevent it. Start with the original 1st order feedback filter in Figure 10.16a and its difference equation.

$$y(n) = a_0 x(n) - b_1 y(n-1) \tag{10.32}$$

10.16.1 Step 1: Take the z Transform of the Difference Equation

This can be done by inspection, using the rules from Section 10.9, or it can be done by pushing the transformed input $X(z)$ through the structure as shown in Figure 10.16b, noting the intermediate nodes, and finding the result $Y(z)$. Using either method, the z Transform of the difference equation is:

$$Y(z) = a_0 X(z) - b_1 Y(z) z^{-1} \tag{10.33}$$

Notice that when we push the input samples $x(n)$ through the filter to form output samples $y(n)$, movement through the delay register changes the indexing from n to $n-1$, but when processing the entire input signal $X(z)$ through the filter to make $Y(z)$, movement through the delay register multiplies it by z^{-1}. Then, we can easily get the difference equation in z.

10.16.2 Step 2: Fashion the Difference Equation into a Transfer Function

Now apply some algebra to convert the transformed difference equation to $H(z)$. The process is always the same: separate the $X(z)$ and $Y(z)$ variables, then form their quotient.

$$Y(z) = a_0 X(z) - b_1 Y(z) z^{-1}$$

Separate variables:

$$Y(z) + b_1 Y(z) z^{-1} = a_0 X(z)$$
$$Y(z)[1 + b_1 z^{-1}] = a_0 X(z)$$
$$Y(z) = \frac{a_0 X(z)}{1 + b_1 z^{-1}} \tag{10.34}$$

From $H(z)$:

$$H(z) = \frac{Y(z)}{X(z)} = \frac{a_0}{1 + b_1 z^{-1}}$$

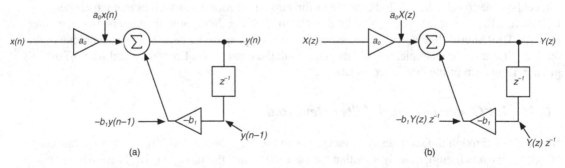

Figure 10.16: (a) Processing samples $x(n)$ through the 1st order feedback structure and (b) processing the input $X(z)$ through the filter allows us to easily find its z Transform.

10.16.3 Step 3: Factor Out a_0 as the Scalar Gain Coefficient

In this case, this step is simple since pulling a_0 out is trivial. However, in more complex filters this requires making substitutions as we did in the last section.

$$H(z) = \frac{a_0}{1 + b_1 z^{-1}}$$

$$= a_0 \frac{1}{1 + b_1 z^{-1}}$$

(10.35)

10.17 The Poles of the Transfer Function

The next step in the sequence is to do a quick estimation of the frequency response using the graphical interpretation method in the z-plane. The pure feed-forward filter we analyzed produced zeros of transmission or zeros at frequencies where its output becomes zero. A pure feedback filter produces *poles* at frequencies where its output becomes *infinite*. We were able to make this happen by applying 100% feedback in the last chapter. For the simple 1st order case, finding the poles is done by inspection.

When the denominator of the transfer function is zero, the output is infinite. The complex frequency where this occurs is the *pole frequency* or *pole*.

Examining the transfer function, we can find the single pole in Equation 10.36.

$$H(z) = a_0 \frac{1}{1 + b_1 z^{-1}}$$

$$= a_0 \frac{1}{1 + \dfrac{b_1}{z}} = a_0 \frac{z}{z + b_1}$$

(10.36)

By re-arranging the transfer function, you can see that the denominator will be zero when $z = -b_1$, so there is a pole at $z = -b_1$. You might also notice something interesting about this transfer function: it has a z in the numerator. If $z = 0$, then this transfer function has a zero at $z = 0$. This zero is called a *trivial zero* because it has no impact on the filter's performance (frequency response). So, you can ignore the zero at $z = 0$. In fact, you can also ignore poles at $z = 0$ for the same reason.

A pole or zero at $z = 0$ is trivial and can be ignored for the sake of analysis since it has no effect on the frequency response.

If you look back at the transfer function in the feed-forward filter, you can see that it also had a trivial pole in Equation 10.37.

$$H(z) = a_0\left[1 + \alpha_1 z^{-1}\right]$$

$$= a_0\left[1 + \frac{\alpha_1}{z}\right] \tag{10.37}$$

pole at $z = 0$!

The poles are plotted in the z-plane in the same manner as the zeros but you use an x to indicate the pole frequency. In the 1st order feedback case, the pole is at $-b_1 + j0$ and so it is a real pole located on the real axis in the z-pane. For this filter let's analyze it with $a_0 = 1.0$ and $b_1 = 0.9$. The measured frequency response of this filter is shown in Figure 10.17a. Let's see how this is estimated first, and then we can do a direct evaluation as before.

10.17.1 Step 4: Estimate the Frequency Response

The single pole is plotted on the real axis at $z = -0.9 + j0$ and a trivial zero at $z = 0 + j0$. This is shown in Figure 10.17b. In the future, we will ignore the trivial zeros or poles.

In the simplest case of only one pole, the method for estimating the frequency response is as follows:

1. Locate each evaluation frequency on the outer rim of the unit circle.
2. Draw a line from the point on the circle to the pole and measure the length of this vector. Do it for each evaluation frequency.
3. The *inverse of the lengths* of the lines will be the *magnitudes* at each frequency in the frequency response, as shown in Equation 10.38.

$$\text{Magnitude} = \frac{1}{L_p} \tag{10.38}$$

a. $L_p = $ length from evaluation frequency to pole

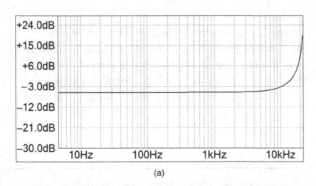

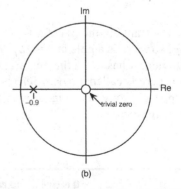

(a)   (b)

Figure 10.17: (a) The measured frequency response of the 1st order feedback structure with *fs* = 44.1 kHz, *a0* = 1.0 and b_1 = 0.9. (b) The pole/zero plot of the algorithm.

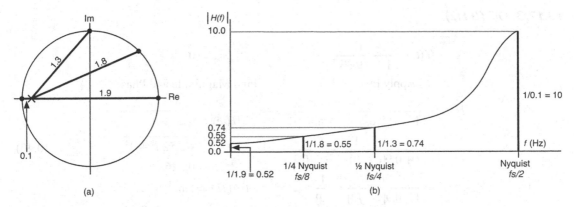

Figure 10.18: (a) Graphical evaluation of the pole/zero plot. (b) The estimated frequency response of our 1st order feedback filter.

So, the mechanism is the same as for the one-zero case, except you take the inverse of the length of the vector. This means that as you near the pole, the vector becomes shorter, but the magnitude becomes larger—exactly opposite of the zero case. The four evaluation frequencies and their vectors are shown in Figure 10.18a and our predicted frequency response is shown in Figure 10.18b. Note the gain at Nyquist is 10.0: convert that to dB and you get +20 dB of gain, which is what we observe in the measured response in Figure 10.19a.

10.17.2 Step 5: Perform Direct Evaluation of the Frequency Response

Now we can evaluate the filter the same way as before, using Euler's equation to separate the real and imaginary components from the transfer function.

Evaluate at the following frequencies:

- DC: 0
- Nyquist: π
- ½ Nyquist: $\pi/2$
- ¼ Nyquist: $\pi/4$

Begin by getting the transfer function in a form to use for all the evaluation frequencies, in Equation 10.39. Also note that $e^{j\omega}$—this is where Euler's equation will come into play.

$$H(z) = a_0 \frac{1}{1 + b_1 z^{-1}}$$

$$a_0 = 1.0 \quad b_1 = 0.9$$

$$H(z) = \frac{1}{1 + 0.9 z^{-1}}$$

$$z = e^{j\omega}$$

(10.39)

10.17.3 DC (0 Hz)

$$H(\omega) = \frac{1}{1 + 0.9e^{-j\omega 1}} \qquad \omega = 0$$

Apply Euler   Find Magnitude and Phase

$$H(\omega) = \frac{1}{1 + 0.9[\cos(\omega) - j\sin(\omega)]}$$

$$= \frac{1}{1 + 0.9[\cos(0) - j\sin(0)]}$$

$$|H(\omega)| = \sqrt{a^2 + b^2}$$

$$= \sqrt{0.526^2 + 0^2} \qquad (10.40)$$

$$= \frac{1}{1 + 0.9[1 - j0]} = \frac{1}{1.9}$$

$$= 0.526$$

$$Arg(H) = \tan^{-1}(b/a)$$

$$= 0.526 - j0$$

$$= \tan^{-1}(0/0.526)$$

$$= 0.0$$

10.17.4 Nyquist (π)

$$H(\omega) = \frac{1}{1 + 0.9e^{-j\omega 1}} \qquad \omega = \pi$$

Apply Euler   Find Magnitude and Phase

$$H(\omega) = \frac{1}{1 + 0.9[\cos(\omega) - j\sin(\omega)]}$$

$$= \frac{1}{1 + 0.9[\cos(\pi) - j\sin(\pi)]}$$

$$|H(\omega)| = \sqrt{a^2 + b^2}$$

$$= \sqrt{10.0^2 + 0^2} \qquad (10.41)$$

$$= 10.0$$

$$= \frac{1}{1 + 0.9[-1 - j0]} = \frac{1}{0.1}$$

$$Arg(H) = \tan^{-1}(b/a)$$

$$= 10 - j0$$

$$= \tan^{-1}(0/10.0)$$

$$= 0.0$$

10.17.5 ½ Nyquist (π/2)

$$H(\omega) = \frac{1}{1 + 0.9e^{-j\omega 1}} \qquad \omega = \pi/2$$

Apply Euler   Find Magnitude and Phase

$$H(\omega) = \frac{1}{1 + 0.9[\cos(\omega) - j\sin(\omega)]}$$

$$|H(\omega)| = \frac{|\text{num}|}{|\text{denom}|}$$

$$= \frac{1}{1 + 0.9[\cos(\pi/2) - j\sin(\pi/2)]}$$

$$|H(\omega)| = \frac{|1|}{|1 - j0.9|} = \frac{1}{\sqrt{1^2 + 0.9^2}} \qquad (10.42)$$

$$= \frac{1}{1 + 0.9[0 - j1]}$$

$$= 0.743$$

$$Arg(H) = Arg(\text{num}) - Arg(\text{denom})$$

$$= \frac{1}{1 - j0.9}$$

$$= \tan^{-1}(0/1) - \tan^{-1}(-0.9/1)$$

$$= 0 - (-42)$$

$$= +42°$$

10.17.6 ¼ Nyquist (π/4)

$$H(\omega) = \frac{1}{1+0.9e^{-j\omega 1}} \qquad \omega = \pi/4$$

Apply Euler

Find Magnitude and Phase

$$H(\omega) = \frac{1}{1+0.9[\cos(\omega)-j\sin(\omega)]}$$

$$|H(\omega)| = \frac{|\text{num}|}{|\text{denom}|}$$

$$= \frac{1}{1+0.9[\cos(\pi/4)-j\sin(\pi/4)]}$$

$$|H(\omega)| = \frac{|1|}{|1.636-j0.636|} = \frac{1}{\sqrt{1.636^2+0.636^2}} \qquad (10.43)$$

$$= \frac{1}{1+0.636-j0.636}$$

$$= 0.57$$

$$= \frac{1}{1.636-j0.636}$$

$$Arg(H) = Arg(\text{num}) - Arg(\text{denom})$$

$$= \tan^{-1}(0/1) - \tan^{-1}(-0.636/1.636)$$

$$= 0 - (-21)$$

$$= +21^\circ$$

Make a special note about how we have to handle the magnitude of a fraction with numerator and denominator. You need to use the two equations in [10.8] to deal with this. The main issue is that the phase is the difference of the *Arg(numerator)* and the *Arg(denominator)*. If the numerator were a complex number instead of 1.0, we would need to take the magnitude of it separately, then divide. The calculated and measured response plots are shown in Figures 10.19a and b. You can see that the phase behaves linearly until it gets near the pole, then behaves nonlinearly. This is not a linear phase filter.

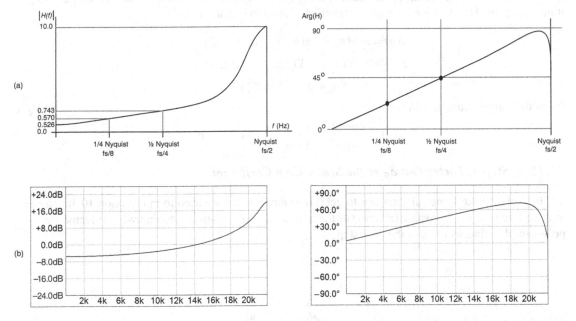

Figure 10.19: (a) The 1st order feedback filter's calculated frequency response (left) and phase response (right). (b) The filter's measured frequency response (left) and phase response (right) with *fs* = 44.1 kHz. Note that the measured frequency response magnitude is in dB, while it is in raw values for the calculated response.

In Figure 10.19 we observe excellent agreement with our evaluation: the response is down −6 dB at DC (0.52) and +20 dB at Nyquist (10), and the phase is 45 degrees at $\pi/2$.

We followed five steps in the direct evaluation of this filter:

1. Take the z Transform of the difference equation.
2. Fashion the difference equation into a transfer function.
3. Factor out a_0 as the scalar gain coefficient.
4. Estimate the frequency response.
5. Perform direct evaluation of the frequency response.

10.18 2nd Order Feed-Forward Filter

Analysis of the 2nd order feed-forward filter proceeds much like the 1st order filters you've seen so far, but there's a bit more math we have to deal with. The topology of a 2nd order feed-forward filter is shown in Figure 10.20.

10.18.1 Step 1 and Step 2: Take the z Transform of the Difference Equation and Fashion it into a Transfer Function

We can combine steps to save time. The difference equation in samples, or z transformed, may be pulled from Figures 10.20a and b respectively as follows.

$$y(n) = a_0 x(n) + a_1 x(n-1) + a_2 x(n-2) \tag{10.44}$$

The z Transform can be taken by inspection, using the rules from Section 10.9, and then we need to get it into the form $Y(z)/X(z)$ for the transfer function, shown in Equation 10.45.

$$y(n) = a_0 x(n) + a_1 x(n-1) + a_2 x(n-2)$$
$$Y(z) = a_0 X(z) + a_1 X(z)z^{-1} + a_2 X(z)z^{-2}$$
$$= X(z)[a_0 + a_1 z^{-1} + a_2 z^{-2}]$$

Form the transfer function $H(z)$

$$H(z) = \frac{\text{output}}{\text{input}} = \frac{Y(z)}{X(z)} = a_0 + a_1 z^{-1} + a_2 z^{-2} \tag{10.45}$$

10.18.2 Step 3: Factor Out a_0 as the Scalar Gain Coefficient

We'll need to make some substitutions to get this in the form we are used to in Equation 10.46. Notice that we force the leading value (z^0 term) of the polynomial to 1.0—this is so that we can factor the polynomial to find its roots in the next step.

$$H(z) = a_0 + a_1 z^{-1} + a_2 z^{-2}$$

$$\text{Let: } \alpha_1 = \frac{a_1}{a_0} \qquad \alpha_2 = \frac{a_2}{a_0} \tag{10.46}$$

$$H(z) = a_0(1 + \alpha_1 z^{-1} + \alpha_2 z^{-2})$$

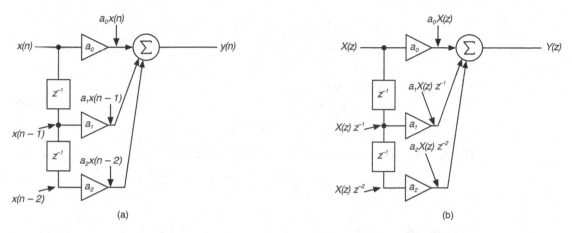

Figure 10.20: (a) The 2nd order feed-forward structure with discrete samples *x(n)* and *y(n)* as input and output. (b) The structure with entire signals *X(z)* and *Y(z)* as input and output.

10.18.3 Step 4: Estimate the Frequency Response

First, this is a pure feed-forward filter so we know there will only be non-trivial zeros; there are no poles to deal with. This transfer function is a 2nd order function because of the z^{-2} term. In fact, this is a quadratic equation. In order to find the poles or zeros, you need to first factor this equation and find the roots. The problem is that this is a complex equation, and the roots could be real, imaginary, or a combination of both. The mathematical break that we get is that our coefficients α_1 and α_2 are *real numbers*. The only way that could work out is if the locations of the zeros are complex conjugates of one another. When you multiply complex conjugates together the imaginary component disappears. So, with a bit of algebra, you can arrive at the deduction shown in Equation 10.47.

$$H(z) = 1 + \alpha_1 z^{-1} + \alpha_2 z^{-2}$$

can be factored as:

$$H(z) = (1 - Z_1 z^{-1})(1 - Z_2 z^{-1})$$

where

(10.47)

$$Z_1 = Re^{j\Theta} = a + jb$$
$$Z_2 = Re^{-j\Theta} = a - jb$$

This analysis results in two zeros, Z_1 and Z_2 located at complex conjugate positions in the z-plane. Figure 10.21a shows an arbitrary conjugate pair of zeros plotted in the z-plane. You can see how they are at complementary angles to one another with the same radii. Remember, any arbitrary point in the z-plane is located at $Re^{j\Theta}$ and the outer rim of the circle is evaluated for $R = 1$ and Θ between $-\pi$ and $+\pi$ (or 0 to 2π).

But how do the complex conjugate pair of zeros at $Re^{j\Theta}$ and $Re^{-j\Theta}$ relate to the coefficients α_1 and α_2? The answer is to just multiply everything out, use Euler's equation, and compare functions as demonstrated in Equations [10.48] and [10.49].

$$H(z) = a_0(1 + \alpha_1 z^{-1} + \alpha_2 z^{-2})$$
$$= a_0(1 - Z_1 z^{-1})(1 - Z_2 z^{-1})$$

where

$$Z_1 = Re^{j\Theta}$$
$$Z_2 = Re^{-j\Theta} \qquad\qquad\qquad \text{[10.48]}$$
$$\text{multiplying } (1 - Z_1 z^{-1})(1 - Z_1 z^{-1})$$

$$(1 - Z_1 z^{-1})(1 - Z_2 z^{-1}) = (1 - Re^{j\Theta} z^{-1})(1 - Re^{-j\Theta} z^{-1})$$

$$= 1 - Re^{j\Theta} z^{-1} - Re^{-j\Theta} z^{-1} + R^2(e^{j\Theta} e^{-j\Theta})z^{-2}$$

noting that $(e^{j\Theta} e^{-j\Theta}) = e^{j\Theta - j\Theta}$
$$= e^0 = 1$$
$$H(z) = 1 - (Re^{j\Theta} + Re^{-j\Theta})z^{-1} + R^2 z^{-2}$$
$$= 1 - R(\cos(\Theta) + j\sin(\Theta) + \cos(\Theta) - j\sin(\Theta))z^{-1} + R^2 z^{-2}$$
$$= 1 - 2R\cos(\Theta)z^{-1} + R^2 z^{-2}$$

compare functions: $\qquad\qquad\qquad\qquad\qquad$ [10.49]
$$H(z) = a_0(1 + \alpha_1 z^{-1} + \alpha_2 z^{-2})$$
$$= a_0(1 - 2R\cos(\Theta)z^{-1} + R^2 z^{-2})$$

then

$$\alpha_1 = -2R\cos(\Theta)$$
$$\alpha_2 = R^2$$

Equations [10.48] and [10.49] show how the coefficients α_1 and α_2 create the zeros at the locations $Re^{j\Theta}$ and $Re^{-j\Theta}$. Once again you see that the coefficients *are* the filter: they determine the locations of the zeros, and these determine the frequency and phase responses of the filter. To demonstrate how to estimate the frequency response using both the graphical interpretation and direct calculation methods we'll need some coefficients to test with. Use the following:

$$a_0 = 1.0$$
$$a_1 = -1.27$$
$$a_2 = 0.81$$

Now, calculate the location of the zeros from Equation 10.49; since $a_0 = 1.0$, then $\alpha_1 = -1.27$ and $\alpha_2 = 0.81$. Start with α_2 then solve for α_1 shown in Equation 10.50. The zeros are plotted in Figure 10.21b.

$$R^2 = \alpha_2 = 0.81$$
$$R = \sqrt{0.81} = 0.9$$

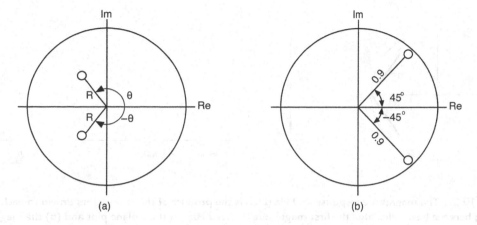

Figure 10.21: (a) An arbitrary conjugate pair of zeros plotted in the z-plane at radius R and angle Θ. (b) The complementary pair of zeros at radii = 0.9 and angles ±45 degrees.

then

$$-2R\cos(\Theta) = -1.27$$
$$2(0.9)\cos(\Theta) = 1.27$$
$$\cos(\Theta) = \frac{1.27}{2(0.9)} \quad\quad (10.50)$$
$$\Theta = \arccos(0.705)$$
$$\Theta = 45°$$

Evaluating the frequency response of the complex pair is similar to before, but with an extra step. When estimating the frequency response with more than one zero, do the following.

1. Locate each evaluation frequency on the outer rim of the unit circle.
2. Draw a line from the point on the circle to *each* zero and measure the length of these vectors. Do it for each evaluation frequency.
3. For each evaluation frequency, the magnitude of the transfer function is the *product of the two vectors* to each zero pair.

Mathematically, this last rule looks like Equation 10.51:

$$\left| H(e^{j\omega}) \right|_{\omega} = a_0 \prod_{i=1}^{N} U_i \quad\quad (10.51)$$

where

N = filter order
U_i = geometric distance from the point ω on the unit circle to the ith pole

Follow the progression in Figures 10.22–10.25 through the four evaluation frequencies, starting at DC (0 Hz). In each figure, the z-plane plot is shown in (a) and a vertical line is placed in the corresponding

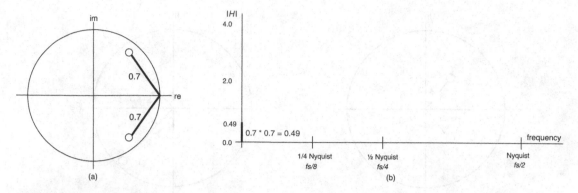

(a)

(b)

Figure 10.22: The magnitude response at 0 Hz (DC) is the product of the two vectors drawn to each zero, or 0.49; here we have calculated the first magnitude line at 0 Hz; (a) the *z*-plane plot and (b) the magnitude response at DC plotted as a vertical line.

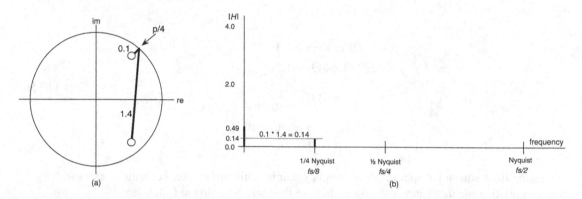

(a)

(b)

Figure 10.23: At ¼ Nyquist, the two vectors multiply out to 0.14; (a) the z-plane plot and (b) the magnitude response at ¼ Nyquist plotted as a vertical line.

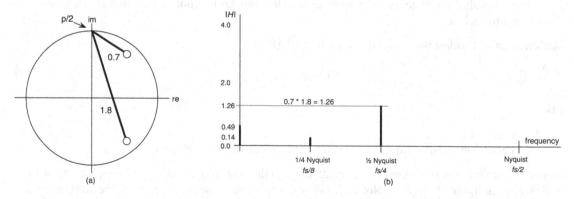

(a)

(b)

Figure 10.24: At ½ Nyquist, the response reaches 1.26 as the vectors begin to stretch out again; (a) the z-plane plot and (b) the magnitude response at ½ Nyquist plotted as a vertical line

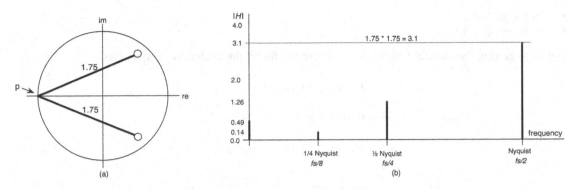

Figure 10.25: At Nyquist, the response reaches a maximum of 3.1 as the vectors stretch out to their longest possible lengths; (a) the z-plane plot and (b) the magnitude response at Nyquist plotted as a vertical line.

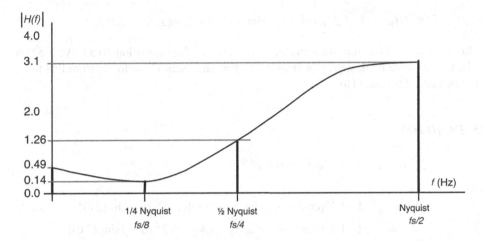

Figure 10.26: The combined response reveals a band-stop (notch) type of filter.

location in the response plots in (b). Finally, put it all together in one plot and sketch the magnitude response in Figure 10.26. The analysis shows a notch filter. The minimum amplitude occurs at the zero frequency, where the vector product is the lowest—this is where the smallest vector is obtained when evaluating on the positive frequency arc.

10.18.4 Step 5: Direct Evaluation

Now we can evaluate the filter the same way as before: using Euler's equation to separate the real and imaginary portions from the transfer function, then finding the magnitude and phase components to generate a plot.

Evaluate at the following frequencies:

* DC: 0
* Nyquist: π

- ½ Nyquist: $\pi/2$
- ¼ Nyquist: $\pi/4$

Begin by getting the transfer function in a form to use for all the evaluation frequencies.

$$H(z) = a_0(1 + \alpha_1 z^{-1} + \alpha_2 z^{-2})$$

$$\text{Where: } \alpha_1 = \frac{a_1}{a_0} \qquad \alpha_2 = \frac{a_2}{a_0}$$

$$H(z) = 1 - 1.27z^{-1} + 0.81z^{-2} \qquad (10.52)$$

$$\text{Let: } z = e^{j\omega}$$

$$H(\omega) = 1 - 1.27e^{-j1\omega} + 0.81e^{-j2\omega}$$

Apply Euler's Equation:

$$H(\omega) = 1 - 1.27[\cos(\omega) - j\sin(\omega)] + 0.81[\cos(2\omega) - j\sin(2\omega)]$$

Now evaluate for each of our four frequencies, starting with DC in Equation 10.53. We are leaving out some of the intermediate math steps now; make sure you understand how to get from line to line and work it out by hand if you need to.

10.18.5 DC (0 Hz)

$$H(\omega) = 1 - 1.27e^{-j1\omega} + 0.81e^{-j2\omega}$$

Apply Euler

$$H(\omega)\big|_{\omega=0} = 1 - 1.27[\cos(\omega) - j\sin(\omega)] + 0.81[\cos(2\omega) - j\sin(2\omega)]$$

$$= 1 - 1.27[\cos(0) - j\sin(0)] + 0.81[\cos(2*0) - j\sin(2*0)]$$

$$= 0.54 + j0$$

Find Magnitude and Phase   (10.53)

$$|H(\omega)| = \sqrt{a^2 + b^2}$$

$$= \sqrt{0.54^2 + 0^2} = 0.54$$

$$Arg(H) = \tan^{-1}(b/a)$$

$$= \tan^{-1}(0/0.54) = 0.0°$$

The exact magnitude is 0.54, which is reasonably close to our estimated value of 0.49 using the z-plane graphical method.

10.18.6 Nyquist (π)

$$H(\omega) = 1 - 1.27e^{-j1\omega} + 0.81e^{-j2\omega}$$

$$H(\omega) = 1 - 1.27e^{-j1\omega} + 0.81e^{-j2\omega}$$

Apply Euler

$$
\begin{aligned}
H(\omega)\big|_{\omega=\pi} &= 1 - 1.27[\cos(\omega) - j\sin(\omega)] + 0.81[\cos(2\omega) - j\sin(2\omega)] \\
&= 1 - 1.27[\cos(\pi) - j\sin(\pi)] + 0.81[\cos(2\pi) - j\sin(2\pi)] \\
&= 3.08 + j0
\end{aligned}
$$

Find Magnitude and Phase

(10.54)

$$
\begin{aligned}
|H(\omega)| &= \sqrt{a^2 + b^2} \\
&= \sqrt{3.08^2 + 0^2} = 3.08 \\
Arg(H) &= \tan^{-1}(b/a) \\
&= \tan^{-1}(0/3.08) = 0.0°
\end{aligned}
$$

The exact magnitude is 3.08, which is close to our estimated value of 3.1 using the z-plane graphical method.

10.18.7 ½ Nyquist (π/2)

$$H(\omega) = 1 - 1.27e^{-j1\omega} + 0.81e^{-j2\omega}$$

Apply Euler

$$
\begin{aligned}
H(\omega)\big|_{\omega=\pi/2} &= 1 - 1.27[\cos(\omega) - j\sin(\omega)] + 0.81[\cos(2\omega) - j\sin(2\omega)] \\
&= 1 - 1.27[\cos(\pi/2) - j\sin(\pi/2)] + 0.81[\cos(\pi) - j\sin(\pi)] \\
&= 0.19 + j1.27
\end{aligned}
$$

Find Magnitude and Phase

(10.55)

$$
\begin{aligned}
|H(\omega)| &= \sqrt{a^2 + b^2} \\
&= \sqrt{0.19^2 + 1.27^2} = 1.28 \\
Arg(H) &= \tan^{-1}(b/a) \\
&= \tan^{-1}(1.27/0.19) = 82°
\end{aligned}
$$

The exact magnitude is 1.28, which is close to our estimated value of 1.26 using the z-plane graphical method.

10.18.8 ¼ Nyquist (π/4)

$$H(\omega) = 1 - 1.27e^{-j1\omega} + 0.81e^{-j2\omega}$$

Apply Euler

$$
\begin{aligned}
H(\omega)\big|_{\omega=\pi/2} &= 1 - 1.27[\cos(\omega) - j\sin(\omega)] + 0.81[\cos(2\omega) - j\sin(2\omega)] \\
&= 1 - 1.27[\cos(\pi/4) - j\sin(\pi/4)] + 0.81[\cos(\pi/2) - j\sin(\pi/2)] \\
&= 0.11 + j0.08
\end{aligned}
$$

(10.56)

Find Magnitude and Phase

$$|H(\omega)| = \sqrt{a^2 + b^2}$$
$$= \sqrt{0.11^2 + 0.08^2} = 0.136$$
$$Arg(H) = \tan^{-1}(b/a)$$
$$= \tan^{-1}(0.08/0.11) = 36°$$

The exact magnitude is 0.136, which is reasonably close to our estimated value of 0.14 using the z-plane graphical method.

10.18.9 z Transform of Impulse Response

This 2nd order feed-forward filter is actually easy to examine for its impulse response. For an impulse stimulus, the impulse response $h(n)$ is: $h(n) = \{1.0, -1.27, 0.81\}$. Taking the z Transform of the impulse response is easy, as shown in Equation 10.57.

$$H(z) = \sum_{n=-\infty}^{n=+\infty} h(n)z^{-n}$$
$$= 1.0z^0 - 1.27z^{-1} + 0.81z^{-2} \tag{10.57}$$
$$= 1 - 1.27z^{-1} + 0.81z^{-2}$$

This is exactly what we expect. This should help you understand two more very important details about pure feed-forward filters.

In a pure feed-forward filters:

- The coefficients $\{a_0, a_1, a_2, \ldots\}$ are the impulse response, $h(n)$.
- The transfer function is the z Transform of the coefficients.

Figure 10.27a shows the frequency response while Figure 10.27b shows the phase response for this example. Both our estimated and direct calculations match very closely with the actual filter's behavior.

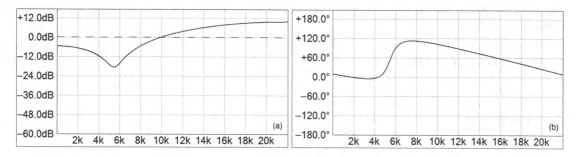

Figure 10.27: Measured responses for the 2nd order feed-forward filter with fs = 44.1 kHz, a_0 = 1.0, a_1 = −1.27, a_2 = 0.81 for (a) frequency response and (b) phase response.

10.19 2nd Order Feedback Filter

Analysis of the 2nd order feedback filter starts with the block diagram and difference equation. Figure 10.28a shows the topology of a 2nd order feedback filter. The difference equation is as follows.

$$y(n) = a_0 x(n) - b_1 y(n-1) - b_2 y(n-2) \qquad (10.58)$$

10.19.1 Step 1, Step 2, and Step 3: Take the z Transform of the Difference Equation to Get the Transfer Function, Then Factor Out a_0 as the Scalar Gain Coefficient

We'll continue to combine steps. Once again, the z Transform can be taken by inspection, using the rules in Section 10.9 or from pushing $X(z)$ through the filter, as shown in Figure 10.28b. Then we need to get it into the form $Y(z)/X(z)$ for the transfer function in Equation 10.59.

$$y(n) = a_0 x(n) - b_1 y(n-1) - b_2 y(n-2)$$

Take the z Transform:

$$Y(z) = a_0 X(z) - b_1 Y(z) z^{-1} - b_2 Y(z) z^{-2}$$

Separate variables:

$$Y(z) + b_1 Y(z) z^{-1} + b_2 Y(z) z^{-2} = a_0 X(z)$$
$$Y(z)[1 + b_1 z^{-1} + b_2 z^{-2}] = a_0 X(z) \qquad (10.59)$$

Form transfer function

$$H(z) = \frac{Y(z)}{X(z)} = \frac{a_0}{1 + b_1 z^{-1} + b_2 z^{-2}}$$

Factor out a_0

$$H(z) = a_0 \frac{1}{1 + b_1 z^{-1} + b_2 z^{-2}}$$

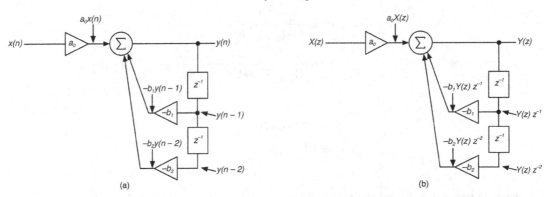

Figure 10.28: The 2nd order feedback structure with (a) discrete samples x(n) and y(n) as input and output and (b) the entire signals X(z) and Y(z) as input and output.

10.19.2 Step 4: Estimate the Frequency Response

Now you can finally see why the feedback filter block diagrams have all the b coefficients negated. This puts the quadratic denominator in the final transfer function in Equation 10.59 in the same polynomial form as the numerator of the feed-forward transfer function, matching. Equation 10.45. So, we can use the same logic to find the poles of the filter: since the coefficients b_1 and b_2 are real values, the poles must be complex conjugates of each other, as derived in Equation 10.60.

$$H(z) = a_0 \frac{1}{1 + b_1 z^{-1} + b_2 z^{-2}}$$

can be factored as:

$$H(z) = a_0 \frac{1}{(1 - P_1 z^{-1})(1 - P_2 z^{-1})}$$

where

$$P_1 = Re^{j\Theta}$$
$$P_2 = Re^{-j\Theta}$$

multiplying out $(1 - P_1 z^{-1})(1 - P_1 z^{-1})$ gives:

$$(1 - Re^{j\Theta} z^{-1})(1 - Re^{-j\Theta} z^{-1})$$
$$= 1 - Re^{j\Theta} z^{-1} - Re^{-j\Theta} z^{-1} + R^2 (e^{j\Theta} e^{-j\Theta}) z^{-2}$$

noting that $(e^{j\Theta} e^{-j\Theta}) = e^{j\Theta - j\Theta} = e^0 = 1$

$$= 1 - (Re^{j\Theta} + Re^{-j\Theta}) z^{-1} + R^2 z^{-2} \qquad (10.60)$$
$$= 1 - R(\cos(\Theta) + j\sin(\Theta) + \cos(\Theta) - j\sin(\Theta)) z^{-1}$$
$$+ R^2 z^{-2}$$
$$= 1 - 2R\cos(\Theta) z^{-1} + R^2 z^{-2}$$

compare functions:

$$H(z) = a_0 \frac{1}{1 + b_1 z^{-1} + b_2 z^{-2}}$$
$$= a_0 \frac{1}{(1 - 2R\cos(\Theta) z^{-1} + R^2 z^{-2})}$$

then

$$b_1 = -2R\cos(\Theta)$$
$$b_2 = R^2$$

This results in two poles, P_1 and P_2, located at complex conjugate positions in the z-plane. Figure 10.29a shows an arbitrary conjugate pair of poles plotted in the z-plane. You can see how they

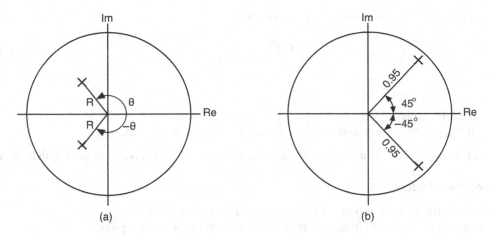

Figure 10.29: (a) An arbitrary conjugate pair of poles plotted in the z-plane at radius R and angle Θ. (b) The complementary pair of zeros at radii = 0.95 and angles ±45 degrees.

are at complementary angles to one another with the same radii. To demonstrate, we'll need some coefficients to test with. Use the following:

$$a0 = 1.0$$
$$b1 = -1.34$$
$$b2 = 0.902$$

Now, calculate the location of the poles from Equation 10.61. Figure 10.29b shows the complex conjugate pair of poles plotted in the z-plane at angles ±45 degrees and radii of 0.95.

$$R^2 = b_2 = 0.902$$
$$R = \sqrt{0.902} = 0.95$$

then

$$-2R\cos(\Theta) = -1.34$$
$$2(0.95)\cos(\Theta) = 1.34 \qquad (10.61)$$
$$\cos(\Theta) = \frac{1.34}{2(0.95)}$$
$$\Theta = \arccos(0.707)$$
$$\Theta = 45°$$

Evaluating the frequency response of the complex pair is similar to before, but with an extra step. When estimating the frequency response with more than one pole, do the following.

1. Locate each evaluation frequency on the outer rim of the unit circle.
2. Draw a line from the point on the circle to *each* pole and measure the length of these vectors. Do it for each evaluation frequency.
3. For each evaluation frequency, the magnitude of the transfer function is the *product of the inverse lengths* of the two vectors to each pole-pair.

Mathematically, this last rule looks like Equation 10.62.

$$\left|H(e^{j\omega})\right|_{\omega} = a_0 \frac{1}{\displaystyle\prod_{i=1}^{N} V_i}$$

(10.62)

where

N = the filter order
V_i = the geometric length from the point (ω) on the unit circle to the ith pole

Thus, the process is the same as with the zeros, except that you take the inverse of the length to the pole.

For feed-forward filters

- The closer the frequency is to the zero, the more attenuation it receives.
- If the zero is *on* the unit circle, the magnitude would go to zero at that point.

For feedback filters

- The closer the evaluation frequency is to the pole, the more gain it receives.
- If a pole is *on* the unit circle, the magnitude would theoretically go to infinity, and it would produce an oscillator, ringing forever at the pole frequency.

In a digital filter:

- Zeros may be located anywhere in the *z*-plane—inside, on, or outside the unit circle—since the filter is always stable and its output can't go lower than 0.0.
- Poles must be located inside the unit circle.
- If a pole is on the unit circle, it produces an oscillator.
- If a pole is outside the unit circle, the filter blows up as the output goes to infinity.

We blew up the 1st order feedback filter as an exercise in Chapter 2—all feedback filters are prone to blowing up when their poles go outside the unit circle. We can now continue with the estimation process for our standard four evaluation frequencies. This time, we'll convert the raw magnitude values into dB. The reason for doing this is that there will be a very wide range of values that will be difficult to sketch if we don't use dB. Follow the evaluation sequence in Figures 10.30–10.33.

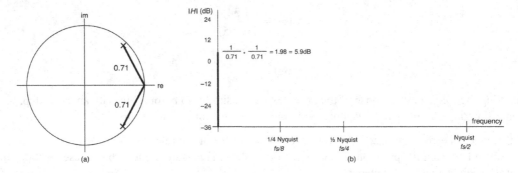

Figure 10.30: The magnitude response at 0 Hz (DC) is the product of the inverse of the two vectors drawn to each zero, or (1/0.71)(1/0.71) = 5.9 dB; (a) the *z*-plane plot and (b) the magnitude response at DC plotted as a vertical line.

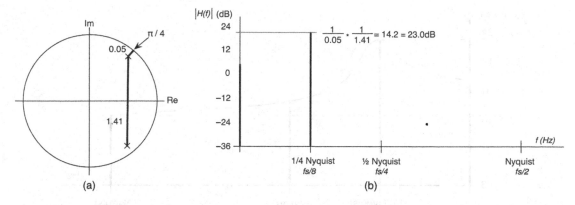

(a)

(b)

Figure 10.31: The magnitude response at π/4 is a whopping +23 dB since the inverse of 0.05 is a large number; (a) the z-plane plot and (b) the magnitude response at π/4 plotted as a vertical line.

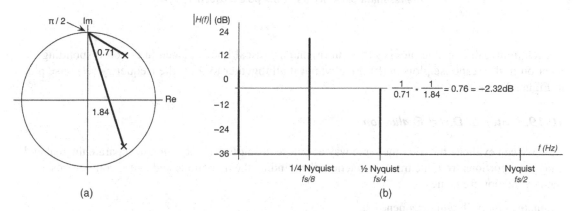

(a)

(b)

Figure 10.32: The magnitude response at π/2 is −2.98 dB; (a) the z-plane plot and (b) the magnitude response at π plotted as a vertical line.

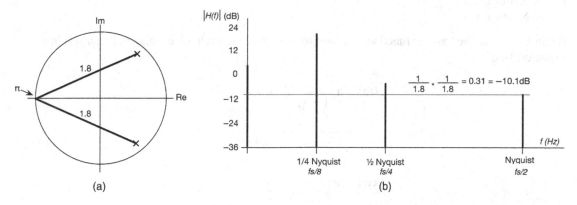

(a)

(b)

Figure 10.33: The magnitude response at π is −10.1 dB; (a) the z-plane plot and (b) the magnitude response at π plotted as a vertical line.

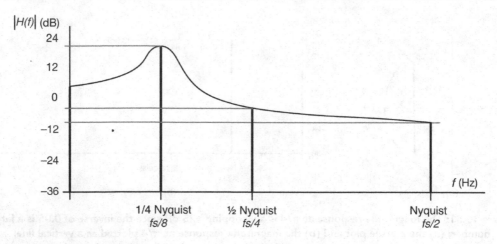

Figure 10.34: The composite magnitude response of the filter shows that it is a resonant low-pass filter; the resonant peak occurs at the pole frequency, $\pi/4$.

In each figure, the z-plane plot is shown in (a) and a vertical line is placed in the corresponding location in the response plots in (b). Finally, put it all together to form the frequency response plot in Figure 10.34.

10.19.3 Step 5: Direct Evaluation

Now we can evaluate the filter the same way as before: using Euler's equation to separate the real and imaginary portions from the transfer function, then finding the magnitude and phase components at each evaluation frequency.

Evaluate at the following frequencies:

- DC: 0
- Nyquist: π
- ½ Nyquist: $\pi/2$
- ¼ Nyquist: $\pi/4$

Begin by getting the transfer function in a form to use for all the evaluation frequencies shown in Equation 10.63.

$$H(z) = a_0 \frac{1}{1 + b_1 z^{-1} + b_2 z^{-2}}$$

$$= \frac{1}{1 - 1.34 z^{-1} + 0.902 z^{-2}} \tag{10.63}$$

Let: $z = e^{j\omega}$

$$H(\omega) = \frac{1}{1 - 1.34 e^{-j1\omega} + 0.902 e^{-j2\omega}}$$

Apply Euler's equation:

$$H(\omega) = \frac{1}{1 - 1.34e^{-j1\omega} + 0.902e^{-j2\omega}}$$

$$H(\omega) = \frac{1}{1 - 1.4[\cos(\omega) - j\sin(\omega)] + 0.902[\cos(2\omega) - j\sin(2\omega)]}$$

Now, evaluate for each of our four frequencies starting with DC.

10.19.4 DC (0 Hz)

$$
\begin{aligned}
H(\omega) &= \frac{1}{1 - 1.34[\cos(\omega) - j\sin(\omega)] + 0.902[\cos(2\omega) - j\sin(2\omega)]} \\[2mm]
&= \frac{1}{1 - 1.34[\cos(0) - j\sin(0)] + 0.902[\cos(2*0) - j\sin(2*0)]} \\[2mm]
&= \frac{1}{1 - 1.34[1 - j0] + 0.902[1 - j0]} \\[2mm]
&= \frac{1}{1 - 1.34 + 0.902} = \frac{1}{0.562 + j0}
\end{aligned}
$$

(10.64)

Finding the magnitude and phase requires examining both the numerator and denominator as with the 2nd order feed-forward case.

$$
\begin{aligned}
H(\omega) &= \frac{1}{0.562 + j0} \\[2mm]
|H(\omega)| &= \frac{|\text{numerator}|}{|\text{denominator}|} = \frac{|1|}{|0.562 + j0|} \\[2mm]
&= \frac{1}{\sqrt{a^2 + b^2}} \\[2mm]
&= \frac{1}{\sqrt{0.562^2}} \\[2mm]
&= 1.78 = 5.00dB
\end{aligned}
$$

(10.65)

$$
\begin{aligned}
Arg(H) &= Arg(Num) - Arg(Denom) \\
&= \tan^{-1}(0/1) - \tan^{-1}(0/0.562) \\
&= 0^\circ
\end{aligned}
$$

Remember that for magnitudes of fractions, you need to take the magnitude of the numerator and denominator separately; and for phase responses, the final *Arg(H)* is the difference of the *Arg(numerator)* and *Arg(denominator)*. The direct evaluation yields 5.00 dB and shows our sketch evaluation was a little off at 5.9 dB.

10.19.5 Exercise

Finish the rest of the direct evaluation calculations on your own. The answers are in Table 10.4.

Figure 10.35a and b show the measured frequency and phase responses for this 2nd order feedback filter.

10.20 1st Order Pole/Zero Filter: The Shelving Filter

The 1st order pole/zero filter consists of a 1st order pole and 1st order zero in the same algorithm. The topology in Figure 10.36a is a combination of feed-forward and feedback components since the structure contains both paths. Audio engineers usually refer to this as a *shelving filter,* while DSP engineers usually call it a *pole/zero (p/z) filter*.

Table 10.4: Exercise answers

Frequency (ω)	$|H(\omega)|$	Arg(H)
Nyquist (π)	−10.2 dB	0.0°
½ Nyquist ($\pi/2$)	−2.56 dB	−85.8°
¼ Nyquist ($\pi/4$)	+23.15 dB	−40.3°

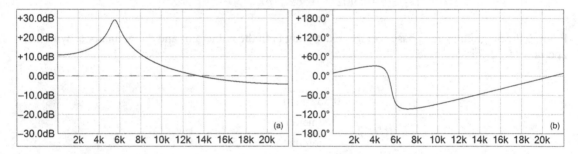

Figure 10.35: Measured responses for the 2nd order feedback filter with fs = 44.1 kHz, a_0 = 1.0, b_1 = −1.34, b_2 = 0.902 for (a) frequency response and (b) phase response.

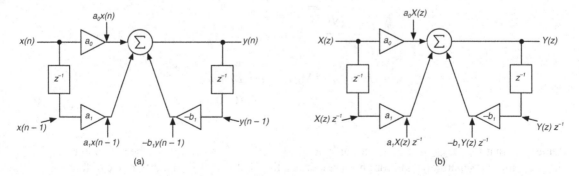

Figure 10.36: The 1st order pole/zero filter with (a) discrete samples $x(n)$ and $y(n)$ as input and output. (b) The entire signals $X(z)$ and $Y(z)$ as input and output.

The difference equation is:

$$y(n) = a_0 x(n) + a_1 x(n-1) - b_1 y(n-1) \tag{10.66}$$

10.20.1 Step 1, Step 2, and Step 3: Take the z Transform of the Difference Equation to Get the Transfer Function, Then Factor Out a_0 as the Scalar Gain Coefficient

Once again, the z Transform can be taken by inspection, using the rules in Section 10.9 or from pushing $X(z)$ through the filter, as shown in Figure 10.36b, and then we need to get it into the form $Y(z)/X(z)$ for the transfer function in Equation 10.67.

$$Y(z) = a_0 X(z) + a_1 X(z)z^{-1} - b_1 Y(z)z^{-1}$$

separate variables:

$$Y(z) + b_1 Y(z)z^{-1} = a_0 X(z) + a_1 X(z)z^{-1}$$

$$Y(z)\left[1 - b_1 z^{-1}\right] = X(z)\left[a_0 + a_1 z^{-1}\right]$$

form the transfer function: $\hspace{6cm}$ (10.67)

$$H(z) = \frac{Y(z)}{X(z)} = a_0 \frac{1 + \alpha_1 z^{-1}}{1 + b_1 z^{-1}}$$

where

$$\alpha_1 = \frac{a_1}{a_0}$$

10.20.2 Step 4: Estimate the Frequency Response

This transfer function has one pole and one zero; both are 1st order. Like the other 1st order cases, we can find the pole and zero by inspection of the transfer function in Equation 10.68.

$$H(z) = a_0 \frac{1 + \alpha_1 z^{-1}}{1 + b_1 z^{-1}}$$

$$= a_0 \frac{1 + \dfrac{\alpha_1}{z}}{1 + \dfrac{b_1}{z}} \tag{10.68}$$

In the numerator, you can see that if $z = -\alpha_1$ the numerator will go to zero and the transfer function will go to zero. In the denominator, you can see that if $z = -b_1$ the denominator will go to zero and the transfer function will go to infinity. Therefore we have a zero at $z = -\alpha_1$ and a pole at $z = -b_1$.

For this example, use the following values for the coefficients:

$$a_0 = 1.0$$
$$a_1 = -0.92$$
$$b_1 = -0.71$$

Then, $\alpha_1 = -0.92$; so we now have a zero at $z = -\alpha_1 = 0.92 + j0$ and a pole at $z = -b_1 = 0.71 + j0$. The pole/zero pair are plotted in Figure 10.37.

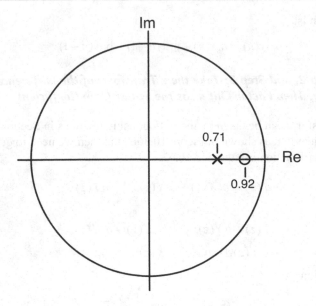

Figure 10.37: The pole and zero are both purely real and plotted on the real axis in the z-plane.

Evaluating the frequency response when you have mixed poles and zeros is the same as before, but you have to implement both magnitude steps.

1. Locate each evaluation frequency on the outer rim of the unit circle.
2. Draw a line from the point on the circle to *each* zero and measure the length of these vectors. Do it for each evaluation frequency.
3. Draw a line from the point on the circle to *each* pole and measure the length of these vectors. Do this for each evaluation frequency.
4. Multiply all the zero magnitudes together.
5. Multiply all the *inverse* pole magnitudes together.
6. Divide the zero magnitude by the pole magnitude for the final result at that frequency.

Mathematically, this last rule looks like Equation 10.69.

$$\left|H(e^{j\omega})\right|_{\omega} = \frac{a_0 \prod\limits_{i=1}^{N} U_i}{\prod\limits_{i=1}^{N} V_i} \tag{10.69}$$

where

    N = the filter order
    U_i = the geometric length from the point (ω) on the unit circle to the ith zero
    V_i = the geometric length from the point (ω) on the unit circle to the ith pole

Equation 10.69 is the final, generalized magnitude response equation for the geometric interpretation of the pole/zero plots in the z-plane for any number of poles and zeros. For completeness, here's the equation for calculating the phase response of a digital filter using the geometric method:

$$Arg(H(e^{j\omega}))\big|_{\omega} = \sum\limits_{i=1}^{N} \theta_i - \sum\limits_{i=1}^{N} \varphi_i \tag{10.70}$$

where

> N = the filter order
> θ_i = the angle between the ith zero and the vector U_i
> φ_i = the angle between the ith pole and the vector V_i

Equations 10.69 and 10.70 together complete the DSP theory for pole/zero interpretation for estimating the frequency and phase responses of any filter with any number of poles and zeros. So, let's do the analysis for this filter. One of the interesting things about it is that by going through the math, you can see the tug of war going on between the pole and zero. Follow the analysis sequence in Figures 10.38–10.41; in each figure, the z-plane plot is shown in (a) and a vertical line is placed in the corresponding location in the response plots in (b). For example, in Figure 10.38a, you can see that the zero wins and the response is down almost 12 dB. Geometrically, you can see this because the zero is closer to the evaluation point, so it has more effect on the outcome. In any filter, the numerator and denominator ultimately fight—in some cases the numerator wins, in other cases the denominator, and sometimes there is a tie, in which case the magnitude of the filter is unity gain for that evaluation frequency. The composite plot is shown in Figure 10.42; note that the y-axis has been zoomed for easier viewing of the shelf.

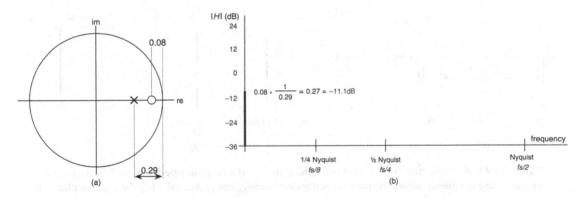

Figure 10.38: The magnitude response at DC is −11.1 dB as the zero wins this fight; (a) the z-plane plot and (b) the magnitude response at DC plotted as a vertical line.

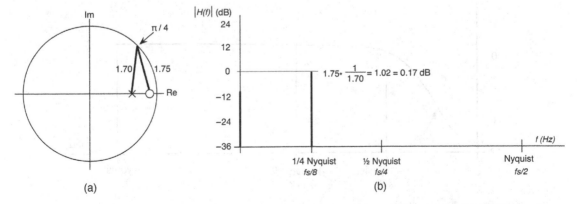

Figure 10.39: The magnitude response at $\pi/4$ is almost unity gain because the pole and zero distances are almost the same. The tug of war ends in stalemate here at 0.17 dB of gain; (a) the z-plane plot and (b) the magnitude response at $\pi/4$ plotted as a vertical line.

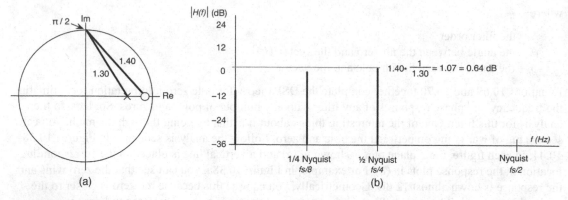

Figure 10.40: With the pole slightly closer to the evaluation frequency, the magnitude response at π/2 picks up a bit to +0.64 dB; (a) the z-plane plot and (b) the magnitude response at π/2 plotted as a vertical line.

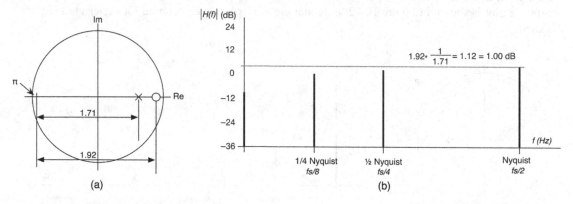

Figure 10.41: At π the pole/zero ratio favors the pole and the response perks up to 1.0 dB; notice that this is the frequency where the pole is clearly dominating, but just barely; (a) the z-plane plot and (b) the magnitude response at π plotted as a vertical line.

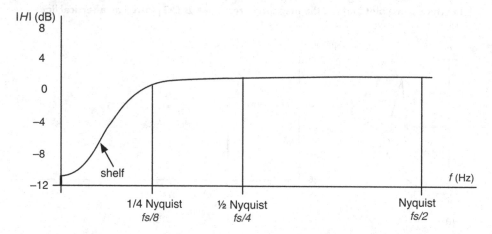

Figure 10.42: The composite frequency response plot shows a −12 dB low shelving filter response— a useful filter in audio.

10.20.3 Step 5: Direct Evaluation

Now we can evaluate the filter the same way as before: using Euler's equation to separate the real and imaginary portions from the transfer function, then finding the magnitude and phase components at each evaluation frequency.

Evaluate at the following frequencies:

- DC: 0
- Nyquist: π
- ½ Nyquist: $\pi/2$
- ¼ Nyquist: $\pi/4$

Begin by getting the transfer function in a form to use for all the evaluation frequencies, as shown in Equation 10.71.

$$H(z) = \frac{a_0 + a_1 z^{-1}}{1 + b_1 z^{-1}}$$

$$= \frac{1 - 0.92 z^{-1}}{1 - 0.71 z^{-1}}$$

Let: $z = e^{j\omega}$

$$H(\omega) = \frac{1 - 0.92 e^{-j1\omega}}{1 - 0.71 e^{-j1\omega}} \tag{10.71}$$

Apply Euler's Equation:

$$H(\omega) = \frac{1 - 0.92 e^{-j1\omega}}{1 - 0.71 e^{-j1\omega}}$$

$$H(\omega) = \frac{1 - 0.92[\cos(\omega) - j\sin(\omega)]}{1 - 0.71[\cos(\omega) - j\sin(\omega)]}$$

Evaluate at our four frequencies.

10.20.4 DC (0 Hz)

$$H(\omega) = \frac{1 - 0.92[\cos(\omega) - j\sin(\omega)]}{1 - 0.71[\cos(\omega) - j\sin(\omega)]}$$

$$= \frac{1 - 0.92[\cos(0) - j\sin(0)]}{1 - 0.71[\cos(0) - j\sin(0)]} = \frac{1 - 0.92[1 - j0]}{1 - 0.71[1 - j0]} \tag{10.72}$$

$$= \frac{1 - 0.92}{1 - 0.71} = \frac{0.08 + j0}{0.29 + j0}$$

$$|H(\omega)| = \frac{|0.08 + j0|}{|0.29 + j0|}$$

$$= \frac{\sqrt{a_{num}^2 + b_{num}^2}}{\sqrt{a_{denom}^2 + b_{denom}^2}}$$

$$= \frac{\sqrt{0.08^2}}{\sqrt{0.29^2}} = 0.276 = -11.2dB \tag{10.73}$$

$$Arg(H) = Arg(Num) - Arg(Denom)$$

$$= \tan^{-1}(0/0.08) - \tan^{-1}(0/0.29) = 0^O$$

10.20.4.1 Exercise

Finish the rest of the direct evaluation calculations on your own. The answers are in Table 10.5.

Figure 10.43 shows the measured frequency response for this shelving filter plotted in (a) linear frequency (that matches our plots here) and with (b) a log frequency axis, which resembles the shelving filter response with which you are most familiar.

Table 10.5: Exercise answers

Frequency (ω)	$|H(\omega)|$	Arg(H)
Nyquist (π)	1.00 dB	0.0
½ Nyquist ($\pi/2$)	0.82 dB	7.23
¼ Nyquist ($\pi/4$)	0.375 dB	16.60

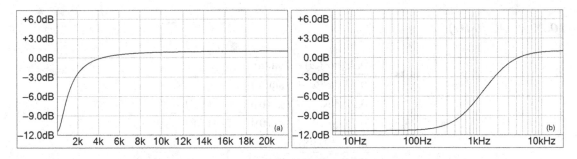

Figure 10.43: The measured (a) linear and (b) log frequency response plots for our pole/zero shelving filter with fs = 44.1 kHz.

10.21 The Biquadratic Filter

The next filter topology to study is the biquadratic (or biquad) structure. The biquad consists of two 2nd order components: a 2nd order feed-forward and a 2nd order feedback filter combined together and shown in Figure 10.44. The resulting transfer function will have two quadratic equations, thus the name. The difference equation is:

$$y(n) = a_0 x(n) + a_1 x(n-1) + a_2 x(n-2) - b_1 y(n-1) - b_2 y(n-2) \tag{10.74}$$

10.21.1 Step 1, Step 2, and Step 3: Take the z Transform of the Difference Equation to Get the Transfer Function, Then Factor Out a_0 as the Scalar Gain Coefficient

$$y(n) = a_0 x(n) + a_1 x(n-1) + a_2 x(n-2) - b_1 y(n-1) - b_2 y(n-2)$$

$$Y(z) = a_0 X(z) + a_1 X(z)z^{-1} + a_2 X(z)z^{-2} - b_1 Y(z)z^{-1} - b_2 Y(z)z^{-2}$$

separate variables:

$$Y(z) + b_1 Y(z)z^{-1} + b_2 Y(z)z^{-2} = a_0 X(z) + a_1 X(z)z^{-1} + a_2 X(z)z^{-2}$$

$$Y(z)[1 + b_1 z^{-1} + b_2 z^{-2}] = X(z)[a_0 + a_1 z^{-1} + a_2 z^{-2}]$$

form transfer function

$$H(z) = \frac{Y(z)}{X(z)} = \frac{a_0 + a_1 z^{-1} + a_2 z^{-2}}{1 + b_1 z^{-1} + b_2 z^{-2}} \tag{10.75}$$

factor out a_0

$$H(z) = a_0 \frac{1 + \alpha_1 z^{-1} + \alpha_2 z^{-2}}{1 + b_1 z^{-1} + b_2 z^{-2}}$$

where

$$\alpha_1 = \frac{a_1}{a_0} \qquad \alpha_2 = \frac{a_2}{a_0}$$

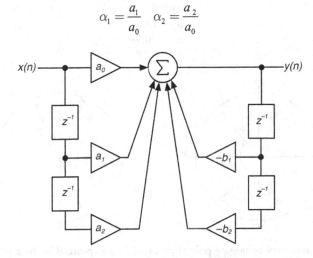

Figure 10.44: The biquad structure combines 2nd order feed-forward and feedback sections.

10.21.2 Step 4: Plot the Poles and Zeros of the Transfer Function

The biquad will produce, at most, a conjugate pair of zeros and conjugate pair of poles from the numerator and denominator respectively. Note that you are not required to use the full 2nd order sections; for example, you might have an algorithm that has a 1st order feed-forward component combined with a 2nd order feedback component. In this case, you just let the a_1 term equal 0.0 to effectively disable its branch. Calculating these pole and zero locations is the same as in the pure 2nd order feed-forward and feedback topologies. All you need to do is plot them in the same unit circle. The transfer function becomes (by simple substitution from previous sections) that of Equation 10.76.

$$H(z) = a_0 \frac{1 - 2R_z \cos(\theta)z^{-1} + R_z^2 z^{-2}}{1 - 2R_p \cos(\phi)z^{-1} + R_p^2 z^{-2}} \tag{10.76}$$

We can plot an arbitrary pair of poles and zeros as shown in Figure 10.45a; each has its own radius, R_z and R_p, and angle θ and φ. The same kind of pole/zero tug-of-war goes on with the biquad, only now there are more competing entities.

Estimating the frequency response is complicated due to the additional poles and zeros but the process is the same:

1. Locate each evaluation frequency on the outer rim of the unit circle.
2. Draw a line from the point on the circle to *each* zero and measure the length of these vectors. Do it for each evaluation frequency.
3. Draw a line from the point on the circle to *each* pole and measure the length of these vectors. Do it for each evaluation frequency.
4. Multiply all the zero magnitudes together.
5. Multiply all the *inverse* pole magnitudes together.
6. Divide the zero magnitude by the pole magnitude for the final result at that frequency.

To demonstrate, use the following coefficients:

$$a_0 = 1.0$$
$$a_1 = 0.73$$
$$a_2 = 1.00$$
$$b_1 = -0.78$$
$$b_2 = 0.88$$

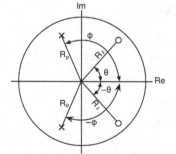 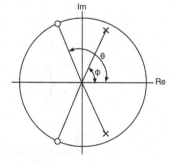

Figure 10.45: (a) An arbitrary conjugate pair of poles and zeros plotted in the z-plane. (b) The pole/zero plot for our biquad example.

We can directly find the pole and zero locations from Equation 10.76 (note that because a_0 is 1.0, you don't have to calculate the α terms). The pairs of poles and zeros are plotted in Figure 10.45b.

The zeros are calculated as follows:

$$a_2 = R_z^2 = 1.00$$
$$R_z = \sqrt{1.00} = 1.00$$

and

$$a_1 = -2R\cos(\Theta) = 0.73 \tag{10.77}$$
$$2(1.00)\cos(\Theta) = -0.73$$
$$\cos(\Theta) = -0.365$$
$$\Theta = \arccos(-0.365)$$
$$\Theta = 111.1^o$$

The poles are calculated as follows:

$$b_2 = R_p^2 = 0.88$$
$$R_p = \sqrt{0.88} = 0.94$$

and

$$b_1 = -2R\cos(\phi) = -0.78 \tag{10.78}$$
$$2(0.94)\cos(\phi) = 0.78$$
$$\cos(\phi) = \frac{0.78}{2(0.94)}$$
$$\phi = \arccos(0.414)$$
$$\phi = 65.5^o$$

The poles and zeros are in relatively close proximity to each other. The zero is directly on the unit circle ($R_z = 1.0$), so we expect a notch to occur at the zero frequency. The pole is near the unit circle but not touching it, so we expect a resonance at the pole frequency. In Figure 10.46a, it's easy to locate the places where the pole or zero very much dominates. In the low frequencies, the pole dominates and

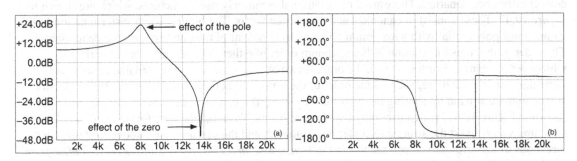

Figure 10.46: The example biquad filter's (a) frequency response and (b) phase response.

at high frequencies the zero dominates. Notice that the sudden phase response shifts in Figure 10.46b correspond to the peak and notch locations. This is an example of a direct z-plane design, where we place a pole and zero pair directly in the z-plane, then calculate the coefficients.

10.21.3 The a_N and b_M Coefficient Naming Conventions

Before proceeding, there is one quick item to cover: how the a_N and b_M coefficients are named in various DSP books. You can see that in this book the numerator coefficients are named with the letter a and the denominator coefficients are named with the letter b. The a_0 coefficient is the filter gain scalar, and there is never a b_0 coefficient. This is a sore topic for some DSP engineers: the reason is that if you buy 10 different DSP books, half of them will use our convention here, and the other half will swap the as and bs, appearing to turn the transfer functions upside down and making the difference equations appear incorrect. The reason that I use the naming convention here is simple: it is exactly how both my analog and digital filtering textbooks named them when I was in college, and it is how I've always named them. It would be nice if all the DSP textbook authors agreed on this, but until then, when looking at other books and papers, make sure you understand the naming convention being used. It is actually fairly simple: if you see a b_0 coefficient, then they are using the "bs in the numerator" convention.

10.22 Other Biquadratic Structures

The structure in Figure 10.44 is called the *direct form* (DF) structure. It directly implements the difference equation, and you can form the output *y(n)* all at once in a single equation. The pseudo-code you might write to implement this structure could look like this:

```
// --- form output directly
yn = a0*xn + a1*x(n-1) + a2*x(n-2)-b1*y(n-1)-b2*y(n-2)

// --- update state registers
x(n-2) = x(n-1)
x(n-1) = xn

y(n-2) = y(n-1)
y(n-1) = yn
```

It turns out that the direct form structure is only one of several structures that will also yield the correct difference equation. There are three additional commonly used structures, which are shown in Figure 10.47. These consist of canonical and transposed forms. The biquadratic transfer function has a numerator polynomial with a maximum polynomial exponent of two, and so does the denominator. The order of the transfer function is the maximum order of either the numerator or denominator— whichever is greatest. For the biquad, the order is two. A *canonical* structure contains exactly the same number of z^{-1} state registers as the order of the transfer function; this means that the numerator and denominator wind up sharing the two-state registers. You can see that in the *canonical form* (CF) shown in Figure 10.47, which uses only two-state registers. To create the transformed version of a structure, you convert the structure's nodes into summers and the summers into nodes, and reverse the directions of the arrows. You can see this in the *transposed direct form* and *transposed canonical form* in Figure 10.47.

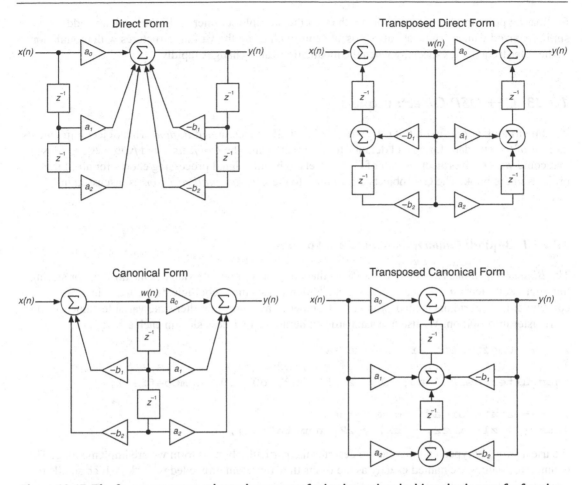

Figure 10.47: The four most commonly used structures for implementing the biquadratic transfer function.

All four of the biquad block diagrams produce the same difference equation at the output. Many DSP engineers classify the structures according to their computational efficiency, memory storage requirements, and sensitivity to round-off errors in the coefficients. The structures themselves—the ordering of the summers and multiplication nodes—can also have an effect on the accuracy of the final calculation for $y(n)$. Higher order polynomials are more sensitive to coefficient errors—so when confronted with a transfer function that is higher than 2nd order, we mathematically split the transfer function into a series connection of 1st and 2nd order structures. A 3rd order filter is realized as a 1st and 2nd order structure in series, while a 10th order transfer function would be realized with a series connection of five 2nd order sections in a row. We should note that all of our C++ FX objects will use double precision coefficients, double precision state registers, and double precision mathematical processing, so we are not as susceptible to these problems as are fixed point DSP systems, still found in many audio applications today.

The direct forms (direct form and transposed direct form) are the simplest to implement but are also the most susceptible to errors. The *transposed canonical form* is generally regarded as the best structure for floating point implementations due to the way the multiple summers split the work and add similarly sized values. There are also pros and cons with using the various structures with modulation of the filter frequency, especially when the modulation value changes rapidly.

10.23 C++ DSP Object: Biquad

It's time to look at our first bit of C++ with our first DSP object named *Biquad*. This object implements all four of the standard forms and defaults to the direct form. The *fxobjects.h* and *fxobjects.cpp* files that come with each project are full of C++ objects to handle signal processing chores for all of the projects in the book. The C++ objects all conform to the *IAudioSignalProcessor*, as discussed in Chapter 8.

10.23.1 Biquad: *Enumerations and Data Structure*

The *Biquad* object has two member variables that are arrays of double values. One array is for storing the filter coefficients a_0, a_1, a_2, b_1, and b_2, in addition to two more coefficients we'll use in more complex filter algorithms named c_0 and d_0. The other array stores the filter state variable values, or the z^{-1} register information. We use two standard enumerations to access slots in each array:

```
// --- coefficient array index values
// a0 = 0, a1 = 1, a2 = 2, b1 = 3, b2 = 4, numCoeffs = 5
enum filterCoeff { a0, a1, a2, b1, b2, c0, d0, numCoeffs };


// --- state array index values
enum { x_z1, x_z2, y_z1, y_z2, numStates };
```

We use a strongly typed enumeration to describe the particular biquad form we are implementing. The enumerated values are named exactly as the forms they represent, preceded with "k" which stands for "constant."

```
// --- type of calculation (algorithm)
enum class biquadAlgorithm { kDirect, kCanonical, kTransposeDirect,
                             kTransposeCanonical };
```

The custom data structure is simple—it only holds the biquad algorithm calculation type variable, using the strongly typed enum:

```
struct BiquadParameters
{
        BiquadParameters () {}

        BiquadParameters& operator=(const BiquadParameters& params) {...}

        // --- one and only variable is a flag for recalculation
        biquadAlgorithm biquadCalcType = biquadAlgorithm::kDirect;
};
```

10.23.2 Biquad: *Members*

Tables 10.6 and 10.7 list the *Biquad* member variables and member functions.

10.23.3 Biquad: *Programming Notes*

Notice that we pass a pointer to an array of biquad coefficients in the *setCoefficients* function, which uses a bulk memory copy to set the array—this means that care must be taken to always use an array whose size matches the state array size, which is *numCoeffs*. Also notice how the *setParameter* argument is identical to the name of the member variable, but with an underscore "_" in front of it; this naming paradigm will be used frequently.

The IAudioSignalProcessor

- **reset**: flushes out the state array to load zeros into the state registers
- **canProcessAudioFrame**: returns false since we only process samples

Table 10.6: *Biquad* **member variables**

Biquad Member Variables		
Type	**Name**	**Description**
double	coeffArray[numCoeffs]	Biquad coefficients
double	stateArray[numCoeffs]	z^{-1} storage registers
BiquadParameters	parameters	Custom object parameters

Table 10.7: *Biquad* **member functions**

Biquad Member Functions		
Returns	**Name**	**Description**
BiquadParameters	*getParameters*	Get the object parameters
void	*setParameters* Parameters: —BiquadParameters _parameters	Set the object parameters
bool	*reset* Parameters: —double sampleRate	Returns true, sample rate not needed
double	*processAudioSample* Parameters: —double xn	Process one audio sample xn through the biquad and return the processed result
void	*setCoefficients* Parameters: —double* coeffs	Set the biquad coefficients via an array of coefficients *coeffs* that is same size as internal array
double*	*getCoefficients*	Returns a pointer to the biquad coefficients array
double*	*getStateArray*	Returns a pointer to the state variable array

All of the work is in the *processAudioSample* function, that takes the input argument *x(n)* and calculates a return variable, *y(n)*. Let's look at the first section that implements the direct form algorithm. For each of the algorithms, we will follow a set pattern:

1. Read variables from the state registers and do whatever is required to formulate the output *y(n)*; this may only be one line of code, or may be many.
2. Check the value of *y(n)* for floating point underflow, using a helper function declared in *fxobjects.h*.
3. Only after the final value for *y(n)* has been prepared do we then update the state registers, pushing in the data in the correct order to preserve the state register relationships.

We begin by comparing the *parameters.biquadCalcType* variable to the strongly typed *enum* constants, and branch accordingly. For the direct form, the sequence is easy:

1. Create *y(n)* all at once, using the current input *x(n)*, the previous inputs and outputs located in the state array, and the coefficients located in the coefficient array.
2. Check for underflow.
3. Update the state registers in sequence, from the oldest to youngest sample value.
4. Notice the use of the enums (*a0, a1*, etc.) to make accessing the coefficients easy and the state array enums (*x_z1, x_z2*, etc.) for accessing those data.
5. Compare the calculation of *yn* with the algorithm in Figure 10.47 and make sure you can see how the line of code properly calculates *y(n)*.
6. Notice how the data are shuffled through the state array.

```
double Biquad::processAudioSample(double xn)
{
    if (parameters.biquadCalcType == biquadAlgorithm::kDirect)
    {
        // y(n) = a0*x(n) + a1*x(n-1) + a2*x(n-2)-b1*y(n-1)
        // -b2*y(n-2)
        double yn = coeffArray[a0] * xn +
                    coeffArray[a1] * stateArray[x_z1] +
                    coeffArray[a2] * stateArray[x_z2] -
                    coeffArray[b1] * stateArray[y_z1] -
                    coeffArray[b2] * stateArray[y_z2];

        // --- 2) underflow check
        checkFloatUnderflow(yn);

        // --- 3) update states
        stateArray[x_z2] = stateArray[x_z1];
        stateArray[x_z1] = xn;

        stateArray[y_z2] = stateArray[y_z1];
        stateArray[y_z1] = yn;

        // --- return value
        return yn;
    }
}
```

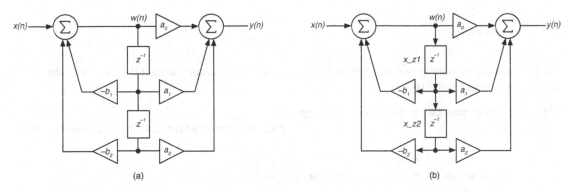

**Figure 10.48: (a) The canonical form includes an intermediate node *w(n)* to be calculated.
(b) The signal flow through the structure.**

Now let's look at the canonical form shown in Figure 10.48a, which is a bit trickier. Notice how the state registers are labeled in Figure 10.48b: the upper z^{-1} register is x_z1 and the lower is x_z2. These are the two index locations used in the state array.

When presented with these algorithms, just remember to follow the steps listed earlier. It can be easy to get your head stuck in an infinite loop when trying to implement these for the first time. Do not be tempted to push any data into any of the state registers until after *y(n)* has been calculated—this will usually cure your infinite loop problem. In Figure 10.48b we've added arrows to show the flow of data and branch values to demonstrate how to calculate them, but it results in an overly decorated figure. We will generally only show arrowheads when they are inputs to summers. For the canonical form, ask yourself what information is needed for the intermediate node *w(n)*. The answer is that *w(n)* is the output of a summer that has three inputs that are all labeled in Figure 10.48b. Make sure you can reconcile how Figure 10.48b translates into this line of code for the *w(n)* node; the minus signs are because the *b* coefficients are negated in the algorithm.

```
else if (parameters.biquadCalcType == biquadAlgorithm::kCanonical)
{
        double wn = xn-(coeffArray[b1]*stateArray[x_z1])
                      -(coeffArray[b2]*stateArray[x_z2]);
```

Now turn your attention to *y(n)*. What do we need to calculate it? Make sure you can figure out that this line of code calculates *y(n)*:

```
double yn = coeffArray[a0]*wn
    + coeffArray[a1]*stateArray[x_z1]
    + coeffArray[a2]*stateArray[x_z2];
```

With *y(n)* calculated, we can then update the states. Take a look at Figure 10.48b. What is the input value for the upper z^{-1} register? The answer is *w(n)*. What is the input value for the lower z^{-1} register? It is the output of the upper register. This means that we can't write *w(n)* into the upper register until we've moved its data into the lower register. That code is here:

```
// --- 3) update states
stateArray[x_z2] = stateArray[x_z1];
stateArray[x_z1] = wn;
```

```
    // --- return value
    return yn;
}
```

The code for the remaining two algorithms follows. Make sure you can connect the code with the block diagrams in Figure 10.47.

```
else if (parameters.biquadCalcType ==
                                biquadAlgorithm::kTransposeDirect)
{
    // --- w(n) = x(n) + stateArray[y_z1]
    double wn = xn + stateArray[y_z1];

    // --- y(n) = a0*w(n) + stateArray[x_z1]
    double yn = coeffArray[a0] * wn + stateArray[x_z1];

    // --- 2) underflow check
    checkFloatUnderflow(yn);

    // --- 3) update states
    stateArray[y_z1] = stateArray[y_z2]-coeffArray[b1] * wn;
    stateArray[y_z2] =-coeffArray[b2] * wn;

    stateArray[x_z1] = stateArray[x_z2] + coeffArray[a1] * wn;
    stateArray[x_z2] = coeffArray[a2] * wn;

    // --- return value
    return yn;
}
else if (parameters.biquadCalcType ==
                                biquadAlgorithm::kTransposeCanonical)
{
    // --- 1) form output y(n) = a0*x(n) + stateArray[x_z1]
    double yn = coeffArray[a0]*xn + stateArray[x_z1];

    // --- 2) underflow check
    checkFloatUnderflow(yn);

    // --- shuffle/update
    stateArray[x_z1] = coeffArray[a1]*xn
                    - coeffArray[b1]*yn + stateArray[x_z2];
    stateArray[x_z2] = coeffArray[a2]*xn-coeffArray[b2]*yn;

// --- return value
return yn;
}
```

Using the Biquad object is straightforward:

1. Reset the object to prepare it for streaming.
2. Set the calculation type in the *BiquadParameters* custom data structure.
3. Call *setParameters* to update the calculation type.
4. Call *processAudioFrame*, passing the input value in and receiving the output value as the return variable.

10.24 Homework

Feedback topology

1. For a given biquad transfer function *H(z)*, it can be shown that the magnitude squared response $|H(z)|^2$ at a given normalized frequency ω such that $(0.0 \leq \omega \leq \pi)$ can be calculated as:

$$|H(z)|^2 = \frac{a_1^2 + (a_0 - a_2)^2 + 2a_1(a_0 + a_2)\cos(\omega) + 4a_0 a_2 \cos^2(\omega)}{b_1^2 + (1 - b_2)^2 + 2b_1(1 + b_2)\cos(\omega) + 4b_2 \cos^2(\omega)} \quad (10.79)$$

Write a C++ function that will calculate the magnitude *H(z)* in dB for a given evaluation frequency ω and a set of biquad coefficients a_0, a_1, a_2, b_1, and b_2. You'll need to take the square root of the result of Equation 10.79 and then convert to dB. If you are good with graphics, write a program to plot the magnitude in dB of any biquad over a range of values from $\omega = 0.0$ to $\omega = \pi$ given the filter coefficients. You might try spacing the various evaluation frequencies evenly over that range, for example 11 evaluation frequencies from 0 to π in $\pi/10$ increments.

2. There are several other algorithms that will implement the biquadratic transfer function. One very interesting algorithm is derived from the wave digital filters of Chapter 12. The block diagram is shown in Figure 10.49.

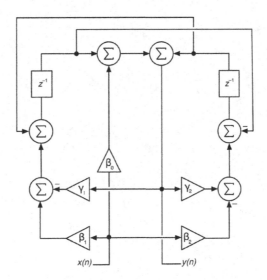

Figure 10.49: The biquad structure derived from 2nd order wave digital filters.

The equations for the coefficients in Figure 10.49 are related to the biquad coefficients a_N and b_M as follows:

$$\beta_0 = a_0$$

$$\beta_1 = a_0 - a_2 - \left(\frac{a_0 - a_1 - a_2}{2}\right)$$

$$\beta_2 = \frac{a_0 - a_1 - a_2}{2}$$

$$\gamma_1 = b_1 + \left(\frac{1 - b_1 - b_2}{2}\right)$$

$$\gamma_2 = \frac{1 - b_1 - b_2}{2}$$

(10.80)

Modify the *Biquad* object to implement this additional algorithm. You can test it when we implement the audio filters in Chapter 11. Make sure you follow the rules about the sequence of operations in the algorithm.

10.25 Bibliography

Ifeachor, E. C. and Jervis, B. W. 1993. *Digital Signal Processing: A Practical Approach*, Chap. 3. Menlo Park: Addison Wesley.

Kwakernaak, H. and Sivan, R. 1991. *Modern Signals and Systems*, Chap. 3. Eaglewood Cliffs: Prentice-Hall.

Moore, R. 1990. *Elements of Computer Music*, Chap. 2. Eaglewood Cliffs: Prentice-Hall.

Oppenheim, A. V. and Schafer, R. W. 1999. *Discrete-Time Signal Processing, 2nd Ed.*, Chap. 3. Eaglewood Cliffs: Prentice-Hall.

Orfanidis, S. 1996. *Introduction to Signal Processing*, Chap. 10–11. Eaglewood Cliffs: Prentice-Hall.

Schlichthärle, D. 2011. *Digital Filters Basics and Design, 2nd Ed.*, Chap. 5. Heidelberg: Springer.

Steiglitz, K. 1996. *A DSP Primer with Applications to Digital Audio and Computer Music*, Chap. 4–5. Menlo Park: Addison Wesley.

CHAPTER 11

Audio Filter Designs: IIR Filters

It's time to put the theory into practice and make some audio filters and EQs. You know that the coefficients of a filter determine its frequency response and other characteristics. But how do you find the coefficients? There are two commonly used methods for calculating the coefficients of IIR filters:

- Direct z-plane design
- Analog filter to digital filter conversion

This chapter uses the following filter naming conventions.

- LPF: Low-pass filter
- HPF: High-pass filter
- BPF: Band-pass filter
- BSF: Band-stop filter
- APF: All-pass filter
- HSF: High shelf filter
- LSF: Low shelf filter
- PEQ: Parametric EQ filter
- GEQ: Graphic EQ system

11.1 Direct z-Plane Design

In this first category of design techniques, you manipulate the poles and zeros directly in the z-plane to create the response you want. You take advantage of the simple equations that relate the coefficients to the pole/zero locations. For the generic biquad, we can relate the coefficients and pole/zero locations with Equations 11.1 and 11.2.

For the numerator:

$$
\begin{aligned}
H(z) &= a_0(1 + \alpha_1 z^{-1} + \alpha_2 z^{-2}) \\
&= a_0(1 - 2R_z \cos(\theta_z)z^{-1} + R_z^2 z^{-2})
\end{aligned}
\tag{11.1}
$$

then

$$
\begin{aligned}
a_0 &= \text{filter gain} \\
\alpha_1 &= -2R_z \cos(\theta_z) \\
\alpha_2 &= R_z^2
\end{aligned}
$$

For the denominator:

$$H(z) = \frac{1}{1 + b_1 z^{-1} + b_2 z^{-2}}$$

$$= \frac{1}{(1 - 2R_p \cos(\theta_p) z^{-1} + R_p^2 z^{-2})}$$

(11.2)

then

$$a_0 = \text{filter gain}$$
$$b_1 = -2R_p \cos(\Theta_p)$$
$$b_2 = R_p^2$$

For the numerator or denominator, the a_1 or b_1 coefficients are in direct control over the angles of the zeros or poles. The distance R to the zeros or poles is determined by either a_1, a_2 or b_1, b_2. For 1st order filters, the coefficients control only the location of the pole and zero on the real axis. There are no conjugate pairs. However, careful placement of the pole and zero can still affect the response of the filter.

11.1.1 Simple Resonator

A resonator is a BPF that can be made to have a very narrow peak. The basic resonator implements a single conjugate pair of poles in a 2nd order feedback topology. Figure 11.1a shows the block diagram of the filter using the direct form structure.

The resonator difference equation is:

$$y(n) = a_0 x(n) - b_1 y(n-1) - b_2 y(n-2)$$

(11.3)

Specify:

- f_c: center frequency
- BW: 3 dB bandwidth or Q: quality factor

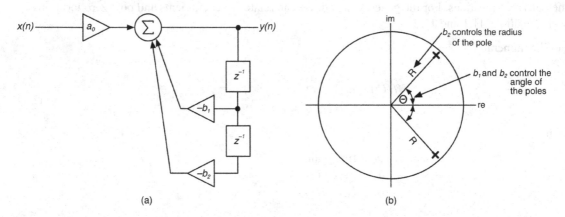

Figure 11.1: The resonator's (a) direct form block diagram and (b) pole/coefficient relationship.

The design equations are as follows.

$$\theta_c = 2\pi f_c / f_s$$
$$BW = f_c / Q$$
$$b_2 = e^{\left(-2\pi \frac{BW}{f_s}\right)}$$
$$b_1 = \frac{-4b_2}{1+b_2}\cos(\theta_c)$$
$$a_0 = (1-b_2)\sqrt{1-\frac{b_1^2}{4b_2}}$$

(11.4)

The resonator works by simple manipulation of the conjugate poles formed with the 2nd order feedback network. The b_2 term controls the distance out to the pole, which makes the resonant peak sharper (when the pole is close to the unit circle) or wider (when the pole is farther away from the unit circle). The b_1 and b_2 terms control the angle of the pole, which controls the center frequency, shown in Figure 11.1b. The a_0 term is used to scale the filter, so its peak output is always normalized to unity gain or 0 dB.

The drawback to this design is that the frequency response is symmetrical only at one frequency, $\pi/2$, when the low frequency and high frequency magnitude vectors are symmetrical. At all other frequencies, the response is shifted asymmetrically. Figure 11.2 shows the resonator with the poles in the right half (a) and left half (b) of the unit circle, which causes the response to be skewed towards the low (a) and high (b) frequencies respectively.

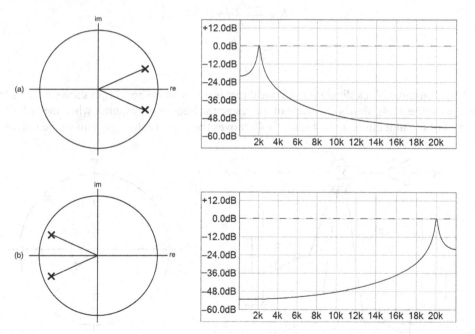

Figure 11.2: (a) When the resonator poles are in the right half of the unit circle, the response is asymmetrical and skewed to the low frequency side. (b) When the resonator poles are in the left half of the unit circle, the response is asymmetrical and skewed to the high frequency side. (The frequency axis is linear to better show the asymmetry.)

11.1.2 Smith-Angell Resonator

The improved design, known as the *Smith-Angell resonator,* reduces the asymmetry problem by adding two zeros into the filter—one at $z = 1$ and one at $z = -1$—in order to pin down DC and Nyquist with a magnitude response of zero. This forces the filter to become somewhat more symmetric (but not entirely) and has the advantage of making the band-pass even more selective in nature. This design is also gain normalized with a_0. As before, the radius is set with b_2 first, then b_1 is calculated using the b_2 coefficient and the desired pole frequency. The filter is not truly normalized to 0.0 dB; there is actually a slight fluctuation of less than 1 dB.

The Smith-Angell resonator difference equation is:

$$y(n) = a_0 x(n) + a_2 x(n-2) - b_1 y(n-1) - b_2 y(n-2) \tag{11.5}$$

Specify:

- f_c: center frequency
- BW: 3 dB bandwidth or Q: quality factor

The design equations are as follows.

$$\theta_c = 2\pi f_c / f_s$$
$$BW = f_c / Q$$
$$b_2 = e^{\left(-2\pi \frac{BW}{f_s}\right)}$$
$$b_1 = \frac{-4b_2}{1+b_2}\cos(\theta_c) \tag{11.6}$$
$$a_0 = 1 - \sqrt{b_2}$$
$$a_2 = -a_0$$

Figure 11.3a shows the block diagram using the direct form structure and 11.3b shows the pole/zero relationship. Figure 11.4a shows the shape of the resonator frequency response when the poles are in the right half of the unit circle while Figure 11.4b moves the poles to the right half of the unit circle.

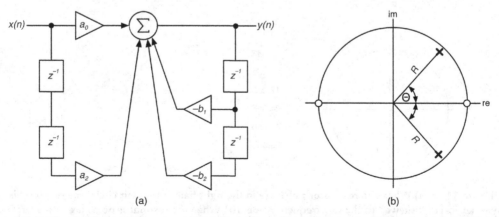

(a)   (b)

Figure 11.3: The Smith-Angell resonator's (a) direct-form block diagram and (b) pole/zero plot showing the two newly added zeros at z = −1 and z = +1.

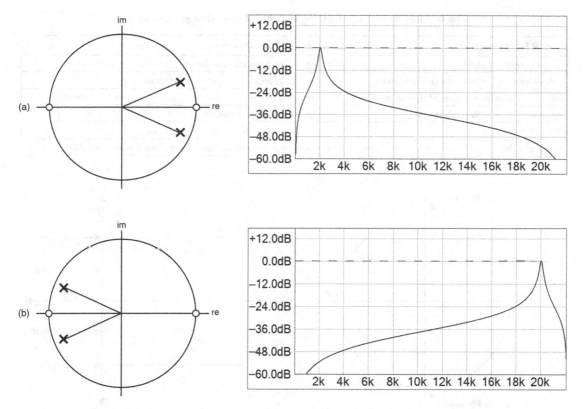

Figure 11.4: The Smith-Angell resonator frequency response at different center frequency settings with (a) f_c = 2 kHz and (b) f_c = 20 kHz. (For both plots f_s = 44.1 kHz; the frequency axis is linear to compare with the simple resonator.)

11.2 Analog Filter to Digital Filter Conversion

A more widely used approach to filter design is to first start with the classic analog designs and then convert them into digital versions. There's good reason to do this because there are many excellent analog designs already done for us. We just need a way to make them work in our digital world. While analog filter design is outside the scope of the book, there are many similarities between the two design worlds. For example, both analog and digital filter design involves a transfer function that you manipulate to produce poles and zeros. They also both use a transform to get from the time domain to the frequency domain. A fundamental difference is that in analog, there is no Nyquist restriction and all frequencies from $-\infty$ to $+\infty$ are included. Analog filters use reactive components such as inductors and capacitors to perform the functions of integrating or differentiating the analog signal. Digital filters use delay elements to produce phase shifts instead. Table 11.1 summarizes the similarities and differences. You can find a more detailed analysis of the two techniques in *Designing Software Synthesizer Plug-Ins in C++*.

In Figure 11.5 you can see the s-plane on the right—it is also a complex plane. The real axis is named the σ *axis* and the imaginary axis is the *$j\omega$ axis*. The $j\omega$ axis is the frequency axis, and it spans $-\infty$ to $+\infty$ rad/sec. The unit circle maps to the portion on the $j\omega$ axis between $-$Nyquist and $+$Nyquist. In order to

Table 11.1: Summary of analog and digital filter design attributes

Digital Filter Design	Analog Filter Design
uses a complex transfer function to relate input/output	uses a complex transfer function to relate input/output
delay elements create phase shift and delay	reactive components perform integration
uses the z Transform (sampled time to frequency)	uses the Laplace Transform (continuous time to frequency)
poles and zeros in the z-plane	poles and zeros in the s-plane
Nyquist limitation	all frequencies from −∞ to +∞ allowed
poles must be inside the unit circle for stable operation	poles must be in the left-hand part of the s-plane for stable operation

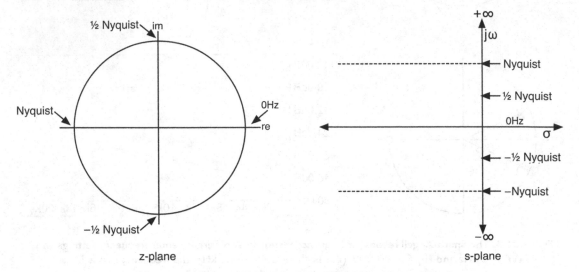

Figure 11.5: The z-plane and s-plane show similar information in different ways.

transform an analog filter design into a digital filter design, we need a mapping system to force the poles and zeros from the s-plane to become poles and zeros in the z-plane. Once we have that, the coefficients that produce those poles and zeros in the digital locations can be calculated. In the analog world, poles that are close to the jω axis will become poles that are close to the unit circle in the digital world.

The problem is that analog poles and zeros can exist anywhere along the jω axis, even at frequencies outside the digital Nyquist zone. It is also common for analog designs to have a pole or a zero at −∞ and +∞ on either the σ and/or jω axes. In Figure 11.6 you can see the problem—the pair of analog zeros close to the jω axis at ½ Nyquist will map to digital zero locations close to the unit circle at the ½ Nyquist angles. What about the pair of poles that are outside the Nyquist frequency in the s-plane— where should they be mapped? The other way to think about the issue is that in the analog s-plane, the entire left-hand side, including all the infinite frequencies, must map to the *interior* of the unit circle in the z-plane shown in Figure 11.7a. In addition, the entire right-hand plane must map to the *exterior* of the unit circle, as shown in Figure 11.7b

So, what we need is a mathematical operation to make the transformation from the s-plane to the z-plane. It must map the infinitely large area of the left-hand side of the s-plane into the unit circle's finite interior area in the z-plane, and the infinite right-hand side of the s-plane to the infinite area

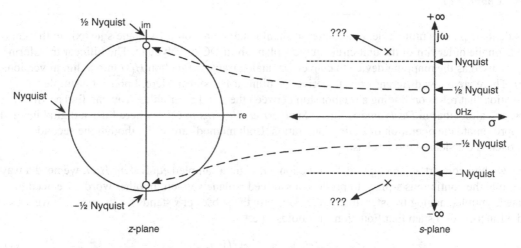

Figure 11.6: Mapping of the analog poles from the *s*-plane to the *z*-plane.

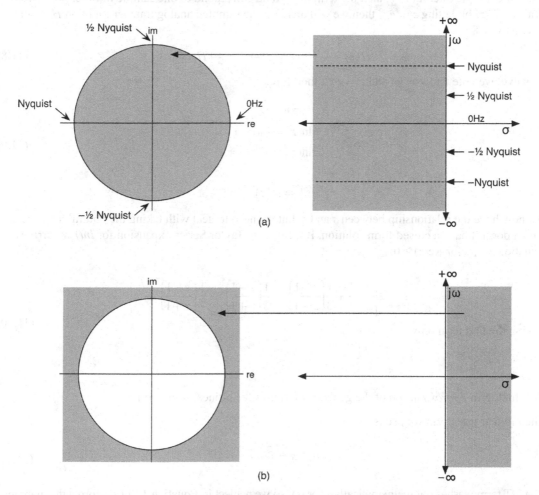

Figure 11.7: To convert an analog filter to a digital filter, (a) the infinitely large left-hand side of the *s*-plane must be mapped into the finite space inside the unit circle in the *z*-plane, while (b) the right-hand plane of the *s*-plane maps to the exterior of the unit circle in the *z*-plane.

outside the z-plane's unit circle. In addition, it should map points on the s-plane's jω axis to the same points on the outer rim of the unit circle in the z-plane from DC to ±Nyquist. The bilinear transform may be used as the mapping device to convert an analog transfer function *H(s)* into a digital version, *H(z)*. There are a few different ways to derive the bilinear transform. Here, I present a simple derivation that relies on finding a relationship between the *s* and *z* variables. You can find a different approach altogether in *Designing Software Synthesizer Plug-Ins in C++* where I use a digital integrator to approximate the operation of analog integration. Both methods are valid, though the second approach is considered more robust.

If we wish to convert an analog transfer function *H(s)* into a sampled equivalent *H(z)*, we need a way to sample the continuous s-plane to produce a sampled z-plane version. In other words, we need to create a sampled analog transfer function $H_s(s)$ where the subscript *s* stands for "sampled." We seek to find a function *g(z)* such that Equation 11.7 holds true.

$$H(s) \xrightarrow{\ s=g(z)\ } H(z) \tag{11.7}$$

Since the sample interval is *T*, then the term $e^{j\omega T}$ would correspond to one sample in time. So, if we evaluate *H(z)* by letting $z = e^{j\omega T}$ then we will arrive at the sampled analog transfer function $H_s(s)$ in Equation 11.8.

$$H(s) \xrightarrow{\ s=g(z)\ } H(z) \xrightarrow{\ z=e^{j\omega T}\ } H_s(s) \tag{11.8}$$

To solve, we note that $s = j\omega$ and $z = e^{j\omega T}$, therefore:

$$z = e^{sT}$$
$$\ln(z) = \ln(e^{sT})$$
$$\ln(z) = sT \tag{11.9}$$

or

$$sT = \ln(z)$$

We now have the relationship between *s* and *z,* but we need to deal with taking the natural log of *z,* which doesn't have a closed form solution. If we use the Taylor Series expansion for *ln()* we arrive at Equation 11.10 for Re(z) > 0.

$$sT = 2\left[\frac{z-1}{z+1} + \frac{1}{3}\left(\frac{z-1}{z+1}\right)^3 - \frac{1}{5}\left(\frac{z-1}{z+1}\right)^5 + \frac{1}{7}\left(\frac{z-1}{z+1}\right)^7 - \cdots \right]$$

Taking the first term only $\tag{11.10}$

$$s = \frac{2}{T}\frac{z-1}{z+1}$$

This first term *approximation* of the general solution is the bilinear transform.

The bilinear transform we use is:

$$s = \frac{z-1}{z+1} \tag{11.11}$$

The 2/T term washes out mathematically later on, so we neglect it. Equation 11.11 performs the mapping by taking values at the infinite frequencies and mapping them to Nyquist. A pole or zero at −∞ or +∞

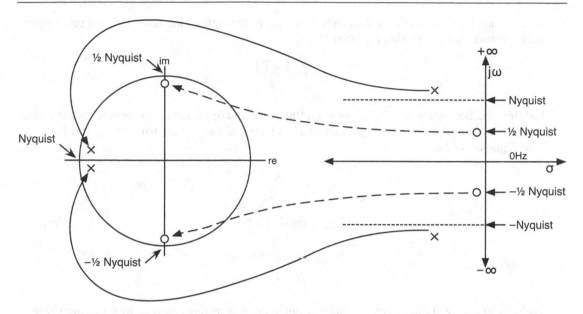

Figure 11.8: The bilinear transform maps the area outside the Nyquist zone on the left-hand plane to an area just inside the circle near Nyquist.

is mapped to a pole or zero at Nyquist. The other frequencies between Nyquist and $-\infty$ or $+\infty$ Hz are squeezed into the little region right around Nyquist, just inside the unit circle as shown in Figure 11.8.

The bilinear transform maps analog frequencies to their corresponding digital frequencies nonlinearly via the *tan* function, as shown in Equation 11.12.

$$\omega_a = \tan\left(\frac{\omega_d T}{2}\right) \tag{11.12}$$

where

$$\omega_a = \text{the analog frequency}$$
$$\omega_d = \text{the mapped digital frequency}$$
$$T = \text{the sample period}$$

The *tan* function is linear at low values of ω but becomes more nonlinear as the frequency increases. At low frequencies, the analog and digital frequencies map linearly. At high frequencies, the digital frequencies become warped and do not map properly to the analog counterparts. This means that a given analog design with a cutoff frequency f_c will generally have the wrong digital cutoff frequency after the conversion. The solution is to pre-warp the *analog* filter so that its cutoff frequency is in the wrong location in the analog domain, but will wind up in the correct location in the digital domain. To pre-warp the analog filter, you just use Equation 11.12 applied to the cutoff frequency of the analog filter. When you combine all the operations you get the bilinear z Transform (BZT). The step-by-step method of conversion is as follows.

1. Start with an analog filter with a normalized transfer function $H(s)$ where "normalized" means the cutoff frequency is set to $\omega = 1\,\text{rad/sec}$; this is the typical way analog transfer functions are described and specified.

2. For LPF and HPF types, choose the cutoff frequency of the digital filter ω_d and get the pre-warped analog corner frequency using Equation 11.13.

$$\omega_a = \tan\left(\frac{\omega_d T}{2}\right) \tag{11.13}$$

For BPF and BSF types, specify ω_{dL} and ω_{dH} (the lower and upper corner frequencies of the digital filter) and calculate the two analog corner frequencies ω_{aL} and ω_{aH} along with the ω_0^2 and W terms with Equation 11.14.

$$\omega_{aL} = \tan\left(\frac{\omega_{dL} T}{2}\right)$$

$$\omega_{aH} = \tan\left(\frac{\omega_{dH} T}{2}\right) \tag{11.14}$$

$$\omega_0^2 = \omega_{aL}\omega_{aH}$$

$$W = \omega_{aH} - \omega_{aL}$$

3. Scale the normalized filter's cutoff frequency out to the new analog cutoff ω_a by replacing s with the appropriate scaling factor in Equation 11.15 in the analog transfer function.

Filter Type	Scaling Factor
LPF	$s = \dfrac{s}{\omega_a}$
HPF	$s = \dfrac{\omega_a}{s}$
BPF	$s = \dfrac{s^2 + \omega_0^2}{Ws}$
BSF	$s = \dfrac{Ws}{s^2 + \omega_0^2}$

$$\tag{11.15}$$

4. Apply the bilinear transform by replacing s with Equation 11.16.

$$s = \frac{z-1}{z+1} \tag{11.16}$$

5. Manipulate the digital transfer function $H(z)$ algebraically to get it into the familiar form so you can identify the coefficients—this is often the most difficult part.

11.2.1 Example

Convert the basic RC analog LPF in Figure 11.9a into a digital LPF. The sample rate is 44.1 kHz and the desired cutoff frequency is 1 kHz.

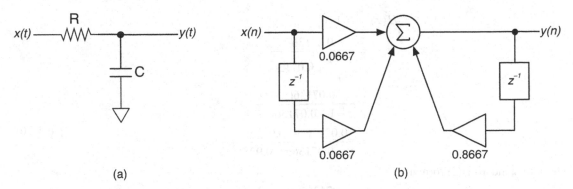

Figure 11.9: (a) A simple analog RC filter circuit. (b) Its digital equivalent.

Step 1: get the normalized analog transfer function *H(s)*.

$$H(j\omega) = \frac{1}{j\omega RC + 1}$$

$$\text{let } s = j\omega$$

Normalize by setting $RC = 1$

$$H(s) = \frac{1}{s+1}$$

(11.17)

Step 2: calculate the pre-warped analog cutoff.

$$f_c = 1\text{kHz}$$
$$\omega_d = 2\pi f_c = 6283.2\text{rad/sec}$$
$$\omega_a = \tan(\frac{\omega_d T}{2}) = \tan\left[\frac{(6283.2)(1/44100)}{2}\right] = 0.07136$$

(11.18)

Step 3: de-normalize the analog transfer function *H(s)* with the appropriate factor.

$$H(s) = \frac{1}{s+1}$$

$$s = \frac{s}{\omega_a}$$

(11.19)

$$H'(s) = \frac{1}{s/\omega_a + 1} = \frac{1}{s/0.07136 + 1}$$

$$= \frac{0.07136}{s + 0.07136}$$

Step 4: apply the BZT. Notice the algebraic manipulation in the last step, dividing by *z* to produce z^{-1} terms and then normalizing so that the first term in the denominator is 1.

$$H(z) = H'(s)\big|_{s=(z-1)/(z+1)}$$

$$= \frac{0.07136}{\dfrac{z-1}{z+1} + 0.07136}$$

$$= \frac{0.07136(z+1)}{z-1+0.07136(z+1)}$$

$$= \frac{0.07136z + 0.07136}{z + 0.07136z - 0.9286} \tag{11.20}$$

Get into standard H(z) format:

$$= \frac{0.07136 + 0.07136z^{-1}}{1 + 0.07136 - 0.9286z^{-1}}$$

$$H(z) = \frac{0.0667 + 0.0667z^{-1}}{1 - 0.8667z^{-1}} = \frac{a_0 + a_1 z^{-1}}{1 + b_1 z^{-1}}$$

Equation 11.20 is in the format that we need with the numerator and denominator properly formed to observe the coefficients. From the transfer function, you can see that:

$$a_0 = 0.0667$$
$$a_1 = 0.0667$$
$$b_1 = -0.8667$$

The difference equation is:

$$y(n) = 0.0667x(n) + 0.0667x(n-1) + 0.8667y(n-1) \tag{11.21}$$

The direct form block diagram shown in Figure 11.9b reveals a pole–zero filter. This makes sense: the original analog filter had a pole at s = −1 and a zero at infinity. The digital equivalent has a pole at $z = 0.8667$ and a zero at $z = -1$ (Nyquist), both on the real axis.

11.3 Audio Biquad Filter Designs

In our last example, we took a fixed analog filter and applied the BZT to convert it to a digital equivalent with the bilinear transform. However, we would like to use the same concept to convert analog filters with arbitrary attributes, such as f_c, Q, or filter gain (boost or cut), into their digital equivalents. Moreover, as audio engineers we would also like the parameters to be continuously adjustable so that the user may control them with GUI elements like knobs or sliders. We would also like to be able to easily modulate the parameters programmatically—for example, we might want to use a low frequency oscillator (LFO) to smoothly modulate the cutoff frequency so that it changes over time.

The good news is that we can apply some fairly straightforward trigonometry and algebra to achieve this goal. In addition, there are multiple approaches to solving this problem mathematically, so you might find two different sets of equations that produce the same frequency response characteristics. There are also multiple topologies for creating the same filters (see Homework problem 4) so you might find different algorithms that combine biquad structures in different ways but which also produce the same result.

The bad news is that the math is not trivial: it will require some work, including the use of trigonometric identities and some complex algebra. Fortunately, DSP engineers worked on this conversion problem very early on in the development of efficient audio algorithms for signal processing. The result is a robust set of filter design equations for just about every classical analog filter type. If you want to learn more, check out Bristow-Johnson's excellent and concise paper (Bristow-Johnson, n.d.) or the set of Motorola application notes (see Motorola, 1991) that constitute some of the earliest work on the subject—you might want to brush up on your trigonometric identities first.

The following classical analog filters are converted to digital and implemented as biquad topologies:

- LPF
- HPF
- BPF
- BSF
- LSF
- HSF
- constant and non-constant Q parametric filters
- 2nd order Butterworth LPF and HPF
- 2nd order Linkwitz-Riley LPF and HPF (good for crossovers)
- 1st and 2nd order APFs

11.3.1 The Audio Biquad Filter Structure

For audio-specific filters, we will use a modified biquad structure that will allow us to blend the dry (unaffected) audio signal with the wet (processed) portion in any ratio we choose. For many of the filter designs, the mix will always be the same: 0% dry and 100% processed. We will add two new coefficients d_0 (dry) and c_0 (processed) to the final topology. You may choose any of the biquad structures from Chapter 10 to implement the signal processing portion. Figure 11.10 shows the audio

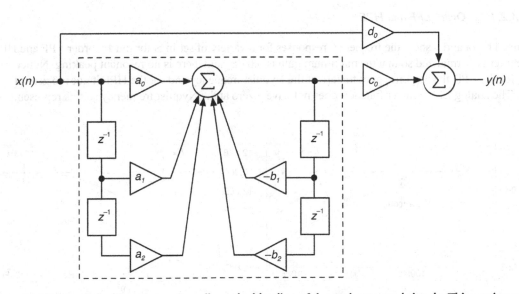

Figure 11.10: The audio biquad structure allows the blending of dry and processed signals, This version uses the direct form (shown in the dashed box), but any of the biquad structures may be substituted in its place.

biquad structure. You can see how the dry and wet signal paths are mixed with the added coefficients. The difference equation is:

$$y(n) = d_0 x(n) + c_0 (a_0 x(n) + a_1 x(n-1) + a_2 x(n-2) - b_1 y(n-1) - b_2 y(n-2)) \quad (11.22)$$

11.3.2 Classical Filters

The following set of filter designs implement the basic analog filter transfer functions that have been around since the earliest days of analog signal processing theory. Sometimes they are called *classical* filters. In all of these filter algorithms, the output is 100% processed and 0% dry so that in each case:

$$c_0 = 1.0$$
$$d_0 = 0.0 \quad (11.23)$$

Due to the trigonometric identities used in the analog filter conversions, and the fact that the BZT requires the *tan* function, the calculations will usually require some kind of trigonometric function: *sin, cos,* and/or *tan.* Of these, the *tan* function has one issue: it is undefined at $\pi/2$. When the design equations require a *tan* function call, we should always check the argument to avoid the undefined condition. We can clamp the argument's value to something very close, such as $0.95\pi/2$. You will see the code to check and clamp this value in the *AudioFilter* object in Section 11.6. These filters are stable across the range of DC to just below Nyquist. Some of the design calculations require a term with $tan(\pi f_c / f_s)$, which will be undefined when $f_c =$ Nyquist.

For our plugin controls, we will always use the full, 10-octave range for the cutoff or center frequencies on the 10-octave span from 20.0 Hz to 20,480 Hz with a sample rate of 44.1 kHz; at this sample rate, the upper limit of 20,480 Hz is about 93% of the Nyquist frequency. If your filters are not stable across this range, double check the coefficient calculations and beware of the *tan* function's problem.

11.3.2.1 1st Order LPF and HPF

Figure 11.11a and b show the frequency responses for a variety of settings for the 1st order LPF and HPF. The filter is normalized so that the maximum gain is 0.0 dB, and there is no resonant peaking. Notice that for the LPF, the response goes to zero at the Nyquist frequency and for the HPF it goes to zero at DC. The analog LPF transfer function does not have a zero at the Nyquist frequency so this represents

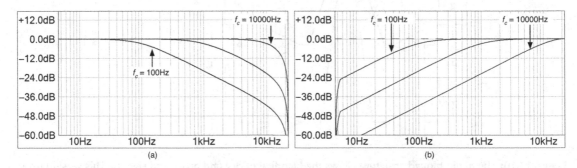

Figure 11.11: (a) The 1st order LPF and (b) HPF with $f_c = $ 100 Hz, 1000 Hz, and 10 kHz.

an error. However, the analog HPF transfer function does have a zero at DC, so its response matches (essentially) perfectly. We will examine the problem with zeros at Nyquist at the end of the chapter and introduce a few more audio filter algorithms.

Specify:

- fc: corner frequency

The design equations are as follows:

$$
\begin{array}{cc}
\text{LPF} & \text{HPF} \\[2mm]
\theta_c = 2\pi f_c / f_s & \theta_c = 2\pi f_c / f_s \\[2mm]
\gamma = \dfrac{\cos(\theta_c)}{1+\sin(\theta_c)} & \gamma = \dfrac{\cos(\theta_c)}{1+\sin(\theta_c)} \\[3mm]
a_0 = \dfrac{1-\gamma}{2} & a_0 = \dfrac{1+\gamma}{2} \\[3mm]
a_1 = \dfrac{1-\gamma}{2} & a_1 = -\left(\dfrac{1+\gamma}{2}\right) \\[3mm]
a_2 = 0.0 & a_2 = 0.0 \\[2mm]
b_1 = -\gamma & b_1 = -\gamma \\[2mm]
b_2 = 0.0 & b_2 = 0.0 \\[2mm]
c_0 = 1.0 & c_0 = 1.0 \\[2mm]
d_0 = 0.0 & d_0 = 0.0
\end{array}
\tag{11.24}
$$

11.3.2.2 2nd Order LPF and HPF

Figure 11.12a and b show the frequency responses for a variety of Q settings for the 2nd order LPF and HPF. The DC gain is normalized to 0.0 dB resulting in peak gains that are greater than unity. A Q of 20 produces a peak gain of ~25 dB or a raw gain of ~18; at this gain, a signal can easily be clipped.

Specify:

- f_c: corner frequency
- Q: quality factor controlling resonant peaking

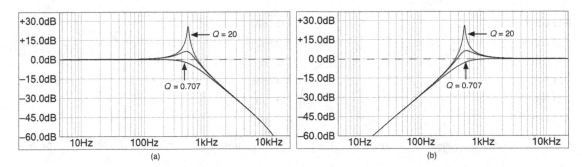

Figure 11.12: (a) The 2nd order LPF and (b) HPF with $f_c = 500$ Hz and Q = 0.707, 2, and 20.

The design equations are as follows.

LPF	HPF

$$\theta_c = 2\pi f_c / f_s \qquad\qquad \theta_c = 2\pi f_c / f_s$$

$$d = \frac{1}{Q} \qquad\qquad\qquad d = \frac{1}{Q}$$

$$\beta = 0.5 \frac{1 - \dfrac{d}{2}\sin(\theta_c)}{1 + \dfrac{d}{2}\sin(\theta_c)} \qquad\qquad \beta = 0.5 \frac{1 - \dfrac{d}{2}\sin(\theta_c)}{1 + \dfrac{d}{2}\sin(\theta_c)}$$

$$\gamma = (0.5 + \beta)\cos(\theta_c) \qquad\qquad \gamma = (0.5 + \beta)\cos(\theta_c)$$

$$a_0 = \frac{0.5 + \beta - \gamma}{2.0} \qquad\qquad a_0 = \frac{0.5 + \beta + \gamma}{2.0}$$

$$a_1 = 0.5 + \beta - \gamma \qquad\qquad a_1 = -(0.5 + \beta + \gamma)$$

$$a_2 = \frac{0.5 + \beta - \gamma}{2.0} \qquad\qquad a_2 = \frac{0.5 + \beta + \gamma}{2.0}$$

$$b_1 = -2\gamma \qquad\qquad\qquad b_1 = -2\gamma$$

$$b_2 = 2\beta \qquad\qquad\qquad b_2 = 2\beta$$

$$c_0 = 1.0 \qquad\qquad\qquad c_0 = 1.0$$

$$d_0 = 0.0 \qquad\qquad\qquad d_0 = 0.0$$

(11.25)

11.3.2.3 2nd Order BPF and BSF

Figure 11.13a and b show the frequency responses for a variety of Q settings for the 2nd order BPF and BSF. The filters are normalized so that the maximum gain is 0.0 dB. Notice once again that the response of the BPF goes to zero at Nyquist—this also represents a slight error from the original analog filter.

Specify:

- f_c: corner frequency
- Q: quality factor controlling width of peak or notch = 1/Bandwidth

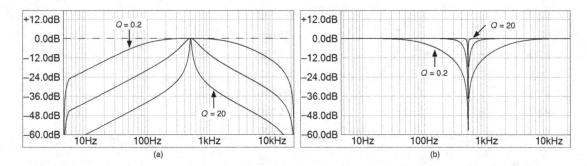

Figure 11.13: (a) The 2nd order BPF and (b) BSF with $f_c = 500$ Hz and Q = 0.2, 2, and 20.

The design equations are as follows.

<table>
<tr><td align="center">BPF</td><td align="center">BSF</td></tr>
</table>

$$K = \tan\left(\frac{\pi f_c}{f_s}\right) \qquad\qquad K = \tan\left(\frac{\pi f_c}{f_s}\right)$$

$$\delta = K^2 Q + K + Q \qquad\qquad \delta = K^2 Q + K + Q$$

$$a_0 = \frac{K}{\delta} \qquad\qquad a_0 = \frac{Q(K^2 + 1)}{\delta}$$

$$a_1 = 0.0 \qquad\qquad a_1 = \frac{2Q(K^2 - 1)}{\delta}$$

$$a_2 = \frac{-K}{\delta} \qquad\qquad a_2 = \frac{Q(K^2 + 1)}{\delta} \qquad (11.26)$$

$$b_1 = \frac{2Q(K^2 - 1)}{\delta} \qquad\qquad b_1 = \frac{2Q(K^2 - 1)}{\delta}$$

$$b_2 = \frac{K^2 Q - K + Q}{\delta} \qquad\qquad b_2 = \frac{K^2 Q - K + Q}{\delta}$$

$$c_0 = 1.0 \qquad\qquad c_0 = 1.0$$

$$d_0 = 0.0 \qquad\qquad d_0 = 0.0$$

11.3.2.4 2nd Order Butterworth LPF and HPF

Butterworth LPFs and HPFs are specialized versions of the ordinary 2nd order HPF. Their Q values are fixed at 0.707, which is the largest value it can assume before peaking in the frequency response is observed. The frequency responses are essentially identical to the general 2nd order filters with a Q of 0.707.

Specify:

* f_c: corner frequency

The design equations are as follows.

<table>
<tr><td align="center">LPF</td><td align="center">HPF</td></tr>
</table>

$$C = \frac{1}{\tan\left(\dfrac{\pi f_c}{f_s}\right)} \qquad\qquad C = \tan\left(\frac{\pi f_c}{f_s}\right)$$

$$a_0 = \frac{1}{1 + \sqrt{2}C + C^2} \qquad\qquad a_0 = \frac{1}{1 + \sqrt{2}C + C^2} \qquad (11.27)$$

$$a_1 = 2a_0 \qquad\qquad a_1 = -2a_0$$

$$a_2 = a_0 \qquad\qquad a_2 = a_0$$

$$b_1 = 2a_0(1 - C^2) \qquad\qquad b_1 = 2a_0(C^2 - 1)$$

$$b_2 = a_0(1 - \sqrt{2}C + C^2) \qquad\qquad b_2 = a_0(1 - \sqrt{2}C + C^2)$$

$$c_0 = 1.0 \qquad\qquad c_0 = 1.0$$

$$d_0 = 0.0 \qquad\qquad d_0 = 0.0$$

11.3.2.5 2nd Order Butterworth BPF and BSF

Butterworth BPFs and BSFs have an adjustable bandwidth that is related to Q. The frequency responses are essentially identical to the general 2nd order BPFs and BSFs.

Specify:

- f_c: corner frequency
- BW: bandwidth of peak or notch $= f_c/Q$

The design equations are as follows.

$$
\begin{array}{ll}
\quad\text{BPF} & \qquad\text{BSF} \\[2mm]
C = \dfrac{1}{\tan(\pi f_c BW\,/\,f_s)} & \quad C = \tan(\pi f_c BW\,/\,f_s) \\[4mm]
D = 2\cos(2\pi f_c\,/\,f_s) & \quad D = 2\cos(2\pi f_c\,/\,f_s) \\[3mm]
a_0 = \dfrac{1}{1+C} & \quad a_0 = \dfrac{1}{1+C} \\[4mm]
a_1 = 0.0 & \quad a_1 = -a_0 D \\[2mm]
a_2 = -a_0 & \quad a_2 = a_0 \\[2mm]
b_1 = -a_0(CD) & \quad b_1 = -a_0 D \\[2mm]
b_2 = a_0(C-1) & \quad b_2 = a_0(1-C) \\[2mm]
c_0 = 1.0 & \quad c_0 = 1.0 \\[2mm]
d_0 = 0.0 & \quad d_0 = 0.0
\end{array}
\tag{11.28}
$$

11.3.2.6 2nd Order Linkwitz-Riley LPF and HPF

2nd order Linkwitz-Riley LPFs and HPFs are designed to have an attenuation of −6 dB at the corner frequency rather than the standard −3 dB. Figure 11.14a shows the Butterworth and Linkwitz-Riley LPF responses while Figure 11.14b zooms in on their cutoff frequencies; their outputs sum exactly and the resulting response is flat. They are often used in crossovers. We will use them for the spectral *Dynamics* plugin later on.

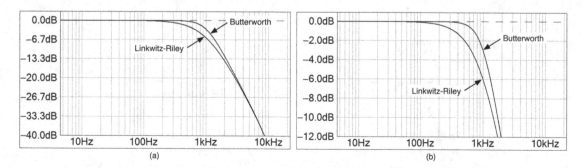

Figure 11.14: A comparison of Butterworth and Linkwitz-Riley filters, each with $f_c = 1$ kHz. (a) The two filters have the same asymptotic slopes for frequencies above the cutoff. (b) Zooming in reveals that the Linkwitz-Riley filter is attenuated by −6 dB at f_c rather than the usual −3 dB of the Butterworth filter.

Specify:

- f_c: corner frequency (−6 dB for Linkwitz-Riley)

The design equations are as follows.

LPF   HPF

$$\text{LPF} \qquad\qquad\qquad\qquad \text{HPF}$$

$$\theta_c = \pi f_c / f_s \qquad\qquad\qquad \theta_c = \pi f_c / f_s$$

$$\Omega_c = \pi f_c \qquad\qquad\qquad\qquad \Omega_c = \pi f_c$$

$$\kappa = \frac{\Omega_c}{\tan(\theta_c)} \qquad\qquad\qquad \kappa = \frac{\Omega_c}{\tan(\theta_c)}$$

$$\delta = \kappa^2 + \Omega_c^2 + 2\kappa\Omega_c \qquad\qquad \delta = \kappa^2 + \Omega_c^2 + 2\kappa\Omega_c$$

$$a_0 = \frac{\Omega_c^2}{\delta} \qquad\qquad\qquad\qquad a_0 = \frac{\kappa^2}{\delta}$$

$$a_1 = 2\frac{\Omega_c^2}{\delta} \qquad\qquad\qquad\qquad a_1 = \frac{-2\kappa^2}{\delta} \qquad\qquad (11.29)$$

$$a_2 = \frac{\Omega_c^2}{\delta} \qquad\qquad\qquad\qquad a_2 = \frac{\kappa^2}{\delta}$$

$$b_1 = \frac{-2\kappa^2 + 2\Omega_c^2}{\delta} \qquad\qquad b_1 = \frac{-2\kappa^2 + 2\Omega_c^2}{\delta}$$

$$b_2 = \frac{-2\kappa\Omega_c + \kappa^2 + \Omega_c^2}{\delta} \qquad b_2 = \frac{-2\kappa\Omega_c + \kappa^2 + \Omega_c^2}{\delta}$$

$$c_0 = 1.0 \qquad\qquad\qquad\qquad c_0 = 1.0$$

$$d_0 = 0.0 \qquad\qquad\qquad\qquad d_0 = 0.0$$

11.3.2.7 1st and 2nd Order APF

APFs have interesting designs that yield equally interesting results. Their frequency responses are perfectly flat from DC to Nyquist. However, the phase responses are the same as those in 1st and 2nd order LPFs. You get all of the phase shift but none of the frequency response change. These filters can be found in crossovers and reverb algorithms. They are also used in the phaser plugin and are the filter kernels for the classic Regalia-Mitra filters (see the Homework section). APFs are designed with pole–zero pairs whose pole/zero radii are reciprocals of one another. For a 1st order APF, the pole lies somewhere on the real axis inside the unit circle at radius R while the zero lies outside the unit circle at radius 1/R as shown in Figure 11.15a. The phase response it produces in Figure 11.15b is identical to that of a 1st order LPF. The 2nd order APF has a conjugate pair of complementary poles and zeros at the reciprocal radii, as shown in Figure 11.16a with the phase response in Figure 11.16b likewise the same as a 2nd order LPF. Figure 11.17 shows the actual phase responses of 1st and 2nd order APFs at a variety of f_c values. Notice that in the 2nd order case, the phase response plot "wraps" at ±180 degrees, a common way of plotting phase since the overall phase shift may be many times 360 degrees; wrapping allows us to plot any phase shift curve within the same ±180 degree range. If you think about the geometric interpretation and analysis of the transfer function, as you move around the unit circle, your vector distance from the pole and zero will always be reciprocal, or 1/each other. The amplitude response is then flat but the phase response does not change because it is not based on the distance to

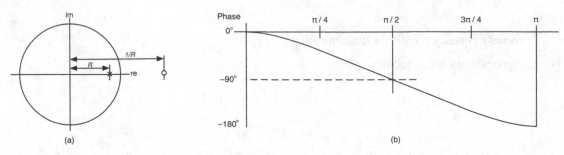

Figure 11.15: The 1st order APF has a flat frequency response but shifts the phase of the signal by −90 degrees at f_c.

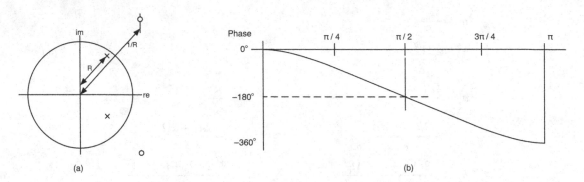

Figure 11.16: The 2nd order APF adds another −90 degrees of phase shift at f_c.

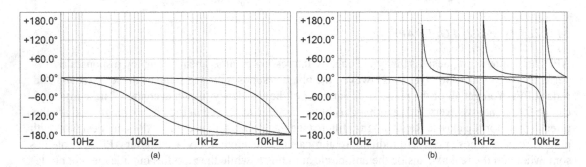

Figure 11.17: The measured phase responses of (a) 1st order and (b) 2nd order APFs with f_c = 100 Hz, 1 kHz, and 10 kHz. Notice the phase wrapping for the 2nd order filters.

the point in question but rather on the angle of incidence of the ray line drawn from the analysis point to the pole or zero.

Specify:

- f_c: corner frequency
- Q: steepness of phase shift at f_c (2nd order only)

The design equations are as follows.

1st Order APF

$$\alpha = \frac{\tan(\pi f_c / f_s) - 1}{\tan(\pi f_c / f_s) + 1}$$

1st Order APF

$$BW = \frac{f_c}{Q}$$

$$\alpha = \frac{\tan(((BW)*\pi)/f_s) - 1}{\tan(((BW)*\pi)/f_s) + 1}$$

$$\beta = -\cos(2\pi f_c / f_s)$$

$$a_0 = \alpha$$
$$a_1 = 1.0$$
$$a_2 = 0.0$$
$$b_1 = \alpha$$
$$b_2 - 0.0$$
$$c_0 = 1.0$$
$$d_0 = 0.0$$

$$a_0 = -\alpha$$
$$a_1 = \beta(1-\alpha)$$
$$a_2 = 1.0$$
$$b_1 = \beta(1-\alpha)$$
$$b_2 = -\alpha$$
$$c_0 = 1.0$$
$$d_0 = 0.0$$

(11.30)

11.3.2.8 1st Order Shelving Filters

Shelving filters are used in many tone controls, especially when there are only two—bass and treble—which are almost always implemented as shelf types. The filters specify a shelving frequency and gain or attenuation value. Figure 11.18a and b show the frequency responses of the LSF and HSF for a variety of shelf gain and attenuation settings. Both filters have a shelf frequency $f_{shelf} = 500$ Hz.

11.3.2.9 Low Shelving

Specify:

- f_c: low shelf frequency
- LF shelf gain/attenuation in dB (boost/cut)

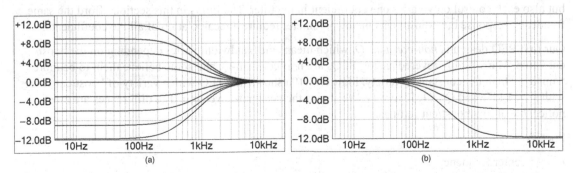

Figure 11.18: 1st order (a) low and (b) high shelf filters with $f_{shelf} = 500$ Hz and shelf gain = −12, −9, −6, −3, 0, 3, 6, 9, and 12 dB.

11.3.2.10 High Shelf

Specify:

- f_c: high shelf frequency
- HF shelf gain/attenuation in dB (boost/cut)

The design equations are as follows.

Low Shelving

$$\theta_c = 2\pi f_c / f_s$$

$$\mu = 10^{ShelfGain(dB)/20}$$

$$\beta = \frac{4}{1+\mu}$$

$$\delta = \beta \tan\left(\theta_c / 2\right)$$

$$\gamma = \frac{1-\delta}{1+\delta}$$

$$a_0 = \frac{1-\gamma}{2}$$

$$a_1 = \frac{1-\gamma}{2}$$

$$a_2 = 0.0$$

$$b_1 = -\gamma$$

$$b_2 = 0.0$$

$$c_0 = \mu - 1.0$$

$$d_0 = 1.0$$

High Shelving

$$\theta_c = 2\pi f_c / f_s$$

$$\mu = 10^{ShelfGain(dB)/20}$$

$$\beta = \frac{1+\mu}{4}$$

$$\delta = \beta \tan\left(\theta_c / 2\right)$$

$$\gamma = \frac{1-\delta}{1+\delta}$$

$$a_0 = \frac{1+\gamma}{2}$$

$$a_1 = -\left(\frac{1+\gamma}{2}\right)$$

$$a_2 = 0.0$$

$$b_1 = -\gamma$$

$$b_2 = 0.0$$

$$c_0 = \mu - 1.0$$

$$d_0 = 1.0$$

(11.31)

11.3.2.11 2nd Order Parametric EQ Filter: Non-Constant-Q

Parametric EQs allow you to adjust the center frequency, Q, and boost or cut, creating any arbitrary bumps or notches in the frequency response. The PEQ filter (also known as a *peaking filter*) is designed to generate symmetrical boost/cut responses. A true digital PEQ not only has independent controls, but also each control only varies one coefficient in the filter. The PEQs in this section afford the same frequency response but adjustments in any parameter require a recalculation of all the coefficients.

Our first PEQ design is *non-constant-Q*, which means the bandwidth varies depending on the boost/cut value. Many highly revered analog PEQs are non-constant-Q, and some believe them to be more "musical" than the constant-Q variety. Figure 11.19a shows this PEQ with a variety of boost/cut values with $f_c = 1$ kHz and Q = 1. Figure 11.19b shows the constant-Q version. The differences are most noticeable at low boost/cut settings.

Specify:

- f_c: center frequency
- Q: quality factor
- gain/attenuation (boost/cut) in dB

The design equations are as follows.

$$\theta_c = 2\pi f_c / f_s$$

$$\mu = 10^{gain(dB)/20}$$

$$\zeta = \frac{4}{1+\mu}$$

$$\beta = 0.5 \frac{1 - \zeta \tan\left(\theta_c / 2Q\right)}{1 + \zeta \tan\left(\theta_c / 2Q\right)}$$

$$\gamma = \left(0.5 + \beta\right)\cos\theta_c \qquad (11.32)$$

$$a_0 = 0.5 - \beta$$

$$a_1 = 0.0$$

$$a_2 = -(0.5 - \beta)$$

$$b_1 = -2\gamma$$

$$b_2 = 2\beta$$

$$c_0 = \mu - 1.0$$

$$d_0 = 1.0$$

11.3.2.12 2nd Order Parametric EQ Filter: Constant-Q

This design creates an almost perfect constant-Q PEQ filter with only a small error for low boost (or cut) values. Figure 11.19b shows the typical frequency response for a variety of boost/cut settings. Compare the −6, −3, +3, and +6 dB curves with Figure 11.19a. Notice that this design specifies $c_0 = 1.0$ and $d_0 = 0.0$, unlike the non-constant-Q PEQ.

Specify:

- f_c: center frequency
- Q: quality factor
- gain/attenuation (boost/cut) in dB

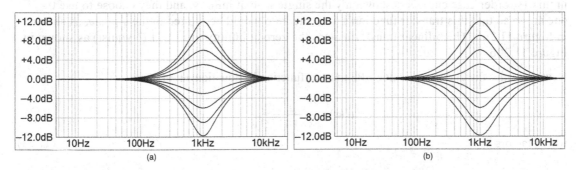

Figure 11.19: Frequency responses of (a) non-constant-Q and (b) constant-Q PEQ filters with *fc* = 1 kHz, Q = 1, and gain = −12, −9, −6, −3, 0, 3, 6, 9, and 12 dB.

The design equations are as follows.

$$K = \tan(\pi f_c / f_s)$$

$$V_o = 10^{gain(dB)/20}$$

$$d_0 = 1 + \frac{1}{Q}K + K^2$$

$$e_0 = 1 + \frac{1}{V_o Q}K + K^2$$

		Boost	Cut
$\alpha = 1 + \dfrac{V_o}{Q}K + K^2$		$a_0 = \dfrac{\alpha}{d_0}$	$a_0 = \dfrac{d_0}{e_0}$
$\beta = 2(K^2 - 1)$		$a_1 = \dfrac{\beta}{d_0}$	$a_1 = \dfrac{\beta}{e_0}$
$\gamma = 1 - \dfrac{V_o}{Q}K + K^2$		$a_2 = \dfrac{\gamma}{d_0}$	$a_2 = \dfrac{\delta}{e_0}$
$\delta = 1 - \dfrac{1}{Q}K + K^2$		$b_1 = \dfrac{\beta}{d_0}$	$b_1 = \dfrac{\beta}{e_0}$
$\eta = 1 - \dfrac{1}{V_o Q}K + K^2$		$b_2 = \dfrac{\delta}{d_0}$	$b_2 = \dfrac{\eta}{e_0}$
		$c_0 = 1.0$	$c_0 = 1.0$
		$d_0 = 0.0$	$d_0 = 0.0$

(11.33)

11.4 Poles and Zeros at Infinity

In the analog transfer function (Equation 11.17) of the bilinear transform example, you can see that there is an analog pole at s = −1 since that would make the transfer function go to infinity and there is a zero at s = ∞ because that will cause *H(s)* to become 0.0. There is also a zero at s = −∞. Interestingly, these two infinity values are in the same location, because the reality is that the σ and jω axes conceptually wrap around an infinitely large sphere and touch each other at ±∞. So, in this 1st order case engineers show only the single zero at infinity, and they choose to use the one at −∞ so that this transfer function's pole and zero would be plotted in the s-plane, as shown in Figure 11.20a. The bilinear transform maps the zero at infinity to z = −1 (Nyquist) as shown in Figure 11.20b.

Next, consider the 2nd order analog LPF transfer function in Equation 11.34.

$$H(s) = \frac{1}{s^2 + (1/Q)s + 1}$$

which factors as

(11.34)

$$H(s) = \frac{1}{(s - P)(s - P^*)}$$

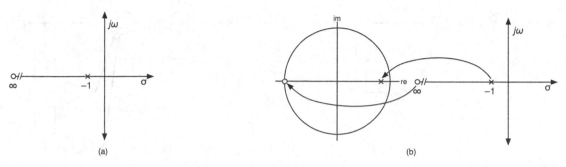

(a)   (b)

Figure 11.20: (a) The pole at $s = -1$ and the zero at $s = \pm\infty$ plotted in the s-plane. (b) The BZT maps real zeros at infinity to the Nyquist frequency in the z-plane.

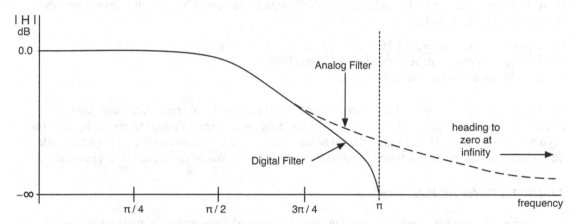

Figure 11.21: The zero at infinity causes an error in the upper part of the frequency that is worst at the Nyquist frequency. This error changes depending on the filter's cutoff frequency.

This transfer function has a pair of conjugate poles at locations P and P* or $(a + bj\omega)$ and $(a - bj\omega)$ as well as a pair of zeros at $\pm\infty$. The BZT maps the poles on the left side of the s-plane to locations inside of the unit circle. Once again, it maps the zeros at $\pm\infty$ to $z = -1$ or the Nyquist frequency.

For both cases, the zero(s) at $z = -1$ will cause the frequency response to drop to 0.0 at Nyquist, where the analog filter would have some finite gain. Figure 11.21 shows the difference between the analog filter and its BZT digital version. There is a small but definite error at high frequencies.

When you think about it, mapping these zeros at infinity to Nyquist makes some sense: Nyquist is the highest possible (legal) frequency in our digital world, and infinity is the highest frequency in the analog domain. Since a BPF is a cascade of low-pass and HPFs, it has the same issue at high frequencies in the frequency response curve. Even with the error in the BZT, it is still an excellent tool for converting existing designs.

There are a few methods available to fix this issue, or at least minimize the error to some extent (Orfanidis, 1997; Massberg, 2011). Both of these methods produce filters with finite gain at Nyquist, but they are also computationally expensive to use if the parameters are to be adjusted in real time or via modulation. The Massberg filter in particular has a very complicated set of design equations when compared to our standard BZT versions here. Fortunately, we have several interesting and easy options

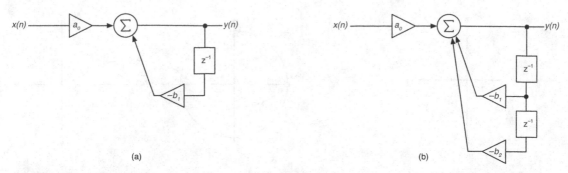

Figure 11.22: The direct-form implementations of (a) 1st order and (b) 2nd order all pole filters.

to create 1st and 2nd order LPF and 2nd order BPFs with finite gain at Nyquist that have reasonably simple designs. These include:

- all-pole 1st and 2nd order LPFs
- Vicanek's analog matched 2nd order LPFs and BPFs
- impulse invariant 1st and 2nd order LPFs

The first two designs are done directly in the z-plane. In both cases, we create APFs that have no zeros in the transfer function. Without zeros, the frequency response can't drop to zero at Nyquist (or anywhere else). Figure 11.22a shows the 1st order while Figure 11.22b shows the 2nd order all pole filter block diagrams in direct form. As with all the other filters, any of the biquad forms may be used.

11.4.1 1st Order All Pole Filter

This simple 1st order filter works by moving the pole back and forth across the real axis between $z = 0$ and $z = +1$. The pole is positioned such that the magnitude of the $H(z)$ transfer function is −3 dB at the desired f_c and 0 dB (unity gain) at DC, representing two degrees of freedom. There are two coefficients in the transfer function, a_0 and b_1 in Equation 11.35, that may be calculated based on these constraints. We will use this filter extensively in our reverb plugin designs in Chapter 17.

$$H(z) = a_0 \left[\frac{1}{1 + b_1 z^{-1}} \right] \tag{11.35}$$

Figure 11.23b shows the 1st order all pole filter in operation with several different values of fc. Notice the finite gain at Nyquist and the severe distortion of the BZT curves in Figure 11.23a from the all-pole version as the cutoff frequency increases.

Specify:

- f_c: the corner frequency

The design equations are as follows.

$$\begin{aligned}
\theta_c &= 2\pi f_c / f_s \\
\gamma &= 2 - \cos(\theta_c) \\
b_1 &= \sqrt{\gamma^2 - 1} - \gamma \\
a_0 &= 1 + b_1
\end{aligned} \tag{11.36}$$

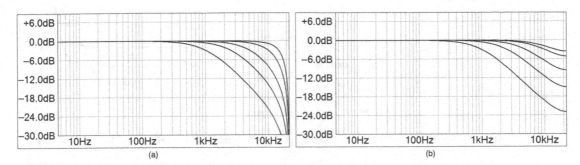

Figure 11.23: The 1st order (a) BZT and (b) all-pole LPF with $f_c = 1$ kHz, 2.5 kHz, 5 kHz, 10 kHz, and 15 kHz.

11.4.2 2nd Order All Pole Filter: The MIDI Manufacturer's Association LPF

The 2nd order all pole filter here is quite unique and comes from the MIDI Manufacturer's Association (MMA). This group specified an early digital software synthesizer known in the DLS Level I and DLS Level II specifications ("DLS" stands for "DownLoadable Sounds"). Part of this specification is for a 2nd order all pole filter that is designed directly in the z-plane, and I will call this the *MMA LPF*. This 2nd order design places a conjugate pair of poles at the desired f_c angle and with a proper radius to produce the desired resonant peak (Q). However, the design become much more interesting with the introduction of two design criteria.

1. The filter's cutoff (−3 dB) frequency f_c is specified in a way such that it falls on an asymptotic cutoff line. The "asymptotic cutoff line will be the line drawn at −12 dB /octave passing through the point one octave below the Nyquist frequency on the filter" and up to the "point of intersection of the DC gain line of the filter." The MMA further clarify that "this method is used rather than the standard −3 dB point because this makes the filter cutoff frequency less sensitive to the height of the resonant peak. Note that in most cases, when calculated in this manner, the difference between the cutoff frequency and the center frequency of the resonant peak (when resonance is present) is very small." (DLS Level 2 Specifications, p.12)
2. The filter's gain is automatically lowered as the filter's Q goes above 0.707—but rather than normalize the filter such that the peak gain is always 0.0 dB (as is done with the 2nd order BPF in this chapter), the filter must attenuate the response by an amount equal to one-half of the resonant peak gain in dB. For example, if the user's Q setting produces a peak gain of +12 dB, the filter should attenuate itself by −6 dB.

The first of these design specs affects only the actual location of the filter's true −3 dB point. Figure 11.24a shows the difference between the MMA LPF and the stock 2nd order digital LPF with the same f_c and Q = 0.707; you can see both the finite gain at Nyquist and the difference in the −3 dB point. However, when the Q rises above 0.707, f_c represents the peak frequency rather than the −3 dB frequency, and it is essentially identical between the filters, but the MMA LPF's gain drops as the resonance is increased as shown in Figure 11.24b.

Why would the MMA specify the filter's gain to drop in response to the increase in Q? There are a couple of reasons: first, the reduction in filter gain helps prevent overload distortion that can occur as a result of the high peak gains that accompany high Q values. Second, this is the same behavior as the famous and highly musical-sounding Moog ladder filter. The Moog ladder filter is analyzed and modeled in detail in *Designing Software Synthesizer Plug-Ins in C++*. The MMA LPF is likewise very musical. It is also easy to disable this gain reduction feature if you wish, or modify it to produce more or less gain reduction based on the filter's Q value. The transfer function is shown in Equation 11.37.

$$H(z) = a_0 \left[\frac{1}{1 + b_1 z^{-1} + b_2 z^{-2}} \right] \tag{11.37}$$

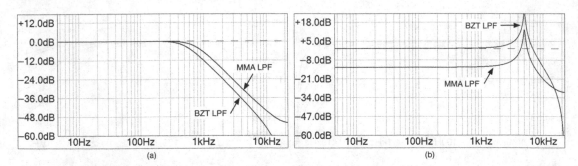

Figure 11.24: The BZT and MMA LPF responses with (a) $f_c = 500$ **Hz and** $Q = 0.707$; **(b)** $f_c = 5$ **kHz and** $Q = 20$ **(where we have filter gain reduction equal to half the resonant peak gain in dB). Notice that in both cases, there is finite gain at Nyquist.**

In order to calculate the filter's a_0 coefficient that controls filter gain, we will need to know the resonant peak gain in dB-based on the filter's Q setting. For our plugins, we will calculate this value on the fly as the Q parameter is changed. You might want to eventually create a lookup table of values to avoid the calculation. Fortunately, this calculation is not overly complex; it is far simpler than either the Orfanidis or Massberg solutions. The equation that relates Q and peak gain is shown in Equation 11.38, and it can then be converted to dB using the standard dB equation. In the design equations that follow, notice the familiar calculations for the b_1 and b_2 coefficients based on the pole radius and angle.

$$\text{Peak Gain} = \frac{Q^2}{\sqrt{Q^2 - 0.25}} \ (Q > 0.707) \tag{11.38}$$

Specify:

- f_c: the frequency of the resonant peak, or the MMA defined −3 dB frequency if the Q is 0.707 or less
- Q: quality factor

The design equations are as follows.

$$\theta_c = 2\pi f_c / f_s$$
$$Q \le 0.707 : resonance = 0$$

$$Q > 0.707 : resonance = 20\log\left[\frac{Q^2}{\sqrt{Q^2 - 0.25}}\right]$$

$$r = \frac{\cos(\theta_c) + \sin(\theta_c)\sqrt{10^{resonance/10} - 1}}{10^{resonance/20}\sin(\theta_c) + 1} \tag{11.39}$$

$$g = 10^{-resonance(dB)/40}$$
$$b_1 = -2r\cos(\theta_c)$$
$$b_2 = r^2$$
$$a_0 = g(1 + b_1 + b_2)$$

The gain reduction g is calculated using the familiar equation $10^{dB/20}$, but with the resonance-in dB value cut in half (which effectively changes the denominator to 40 in the exponent), and negated (−), which creates an attenuation value. To disable the gain reduction with increased Q, just let $g = 1.0$. Figure 11.25a shows the MMA LPF with $f_c = 2$ kHz and various Q values; Figure 11.25b shows the same filter with $f_c = 3$ kHz with gain reduction disabled. In Figure 11.25b, the peak gain is ~32 dB or a gain of almost 40, which will certainly distort most signals with frequency components near the peak frequency. Both version preserve finite gain at Nyquist.

11.4.3 Vicanek's Analog Matched Magnitude 2nd Order LPF

Vicanek formulated several analog matched filter designs that are simple to implement, including 2nd order LPF and BPF designs. These designs involve matching analog and digital transfer function magnitudes at multiple frequencies including DC, Nyquist, and the cutoff frequency f_c. In addition, the designs have a hybrid quality: they use pole locations taken from the *impulse invariant* filter design method in Section 11.5 combined with one or more zeros; for the LPF, one additional real zero is located just inside the unit circle at the Nyquist frequency, but not right on it. This gives the filter a finite gain at the Nyquist frequency. There are two variations—a tight fit and loose fit—which involve strictness and degrees of freedom in the curve matching strategy. The difference between the two fits is negligible at almost all frequencies, except those extremely close to Nyquist.

The LPF version uses a conjugate pair of poles along with one zero located on the real axis. Both designs share the same exact pole locations (and therefore the denominator b_N coefficients) and only differ in the calculation for the zero, which affects the a_0 and a_1 coefficients—the a_2 coefficient is 0.0 since there is only a single 1st order zero to implement. This 2nd order LPF does not implement gain reduction with increasing Q like the MMA LPF, but you could certainly add this functionality yourself. Figure 11.26

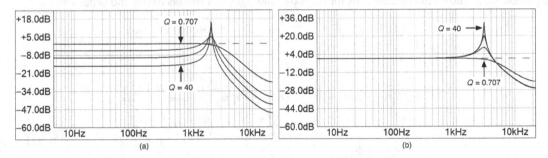

Figure 11.25: The MMA LPF in (a) normal operation with $f_c = 2$ kHz and (b) with $f_c = 3$ kHz and gain reduction disabled; for both plots $Q = 0.707$, 3, 10, and 40.

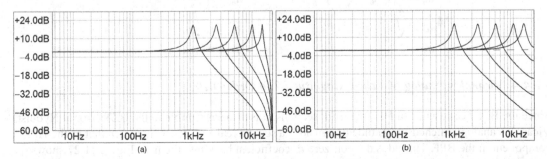

Figure 11.26: 2nd order LPF responses with (a) the standard BZT and (b) Vicanek's "tight fit" matched magnitude implementations. In both plots $Q = 10$ and $f_c = 1$ kHz, 2.5 kHz, 5 kHz, 10 kHz, and 15 kHz.

shows how this design compares with the normal BZT filter. Figure 11.26a shows the high frequency details of the BZT LPF; Figure 11.26b shows the loose and tight fit 2nd order Vicanek LPFs.

The design equations for Vicanek's analog-matched-magnitude filters use a common set of equations—two of which are for the b_M coefficients—which set the pole locations. These are used for the other designs as well, and are shown in Equation 11.40.

Vicanek's Analog-Magnitude-Matched b_M Coefficientswc

$$\omega_c = \frac{2\pi f_c}{f_s} \quad q = \frac{1}{2Q}$$

$$b_1 = \begin{cases} -2e^{-q\omega_c}\cos\left(\left(\sqrt{1-q^2}\right)\omega_c\right) & q \le 1 \\ -2e^{-q\omega_c}\cosh\left(\left(\sqrt{q^2-1}\right)\omega_c\right) & q > 1 \end{cases}$$

$$b_2 = e^{-2q\omega_c}$$

(11.40)

Other Required Variables

$$\Phi_0 = 1 - \sin^2\left(\frac{\omega_c}{2}\right) \quad \Phi_1 = \sin^2\left(\frac{\omega_c}{2}\right) \quad \Phi_2 = 4\Phi_0\Phi_1$$

$$B_0 = (1 + b_1 + b_2)^2 \quad B_1 = (1 - b_1 + b_2)^2 \quad B_2 = -4b_2$$

Equation 11.41 shows the calculations for the loose and tight fit versions of the 2nd order LPF.

Tight Fit

$$A_0 = B_0$$

$$R_1 = (B_0\Phi_0 + B_1\Phi_1 + B_2\Phi_2)Q^2$$

$$A_1 = \frac{(R_1 - A_0\Phi_0)}{\Phi_1}$$

$$a_0 = \frac{1}{2}\left(\sqrt{A_0} + \sqrt{A_1}\right)$$

$$a_1 = \sqrt{A_0} - a_0$$

$$a_2 = 0$$

Loose Fit

$$f_0 = \frac{\omega_c}{\pi}$$

$$r_1 = \frac{(1 - b_1 + b_2)f_0^2}{\sqrt{\left(1 - f_0^2\right)^2 + \frac{f_0^2}{Q^2}}}$$

$$r_0 = 1 + b_1 + b_2$$

$$a_0 = \frac{(r_0 + r_1)}{2}$$

$$a_1 = r_0 - a_0$$

$$a_2 = 0$$

(11.41)

11.4.4 Vicanek's Analog Matched Magnitude 2nd Order BPF

Since the BPF is a cascade of HPF and LPF sections, it is also susceptible to the BZT error at Nyquist. Vicanek's analog-matched-magnitude design adds a second real zero at DC to match that of the HPF component of the BPF. This adds the non-zero a_2 coefficient back into the mix. Figure 11.27 shows the difference between (a) the BZT and (b) analog matched versions that include the finite gain at Nyquist. As with the LPF version, the difference between the tight and loose versions is negligible and manifests

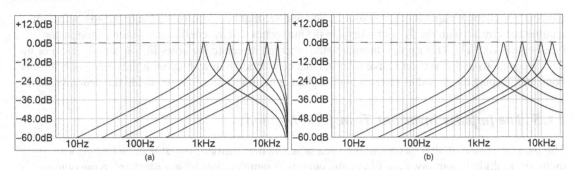

Figure 11.27: 2nd order BPF responses with (a) the standard BZT and (b) Vicanek's "tight fit" matched magnitude implementations, In both plots $Q = 10$ and $f_c = 1$ kHz, 2.5 kHz, 5 kHz, 10 kHz, and 15 kHz.

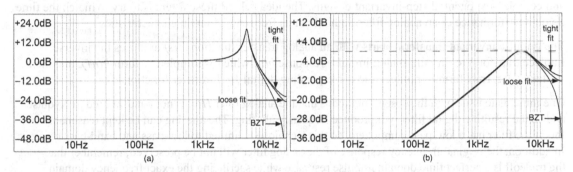

Figure 11.28: High frequency differences between the BZT and Vicanek's loose and tight fit analog magnitude matched versions for $f_c = 5$ kHz and (a) 2nd order LPF and (b) 2nd order BPF.

itself only at frequencies very near Nyquist. Figure 11.28 shows the high frequency details and differences between the BZT, loose fit, and tight fit versions of the 2nd order (a) LPF and (b) BPF.

Equation 11.42 shows the calculations for the loose and tight fit versions of the 2nd order BPF.

Tight Fit

$$R_1 = B_0\Phi_0 + B_1\Phi_1 + B_2\Phi_2$$

$$R_2 = -B_0 + B_1 + 4(\Phi_0 - \Phi_1)B_2$$

$$A_2 = \frac{R_1 - R_2\Phi_1}{4\Phi_1^2}$$

$$A_1 = R_2 + 4(\Phi_1 - \Phi_0)A_2$$

$$a_1 = -\frac{1}{2}\sqrt{A_1}$$

$$a_0 = \frac{1}{2}\left(\sqrt{A_2 + a_1^2} - a_1\right)$$

$$a_2 = -a_0 - a_1$$

Loose Fit

$$f_0 = \frac{\omega_c}{\pi}$$

$$r_0 = \frac{(1 + b_1 + b_2)}{\pi f_0 Q}$$

$$r_1 = \frac{(1 - b_1 + b_2)\left(\dfrac{f_0}{Q}\right)}{\sqrt{\left(1 - f_0^2\right)^2 + \dfrac{f_0^2}{Q^2}}}$$

$$a_1 = \frac{-r_1}{2}$$

$$a_0 = \frac{(r_0 - a_1)}{2}$$

$$a_2 = -a_0 - a_1$$

(11.42)

Note: if you check Vicanek's original source, you will see that he conforms to the reversed coefficient naming convention with the b_N coefficients in the numerator and a_M coefficients in the denominator (see Section 10.21.1). This ripples through to the other intermediate variables like A_0 and B_1, etc. For the design equations in this section I have reversed that convention to match up with the convention for this book and keeping all As and as referencing the numerator and Bs and bs referencing the denominator.

11.5 The Impulse Invariant Transform Method

The *impulse invariant* filter design method is a subset of the family of *input invariant transform* methods. In digital input invariant filters, the outputs at sample times nT are identical to the outputs of the corresponding analog filter when stimulated with the same input. This family includes impulse-invariant, step-invariant, and ramp-invariant types; other variations are possible including sine, cosine, and complex-exponential step-invariant designs. The idea behind these filters is to try to match the time domain performance of an analog filter, rather than its frequency domain characteristics.

The impulse invariant transform starts with an analog transfer function *H(s)* and uses the inverse Laplace transform to find the analog impulse response *h(t),* which is sampled to form *h(nT),* and then we take the *z* Transform to produce the corresponding digital *H(z)* transfer function.

One immediate issue with the design method is that the analog filter *H(s)* is not band-limited to Nyquist, and so the resulting digital filter will produce frequency aliasing. The aliasing can be kept reasonably low if the filter has a low-pass component, so this technique is limited to low-pass and band-pass types. In addition, the original frequency response of the analog filter will not be preserved; remember that the tradeoff is a perfect time domain impulse response while sacrificing the exact frequency domain performance. For filter orders greater than one, there is going to be some significant math effort required to perform the inverse Laplace transform on the original analog filter *H(s)*. The impulse invariant method also has strict requirements regarding the order of the numerator and denominator. The LPF impulse invariant filters are also examples of all-pole filters along with those in Sections 11.4.1 and 11.4.2.

11.5.1 Impulse Invariant 1st Order LPF

For a simple 1st order LPF based on a traditional RC circuit in Figure 11.9a, the impulse invariant design is not too difficult and produces a frequency response that is erroneous only at very high frequencies near Nyquist, with low aliasing. If you have had exposure to basic analog circuit design, then these equations will be reasonably familiar. The digital design equations are also fairly straightforward. For this design, we can start with the analog impulse response of a simple RC LPF, and then find the discrete version. The analog impulse *h(t)* response is a well-known exponentially decaying function based on the RC time constant of the circuit.

$$H(s) = \frac{1}{1 + s/\omega_c}$$

$$\omega_c = \frac{1}{RC}$$

$$h(t) = \omega_c e^{-\omega_c t} u(t)$$

(11.43)

where u is the unit impulse function

$$h(n) = Th(t)\big|_{t=nT}$$
$$= \omega_c Te^{-\omega_c nT} u(n)$$

then

$$H(z) = \frac{\omega_c T}{1 - e^{-\omega_c T} z^{-1}}$$

Examination of the transfer function in Equation 11.43 allows us to set the two coefficients a_0 (numerator) and b_1 (the denominator's z^{-1} coefficient) noting that the filter's overall gain will be dependent on the cutoff frequency ω_c. We can easily normalize the transfer function by scaling a_0 appropriately. The design equations are then found with Equation 11.44. As a 1st order system, the other coefficients a_1, a_2, and b_2 are all set to 0.0.

$$T = \frac{1}{f_s}$$
$$\omega_c = 2\pi f_c$$
$$a_0 = 1 - e^{-\omega_c T} \qquad\qquad (11.44)$$
$$b_1 = -e^{-\omega_c T}$$

Figure 11.29a shows the frequency response of the filter with various settings of f_c while Figure 11.29b shows the very high frequency details comparing the BZT, one-pole LPF, and impulse invariant designs. The impulse invariant version has a slight error in the high frequency response while the one-pole and BZT versions have the correct $f-3dB$ frequency value of 10 kHz.

11.5.2 Impulse Invariant 2nd Order LPF

The design method can be extended to higher order filters, albeit with some deeper math. One design tactic to take from the start is to use a normalized sample rate of 1 Hz, with a period $T = 1$. This manifests itself in the design equations by normalizing the analog cutoff frequency f_c by f_s and it

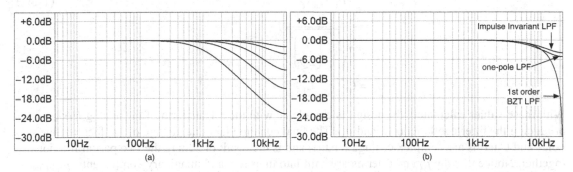

Figure 11.29: (a) The frequency response of the 1st order impulse invariant LPF with $f_c = 1$ kHz, 2.5 kHz, 5 kHz, 10 kHz, and 15 kHz. (b) A comparison of the impulse invariant, 1-pole LPF, and 1st order BZT LPF with $f_c = 10$ kHz.

allows the filter gain-scaling factor to be calculated automatically during the design process, as well as simplifying the math a bit. In Equation 11.45, this f_s normalized f_c is converted into ω_c.

For a 2nd order analog LPF, the frequency scaled analog $H(s)$ may be decomposed with partial fraction expansion where p_1 and p_2 are the analog poles as in Equation 11.45.

$$H(s) = \frac{\omega_c{}^2}{s^2 + \dfrac{\omega_c}{Q}s + \omega_c{}^2} = \frac{\omega_c{}^2}{(s-p_1)(s-p_2)} = \frac{C_1}{s-p_1} + \frac{C_2}{s-p_2}$$

$$\omega_c = \frac{2\pi f_c}{f_s}$$

(11.45)

For real filter coefficients, the poles are guaranteed to be complex conjugates of one another, and this forces the two partial fraction C numerators to also be complex conjugates. This cuts the amount of work in half, as we only need to calculate one pole (p_1) and one numerator term C_1, then the conjugates may be easily found by negating the imaginary component.

The relationship between the pole locations of a 2nd order analog $H(s)$ and the filter cutoff frequency and Q is well known (see any good analog filtering textbook), so we may find the pole locations as follows.

$$p_1 = a + jb$$

$$a = \frac{-\omega_c}{2Q} \qquad b = \omega_c \sqrt{1 - \left(\frac{1}{2Q}\right)^2}$$

(11.46)

$$p_2 = p_1{}^* = a - jb$$

The C_1 term is found using standard partial fraction expansion techniques.

$$C_1 = \frac{\omega_c{}^2}{(s-p_1)(s-p_1{}^*)}(s-p_1{}^*)\Bigg|_{s=p_1{}^*}$$

$$= \frac{\omega_c{}^2}{2b} j$$

(11.47)

$$= \frac{\omega_c}{2\sqrt{1 - \left(\dfrac{1}{2Q}\right)^2}} j$$

$$C_2 = C_1{}^* = -C_1$$

With these values calculated, the design is completed with the z Transform of the analog $H(s)$, producing the filter coefficients a_N and b_M. The final equation takes advantage of the fact that the poles and C terms are complex conjugate pairs, and is the result of adding the two partial fractions together. Notice that the p_1 and C terms are split into their real and imaginary components, p_{re}, p_{im} and C_{re}, C_{im}.

$$H(s) = \frac{C_1}{s - p_1} + \frac{C_1^*}{s - p_1^*} \rightarrow H(z) = \frac{C_1}{1 - e^{p_1 T} z^{-1}} + \frac{C_1^*}{1 - e^{p_1^* T} z^{-1}}$$

$$p_{re} = \frac{-\omega_c}{2Q} \qquad p_{im} = \omega_c \sqrt{1 - \left(\frac{1}{2Q}\right)^2}$$

$$C_{re} = 0 \qquad C_{im} = \frac{\omega_c}{2\sqrt{1 - \left(\frac{1}{2Q}\right)^2}} \qquad (11.48)$$

$$H(z) = \frac{2C_{re} - 2\left[C_{re}\cos(p_{im}T) + C_{im}\sin(p_{im}T)\right]e^{p_{re}T} z^{-1}}{1 - 2\cos(p_{im}T)e^{p_{re}T} z^{-1} + \left(e^{p_{re}T}\right)^2 z^{-2}}$$

Examining the resulting transfer function in Equation 11.48 and noting that we forced the T term to 1.0 early in the design lets us read off the biquad coefficients directly as follows.

$$T = 1$$
$$a_0 = 2C_{re} = 0$$
$$a_1 = -2\left[C_{re}\cos(p_{im}) + C_{im}\sin(p_{im})\right]e^{p_{re}}$$
$$a_2 = 0 \qquad (11.49)$$
$$b_1 = -2\cos(p_{im})e^{p_{re}}$$
$$b_2 = \left(e^{p_{re}}\right)^2$$

Figure 11.30a shows the frequency response of the impulse invariant 2nd order LPF with a variety of cutoff frequencies while Figure 11.30b compares the high frequency response with the standard 2nd order BZT LPF and Vicanek's 2nd order LPF. At very high f_c values, there is a significant error in the frequency response.

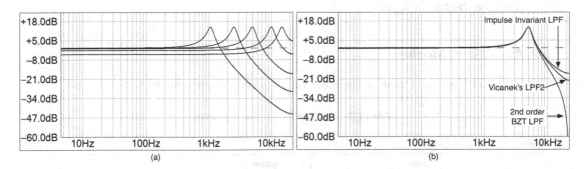

Figure 11.30: (a) The frequency response of the 2nd order impulse invariant LPF with f_c = 1 kHz, 2.5 kHz, 5 kHz, 10 kHz, and 15 kHz and Q = 5. (b) A comparison of the 2nd order impulse invariant, Vicanek, and BZT LPF with f_c = 5 kHz.

11.6 C++ *Effect Object:* AudioFilter

The *fxobjects.h* and *fxobjects.cpp* files include the implementation of a C++ object *AudioFilter* that creates all of the filters in this chapter. You can easily support more filter types if you wish, including those in the Homework section. Like the other C++ objects in the book, it is derived from *IAudioSignalProcessor* and implements the familiar functions *reset*, and *processAudioSample*. The *AudioFilter* object can process one channel of audio.

11.6.1 AudioFilter: *Enumerations and Data Structure*

The filter algorithm is encoded as a strongly typed enumeration and is the only constant required. It is declared at the top of the *fxobjects.h* file and listed here (you can see that the enumeration strings encode the filter types in Sections 11.3–11.5, e.g. *kButterHPF2* is the 2nd order Butterworth HPF).

```
enum class filterAlgorithm {
    kLPF1P, kLPF1, kHPF1, kLPF2, kHPF2, kBPF2, kBSF2,
    kButterLPF2, kButterHPF2, kButterBPF2,kButterBSF2, kMMALPF2,
    kMMALPF2B, kLowShelf, kHiShelf, kNCQParaEQ, kCQParaEQ,
    kLWRLPF2, kLWRHPF2, kAPF1, kAPF2, kResonA, kResonB, kMatchLP2A,
    kMatchLP2B, kMatchBP2A, kMatchBP2B, kImpInvLP1, kImpInvLP2};
```

The object is updated via the custom structure *AudioFilterParameters* that contains members for the fundamental controls f_c, Q, boost/cut, and algorithm.

```
struct AudioFilterParameters
{
    AudioFilterParameters(){}
    AudioFilterParameters& operator=(const AudioFilterParameters&
                                     params){. . .}
    // --- individual parameters
    filterAlgorithm algorithm = filterAlgorithm::kLPF1;
    double fc = 100.0;
    double Q = 0.707;
    double boostCut_dB = 0.0;
};
```

11.6.2 AudioFilter: *Members*

Tables 11.2 and 11.3 list the *AudioFilter* member variables and member functions.

Table 11.2: *AudioFilter* member variables

AudioFilter Member Variables		
Type	**Name**	**Description**
AudioFilterParameters	*parameters*	Filter parameters, f_c, Q, boost/cut and algorithm
Biquad	*biquad*	Biquad calculator object
double	*coeffArray[numCoeffs]*	array to store filter coefficients
double	*sampleRate*	current sample rate

Table 11.3: *AudioFilter* member functions

AudioFilter Member Functions		
Returns	**Name**	**Description**
AudioFilterParameters	*getParameters*	Get all parameters at once
void	*setParameters* Parameters: —*AudioFilterParameters* _parameters	Set all parameters at once
bool	*reset* Parameters: —*double* sampleRate	Flushes biquad state variables
double	*processAudioSample* Parameters: —*double* xn	Process input xn through the audio filter
void	*setSampleRate* Parameters: —*double* _sampleRate	Set sample rate
bool	*calculateFilterCoeffs*	Update filter coefficients, called from setters below

11.6.3 AudioFilter: *Programming Notes*

The *AudioFilter* object implements the specialized biquad topology in Figure 11.10 and is fairly straightforward.

- It contains a *Biquad* member object to perform the biquad filter calculation.
- It contains an array of coefficients. The a_N and b_N coefficients are passed to the *Biquad* member object for use, while the c_0 and d_0 coefficients are used in the *processAudioSample* function to mix the processed and dry signal paths according to the algorithm being used.
- The *IAudioSignalProcessor* interface is used for common reset/process operations.

Using the *AudioFilter* object is straightforward:

1. Reset the object to prepare it for streaming.
2. Set the calculation type in the *AudioFilterParameters* custom data structure.
3. Call *setParameters* to update the calculation type.
4. Call *processAudioSample*, passing the input value in and receiving the output value as the return variable.

The majority of the work in this object happens in the *calculateFilterCoeffs* function, which decodes the filter algorithm and then calculates the filter coefficients using the same equations from this chapter. The object implements about 30 different filter algorithms.

Function: *calculateFilterCoeffs*
Details:

- At the top of the function, we clear out and initialize the coefficients and then we grab the filtering variables from the *AudioFilterParameters* structure.
- Next, we decode the algorithm variable and process accordingly.

We will look at the two pieces of code that calculate the 1st and 2nd order LPF coefficients. Go through the two calculations listed here, compare them with the equations in Sections 11.3.2.1 and 11.3.2.2, and make sure you understand how the C++ code here implements those filter coefficient calculations.

```cpp
bool AudioFilter::calculateFilterCoeffs()
{
    // --- clear coeff array
    memset(&coeffArray[0], 0, sizeof(double)*numCoeffs);

    // --- set default pass-through
    coeffArray[a0] = 1.0;
    coeffArray[c0] = 1.0;
    coeffArray[d0] = 0.0;

    // --- grab these variables like the book
    filterAlgorithm algorithm = audioFilterParameters.algorithm;
    double fc = audioFilterParameters.fc;
    double Q = audioFilterParameters.Q;
    double boostCut_dB = audioFilterParameters.boostCut_dB;

    if (algorithm == filterAlgorithm::kLPF1)
    {
        // --- see book for formulae
        double theta_c = 2.0*kPi*fc / sampleRate;
        double gamma = cos(theta_c) / (1.0 + sin(theta_c));

        // --- update coeffs
        coeffArray[a0] = (1.0-gamma) / 2.0;
        coeffArray[a1] = (1.0-gamma) / 2.0;
        coeffArray[a2] = 0.0;
        coeffArray[b1] = -gamma;
        coeffArray[b2] = 0.0;

        // --- update on calculator
        biquad.setCoefficients(coeffArray);

        // --- we updated
        return true;
    }
    else if (algorithm == filterAlgorithm::kLPF2)
    {
        // --- see book for formulae
        double theta_c = 2.0*kPi*fc / sampleRate;
        double d = 1.0 / Q;
        double betaNumerator = 1.0-((d / 2.0)*(sin(theta_c)));
        double betaDenominator = 1.0 + ((d / 2.0)*(sin(theta_c)));

        double beta = 0.5*(betaNumerator / betaDenominator);
        double gamma = (0.5 + beta)*(cos(theta_c));
        double alpha = (0.5 + beta-gamma) / 2.0;

        // --- update coeffs
        coeffArray[a0] = alpha;
        coeffArray[a1] = 2.0*alpha;
```

```
        coeffArray[a2] = alpha;
        coeffArray[b1] = -2.0*gamma;
        coeffArray[b2] = 2.0*beta;

        // --- update on calculator
        biquad.setCoefficients(coeffArray);

        // --- we updated
        return true;
    }
```

Function: *setParameters*

Details: the *setParameters* function checks to see if parameters have changed, checks the Q value for validity, then calls the *updateFilterCoeffs* function to calculate the coefficients.

```
void setParameters(const AudioFilterParameters& parameters)
{
    if (audioFilterParameters.algorithm != parameters.algorithm ||
        audioFilterParameters.boostCut_dB != parameters.boostCut_dB ||
        audioFilterParameters.fc != parameters.fc ||
        audioFilterParameters.Q != parameters.Q)
    {
        // --- save new params
        audioFilterParameters = parameters;
    }
    else
        return;

    // --- don't allow 0 or (-) values for Q
    if (audioFilterParameters.Q <= 0)
        audioFilterParameters.Q = 0.707;

    // --- update coeffs
    calculateFilterCoeffs();
}
```

Function: *processAudioSample*

Details: the *processAudioSample* function is simple; the *Biquad* member object does the filtering calculation and the *AudioFilter* object mixes its output according to the c_0 and d_0 values.

```
virtual double processAudioSample(double xn)
{
    // --- let biquad do the grunt-work
    //
    // output = (dry) + (processed): x(n)*d0 + y(n)*c0
    return coeffArray[d0] * xn +
           coeffArray[c0] * biquad.processAudioSample(xn);
}
```

11.7 *Chapter Plugin:* IIRFilters

As an example of using the *AudioFilter* object, we will create the *IIRFilters* plugin that implements all 30 of this chapter's filter algorithms in one compact plugin. Figure 11.31 shows the GUI and block diagram for this simple plugin. The filter type control uses a string-list parameter to display strings to the user that correspond to the strongly typed *filterAlgorithm* enumeration. The user adjusts the controls for each filter; note that not all GUI controls are active for all filter types (e.g. 1st order filters have no Q control).

11.7.1 IIRFilters: *GUI Parameters*

The GUI parameter table is shown in Table 11.4—use it to declare your GUI parameter interface for the plugin framework you are using, or alternatively roll your own API-specific parameters if you want to write the plugin "from scratch." If you are using ASPiK, remember to first create your enumeration of control ID values at the top of the *plugincore.h* file and then use automatic variable binding to connect the linked variables to the GUI parameters.

ASPiK: top of *plugincore.h* file:

```
enum controlID {
        filterFc_Hz = 0,
        filterQ = 1,
        boostCut_dB = 2,
        filterType = 3
};
```

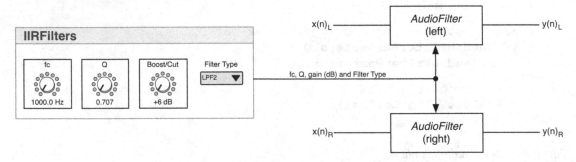

Figure 11.31: The *IIRFilters* plugin's GUI ((shown here on the left) is used to feed parameter information to two *AudioFilter* objects, one for each channel.

TABLE 11.4 The *IIRFilter* plugin's GUI parameter and linked-variable list

Control Name	Units	Min/max/default or string-list	Taper	Linked Variable	Linked Variable Type
fc	Hz	20 / 20480 / 1000	1V/oct	*filterFc_Hz*	*double*
Q	-	0.707 / 20 / 0.707	linear	*filterQ*	*double*
Boost/Cut	dB	−20 / +20 / 0	linear	*boostCut_dB*	*double*
Filter Type		string-list follows this table	-	*filterType*	*int*

The filter type *string list* is a series of strings that exactly matches the sequence of the strongly typed enumeration for *filterAlgorithm*—in reality it is the same list but with the leading "k" removed:

"LPF1P, LPF1, HPF1, LPF2, HPF2, BPF2, BSF2, ButterLPF2, ButterHPF2, ButterBPF2,ButterBSF2, MMALPF2, MMALPF2B, LowShelf, HiShelf, NCQParaEQ, CQParaEQ, LWRLPF2, LWRHPF2, APF1, APF2, ResonA, ResonB, MatchLP2A, MatchLP2B, MatchBP2A, MatchBP2B, ImpInvLP1, ImpInvLP2"

11.7.2 IIRFilters: Object Declarations and Reset

The *IIRFilters* plugin uses two *AudioFilter* members to provide the filtering for the left and right channels of a stereo plugin. To support more channels, just add more member objects. In your plugin's processor (*PluginCore* for ASPiK) declare two members plus the *updateParameters* method:

```
// --- in your processor object's .h file
#include "fxobjects.h"

// --- in the class definition
protected:
    AudioFilter leftAudioFilter;
    AudioFilter rightAudioFilter;
    void updateParameters();
```

The filtering objects require the current DAW sample rate to properly calculate their coefficients. Call the *reset* function and pass in the current sample rate in your plugin framework's prepare-for-audio-streaming function. In this example, *resetInfo.sampleRate* holds the current DAW sample rate.

ASPiK: *reset( )*
```
    // --- reset the filters
    leftAudioFilter.reset(resetInfo.sampleRate);
    rightAudioFilter.reset(resetInfo.sampleRate);
```

11.7.3 IIRFilters: GUI Parameter Update

Update the left and right channel objects at the top of the buffer processing loop. Gather the values from the GUI parameters and apply them to the custom data structure for each of the filter objects by using the *getParameters* method to get the current parameter set, and then overwrite only the parameters your plugin actually implements. Do this for all channels you support. For ASPiK we use the helper *convertIntToEnum* to convert the user's selected value (a zero-indexed integer) to the corresponding strongly typed enumeration as an integer. Notice that since we want both filters to share identical parameters, we just reuse the parameter structure and set the right channel with it.

ASPiK: *updateParameters( )* function

```
// --- update left filter with GUI parameters
AudioFilterParameters filterParams = leftAudioFilter.getParameters();

// --- alter param values
filterParams.fc = filterFc_Hz;
filterParams.Q = filterQ;
```

```
filterParams.boostCut_dB = boostCut_dB;
filterParams.algorithm = convertIntToEnum(filterType, filterAlgorithm);

// --- set on objects
leftAudioFilter.setParameters(filterParams);
rightAudioFilter.setParameters(filterParams);
```

11.7.4 IIRFilters: *Process Audio*

Add the code to process the plugin object's input audio samples through each filter and write the output samples. The ASPiK code is shown here for grabbing the input samples and writing the output samples; use the method that your plugin framework requires to read and write these values.

ASPiK: *processAudioFrame( )* function

```
// --- read input (adapt to your framework)
double xnL = processFrameInfo.audioInputFrame[0]; //< input sample L
double xnR = processFrameInfo.audioInputFrame[1]; //< input sample R

// --- process the audio to produce output
double ynL = leftAudioFilter.processAudioSample(xnL);
double ynR = rightAudioFilter.processAudioSample(xnR);

// --- write output (adapt to your framework)
processFrameInfo.audioOutputFrame[0] = ynL; //< output sample L
processFrameInfo.audioOutputFrame[1] = ynR; //< output sample R
```

Rebuild and test the plugin. Notice especially any sonic differences you may hear between the standard bilinear LPFs, and those that have a finite gain at Nyquist. Have a listening contest to see just how much of that high frequency content you can really hear.

11.8 Homework

1. Modify the BPF filter to add a post-gain stage to provide a user controlled boost. Modify the BPF so that the boost is related to the Q value with sharper peaks getting more boost. The mathematical relationship between Q and boost is up to you.
2. Modify the MMA LPF so that the filter gain is ¼ of the resonant peak gain in dB. Then, modify the plugin to allow the user to specify the gain reduction ratio from 0.1 to 1.0 of the peak gain in dB using a continuous GUI control (knob or slider).
3. The analog transfer function of a 2nd order LPF is:

$$H(s) = \frac{1}{s^2 + (1/Q)s + 1}$$

An analog LPF has the following specifications: $Q = 1, f_c = 1 \text{ kHz}, f_s = 44.1 \text{ kHz}$

Apply the bilinear transform and some algebra to find the coefficients and create the digital equivalent.

(Answer: $a_0 = 0.0047$, $a_1 = 0.0095$, $a_2 = 0.0047$, $b_1 = -1.8485$, $b_2 = 0.8673$)

4. Regalia-Mitra (R-M) filters are some of the earliest audio-specific filter designs. At the core of the R-M topology is an APF. The APF may be 1st or 2nd order. The output of the APF is both added to

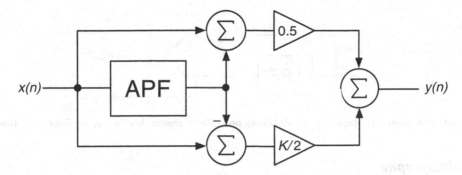

Figure 11.32: The Regalia-Mitra filter topology

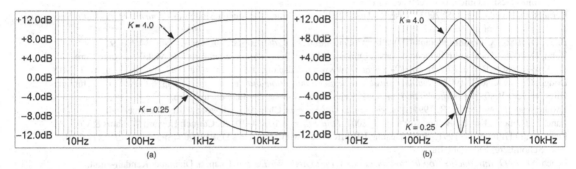

Figure 11.33: (a) The 1st order and (b) 2nd order Regalia-Mitra filters create high shelf and parametric EQ responses. In both filters, f_c = 500 Hz and K = 0.25, 0.4, 0.65, 1.0 (flat), 1.6, 2.5 and 4.0 and Q = 1.0 for the 2nd order version.

and subtracted from the dry signal, then those outputs are mixed using two coefficients: one fixed at 0.5 and the other variable with $K/2$ as shown in Figure 11.32.

When the APF is a 1st order filter, the response is a high shelf type. The APF's f_c controls the shelf frequency and K controls the boost/cut, shown in Figure 11.33a. When K = 0.0, the filter becomes a pure low-pass type.

When the APF is a 2nd order filter, the response is that of a PEQ. The APF's f_c controls the center frequency, its Q controls the width of the peak/notch, and K controls the boost/cut, shown in Figure 11.33b. When K = 0.0, the filter becomes a perfect notch type.

Modify the *AudioFilter* C++ object to include the R-M filters. You will need to add a new coefficient K and provide a mechanism for the user to adjust it from 0.0 to 10.0.

5. We can take advantage of some simple mathematical relationships between the LPF, HPF, and APF to create a multi-filter that can implement all three types simultaneously with only a single LPF core, shown in Figure 11.34. The filters share the same order with the LPF core; when it is 1st order, the other responses are also 1st order, etc. Note: this topology will not work properly with the all-pole LPFs from Sections 11.5.1–11.5.2.

    Create a new C++ object named *MultiFilter* and derive it from the existing *AudioFilter* object to implement this useful filter. Include both 1st and 2nd order types.

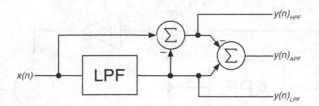

Figure 11.34: The multi-filter topology requires only one LPF to create three different filters, as shown here.

11.9 Bibliography

Allred, R. 2003. *Second-Order IIR Filters Will Support Cascade Implementations: 5 Part Digital Audio Application Tutorial*. EE Times Design Guide, www.eetimes.com.

Berners, D. P. and Abel, J. S. 2003. "Discrete-Time Shelf Filter Design for Analog Modeling." *Journal of the Audio Engineering Society*, Preprint 5939.

Bohn, D. 1986. "Constant-Q Graphic Equalizers." *Journal of the Audio Engineering Society*, Vol. 34, No. 9.

Bohn, D. 2008. "Bandwidth in Octaves Versus Q in Band Pass Filters." *Application Note 170*, Mukilteo: Rane Corp.

Bristow-Johnson, R. n.d. *RBJ Audio-EQ-Cookbook*, www.musicdsp.org/showone.php?id=197, Accessed August 1, 2018.

Dodge, C. and Jerse, T. 1997. *Computer Music Synthesis, Composition and Performance*, Chap. 6. New York: Schirmer.

Giles, M., ed. 1980. *The Audio/Radio Handbook*. Santa Clara: National Semiconductor Corp.

Kwakernaak, H. and Sivan, R. 1991. *Modern Signals and Systems*, Chap. 9. Eaglewood Cliffs: Prentice-Hall.

Lane, J., Datta, J., Karley, B., and Norwood, J. 2001. *DSP Filters*, Chap. 4–10, 20. Carmel: Howard W. Sams & Co.

Lane, J. and Hillman, G. 1991. "Implementing IIR/FIR Filters with Motorola's DSP56000/DSP56001." *APR7/D Rev1*. Schomberg: Motorola, Inc.

Leach, M. 1999. *Introduction to Electroacoustics and Audio Amplifier Design*, Chap. 6. Dubuque: Kendall-Hunt.

Lindquist, C. 1977. *Active Network Design*, Chap. 2. Miami: Steward and Sons.

Lindquist, C. 1999. *Adaptive and Digital Signal Processing*, Chap. 5. Miami: Steward and Sons.

Massberg, M. 2011. "Digital Low-Pass Filter Design with Analog-Matched Magnitude Response." *Journal of the Audio Engineering Society*, Preprint 8551.

MIDI Manufacturer's Association. 1999. *Downloadable Sounds Level 2*, www.midi.org/specifications-old/item/dls-level-2-specification, Accessed August 1, 2018.

Moore, R. 1990. *Elements of Computer Music*, Chap. 2. Eaglewood Cliffs: Prentice-Hall.

Motorola, Inc. 1991. "Digital Stereo 10-Band Graphic Equalizer Using the DSP56001." *APR2/D*. Schomberg: Motorola, Inc.

Oppenheim, A. V. and Schafer, R. W. 1999. *Discrete-Time Signal Processing, 2nd Ed.*, Chap. 7. Eaglewood Cliffs: Prentice-Hall.

Orfanidis, Sophacles. 1996. *Introduction to Signal Processing*, chapters 10–11. Eaglewood Cliffs: Prentice-Hall.

Smith, J. O. and Angell, J. B. 1982. "A Constant Gain Digital Resonator Tuned by a Single Coefficient." *Computer Music Journal*, Vol. 4, No. 4.

Vicanek, M. *Matched Second Order Digital Filters*, http://vicanek.de/articles/BiquadFits.pdf, Accessed August 1, 2018.

Zölzer, U. 2011. *Digital Audio Effects, 2nd Ed.*, Chap. 2. West Sussex: J. Wiley & Sons.

CHAPTER 12

Audio Filter Designs: Wave Digital and Virtual Analog

Wave digital filters (WDFs) are designed to simulate analog electronic circuits, and do not use any biquadratic structures. The idea dates to 1970, but the seminal paper published on the topic didn't arrive until Fettweis's "Wave Digital Filters: Theory and Practice" in 1986. This 327-page document is a highly detailed treatise on the subject. If you are at all interested in this topic, it is the first document you should get. There are more references in the chapter bibliography. The interest in WDFs seems to come and go; however, a recent release of a WDF simulation software package called RT-WDF in 2016 has brought it back into the spotlight. The RT-WDF software can be used to simulate many electronic, mechanical and acoustic circuits with components connected in highly complex fashions. The software is easy to use, but the learning curve for converting the circuit into the functional blocks that are encoded is steep: the circuit is converted into a SPQR tree, which then becomes the WDF representation. That SPQR conversion is not trivial and involves graph theory. For this chapter, we've developed a WDF C++ library of objects specifically suited to modeling analog RLC filter circuits called *ladder filters*. The ladder filter library contains objects representing all of the WDF components you need to create ladder filters, including very high-order filters. The library can be extended in the future but if you decide to go deeper, you might want to think about investing some time in SPQR tree reduction and check out the RT-WDF package.

Zavalishin's *Virtual Analog* (VA) filters represent another class of digital filter designs aimed at simulating analog circuits that use a technique known as *digital integrator replacement*. These filters work particularly well when their parameters are sharply modulated, making them suitable for synthesizer applications. In this chapter, we will focus three particular algorithms to add to our collection of filter designs and make some tweaks to the original designs to generate some unique filter variations.

12.1 Wave Digital Filters

The idea behind WDFs is to simulate the signal flow through analog circuit components: Resistors (R), Capacitors (C), and Inductors (L) and use this simulation to convert an input $x(n)$ into an output $y(n)$. WDFs can be used to simulate many types of components such as tubes, transistors, op-amps, and transformers. For our analog RLC simulation filters we'll stick with just the three basic components. Whereas the VA filters may be designed to simulate signal flow through an analog block diagram, or a signal flow graph, the WDFs simulate the signal flow through the physical components and interconnections between the components. The first step involves two-port network theory and *scattering parameters*, also known as *s-parameters*.

12.1.1 Scattering Parameters and WDFs

In electronics, a two-port network is any circuit with an input port and an output port. Typically the input and output ports of the two-port "box" each consist of a two-wire (or two-pin) interface as shown in Figure 12.1a. If you understand electronics, you can think of these as the signal and ground connections. Two-port theory allows us to fully and easily characterize and describe the behavior of the circuit inside the two-port box without knowing anything about the actual circuit itself. The two-port box is sometimes thought of as a "black box" where we can't see the circuit inside to glean any information about its operation. In Figure 12.1a, we make the traditional measurements of input and output voltages across the ports, and input and output currents through the ports, producing four basic measurements: V_1, V_2, I_1, and I_2. We can then make measurements with each port in an open or short circuit connection. Applying the basic laws of electronics formulates the set of equations in Equation 12.1 that relate the measurements.

$$\begin{bmatrix} V_1 \\ V_2 \end{bmatrix} = \begin{bmatrix} Z_{11} & Z_{12} \\ Z_{21} & Z_{22} \end{bmatrix} \begin{bmatrix} I_1 \\ I_2 \end{bmatrix} \qquad Z_{11} = \left. \frac{V_1}{I_1} \right|_{I_2=0} \tag{12.1}$$

The matrix of Z terms is called the *impedance matrix*, and the values represent every possible combination of impedance (voltage/current) that may be calculated on each port with the ports alternatively opened ($I_{port} = 0$)—so Z_{11} is the impedance using V_1 and I_1 with port number 2 open. This produces a family of two equations after multiplying the matrix out. A problem that arises in microwave frequency circuits is that these voltage and current measurements are extremely difficult to measure as the measurement devices themselves affect the network at such high frequencies. Microwave (or RF) engineers use a different two-port analysis technique based on incident and reflected waves. At microwave frequencies, the transmission lines (cables) and the microwave components themselves have an effect on the transmission of the signal. This causes parts of the high frequency signals to be reflected back from the ports, as shown in Figure 12.1b. The incident waves are labeled a_N and the reflected waves are labeled b_N. The amplitudes of these waveforms can be used to characterize the circuit in terms of its scattering parameters, which refer to the ability of the waveforms to reflect or scatter. They are shown in Figure 12.1b with their own directional arrows with the incident wave entering the top pin and the reflected wave exiting the

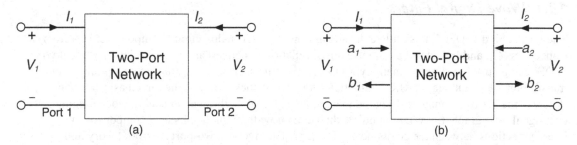

Figure 12.1: (a) A two-port network with voltage and current being measured at each port. (b) A two-port network with incident (a_1, a_2) and reflected (b_1, b_2) waves being measured.

bottom pin of each port. The relationship between these signals and the scattering parameters S_{XY} is shown in Equation 12.2.

$$\begin{bmatrix} b_1 \\ b_2 \end{bmatrix} = \begin{bmatrix} S_{11} & S_{12} \\ S_{21} & S_{22} \end{bmatrix} \begin{bmatrix} a_1 \\ a_2 \end{bmatrix} \tag{12.2}$$

In the 1970s, analyzers became available that allowed easy and accurate measurement of the scattering parameters, and microwave electronics really took off; now microwave satellite dishes are commonplace.

12.1.2 Simulating WDF Components

If we consider a transmission line (think satellite-TV cable) connected to a single one-port component (R, L, or C), we can characterize the structure with an incident voltage V_i, a reflected voltage V_r, and a series current I that drives in input impedance Z. Because we take the actual transmission line itself into consideration, we note that it has a characteristic impedance that we call the *reference impedance* or Z_o. You may have seen microwave cables with the marking "50 ohms"—that is the standard RF reference impedance, Z_o. With WDF designs, we want to digitally simulate the incident and reflected waves through an electronic component (resistor, capacitor, or inductor) and we can formulate this in any domain we wish—time or frequency. Figure 12.2 shows several examples in (a) time, (b) the s-domain and (c) the z-domain. In our DSP designs, we set our digital transfer function $H(z)$ equal to $Y(z)/X(z)$ or output/input. We can form a similar transfer function ρ that relates the reflected digital signal B (output signal) to the incident digital signal A (input signal) shown in Figure 12.2d. This special transfer function $\rho = B/A$ is called the *reflection coefficient*.

Looking at the incident and reflected voltages A and B in Figure 12.2d we can write that the incident voltage A equals the input voltage V plus whatever voltage is dropped across the transmission line's impedance Z_o. The reflected voltage will be the same but subtracting the voltage drop as shown in Equation 12.3.

$$A = V + IZ_0$$
$$B = V - IZ_0 \tag{12.3}$$

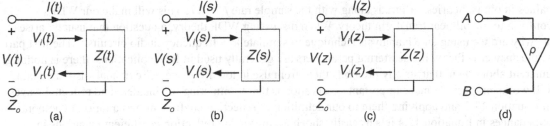

Figure 12.2: A representation of some electrical component (or components) with impedance Z, connected via a transmission line with *reference impedance* or Z_o in the following domains: (a) time, (b) analog frequency, and (c) digital frequency. (d) A compact version of (c) with incident wave A, reflected wave B, and reflection coefficient ρ.

Evaluating the analog transmission coefficient in the frequency domain as $B(s)/A(s)$ and knowing that at the termination (component) that $V = IZ$, we can then write:

$$\rho(s) = \frac{B(s)}{A(s)} = \frac{V_r(s)}{V_i(s)} = \frac{-I_r(s)}{I_i(s)} = \frac{\dfrac{V(s)}{I(s)} - Z_o}{\dfrac{V(s)}{I(s)} + Z_o} = \frac{\dfrac{Z(s)}{Z_o(s)} - 1}{\dfrac{Z(s)}{Z_o(s)} + 1}$$

(12.4)

$$s = j\omega$$

We can then apply the basic electronic impedance equations for the passive components and formulate reflections coefficients for each one, replacing the reference impedance Z_o with a reference resistance R_o that is going to be different for each component. This R_o value will be called the *component resistance* and will be key in later calculations.

$$Z_R = R \qquad \text{resistor (R)}$$
$$Z_L = j\omega L = sL \qquad \text{inductor (L)}$$
$$Z_C = \frac{1}{j\omega C} = \frac{1}{sC} \qquad \text{capacitor (C)}$$

$$\rho_R(s) = \frac{\dfrac{R}{R_o} - 1}{\dfrac{R}{R_o} + 1} \qquad \rho_L(s) = \frac{\dfrac{sL}{R_o} - 1}{\dfrac{sL}{R_o} + 1} \qquad \rho_C(s) = \frac{\dfrac{(1/j\omega C)}{R_o} - 1}{\dfrac{(1/j\omega C)}{R_o} + 1}$$

(12.5)

Digitizing the transmission coefficient requires an analog-to-digital transform of some kind that can convert values in $s = j\omega$ to values in $z = e^{j\omega}$. We observe that all three of the reflection coefficient equations have the same form as the BZT of $(x-1)/(x+1)$.

$$s = \frac{2}{T} \frac{z-1}{z+1}$$

(12.6)

This creates a happy coincidence for us as we can then replace the component reflection coefficients with digitized versions, and create easy-to-calculate component resistances from the actual component values in ohms, henries, or farads along with the sample rate $f_s = 1/T$. This will make our WDFs conform to the bilinear transform theory. When discussing WDF theory, a question that usually arises is "why are we using an RF analysis technique to simulate low frequency audio circuits?" The first part of the answer is that while scattering parameters are generally used in RF applications, there is nothing inherent about them that strictly prohibits them from use in low frequency circuit analysis—the math still works properly. A more important observation is that formulating the incident and reflected waves in Equation 12.3 and applying them to our definition of reflection coefficient ρ as a ratio of component impedances in Equation 12.4 lets us easily shoehorn the analog reflection coefficient equations in Equation 12.5 into the bilinear transform equations in Equation 12.6. This allows us to digitize each analog component's reflection coefficient. Those digitized component simulations are listed next. You should note that what is being digitized here is the component's reflection coefficient ρ and not the actual component itself, though like the literature, I reference the components as being digitized, naming them "WDF Resistor" and the like.

12.1.2.1 WDF Resistor

For a resistor of value R (ohms), we can write the reflection coefficient as:

$$\rho_R(z) = k = \frac{R/R_o - 1}{R/R_o + 1}$$

$$R_o = \text{arbitrary scaling factor}$$

(12.7)

Equation 12.7 shows that the digitized reflected wave out of a resistor is a scaled version of the incident wave entering the resistor. Because R_o is an arbitrary scaling factor, we often make it equal to the given resistance R. In this way, all of the incident wave energy is absorbed in the resistor and none of it is reflected back as $k - 0$ such that it becomes an energy sink with no reflected output waveform.

12.1.2.2 WDF Inductor

For an inductor of value L (henries) we can combine Equations 12.5 and 12.6 to write:

$$\rho_L = -z^{-1}$$

$$if$$

$$R_o(L) = 2Lf_s$$

(12.8)

Equation 12.8 shows that the digitized reflected wave out of an inductor is an inverted and delayed version of the incident wave entering the inductor. The inductor's component resistance $R_o = 2Lf_s$.

12.1.2.3 WDF Capacitor

For a capacitor of value C (farads) we can combine Equations 12.5 and 12.6 to write:

$$\rho_C = z^{-1}$$

$$if$$

$$R_o(C) = \frac{1}{2Cf_s}$$

(12.9)

Equation 12.9 shows that the digitized reflected wave out of a capacitor is delayed version of the incident wave entering the capacitor. The capacitor's component resistance $R_o = 1/2Cf_s$. To distinguish between the different component resistances, we will place the components in parentheses so that $R_o(L)$ is the component resistance for an inductor, while $R_o(LC)$ is the component resistance of a series or parallel combination of inductor and capacitor.

Equations 12.7–12.9 allow us to simulate R, L, and C using the one-port block diagrams in Figures 12.3a–c respectively. Notice the equivalent diagram for the resistor, which shows an energy sink as a kind of dead-end input, with the "0" on the reflected output branch to signify that there is no reflected

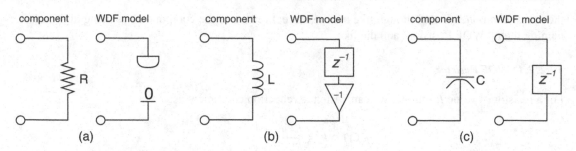

Figure 12.3: The analog electronic component and WDF digitized equivalent for
(a) resistor, (b) inductor and (c) capacitor.

component. With digital versions of each of the three components, we can then turn our attention to simulating the interconnections between components—and this is where the component resistances come into play.

Our analog circuits are going to contain at least some components in series and parallel, and we'll often find repeated combinations in the ladder filter designs.

12.1.3 Simulating WDF Component Interconnections

The WDF components may be connected in any combination of series and parallel topologies. Those familiar with passive analog circuits are familiar with the ladder shapes of these circuits. The interconnecting blocks are called *adaptors*. The simulation of the interconnections of the components is just as important as the simulation of the components themselves. In fact, you'll never see any of the simulated components in Figure 12.3 hanging out alone—they will always be connected to adaptors, which will always be connected to something else for a meaningful circuit. The adaptors we will use for our analog ladder filter simulations all have three ports, that is, they are *three-port networks*. The input port is on the left and named *port 1*; the output port is on the right and named *port 2*, just as the two-port networks in Figure 12.1. The third port is at the top (or bottom) of the adaptor and is where we will connect our electronic component models, or alternatively, other branches of series or parallel components. The adaptors we will use are divided into two groups: series and parallel. Series adaptors connect components in series while parallel adaptors connect components in parallel. Figure 12.4 shows the (a) series and (b) parallel adaptors. Each port has an associated port impedance (resistance) labeled *R1*, *R2* and *R3*.

For our WDF ladder filter library, we will add a few more constraints as shown in Figure 12.4c; for port 3, which is always connected to a component, the port resistance *R3* will always be the connected R_o component resistance in Equations 12.7–12.9. The incident waves will flow through the top port pins and the reflected waves will flow through the bottom port pins.

The adaptors have a difficult and important job—they must propagate the incident and reflected waves through the circuit (port 1 and port 2) and its component (port 3) taking into account all of the port impedances and signal directions. Figure 12.5 shows three circuits—(a) series components, (b) parallel components, and (c) two series and one parallel component, along with the adaptors used to connect them. Notice how the internal component series or parallel connections are hidden by the adaptor blocks, which are all connected in series. The signal flow will always be from left (input) to right (output).

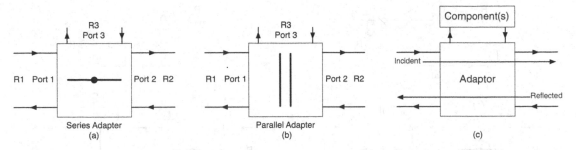

Figure 12.4: As shown here, series (a) and parallel (b) adaptors have three ports. (c) For our WDF ladder filter library, we will constrain our adaptors to make component connections to Port 3.

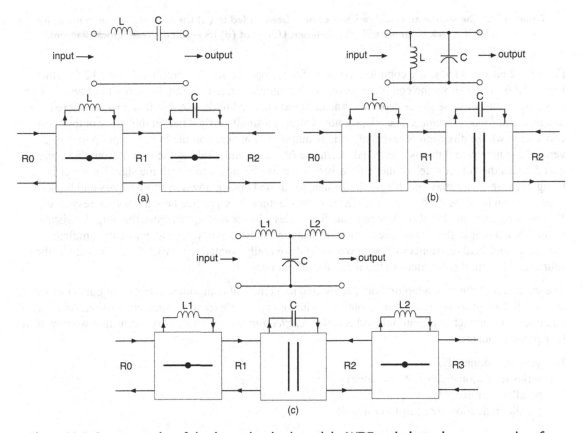

Figure 12.5: Some examples of simple passive circuits and the WDF equivalent adaptor connections for (a) series LC, (b) parallel LC, and (c) series L1 with parallel C1 and series L2.

While Figure 12.5 shows some basic component connections, it doesn't show a complete passive circuit. A complete passive circuit technically consists of a driving voltage source, with inherent source resistance R_s and a load resistance R_L into which the circuit delivers its power. These two resistances turn out to be extremely important in our ladder filter designs.

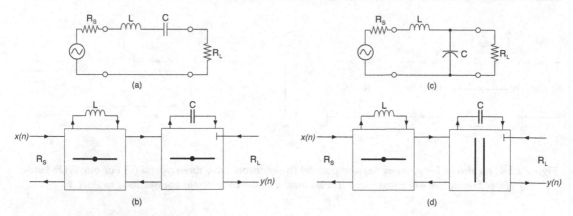

Figure 12.6: The source and load resistances have been added to (a) the series LC circuit and (b) its resulting WDF block diagram and (c) a different LC circuit (d) its resulting WDF block diagram.

Figures 12.6a and b show the complete version of the simple series LC circuit in Figure 12.5a while Figure 12.6c and d show the complete version of the simple series-L-parallel-C circuit in Figure 12.5b. First, notice the way the series capacitor's adaptor has changed to denote that it is now connected to a *terminal load resistance*. The adaptor now features a small vertical bar on the upper output port connection whose direction is reversed, and the output $y(n)$ appears on the lower output port pin. The vertical bar means that this is a terminal resistance (the dead end). Taking the output from the lower port denotes the "other side" of the load resistor—that is, the side into which the signal's power is being delivered. You may see this particular adaptor turned upside down in some papers and books, depending on how the authors want to define their adaptors. Next, notice how the source resistance R_s now appears as the input resistance to the first series adaptor and R_L becomes the output resistance of the last adaptor in the chain. These are significant in that we do not need to explicitly simulate the source and load resistances (though we could if we really wanted to); instead we use them as the source and terminal resistances of the input and output adaptors.

The inclusion of the new adaptor that is connected to a terminal component means that our set of series and parallel adaptors comes in two variations: with and without a terminal load connection. An adaptor that does not connect to a terminal load is called a *reflection free* adaptor. This means that we now have four possible adaptors:

1. series (terminated)
2. series reflection-free (un-terminated)
3. parallel (terminated)
4. parallel reflection-free (un-terminated)

12.2 WDF Adaptors

Designing the adaptors involves more work back in the analog circuit and scatter parameter areas that is not trivial. Each adaptor must take into account the type of connection, the input and output resistances, and the connected component's input and output signals, along with routing the incident and reflected signal paths properly. If you are really interested in diving deeper into the adaptor theory (and be forewarned, this can be a really deep rabbit hole to dive into), check out the Fettweis source first, then tackle some of the more modern papers. Within the various literature are alternate versions of the adaptors, and you can also find splitter and combiner adaptors, as well as adaptors with more than three ports. We will first present the standard series adaptors and their equations, then the parallel adaptors. You will also notice that the adaptors are adorned with many signal flow arrows to help you understand the flow in the adaptor as well as how the resulting C++ code will implement them.

12.2.1 Series Adaptors

The two series adaptors—one normal (terminated) and the other reflection-free (non-terminated)—are shown in compact and detailed formats in Figure 12.7a–d. These are typical digital signal block diagrams with summers, coefficients, and nodes just like the other diagrams we've studied.

You'll notice that the reflection-free version is simpler, with only one coefficient B to calculate, while the normal terminated version requires two coefficients, $B1$ and $B3$, numbered this way to keep consistent with the literature. The generalized equations for the series adaptor coefficients are shown here.

Series Adaptor (may be terminated)   Series Reflection-Free Adaptor

$$B1 = \frac{2R1}{R1 + R3 + R2}$$

$$R2 = R1 + R3$$

$$B3 = \frac{2R2}{R1 + R3 + R2}$$

$$B = \frac{R1}{R1 + R3}$$

$$\text{terminated:} \quad R2 = R_L$$

(12.10)

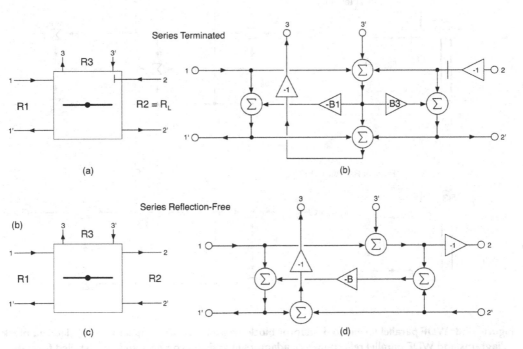

Figure 12.7: WDF series terminated adaptor block diagrams as (a) compact and (b) detailed block diagrams and WDF series reflection-free adaptors in their (c) compact and (d) detailed formats.

12.2.2 Parallel Adaptors

The two parallel adaptors—one normal and the other reflection-free—are shown in compact and detailed formats in Figure 12.8a–d. As with the series versions, the reflection-free structure is somewhat simpler.

The equations for the parallel adaptor coefficients are as follows.

Parallel Adaptor (may be terminated)   Parallel Reflection-Free Adaptor

$$A1 = \frac{2G1}{G1 + G3 + G2} \qquad\qquad R2 = \frac{1}{G1 + G3}$$

$$A3 = \frac{2G2}{G1 + G3 + G2} \qquad\qquad A = \frac{G1}{G1 + G3} \qquad (12.11)$$

$$G1 = \frac{1}{R1} \quad G3 = \frac{1}{R3} \quad G2 = \frac{1}{R2}$$

$$\text{terminated:} \quad G2 = G_L = \frac{1}{R_L}$$

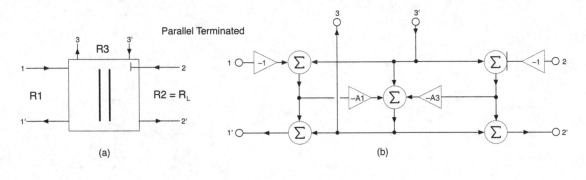

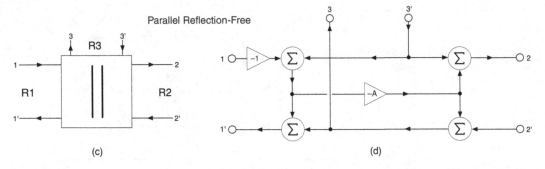

Figure 12.8: WDF parallel terminated adaptor block diagrams as (a) compact and (b) detailed block diagrams and WDF parallel reflection-free adaptors in their (c) compact and (d) detailed formats.

12.2.3 More Component Combinations

When designing analog ladder filters, we often wind up with circuits that include combination connections, as shown in Figure 12.9. In Figure 12.9a there are three parallel structures in series, while in Figure 12.9b there are three parallel structures in parallel. There can also be combinations of two components in series but in parallel to the larger structure, as shown in Figure 12.9c, and we may also have resistors sprinkled in the mixture as in Figure 12.9d.

To facilitate these common types of connection for higher order ladder filters while keeping with the port conventions in Figure 12.4c, we may combine models of series and parallel components. Figure 12.10 shows the combination models for the following component combinations: (a) and (b) series and parallel *LC*, (c) and (d) series and parallel *RL*, (e) and (f) series and parallel *RC*.

Each combined model contains two-state registers and one or more coefficients involving the *K* value. The component resistances R_o and the *K* coefficient calculations are shown in Equation 12.12 and have been fashioned to work properly within our ladder filter WDF library. In some cases, the component's conductance $G_o = 1/R_o$ is used.

$$
\begin{array}{cc}
\text{Series LC} & \text{Parallel LC}
\end{array}
$$

$$
\begin{array}{ccr}
R_o(LC) = R_o(L) + G_o(C) & R_o(LC) = R_o(C) + G_o(L) & (12.12) \\[6pt]
K = \dfrac{1 - R_o(L)G_o(C)}{1 + R_o(L)G_o(C)} & K = \dfrac{R_o(C)G_o(L) - 1}{R_o(C)G_o(L) + 1} &
\end{array}
$$

$$
\begin{array}{cc}
\text{Series RL} & \text{Parallel RL}
\end{array}
$$

$$
\begin{array}{ccr}
R_o(RL) = R_o(R) + R_o(L) & R_o(RL) = \dfrac{1}{G_o(R) + G_o(L)} & (12.13) \\[10pt]
K = \dfrac{R_o(R)}{R_o(R) + R_o(L)} & K = \dfrac{G_o(R)}{G_o(R) + G_o(L)} &
\end{array}
$$

$$
\begin{array}{cc}
\text{Series RC} & \text{Parallel RC}
\end{array}
$$

$$
\begin{array}{ccr}
R_o(RC) = R_o(R) + R_o(C) & R_o(RC) = \dfrac{1}{G_o(R) + G_o(C)} & (12.14) \\[10pt]
K = \dfrac{R_o(R)}{R_o(R) + R_o(C)} & K = \dfrac{G_o(R)}{G_o(R) + G_o(C)} &
\end{array}
$$

The ladder filter WDF library contains C++ objects for all of these component pairs, plus the single components, plus all of the adaptors so you can create a wide array of ladder filter plugins.

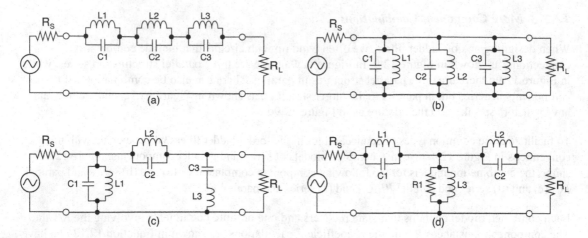

Figure 12.9: (a) Three parallel structures in series. (b) Three parallel structures in parallel. (c) Parallel sets of series structures. (d) The addition of resistors in the branches.

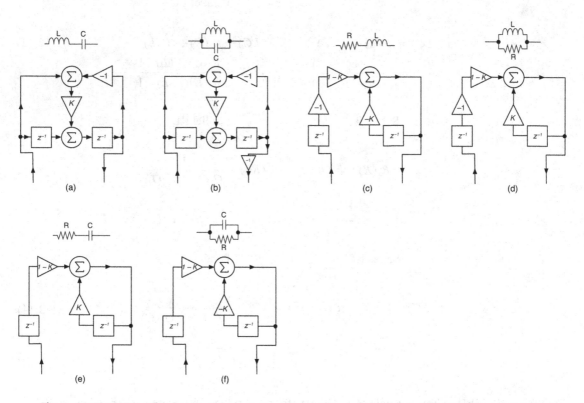

Figure 12.10: WDF reflection models for (a) series LC (b) parallel LC, (c) series RL, (d) parallel RL, (e) series RC, and (f) parallel RC combinations.

12.2.4 Signal Flow Through a WDF Circuit

Before continuing further, let's discuss the signal flow through a pair of connected adaptors simulating the simple series LC component connection in Figure 12.6 and terminating in a load resistance R_L. Figure 12.11 shows (a) the circuit, (b) simplified WDF block diagram, and (c) the detailed WDF circuit. Notice the node labeled N4 in the center of Figure 12.8c—this is the output of the first adaptor and the input to the second adaptor.

Processing the input sample $x(n)$ through the pair of adaptors to form the output $y(n)$, then continuing by processing the reflected signal on the lower port back through the diagram in reverse to set the state register values will proceed as follows.

1. Form the first node value *N1* by reading the state register and inverting it from port 3, then sum that with the input $x(n)$; inverting this value produces the output *N4* (i.e. *N4* = −*N1*).
2. Form the output of the second adaptor $y(n)$ next; it is the output of the summer that processes *(−B3)(N8)* + *N7* and *N7* is the dead-end input (0.0) while *N8* is the sum of the adaptor input *N4* plus the value from the state register in port 3.
3. Now that $y(n)$ has been formed, we reverse direction, reflect it, and head backwards through the adaptor, finding the reflected output port value *N5* = *N4* − *(B1)(N8)*. With that we can calculate and set the value in the state register, *N6* = − *(N5* + *N8* + $y(n)$*)*.
4. The reflected signal *N5* then propagates backwards into the first adaptor as *N2*; we can then form the input to the state register *N3* = − *(N2* + *(x(n)* − *B(N1* + *N2))*.
5. Notice that since we are back at the input adaptor, there is no upstream adaptor to send the reflected signal to—we are done.

You can see that while this does look somewhat complicated, it is ultimately a matter of keeping track of the incident signal, forming the output, then reflecting the signal back upstream to calculate and

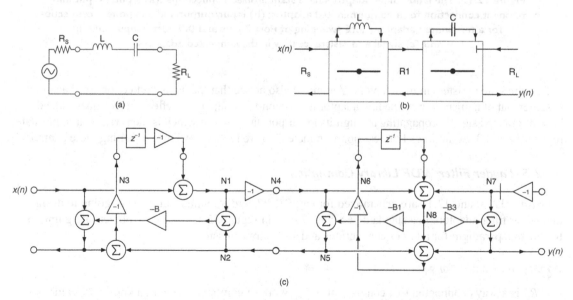

(c)

Figure 12.11: (a) A simple series LC circuit, (b) the simplified WDF adaptor block diagram, and (c) the full WDF connection graph.

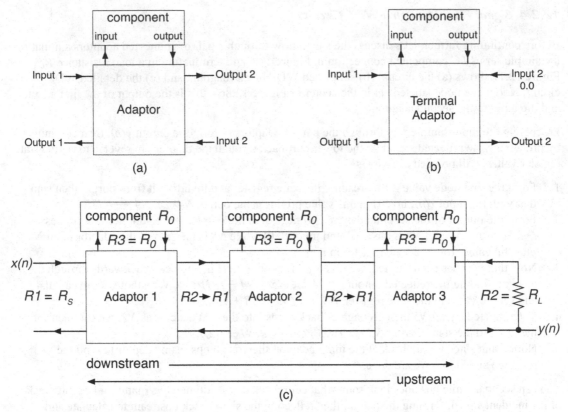

Figure 12.12: The ladder filter adaptors have standardized connections: (a) input/output and component connection for a non-terminated adaptor; (b) input/output and component connection for a terminated adaptor (note swapping of Port 2 pins and 0.0 Port 2 input value); and (c) signal and resistance flow in the connected adaptors.

update the state registers in the adaptors. You might also notice that the name "reflection free" adaptor is somewhat of a misnomer as the first adaptor that is non-terminated and reflection free does indeed have a reflected signal propagating through its lower port line, part of which is then written into the state register in port 3. Sometimes I prefer "non terminated" to "reflection free" when naming these adaptors.

12.2.5 Ladder Filter WDF Library Conventions

Equations 12.10 and 12.11 are generalized for any $R1$, $R2$, and $R3$ values, but we are going to design the ladder filter adaptors in a standardized way, shown in Figure 12.12. Our C++ objects are going to follow this paradigm for adaptor connections and signal transmission.

Connection Conventions

- $R3$ is always connected to a component (or network of components) with a known R_o value, so $R3$ always equals the component R_o; parallel adaptors convert $R3$ to the conductance $G3 = 1/R3$ for the coefficient calculation.

- The port 2 output resistance $R2$ of an *upstream* adaptor becomes $R1$ for the *downstream* adaptor. Parallel adaptors further process $R1$ into the admittance value $G1$.
- $R2$ is the sum of $R1 + R3$ for series adaptors. Parallel adaptors convert the resistances to conductances first so $G2$ is the sum of $G1 + G3$.
- For terminal adaptors, the $R2$ value is equal to the terminal resistance and parallel adaptors further process this into $G2 = GL = 1/R_L$.
- Components are one-port networks that only have an *input* and *output* value; these may be replaced with the component pairs in Figure 12.10.

Initialization

- Adaptor 1 is initialized with the circuit's source resistance R_s. It then calculates its $R1$ value from that. Parallel adaptors further process $R1$ into the admittance value $G1$.
- Adaptor 1 calculates its $R2$ value as $R2 = R1 + R3$ (or $G2 = G1 + G3$).
- Adaptor 1 then calls the initialization function on the downstream adaptor, delivering $R2$ as the source resistance for the next adaptor.
- Adaptor 2 follows suit, calculates its R2 value as $R2 = R1 + R3$, and initializes its coefficients with the appropriate values from the known $R1$ and $R3$.
- Adaptor 2 then calls the initialization function on the downstream adaptor, delivering $R2$ as the source resistance for the next adaptor. The process repeats until we get to the terminal adaptor; once it calculates its coefficients, the initialization is complete.

Operation

- Operation begins when you push the value of $x(n)$ into the first adaptor by calling its *setInput1( )* function. Adaptor 1 calculates the value for its output port, *Ouput 2*, then delivers it to the downstream Adaptor 2 by calling its *setInput1( )* function.
- Each adaptor in turn calculates its *output 2* value, and then delivers it to the attached *downstream* adaptor until the terminal adaptor is reached.
- The terminal adaptor calculates the final $y(n)$ as its *output 2* value—this is where the output variable is stored and located for future use.
- The terminal adaptor uses $y(n)$ along with other values to calculate its reflected output value, *output 1,* and then it calls the *setInput2( )* function on the *upstream* adaptor, sending the reflected signal backwards through the structure.
- The operation is complete when adaptor 1 receives the last reflected signal and updates the component's state register value; at this point, we can pick up $y(n)$ from the terminal adaptor's *output 2.*

With the C++ objects ready and our programming convention set, there are a few more items to discuss before implementing a ladder filter.

12.2.6 Filter Source/Termination Impedance Matching

Referring back to Figure 12.6, we see that our passive filtering circuits necessarily contain a source resistance R_S and a load resistance R_L. Passive filter design is challenging because of the way that the impedances of the filter sections affect each other. For example, in Figure 12.9a, each successive parallel LC pair's impedance becomes the load for the previous section. Complicating this is the fact that the source has a resistance, and ultimately we plan on delivering the final signal into some load resistance.

The values of all of the components affect each other. This has led to the notion of passive filter circuits with matched impedances. This means that the source and load resistances are equal, or $R_S = R_L$. Adding this constraint makes the filter design at least a bit simpler; however, we need to understand the ramifications of the matched source and load impedance on the performance of the internal passive filter. Figure 12.13a shows a resistor divider circuit with identical values for each resistor. The attenuation through the resistor divider is 0.5 or −6 dB. This means that when we embed the passive filter circuit inside the resistor divider as in Figure 12.13b, we will automatically lose half of the input signal to the matched source and load impedance circuit. The WDF equations are already set up to compensate for this condition, *boosting* the output by +6 dB as in Figure 12.13c. This means that when we implement the ideal RLC filters as WDF structures where the source impedance is zero and the load is an open circuit, output will be +6 dB larger that it should be as shown in Figure 12.13d—this is not a big issue, but we will look out for it.

12.2.7 Bilinear Transform Frequency Warping

Another issue is that we derived the filter component reflection coefficients in a matched equation to the bilinear transform. This means that WDFs are types of bilinear transform filters and have the same issues. One of those is the frequency-warping problem where the analog frequencies do not exactly map to the digital frequencies. The WDF component equations do not take the frequency warping into account and so they do not attempt to correct for this error. You might say that the bilinear frequency warping is *baked in to* the WDF equations. We'll handle this issue in two different ways depending on our filter design technique.

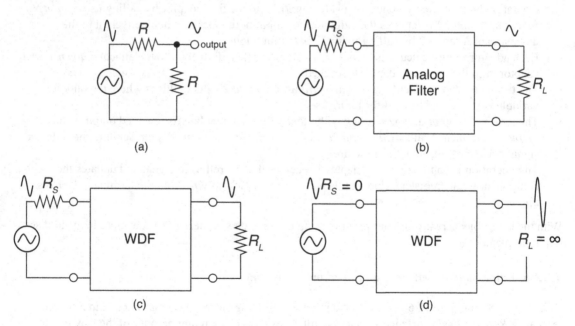

Figure 12.13: (a) A resistor divider with identical resistors cuts the input signal in half.
(b) Embedding the filter circuit inside of the resistor divider causes a signal loss at the output.
(c) The WDF equations are set up with this assumption and make up for the loss.
(d) When implementing ideal filters we will need to attenuate the output.

12.3 Designing Digital Ladder Filters With the WDF Library

The ladder filter WDF library is a set of C++ objects that you incorporate into an outer C++ container object. The outer container will inherit from *IAudioSignalProcessor* and expose the inherited *reset* and *processAudioSample* functions so it may be integrated with the rest of our FX C++ objects easily, and will snap into the *PluginCore* without much effort. These objects inherit from a special interface for adaptors called *IComponentAdaptor*. The interface is used internally to allow the adaptor objects to send data to and from one another's input and output ports. You should not need to deal with it unless you would like to modify the library to extend its functionality. The library consists of the following objects that are included in *fxobjects.h*.

12.3.1 Components and Component Combinations

These objects implement WDF one-port components. The names and implemented components are listed in Table 12.1.

12.3.2 Adaptors

These objects implement WDF 3-port adaptors. The names and adaptors are listed in Table 12.2. The adaptors implemented use the constraints and conventions in Section 12.2.5.

**Table 12.1: WDF library objects that implement *R*, *L*,
and *C* components or combinations**

WDF Object Type	WDF Object Name
Individual Components	*WdfResistor* *WdfCapacitor* *WdfInductor*
LC Combinations	*WdfSeriesLC* *WdfParallelLC*
RL Combinations	*WdfSeriesRL* *WdfParallelRL*
RC Combinations	*WdfSeriesRC* *WdfParallelRC*

Table 12.2: WDF library objects that implement standard adaptors

WDF Object Name	Description
WdfAdaptorBase	Provides convenient functions for connecting components to adaptors and adaptors to each other; all adaptors derive from this base class
WdfSeriesAdaptor	Adaptor for series component connection that is considered reflection free or non-terminated
WdfSeriesTerminatedAdaptor	Adaptor for series component connection that is terminated in a load R_L
WdfParallelAdaptor	Adaptor for parallel component connection that is considered reflection free or non-terminated
WdfParallelTerminatedAdaptor	Adaptor for parallel component connection that is terminated in a load R_L

The bulk of the programming work is inside the component and adaptor objects; they implement all of the WDF equations from this chapter. You use them to design digital ladder filters by combining them together.

12.3.3 WDF Ladder Filter Design: 3rd Order Butterworth LPF

For our first design, we will do a simple textbook 3rd order Butterworth low-pass ladder filter. For these fixed ladder filter designs, we will be using a free passive filter design program called Elsie (sounds like "LC"). While designed for amateur radio users, it will also design correct passive audio frequency filters as well, in numerous topologies and polynomial varieties. The software is available at www.tonnesoftware.com/elsie.html. For the first ladder filter, we designed a 3rd order Butterworth LPF with $f_c = 1$ kHz and then redesigned it for $f_c = 10$ kHz. The circuit is the same topology, but the component values are different. The source and load impedance are 600 ohms and the circuit along with its WDF equivalent are shown in Figure 12.14a and b, respectively.

The component values are shown in Table 12.3.

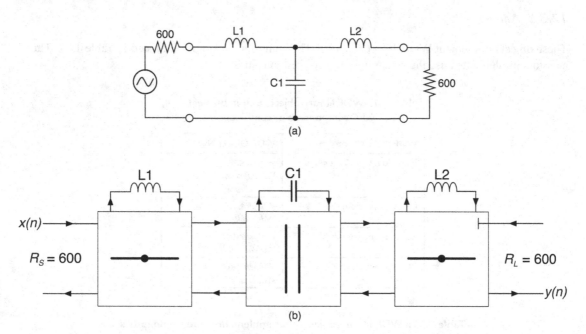

Figure 12.14: The 3rd order Butterworth LPF (a) circuit and (b) its WDF equivalent.

Table 12.3: Component values for the 3rd order Butterworth LPF
at different f_c values

$f_c = 1$ kHz	$f_c = 10$ kHz
L1 = 95.49 mH = 95.49e−3	L1 = 9.549 mH = 9.549e−3
C1 = 0.5305 μF = 0.5305e−6	C1 = 0.05305 μF = 0.05305e−6
L2 = 95.49 mH = 95.49e−3	L2 = 9.549 mH = 9.549e−3
$R_s = R_L = 600$ ohms	Rs = $R_L = 600$ ohms

The WDF library makes coding this object very easy. I'll show this object in full code. Begin by creating a C++ object that will encapsulate the WDF design and inherit from *IAudioSignalProcessor*. Declare a function to create the WDF circuit and call it from the constructor.

```
class WDFButterLPF3 : public IAudioSignalProcessor
{
public:
        WDFButterLPF3(void) { createWDF(); } /* C-TOR */
        ~WDFButterLPF3(void) {} /* D-TOR */
```

Next, give the object three member variables—one for each adaptor (series reflection-free, parallel, series-terminated).

```
protected:
        // --- three adapters
        WdfSeriesAdapter              seriesAdapter_L1;
        WdfParallelAdapter            parallelAdapter_C1;
        WdfSeriesTerminatedAdapter    seriesTerminatedAdapter_L2;
};
```

This function shows you how to assemble the WDF circuit. First, add components to the adaptors. A strongly typed enum is used to distinguish the components. The choices reflect each of the models:

```
enum class wdfComponent { R, L, C, seriesLC, parallelLC,
                          seriesRL, parallelRL, seriesRC, parallelRC };
```

In this WDF circuit, we have two L components and one C component whose values are initialized with the components in Table 12.3.

```
void createWDF()
{
// --- set adapter components
seriesAdapter_L1.setComponent(wdfComponent::L, 95.49e-3);
parallelAdapter_C1.setComponent(wdfComponent::C, 0.5305e-6);
seriesTerminatedAdapter_L2.setComponent(wdfComponent::L, 95.49e-3);
```

Connect the adapters together in sequence using the *WdfAdapterBase* method. Notice how *L1* connects to *C1*, then *C1* connects to *L2*.

```
// --- connect adapters: L1 -> C1
WdfAdapterBase::connectAdapters(&seriesAdaptor_L1,
                                &parallelAdapter_C1);
```

```
// --- C1 -> L2
WdfAdapterBase::connectAdapters(&parallelAdapter_C1,
                                &seriesTerminatedAdapter_L2);
```

Set the source resistance of the first adaptor to 600 ohms and set the terminal resistance of the last adaptor to 600 ohms.

```
// --- set source resistance
seriesAdapter_L1.setSourceResistance(600.0);

// --- set terminal resistance
seriesTerminatedAdapter_L2.setTerminalResistance(600.0);
```

There are only two functions left. The *reset* function resets the adaptors. If adaptors have components connected at port 3, they will also reset themseleves and recalculate the component values based on the sample rate; this also flushes component state registers. The crucial line of code is the last one, which initializes the chain of adaptors that are connected to the first one in the series (note: for complex WDF circuits, you may have multiple, parallel sets of adaptor chains—that is OK, just initialize the first adaptor in each chain).

```
virtual bool reset(double _sampleRate)
{
      // --- reset WDF components (flush state registers)
      seriesAdapter_L1.reset(_sampleRate);
      parallelAdapter_C1.reset(_sampleRate);
      seriesTerminatedAdapter_L2.reset(_sampleRate);

      // --- intialize the chain of adapters
      seriesAdapter_L1.initializeAdapterChain();

      return true;
}
```

The *processAudioSample* function pushes the audio data into the incident input port of the first adaptor (input 1) that calculates its output and pushes the incident signal downstream. The terminating adaptor will generate *y(n)* and then push the reflected signal backwards through the structure. The output value is then picked up from the terminal adaptor's Output 2 pin. Notice that the entire function requires only two lines of code!

```
// --- process audio: run the filter
virtual double processAudioSample(double xn)
{
      // --- push audio sample into series L1
      seriesAdapter_L1.setInput1(xn);

      // --- output is at terminated L2's output2
      return seriesTerminatedAdapter_L2.getOutput2();
}
```

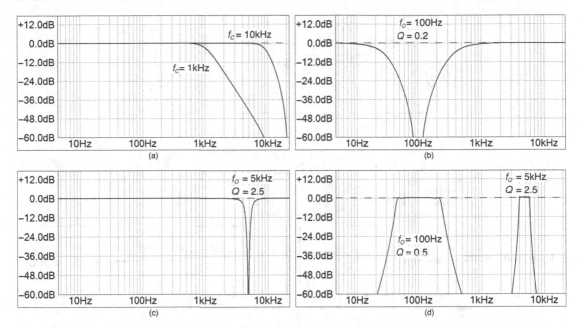

Figure 12.15: Frequency response plots for WDF ladder filters for (a) 3rd order Butterworth LPF with f_c = 1 kHz and f_c = 10 kHz; (b) 3rd order Bessel BSF with f_o = 100 Hz and Q = 0.2; (c) 3rd order Bessel BSF with f_o = 5 kHz and Q = 2.5, (d) 6th order Constant-K BPF with f_o = 100 Hz and Q = 0.5 and f_o = 5 kHz and Q = 2.5.

The resulting WDF filter responses are shown in Figure 12.15a, with both f_c curves plotted together. The bilinear transform warping caused frequency errors in the result, as expected. The measured error at the cutoff frequency for each f_c was as follows.

- f_c = 1 kHz: error = −0.65 dB
- f_c = 10 kHz: error = −6.78 dB

To show how to create more complex ladder filters, the next Sections 12.3.2 through 12.3.4 show the analog circuit, WDF block diagram and highlighted C++ code for each circuit. You can create very high-order filters with ease using the WDF library.

12.3.4 WDF Ladder Filter Design: 3rd Order Bessel BSF

Elsie was used to design a 3rd order Bessel BSF circuit that is shown in Figure 12.16a. Notice that the sequence of adaptors in Figure 12.16b has not changed from the previous design: only the components being modeled were changed, and here we have series and parallel sets of them. In this case, the series and parallel structures attached to port 3 are replaced with their equivalent structures in Figure 12.10 and accompanying Equations 12.12–12.14. The resulting WDF filter responses are shown in Figure 12.15b and c.

The component values are listed in Table 12.4.

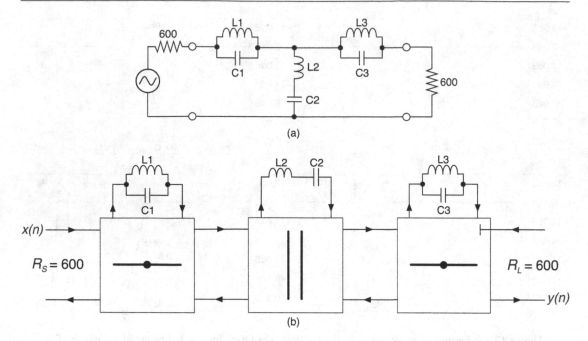

Figure 12.16: The 3rd order Bessel BSF (a) circuit and (b) its WDF equivalent.

Table 12.4: Component values for the 3rd order Bessel BSF at different f_c and Q values; for this filter, $Q = f_c$/bandwidth (Hz)

f_o = 100 Hz Q = 0.2	f_o = 5 kHz Q = 2.5
L1 = 10.5202 H	L1 = 16.8327 mH
C1 = 0.240772 µF	C1 = 0.060193 µF
L2 = 196.71 mH	L2 = 49.1978 mH
C2 = 2.8717 µF	C2 = 0.02059 µF
L3 = 1.61097 H	L3 = 2.57755 mH
C3 = 1.57237 µF	C3 = 0.393092 µF
R_s = R_L = 600 ohms	R_s = R_L = 600 ohms

The C++ code for assembling the structure and initializing the component variables to those for f_c = 100 Hz is shown next. Note that the components are now doubled in series or parallel pairs, and their initializations are done with a single function where the order of the arguments matches the syntax of the function. The adaptor declarations are first.

```
protected:
    // --- three adapters
    WdfSeriesAdapter            seriesAdapter_L1C1;
    WdfParallelAdapter          parallelAdapter_L2C2;
    WdfSeriesTerminatedAdapter  seriesTerminatedAdapter_L3C3;
```

Wave Digital and Virtual Analog Designs **323**

The code for assembling the WDF circuit is as follows

```
// --- fo = 5kHz
//     BW = 2kHz or Q = 2.5
seriesAdapter_L1C1.setComponent(wdfComponent::parallelLC,
                           16.8327e-3, 0.060193e-6); /* L,C */


parallelAdapter_L2C2.setComponent(wdfComponent::seriesLC,
                           49.1978e-3, 0.02059e-6); /* L,C */


seriesTerminatedAdapter_L3C3.setComponent(wdfComponent::parallelLC,
                              2.57755e-3, 0.393092e-6);


// --- connect adapters
WdfAdapterBase::connectAdapters(&seriesAdapter_L1C1,
                           &parallelAdapter_L2C2);


WdfAdapterBase::connectAdapters(&parallelAdapter_L2C2,
                           &seriesTerminatedAdapter_L3C3);


// --- set source & terminal resistances
seriesAdapter_L1C1.setSourceResistance(600.0);
seriesTerminatedAdapter_L3C3.setTerminalResistance(600.0);
```

The *reset* and *processAudioFrame* functions are as follows.

```
virtual bool reset(double _sampleRate)
{
     // --- reset WDF components
     seriesAdapter_L1C1.reset(_sampleRate);
     parallelAdapter_L2C2.reset(_sampleRate);
     seriesTerminatedAdapter_L3C3.reset(_sampleRate);

     // --- intialize the chain of adapters
     seriesAdapter_L1C1.initializeAdapterChain();

     return true;
}
// --- process audio: run the filter
virtual double processAudioSample(double xn)
{
     // --- push audio sample into series L1C1
     seriesAdapter_L1C1.setInput1(xn);

     // --- output is at terminated L3C3's output2
     return seriesTerminatedAdapter_L3C3.getOutput2();
}
```

12.3.5 WDF Ladder Filter Design: 6th Order Constant-K BPF

Elsie was used to design a 6th order Constant-K BPF circuit shown in Figure 12.17a with its WDF equivalent structure 12.17b. Although this looks very complex, it is really a simple repeating set of similar structures.

The resulting WDF filter responses are shown in Figure 12.15d, and the component values are listed in Table 12.5.

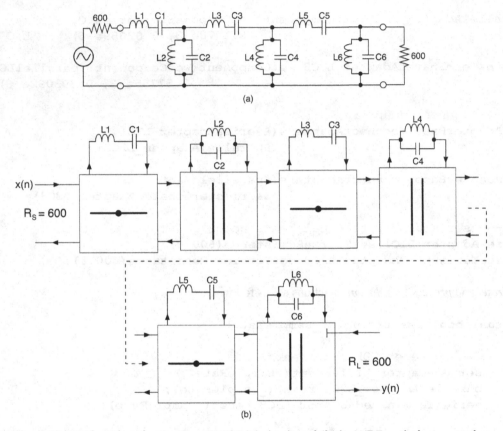

Figure 12.17: The 6th order Constant-K BPF (a) circuit and (b) its WDF equivalent; note that the WDF structure is broken into two pieces to facilitate printing and that the dotted line is usually not shown in these multi-line WDF block diagrams.

Table 12.5: Component values for the 6th order Constant-K BPF at different f_c and Q values; for this filter, $Q = f_c$/bandwidth (Hz)

f_o = 100 Hz Q = 0.5		f_o = 5 kHz Q = 2.5	
L1 = 477.465 mH	L4 = 954.93 mH	L1 = 47.7465 mH	L4 = 3.81972 mH
C1 = 5.30516 µF	C4 = 2.65258 µF	C1 = 0.02122 µF	C4 = 0.265258 µF
L2 = 954.93 mH	L5 = 954.93 mH	L2 = 3.81972 mH	L5 = 95.493 mH
C2 = 2.65258 µF	C5 = 2.65258 µF	C2 = 0.265258 µF	C5 = 0.01061 µF
L3 = 954.93 mH	L6 = 1.90986 H	L3 = 95.493 mH	L6 = 7.63944 H
C3 = 2.65258 µF	C6 = 1.32629 µF	C3 = 0.01061 µF	C6 = 0.132629 µF
$R_s = R_L$ = 600 ohms		$R_s = R_L$ = 600 ohms	

The C++ code for the assembly of the circuit is shown here.

```
// --- fo = 5 kHz
//     BW = 2kHz or Q = 2.5
seriesAdapter_L1C1.setComponent(wdfComponent::seriesLC, 47.7465e-3, 0.02122e-6);
parallelAdapter_L2C2.setComponent(wdfComponent::parallelLC, 3.819e-3, 0.26525e-6);

seriesAdapter_L3C3.setComponent(wdfComponent::seriesLC, 95.493e-3, 0.01061e-6);
parallelAdapter_L4C4.setComponent(wdfComponent::parallelLC, 3.819e-3, 0.26525e-6);

seriesAdapter_L5C5.setComponent(wdfComponent::seriesLC, 95.493e-3, 0.01061e-6);
parallelTerminatedAdapter_L6C6.setComponent(wdfComponent::parallelLC,
                              7.63944e-3, 0.132629e-6);

// --- connect adapters
WdfAdapterBase::connectAdaptors(&seriesAdapter_L1C1, &parallelAdapter_L2C2);
WdfAdapterBase::connectAdapters(&parallelAdapter_L2C2, &seriesAdapter_L3C3);
WdfAdapterBase::connectAdapters(&seriesAdapter_L3C3, &parallelAdapter_L4C4);
WdfAdapterBase::connectAdapters(&parallelAdapter_L4C4, &seriesAdapter_L5C5);
WdfAdapterBase::connectAdapters(&seriesAdapter_L5C5,
                              &parallelTerminatedAdapter_L6C6);

// --- set source/load resistances
seriesAdapter_L1C1.setSourceResistance(600.0); // Ro = 600
parallelTerminatedAdapter_L6C6.setTerminalResistance(600.0);
```

12.3.6 WDF Ladder Filter Design: Ideal 2nd Order RLC Filters

You might have noticed that for the preceding fixed ladder filter designs, changes to the cutoff frequency or Q caused all of the component values to change. For the Butterworth 3rd order LPF you could use frequency-scaling techniques to create a tunable filter whose f_c could then be changed (see the Homework section). But there are a set of ideal RLC filters that form the basis for many important analog EQ designs, including the Pultec® passive EQ. One issue with these filters is that they are ideal in the sense that the input impedance $R_s = 0$ and the load impedance $R_L =$ infinite. The infinite load impedance means that the filter terminates in an open circuit, or that the output node must be measured under open circuit conditions. The four ideal 2nd order RLC filters are shown in Figure 12.8.

Interestingly, all of the filters share identical and simple design equations:

$$f_o = \frac{1}{2\pi\sqrt{LC}} \qquad Q = \frac{1}{R}\sqrt{\frac{L}{C}} \tag{12.15}$$

For the LPF and BPF, f_o is the cutoff frequency, while for BPF and BSF, it is the center frequency. For LPF and BPF, Q is the quality or resonant peaking factor, while for BPF and BSF, $Q = f_o/BW$ where BW is the −3 dB bandwidth. This means that we could concoct a WDF version of the filters with

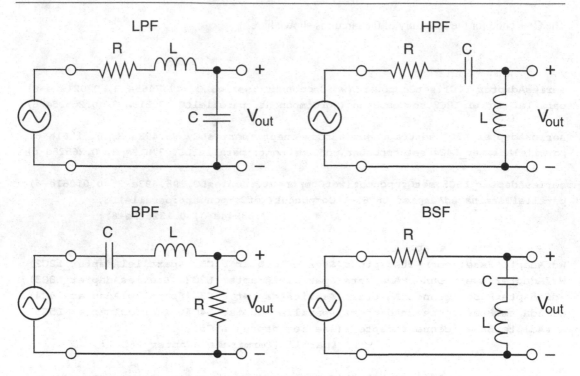

Figure 12.18: The four ideal 2nd order RLC filters; notice the location of the output voltage.

tunable f_o and Q values. If we hold the capacitor (C) value constant, we can calculate the values of the L and R components with Equation 12.16.

Let C = constant

$$L = \frac{1}{C(2\pi f_o)^2} \qquad R = \frac{1}{Q}\sqrt{\frac{L}{C}} \qquad (12.16)$$

Our WDF ladder filter library adaptors are already set up to allow for a zero R_s and infinite R_L, so we can design a plugin that exposes just the f_o and Q controls to the user, then modify the component values based on the user selections. This is about as close as you can get to the way some hardware EQs work. For example, on the Pultec EQ, the filter frequency controls are not potentiometers, but multi-position switches instead. Turning these controls switches various R, L, and C values in and out of the filter circuits to change the parameters. In that design, a tube amplifier is used post-EQ as a gain make-up stage since the filters are passive.

Converting the ideal 2nd order RLC low-pass filter for use with our WDF library yields two different designs with identical results. In one version we use separate adaptors for each component, and in the other we combine the series R and L together into a combination component, noting that we have full capability to alter either the R or L values, or both in the mode during operation. The circuit and WDF block diagrams are shown in Figure 12.19a–d for all four ideal RLC filters. We'll be using the shortened version in Figure 12.19a for the LPF example.

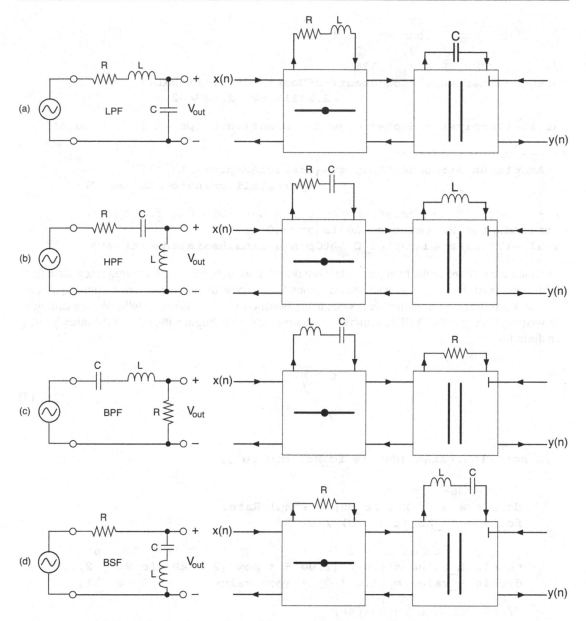

Figure 12.19: Ideal RLC 2nd order circuits and their corresponding WDF structures for (a) LPF, (b) HPF, (c) BPF, and (d) BSF; in all cases $R_s = 0$ and R_L = infinite (open circuit).

The C++ code for creating the pure RLC LPF WDF is simple and shown here. We only need two adaptors: one for the series RL and the other for the parallel C. We will provide a function to allow the user to set the f_c and Q controls, which will in turn alter the underlying component values. The adaptors will update their coefficients based on the new values. The code for initializing the WDF follows; notice how the source and load resistances are set in the last two lines of code.

```
//      initial values for fc = 1kHz Q = 0.707
//      Holding C Constant at 1e-6
//             L = 2.533e-2
//             R = 2.251131e2
seriesAdapter_RL.setComponent(wdfComponent::seriesRL,
                          2.251131e2, 2.533e-2);

parallelTerminatedAdapter_C.setComponent(wdfComponent::C, 1.0e-6);

// --- connect adapters
WdfAdapterBase::connectAdapters(&seriesAdaptor_RL,
                              &parallelTerminatedAdapter_C);

// --- set source resistance to 0 and set load flag for open ckt
seriesAdapter_RL.setSourceResistance(0.0);
parallelTerminatedAdapter_C.setOpenTerminalResistance(true);
```

The cooking equation for the filter parameters requires that we either live with the frequency error that the bilinear transform creates, or undo the erroneous frequency with a frequency un-warping equation. We'll de-warp the user's chosen filter f_c value using Equation 12.17 (Lindquist 1989). We are calling it "de-warping" because the WDF equations were not pre-warped to begin with, as with the other bilinear transform filters.

$$\theta = \frac{\pi f_c}{f_s}$$

$$f_c' = f_c \frac{\tan(\theta)}{\theta}$$

(12.17)

```
void setFilterParams(double fc_Hz, double Q)
{
      // --- de-warp fc
      double arg = (kPi*fc_Hz) / sampleRate;
      fc_Hz = fc_Hz*(tan(arg) / arg);

      // --- calculate new component values; C = 1e-6 constant
      double L_value = 1.0 / (1.0e-6 * pow((2.0*kPi*fc_Hz), 2.0));
      double R_value = (1.0 / Q) * (pow(value / 1.0e-6, 0.5));

      // --- update on adapter
      seriesAdapter_RL.setComponentValue_RL(R_value, L_value);
}
```

The frequency response plots for the WDF ideal RLC LPF are shown in Figure 12.20; (a) and (b) show non-resonant versus resonant filters with corrected cutoff frequencies while (c) and (d) show that the frequency warping from the bilinear transform error becomes worse at high frequencies.

The rest of the ideal RLC filters were all implemented using almost identical C++ object code and the appropriate component combinations as needed. These projects are all included in the ASPiK SDK for your reference. The plots for the WDF ideal RLC HPF, BPF, and BSF (2 variations) are shown in Figures 12.21a–d respectively. All filters are stable over the 10-octave range from 20.0 to 20,480 Hz.

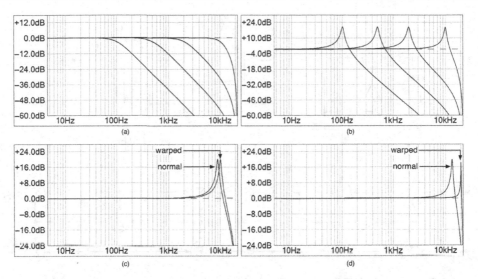

Figure 12.20: Frequency response plots for the WDF ideal RLC LPF (a) f_c = 100 Hz, 500 Hz, 2 kHz, 10 kHz and Q = 0.707 with frequency de-warping, (b) f_c = 100 Hz, 500 Hz, 2 kHz, 10 kHz and Q = 10 with frequency de-warping, (c) f_c = 10 kHz and Q = 10 normal (bilinear error) and warped (corrects error), (d) f_c = 20 kHz and Q = 10 normal (bilinear error) and warped (corrects error).

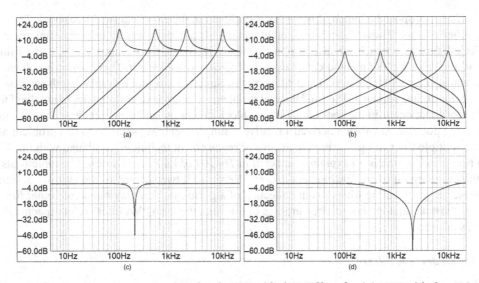

Figure 12.21: Frequency response plots for the WDF ideal RLC filtersfor (a) HPF with f_c = 100 Hz, 500 Hz, 2 kHz, 10 kHz and Q = 10, (b) BPF with f_c = 100 Hz, 500 Hz, 2 kHz, 10 kHz and Q = 10, (c) BSF with f_c = 200 Hz and Q = 2 and (d) BSF with f_c = 200 Hz and Q = 0.2

You now have a great C++ library of WDF objects for implementing scores of different ladder filter designs, including extremely complex high-order filters, and you now have a strategy for implementing ideal filters and tunable variations. Before leaving the topic, you might want to look at a diagram that shows a complete filter with all blocks exposed and using a combination series *RL* component; we typically use the compact versions as in Figure 12.19. Figure 12.22 shows the complete WDF structure for the ideal RLC LPF used in our plugin here. The C++ objects do all the work.

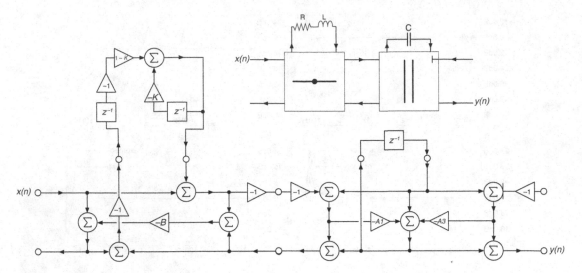

Figure 12.22: The complete WDF structure for the ideal RLC LPF; the compact form is in the upper right.

12.4 Zavalishin's Virtual Analog Filters

Zavalishin (2012) proposed a bunch of new filter algorithms in the excellent (and free) book *The Art of VA Filter Design*. These algorithms attempt to simulate analog filters by implementing their analog block diagrams—which can also be done using analog signal flow graphs. At the core of this approach is a relatively older method called *digital integrator replacement*. This method involves taking an analog filter's block diagram and converting it directly to digital by replacing the analog integrators with digital versions. The fundamental problems that arose in early attempts involved the delay-free loops that occur in the continuous analog block diagrams, and these prevented the designs from becoming popular. Zavalishin derived a relatively simple algebraic solution to this problem and formulated multiple filter designs using the integrator replacement technique combined with his delay-free loop resolution, which is all explained in detail in *The Art of VA Filter Design*. The topic is also covered in *Designing Software Synthesizers in C++*. Here, we'll tweak some of the existing designs and add these very interesting filters to our collection of algorithms. You may run across other filters described as "virtual analog" but for the purposes of this text, the term "VA filter" refers specifically to Zavalishin's designs.

12.4.1 1st Order VA Filters

The block diagram for the 1st order analog *RC* low-pass filter shown in Figure 11.9a is shown in Figure 12.23a. You can see that it includes some familiar components like summers and scalar multipliers (coefficients). It also includes an analog integrator block and a continuous feedback loop. In the digital integrator replacement method, you swap out the analog integrator for a digital equivalent, making any changes to coefficients as needed. Zavalishin's designs not only swap the integrator, but also fix the delay-free loop in the block diagram.

The integrator in the dotted box in Figure 12.23b is one of several variations on the bilinear integrator. This integrator is covered in detail in *The Art of VA Filter Design,* so please refer to that text for

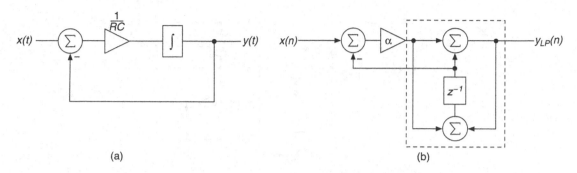

Figure 12.23: (a) Analog block diagram for a 1st order RC low-pass filter. (b) Zavalishin's VA equivalent (the digital integrator is shown in the dotted box).

elaboration. Swapping the analog integrator with a digital bilinear integrator converts the original analog filter into a digital bilinear transform design. It also preserves the original ordering of the mathematical components in the analog block diagram—the coefficient multiplier is in the same location in both versions in Figure 12.23. This means that these designs have the benefits of the bilinear transform, including stability, and mapping the analog $j\omega$ axis to the rim on the unit circle in the z-plane. They also suffer from the same issue with a gain of zero at Nyquist for the low-pass variety.

There are multiple aspects to like about this design: there is only one coefficient to calculate and only one state register to update. The filter performs extremely well when its f_c is modulated drastically. In addition, this design also lends itself easily to producing a multi-filter, in which multiple filter types are implemented in one algorithm. In this case, we can easily generate 1st order low-pass, high-pass, and all-pass filters from the single low-pass filter in Figure 12.24. One annoyance with these filters is that although the name includes "virtual analog," the low-pass version does not have a matched analog frequency response due to the zero at Nyquist issue discussed in Chapter 11. I've modified this algorithm to give us a better analog matching gain at Nyquist using the sub-circuit in the top of Figure 12.24a. For the $y'_{LP}(n)$ output, we blend in some of the HPF response, whose gain at Nyquist is fixed at 1.0. It turns out that for this design, the alpha coefficient may be used directly as a scaling factor. The responses at Nyquist are so close to the one-pole LPF from Chapter 11 that there is only a tiny discernable difference at the highest of f_c settings. I've also calculated the exact analog magnitude at the Nyquist frequency and the error ranges from −12.4% to +1.1% across the range from 100 Hz to 10 kHz. In any event, I feel that this is an improvement over the ordinary design that does not require oversampling or other high CPU usage methods. Figures 12.24b and c show a comparison of the two LPF outputs while Figures 12.24d and e show the HPF and APF respnoses of this multiple output structure.

The equations for calculating the 1st order VA filter coefficient are as follows.

$$T = 1/f_s$$

$$\omega_a = \frac{2}{T}\tan\left(\frac{2\pi f_c T}{2}\right)$$

$$\alpha = \frac{\omega_a T}{2}$$

(12.18)

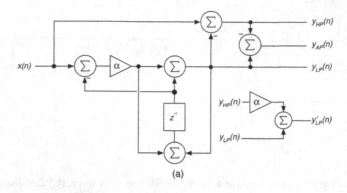

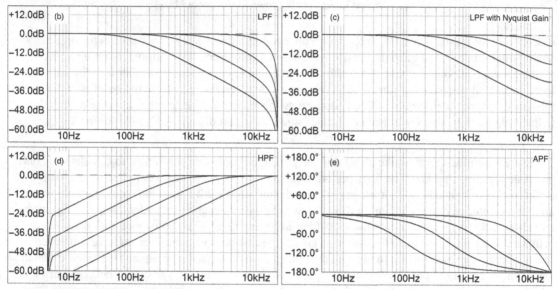

Figure 12.24: The 1st order VA multi-filter (a) block diagram along with its (b) normal and (c) Nyquist gain adjusted LPF responses and (d) HPF and (e) APF responses; note that the HPF rapid magnitude descents at the left of the plot are due to its zero of transmission at DC while the APF response plots phase shift rather than frequency magnitude.

12.4.2 2nd Order State Variable VA Filter

Zavalishin extends his method to the 2nd order case by simulating the analog state Variable filter (SVF). This well-studied topology is a multi-filter design that produces 2nd order LPF, HPF, BPF, and BSF responses from one topology, and there are numerous analog circuit variations. The digital version is sometimes called the "Chamberlin Filter" due to its appearance in Chamberlin's seminal book *Musical Applications of Microprocessors* from 1985. This is a great book to acquire if you can hunt one down. The fundamental flaw in Chamberlin's design was once again a delay-free loop. Zavalishin mitigates this issue using his delay-free loop resolution technique, producing a compact and simple-to-implement block diagram shown in Figure 12.25a.

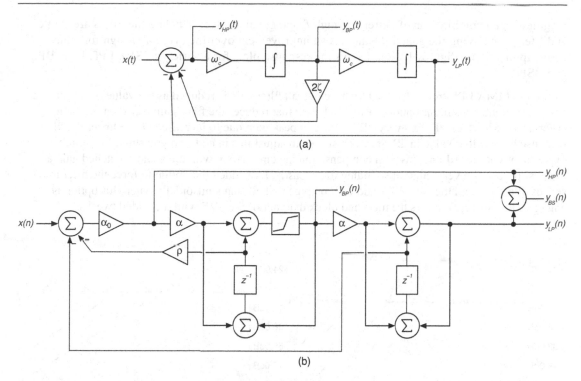

Figure 12.25: (a) Analog block diagram for the 2nd order state variable filter. (b) Zavalishin's VA equivalent; you can spot the two integrator sub-circuits and the additional output for the band-stop filter.

In Figure 12.25b you can see the block diagram for a nonlinear processor inserted into the first integrator circuit; the symbol inside it looks like a bent "Z" shape. As with all our other 2nd order LPF and HPF algorithms, this filter can produce very high gains for high Q settings. For $Q = 20$, the peak gain is about +26 dB—which is quite high. For this particular design we are going to add another tweak and allow the Q to become infinite. When this happens, the filter will self-oscillate, a trait generally desirable for synthesizer filter designs. Under self-oscillating conditions, when the source material contains frequencies at or near the filter's f_c settings, the oscillating output can grow very large, easily distorting the [−1, +1] range of our digital audio input/output signal. The methods to try to mitigate the obnoxiously high signal components always involve a nonlinear processor of some kind. Zavalishin's book (and most of the literature) concentrates on using a waveshaper for this nonlinear element, in part because some of the original analog versions, including the Oberheim SEM® filter, contained a nonlinear distortion circuit using diodes to try to tame the overbearing filter output at high Q or self-oscillation. We also use them exclusively in the filter designs in *Designing Software Synthesizers in C++*. We'll use a simple waveshaper equation (see Chapter 19) and you may want to use the oversampling techniques in Chapter 22 to mitigate any aliasing that forms.

As with the 1st order version, the LPF suffers from the zero-gain-at-Nyquist problem. We have addressed this issue with another tweak to provide the correct gain at Nyquist: in this case, the gain at Nyquist exactly matches that of the analog equivalent, however, there is a small error in the peak magnitude at a combination of extremely high f_c and Q values. Nevertheless the filters are stable and I feel that having the gain at Nyquist is an improvement over the original design for many audio applications. Figures 12.26a–f show the response plots for a variety of SVF LPF, HPF, BPF and BSFs.

Recall the MMA LPF from Chapter 11 whose overall filter gain is reduced as the value of Q increases. I've implemented a similar option for the SVF LPF that reduces the filter gain based on the same criteria as the MMA LPF: for every dB increase in peak gain due to increased Q, we lower the filter response by half that value in dB. It is also simple to adjust this in code, so you should definitely experiment with it. Adding this gain compensation in combination with the analog matched gain at Nyquist produces a very musical sounding filter. Lastly, I've added the option to force the filter into self-oscillation—and the LPF, HPF and BPF outputs will all ring sinusoidally when this option is engaged. Figure 12.27 shows the modified block diagram of the SVF with the added tweaks.

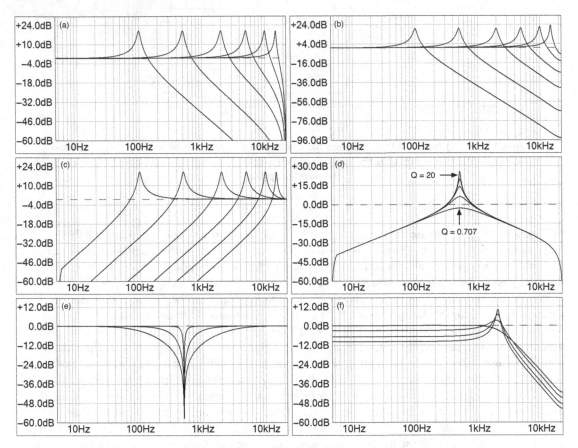

Figure 12.26: Frequency response plots for the 2nd order VA SVF: (a) normal LPF, (b) LPF analog matched gain at Nyquist, (c) HPF, (d) BPF with f_c = 500 Hz and various Q values, (e) BSF with f_c = 500 Hz and various Q values and (f) LPF with analog matched gain at Nyquist and Moog-style gain compensation with increasing Q. For (a) through (c) the values for f_c are 100 Hz, 500 Hz, 2 kHz, 5 kHz, 10 kHz, 15 kHz with Q = 20.

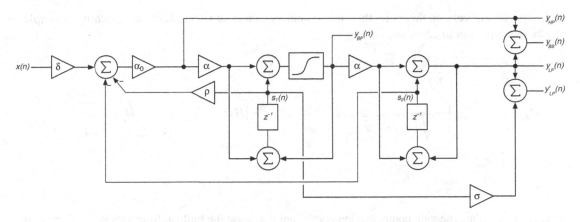

Figure 12.27: The modified VA SVF with added outputs and coefficients.

The equations for calculating the basic SVF coefficients are in several parts. First, we have the original Zavalishin defined calculations for α and ρ shown in Equation 12.19. Notice that self-oscillation is achieved when Q is infinite, which implies that the value for R is zero. You may recognize R as the traditional damping coefficient ζ in analog filter theory books.

$$T = 1/f_s$$

$$\alpha = \frac{2}{T}\tan\left(\frac{2\pi f_c T}{2}\right)$$

$$R = \zeta = \begin{cases} 0.0 & Q = \infty \\ \dfrac{1}{2Q} & Q \neq \infty \end{cases} \qquad (12.19)$$

$$\alpha_o = \frac{1}{1 + 2R\alpha + \alpha^2}$$

$$\rho = 2R\alpha$$

In order to correct the gain at Nyquist, we observe that the value residing in the first integrator's state register labeled $s_1(n)$ in Figure 12.27 is actually a mixture of the current band-pass output plus the high-pass output scaled by α. We can take advantage of the fact that the HPF gain at Nyquist is easily calculated (and mostly equals 1.0 except for very high combinations of f_c and Q). We can then use a similar trick of blending in this "correction" signal knowing the values of α and the desired gain at Nyquist. This works very well up until just before Nyquist where we observe a high frequency offset that is due to the fact that the internal HPF + BPF signal we are using has the same value of Q applied to it. You can see this slight increase in peak gain in Figure 12.26b. The Nyquist gain still matches the analog version and the filter is still stable right up to f_c = Nyquist. I have never been able to hear a difference of a few dB of peaking at 15–20 kHz between the two versions. To calculate the analog matching coefficient σ in Figure 12.25, we use Equation 12.20, evaluating the analog matching transfer functions at f = Nyquist. The equations for M_{LPF} and M_{HPF} are the standard equations to calculate the magnitude of the analog filter $H(s)$ for 2nd order LPF and HPF respectively. In this ratio, their denominators cancel so we are left with a relatively simple coefficient equation to calculate.

$$f_o = \frac{f}{f_c} \qquad \zeta = \frac{1}{2Q}$$

$$M_{LPF} = |H(s)_{LPF}| = \frac{1}{\sqrt{\left(1 - f_o^2\right)^2 + 4\zeta^2 f_o^2}} \qquad M_{HPF} = |H(s)_{HPF}| = \frac{f_o^2}{\sqrt{\left(1 - f_o^2\right)^2 + 4\zeta^2 f_o^2}} \qquad (12.20)$$

$$\sigma = \frac{M_{LPF}}{\alpha M_{HPF}}\bigg|_{f = Nyquist} = \frac{1}{\alpha f_o^2}$$

Finally, to calculate the gain compensation coefficient δ, we use the built-in functions in *fxobjects.h* to easily convert Q and gain values. The C++ code for computing the compensation gain reduction value that is then applied to the filter input signal $x(n)$ is shown next. Note that you can easily change the logic to reduce gain by more or less than half of the peak value in dB. See the two functions *dBTo_Raw* and *dBPeakGainFor_Q*, which are very basic.

```
double peak_dB = dBPeakGainFor_Q(Q);
if (peak_dB > 0.0)
{
        double halfPeak_dBGain = dBTo_Raw(-peak_dB / 2.0);
        xn *= halfPeak_dBGain;
}
```

12.5 C++ DSP Object: ZVAFilter

In the next chapter's modulated filter plugin, we will use a 2nd order virtual analog SVF filter. I've created an object that implements the following Zavalishin VA filters:

* 1st order LPF and 1st order HPF
* 1st order APF
* 2nd order SFV (LPF, HPF, BPF, BSF)

Like the other C++ objects in the book, it is derived from *IAudioSignalProcessor* and implements the familiar functions *reset*, and *processAudioSample*. The *ZVAFilter* object can process one channel of audio.

12.5.1 ZVAFilter: *Enumerations and Data Structure*

The filter algorithm is encoded as a strongly typed enumeration and is the only constant required. It is declared at the top of the *fxobjects.h* file and listed next—compare the enumeration strings with the block diagrams in Figures 12.23 and 12.24.

```
enum class vaFilterAlgorithm {kLPF1, kHPF1, kAPF1,
                      kSVF_LP, kSVF_HP, kSVF_BP, kSVF_BS};
```

The object is updated via the custom structure *ZVAFilterParameters* that contains members not only for the fundamental controls (f_c, Q, algorithm), but also for numerous features introduced in Section 12.4.2 such as gain compensation, self-oscillation, and nonlinear processing.

```
struct ZVAFilterParameters
{
        ZVAFilterParameters() {}
        ZVAFilterParameters& operator=(. . .){. . .}

        // --- individual parameters
        vaFilterAlgorithm filterAlgorithm = vaFilterAlgorithm::kSVF_LP;
        double fc = 1000.0;
        double Q = 0.707;
        double filterOutputGain_dB = 0.0; // usually not used
        bool enableGainComp = false;
        bool matchAnalogNyquistLPF = false;
        bool selfOscillate = false;
        bool enableNLP = false;
};
```

12.5.2 ZVAFilter: *Members*

Tables 12.6 and 12.7 list the *ZVAFilter* member variables and member functions.

12.5.3 ZAFilter: *Programming Notes*

The *ZVAFilter* object implements the seven basic virtual analog filters along with the gain compensation and nonlinear processing (NLP) options that we discussed in Chapter 12. We use a simple array to store the two z^{-1} state data values and follow the design equations directly.

Using the *ZVAFilter* object is straightforward:

1. Reset the object to prepare it for streaming.
2. Set the GUI control information in the *ZVAFilterParameters* custom data structure.
3. Call *setParameters* to update the calculation type.
4. Call *processAudioSample*, passing the input value in and receiving the output value as the return variable.

Table 12.6: *ZVAFilter* **member variables**

ZVAFilter Member Variables		
Type	**Name**	**Description**
ZVAFilterParameters	*parameters*	Filter parameters, f_c, Q, boost/cut and algorithm
double	*integrator_z[2]*	Array to store two z^{-1} state register values
double	*alpha0*	The alpha0 filter coefficient for input scaling and delay-free loop resolution
double	*alpha*	The alpha filter coefficient
double	*rho*	The rho filter coefficient
double	*analogMatchSigma*	Variable used for the Nyquist analog matching feature
double	*sampleRate*	Current sample rate

Table 12.7: *ZVAFilter* **member functions**

ZVAFilter Member Functions		
Returns	**Name**	**Description**
ZVAFilterParameters	*getParameters*	Get all parameters at once
void	*setParameters* Parameters: —*ZVAFilterParameters* _parameters	Set all parameters at once
bool	*reset* Parameters: —*double* sampleRate	Flush integrators
double	*processAudioSample* Parameters: —*double* xn	Process input xn through the audio filter
void	*setSampleRate* Parameters: —*double* _sampleRate	Set sample rate
bool	*calculateFilterCoeffs*	Update filter coefficients

The majority of the work in this object happens in the *calculateFilterCoeffs* function, which decodes the filter algorithm and then calculates the filter coefficients using Equations 12.18 and 12.19; as an exercise, you should compare the C++ code with the parameter calculations.

Function: *calculateFilterCoeffs*

Details: For 1st order filters, calculate the alpha coefficient, while for 2nd order filters, add the alpha0 and rho coefficients; calculate the analog matching coefficient.

```
void calculateFilterCoeffs()
{
        double fc = zvaFilterParameters.fc;
        double Q = zvaFilterParameters.Q;
        vaFilterAlgorithm filterAlgorithm =
                        zvaFilterParameters.filterAlgorithm;

        // --- normal Zavalishin SVF calculations here
        //     prewarp the cutoff- these are bilinear-transform filters
        double wd = kTwoPi*fc;
        double T = 1.0 / sampleRate;
        double wa = (2.0 / T)*tan(wd*T / 2.0);
        double g = wa*T / 2.0;

        // --- for 1st order filters:
        if (filterAlgorithm == vaFilterAlgorithm::kLPF1 ||
            filterAlgorithm == vaFilterAlgorithm::kHPF1 ||
            filterAlgorithm == vaFilterAlgorithm::kAPF1)
```

```
      {
            // --- calculate alpha
            alpha = g / (1.0 + g);
      }
      else // state variable variety
      {
            // --- note R is the traditional analog damping factor zeta
            double R = zvaFilterParameters.selfOscillate ? 0.0:
                                                1.0 / (2.0*Q);

            alpha0 = 1.0 / (1.0 + 2.0*R*g + g*g);
            alpha = g;
            rho = 2.0*R + g;

            // --- sigma for analog matching version
            double f_o = (sampleRate / 2.0) / fc;
            analogMatchSigma = 1.0 / (alpha*f_o*f_o);
      }
}
```

Function: *setParameters*

Details: The *setParameters* function checks to see if parameters have changed, checks the Q value for
   validity, then calls the *updateFilterCoeffs* function to calculate the coefficients:

```
void setParameters(const ZVAFilterParameters& params)
{
      if(params.fc != zvaFilterParameters.fc ||
         params.Q != zvaFilterParameters.Q ||
         params.selfOscillate != zvaFilterParameters.selfOscillate ||
         params.matchAnalogNyquistLPF !=
                  zvaFilterParameters.matchAnalogNyquistLPF)
            calculateFilterCoeffs();

      zvaFilterParameters = params;
}
```

Function: *processAudioSample*

Details: In the *processAudioSample* function note the branching to set the output value. As a state variable
   filter, the 2nd order version produces three of the output types directly. Notice how we scale the filter
   gain knowing the Q value, from which we can directly calculate the expected peak gain using the helper
   function *dBPeakGainFor_Q*. You may easily experiment with this gain control here.

```
virtual double processAudioSample(double xn)
{
      // --- with gain comp enabled, we reduce the input by
      //      half the gain in dB at resonant peak
      vaFilterAlgorithm filterAlgorithm =
                  zvaFilterParameters.filterAlgorithm;
```

```
bool matchAnalogNyquistLPF =
        zvaFilterParameters.matchAnalogNyquistLPF;

if (zvaFilterParameters.enableGainComp)
{
        double peak_dB = dBPeakGainFor_Q(zvaFilterParameters.Q);
        if (peak_dB > 0.0)
        {
                double halfPeak_dBGain = dB2Raw(-peak_dB / 2.0);
                xn *= halfPeak_dBGain;
        }
}

// --- for 1st order filters:
if (filterAlgorithm == vaFilterAlgorithm::kLPF1 ||
        filterAlgorithm == vaFilterAlgorithm::kHPF1 ||
        filterAlgorithm == vaFilterAlgorithm::kAPF1)
{
        // --- create vn node
        double vn = (xn -integrator_z[0])*alpha;

        // --- form LP output
        double lpf = ((xn-integrator_z[0])*alpha) +
                        integrator_z[0];

        double sn = integrator_z[0];

        // --- update memory
        integrator_z[0] = vn + lpf;

        // --- form the HPF = INPUT = LPF
        double hpf = xn - lpf;

        // --- form the APF = LPF - HPF
        double apf = lpf - hpf;

        // --- set the outputs
        // --- see code
}

// --- form the HP output first
double hpf = alpha0*(xn - rho*integrator_z[0] - integrator_z[1]);

// --- BPF Out
double bpf = alpha*hpf + integrator_z[0];
if (zvaFilterParameters.enableNLP)
        bpf = softClipWaveShaper(bpf, 1);
```

```
// --- LPF Out
double lpf = alpha*bpf + integrator_z[1];

// --- BSF Out
double bsf = hpf + lpf;

// --- finite gain at Nyquist; slight error at VHF
double sn = integrator_z[0];

// update memory
integrator_z[0] = alpha*hpf + bpf;
integrator_z[1] = alpha*bpf + lpf;

double filterOutputGain = pow(10.0,
        zvaFilterParameters.filterOutputGain_dB / 20.0);

// return our selected type
if (filterAlgorithm == vaFilterAlgorithm::kSVF_LP)
{
    if (matchAnalogNyquistLPF)
        lpf += analogMatchSigma*(sn);
    return filterOutputGain*lpf;
}
else if (filterAlgorithm == vaFilterAlgorithm::kSVF_HP)
    return filterOutputGain*hpf;
else if (filterAlgorithm == vaFilterAlgorithm::kSVF_BP)
    return filterOutputGain*bpf;
else if (filterAlgorithm == vaFilterAlgorithm::kSVF_BS)
    return filterOutputGain*bsf;

// --- unknown filter
return filterOutputGain*lpf;
}
```

12.6 C++ DSP Objects: WDF Ladder Filter Library

In Section 12.2 we listed the WDF component and adaptor objects that are implemented in the *fxobjects.h* and *fxobjects.cpp* files. For examples we also included a set of WDF classes that implement the fixed ladder filter designs in Section 12.2.1–12.2.3, parts of which are shown in text. We also included a set of pure RLC WDF classes that implement the four resonant RLC filter circuits from Section 12.3.4. All objects inherit the *IAudioSignalProcessor* interface and feature identically named member functions across all objects. The only real differences are the class names and the filter implementations inside. Table 12.8 shows the list of WDF filter objects.

The *WDFButterLPF3* fixed filter object is documented completely and the other fixed designs have their individual code chunks printed as well, including the pure RLC LPF example. Since the chapter plugin design project implements pure RLC WDF filters and the objects share identical function names, we will document the *WDFIdealRLCLPF* here.

Table 12.8: WDF ladder filter objects

WDF Object Name	Description
WDFButterLPF3	Implements the 3rd order Butterworth LPF (Section 12.3.1) as a WDF ladder filter
WDFBesselBSF3	Implements the 3rd order Bessel BSF (Section 12.3.2) as a WDF ladder filter
WDFConstKBPF6	Implements the 6th order Constant-K BPF (Section 12.3.3) as a WDF ladder filter
WDFIdealRLCLPF	Implements the ideal Pure RLC LPF (Section 12.3.4) as a WDF ladder filter
WDFIdealRLCHPF	Implements the ideal Pure RLC HPF (Section 12.3.4) as a WDF ladder filter
WDFIdealRLCBPF	Implements the ideal Pure RLC BPF (Section 12.3.4) as a WDF ladder filter
WDFIdealRLCBSF	Implements the ideal Pure RLC BSF (Section 12.3.4) as a WDF ladder filter

12.6.1 WDFIdealRLCxxx: *Enumerations and Data Structure*

The objects are updated via the custom structure *WDFParameters* that contains members for the fundamental controls f_c, Q, and boost/cut, as well as a flag for un-warping the resonant frequency.

```
struct WDFParameters
{
    WDFParameters() {}
    WDFParameters& operator=(const WDFParameters& params){. . .}

    // --- individual parameters
    double fc = 100.0;
    double Q = 0.707;
    double boostCut_dB = 0.0;
    bool frequencyWarping = true;
};
```

12.6.2 WDFIdealRLCxxx: *Members*

Tables 12.9 and 12.10 list the *WDFIdealRLCxxx* member variables and member functions. These are consistent across all objects and have identical behaviors.

12.6.3 WDFIdealRLCxxx: *Programming Notes*

The *WDFIdealRLCxxx* objects implement the specialized WDF adaptor connections and simulate the RLC components based on the users selected values for f_c and Q.

- The WDF component and adaptor objects do very simple and specialized functions.
- The static base class method for connecting the adaptors together handles the trading of interface pointers.
- The RLC components are updated in the *setParameters* method that uses information about the f_c, Q, and frequency un-warping that the user has set.
- The *processAudioSample* functions push an audio sample into the first adaptor in the chain, and extract the output sample from the desired location in the last adaptor in the chain.

Table 12.9: *WDFIdealRLCxxx* **member variables**

WDFIdealRLCxxx Member Variables		
Type	**Name**	**Description**
WDFParameters	*wdfParameters*	Object parameters
WdfSeriesAdaptor	*seriesAdaptor_XY*	Series WDF adaptor for components XY (e.g. RL, RC, RC)
WdfParallelAdaptor	*parallelAdaptor_XY*	Parallel WDF adaptor for components XY (e.g. RL, RC, RC)
double	*sampleRate*	current sample rate

Table 12.10: *WDFIdealRLCxxx* **member functions**

WDFIdealRLCxxx Member Functions		
Returns	**Name**	**Description**
WDFParameters	*getParameters*	Get all parameters at once
void	*setParameters* Parameters: —*WDFParameters* _wdfParameters	Set all parameters at once
bool	*reset* Parameters: —*double* sampleRate	Initialize all WDF components and initialize adaptor chain
double	*processAudioSample* Parameters: —*double* xn	Process input xn through the WDF structure
void	createWDF	Create the WDF adaptor structure for the circuit; may be called more than once and will destroy/rebuild the circuit as needed

The reset function calls the reset method on the components and then initializes the adaptors by calling the *initializeAdaptorChain* method on the first adaptor in the series. This function propagates initialization calls through the rest of the objects. Your WDF circuit may contain parallel branches of adaptors; you call the *initializeAdaptorChain* method on the first adaptor of any branch.

```
virtual bool reset(double _sampleRate)
{
    sampleRate = _sampleRate;

    // --- reset WDF components (flush state registers)
    seriesAdapter_RL.reset(_sampleRate);
    parallelTerminatedAdapter_C.reset(_sampleRate);

    // --- intialize the chain of adapters
    seriesAdapter_RL.initializeAdapterChain();
    return true;
}
```

The *createWDF* method attaches the adaptors together as discussed in Section 12.2.5 and is called once from the constructor. However, the function is designed so that it may be called more than once, destroying, and re-creating the structures a needed.

```
void createWDF()
{
        //      Holding C Constant at 1e-6
        //               L = 2.533e-2
        //               R = 2.251131 e2
        // --- initial values, will be changed by user interaction
        seriesAdapter_RC.setComponent(wdfComponent::seriesRC,
                                2.251131e2, 1.0e-6);
        parallelTerminatedAdapter_L.setComponent(wdfComponent::L,
                                2.533e-2);


        // --- connect adapters
        WdfAdapterBase::connectAdapters(&seriesAdaptor_RC,
                                &parallelTerminatedAdapter_L);


        // --- set source resistance
        seriesAdapter_RC.setSourceResistance(0.0); // --- Rs = 600 to 0


        // --- set open ckt termination
        parallelTerminatedAdapter_L.setOpenTerminalResistance(true);
}
```

Finally, the *processAudioSample* function processes the input sample through the structure. Make sure to check out the *createWDF* and *processAudioSample* for each of the pure RLC objects so you may understand how the connections are made for the four different analog circuits.

```
virtual double processAudioSample(double xn)
{
        // --- push audio sample into series L1
        seriesAdapter_RC.setInput1(xn);

        // --- output is at terminated L2's output2
        //      note compensation scaling by -6dB = 0.5
        //      because of WDF assumption about Rs and Rload
        return 0.5*parallelTerminatedAdapter_L.getOutput2();
}
```

12.7 *Chapter Plugin:* RLCFilters

As an example of using the *WDFIdealRLCxxx* objects, we will assemble a plugin that contains all four of the pure RLC WDF filter components. We will need objects for each filter and each channel, and we will need to decode the user selection and route the audio into and out of the appropriate object.

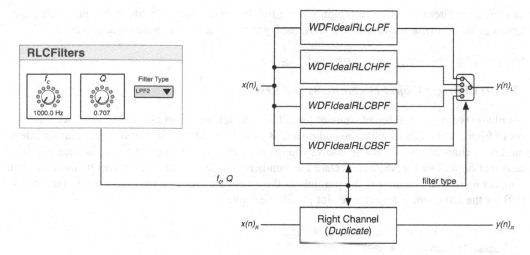

Figure 12.28: The *RLCFilters* plugin GUI and block diagram.

The GUI and block diagram are shown in Figure 12.28. In the example plugin, we will process only the filter that the user has selected. For homework you may improve on the plugin by using a different approach to running the filters.

> We will demonstrate using the virtual analog *SVF DSP* object in a plugin in the next chapter when we tackle modulated filter designs.

12.7.1 RLCFilters: *GUI Parameters*

The GUI parameter table is shown in Table 12.11—use it to declare your GUI parameter interface for your chosen plugin framework. If you are using ASPiK, remember to first create your enumeration of control ID values at the top of the *plugincore.h* file, then use automatic variable binding to connect the linked variables to the GUI parameters.

ASPiK: top of *plugincore.h* file

```
enum controlID {
    filterFc_Hz = 0,
    filterQ = 1,
    filterType = 2
};
```

Table 12.11: The *RLCFilter* plugin's GUI parameter and linked-variable list

Control Name	Units	Min/max/default or string-list	Taper	Linked Variable	Linked Variable Type
fc	Hz	20 / 20480 / 1000	1V/oct	*filterFc_Hz*	*double*
Q	-	0.707 / 20 / 0.707	linear	*filterQ*	*double*
Filter Type		string-list follows this table	-	*filterType*	*int*

We declare four filter types for this plugin's string list. If you are using ASPiK, note that placing the underscores in the string-list results in a GUI display that replaces them with whitespaces:

"RLC_LPF, RLC_HPF, RLC_BPF, RLC_BSF"

12.7.2 RLCFilters: *Object Declarations and Reset*

In this plugin we'll use a different strategy for holding the left and right channel filters—we will use arrays of filters. This will simplify the coding and is a bit more advanced than declaring independent left and right channel objects; it also allows you to easily change the channel count. In your plugin's processor (*PluginCore* for ASPiK), declare the members plus the *updateParameters* method. Note that we support up to two channels in the plugins, so the arrays are sized accordingly. We will use array slot [0] for the left channel object, and slot [1] for the right:

```
// --- in your processor object's .h file
#include "fxobjects.h"

// --- in the class definition
protected:
      WDFIdealRLCLPF rlcLPF[2];
      WDFIdealRLCHPF rlcHPF[2];
      WDFIdealRLCBPF rlcBPF[2];
      WDFIdealRLCBSF rlcBSF[2];
      void updateParameters();
```

The filtering objects require the current DAW sample rate to properly calculate their coefficients. Call the *reset* function and pass in the current sample rate in your plugin framework's prepare-for-audio-streaming function. In this example, *resetInfo.sampleRate* holds the current DAW sample rate. Notice the use of the for-loop to simplify coding.

ASPiK: *reset( )*

```
// --- reset components
for (unsigned int i = 0; i < 2; i++)
{
      rlcLPF[i].reset(resetInfo.sampleRate);
      rlcHPF[i].reset(resetInfo.sampleRate);
      rlcBPF[i].reset(resetInfo.sampleRate);
      rlcBSF[i].reset(resetInfo.sampleRate);
}
```

12.7.3 RLCFilters: *GUI Parameter Update*

Update the left and right channel objects at the top of the buffer processing loop. Gather the values from the GUI parameters and apply them to the custom data structure for each of the filter objects by using the *getParameters* method to get the current parameter set, and then overwrite only the parameters your plugin actually implements. Do this for all channels you support. Notice that since we want all filters to share identical parameters, we just reuse the parameter structure repeatedly.

ASPiK: *updateParameters( )* function

```
// --- all objects share same params, so get first
WDFParameters params = rlcLPF[0].getParameters();

// --- update with our GUI parameter variables
params.fc = filterFc_Hz;
params.Q = filterQ;

// --- apply to all filters
for (int i = 0; i < 2; i++)
{
    rlcLPF[i].setParameters(params);
    rlcHPF[i].setParameters(params);
    rlcBPF[i].setParameters(params);
    rlcBSF[i].setParameters(params);
}
```

12.7.4 RLCFilters: *Process Audio*

Add the code to process the plugin object's input audio samples through each filter and write the output samples. The ASPiK code is shown here for grabbing the input samples and writing the output samples; use the method that your plugin framework requires to read and write these values. In this design, we will decode the user's selection for the filter type, and then process only the input sample through the appropriate filter. The *filterType* variable encodes the user selection as an integer. For ASPiK, we use a helper function to compare the value to a pre-stored strongly typed enumeration. Use the method for your framework to make the string-list comparisons. We will show one channel of operation here; duplicate the code for the other channels. Remember that your right channel objects are in array location 1 (e.g. *rlcLPF[1]*).

```
// --- in your processing object's audio processing function
double xnL = processFrameInfo.audioInputFrame[0]; //< input sample L

// --- setup for left output
double ynL = 0.0;

// --- choose filter to process
if (compareIntToEnum(filterType, filterTypeEnum::RLC_LPF))
    ynL = rlcLPF[0].processAudioSample(xnL);
else if (compareIntToEnum(filterType, filterTypeEnum::RLC_HPF))
    ynL = rlcHPF[0].processAudioSample(xnL);
if (compareIntToEnum(filterType, filterTypeEnum::RLC_BPF))
    ynL = rlcBPF[0].processAudioSample(xnL);
if (compareIntToEnum(filterType, filterTypeEnum::RLC_BSF))
    ynL = rlcBSF[0].processAudioSample(xnL);

// --- write output
processFrameInfo.audioOutputFrame[0] = ynL; //< output sample L
```

Rebuild and test the plugin; play around with changing the filter type while audio streams. If you hear any pops or clicks, check out the Homework section for a solution to minimize clicking when changing the filter type.

12.8 Homework

1. **Tunable 3rd Order Butterworth LPF**: if you look at Table 12.3 you might notice that the L and C component values change by a factor of 10 when the cutoff frequency f_c changes by a factor of 10. This particular design lends itself to being tunable through frequency scaling. In fact, making high-order tunable filters is something that the WDF ladder filters excel at, as long as you can get the tuning equations figured out. In the case of the 3rd order Butterworth filter, we can use an analog filter design technique called *frequency scaling*. You first start by designing the filter with a cutoff frequency of $f_c = 1$ Hz, called the *normalized cutoff frequency*. For this particular design, we can easily find the new component values by simply dividing the normalized values by the user-set cutoff frequency. For a fs = 44.1 kHz, the normalized values are listed in Equation 12.21.

$$R(f_c) = \frac{R(1Hz)}{f_c} \quad L(f_c) = \frac{L(1Hz)}{f_c} \quad C(f_c) = \frac{C(1Hz)}{f_c} \tag{12.21}$$

$$f_s = 44.1kHz$$
$$L1(1Hz) = 95.493H \quad C1(1Hz) = 530.51\mu F \quad L2(1Hz) = 95.493H \tag{12.22}$$

Redesign the C++ object in Section 12.3.1 to make it tunable and allow the user to adjust the f_c with a GUI control on your plugin. Figure 12.29 shows the various frequency responses you can expect to achieve.

2. Combine the *WDF Ideal RLC* filter objects into one aggregate object that houses all four types; re-factor the *RLCFilters* project using this object instead of the set of arrays.

3. The *RLCFilters* has an annoyance: sometimes when the user changes filter types, a louder-than-usual click occurs. This is because we select only the chosen filter to process audio through. When the user changes the filter type while audio is streaming, we abruptly switch to a different filter.

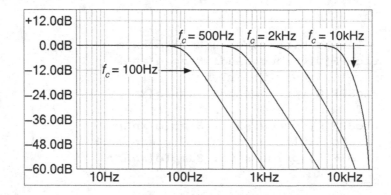

Figure 12.29: Responses of the tunable WDF 3rd order Butterworth LPF.

That filter's state registers contain old data and this can sometimes cause a popping sound. Change the design structure of the *RLCFilters* plugin so that all filters are always running all the time. Then, decode the filter type and select only the outputs from the proper filters (left and right) to send to the output. This keeps the state data in each filter updated correctly, but at a cost of CPU cycles.

4. Use the *WDF* library to implement the 1st order filters shown in Figure 12.30 and make them tunable for various values of f_c and boost/cut (shelving filters).

The equations for the filters are as follows.

LPF and HPF

$$f_c = \frac{1}{2\pi RC} \tag{12.23}$$

Low Shelf	High Shelf
f_{shelf} = shelf break frequency	f_{shelf} = shelf break frequency
$R2$ = choose arbitrary R2	$R2$ = choose arbitrary R2
$gain = 10^{(-shelf\_gain\_dB/20)}$	$gain = 10^{(-shelf\_gain\_dB/20)}$
$R1 = \dfrac{R2(1\text{-gain})}{gain}$	$R1 = \dfrac{R2(1\text{-gain})}{gain}$
$C1 = \dfrac{1}{2\pi f_{shelf} R1}$	$C1 = \dfrac{1}{2\pi f_{shelf} R1}$

$$\tag{12.24}$$

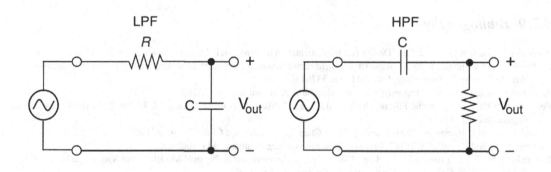

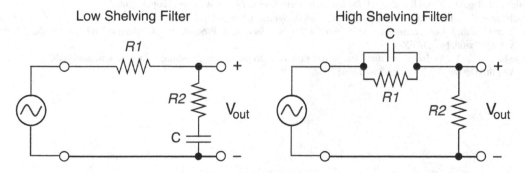

Figure 12.30: Four 1st order passive analog filters to implement as WDF plugins.

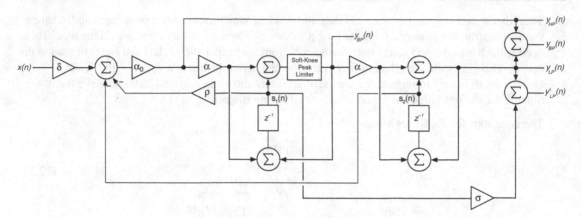

Figure 12.31: Replacement of the nonlinear waveshaper with a soft-knee compressor in the SVF

5. Replace the nonlinear waveshaper in the SVF with a peak limiter as shown in Figure 12.31. The peak limiter is designed using the same technique as you'll find in Chapter 18, but it is mostly a fixed implementation—we don't allow the user to mess with it. When using a soft-knee curve, the peak limiter behaves much like a dynamic waveshaper. The *fxobjects.h* file contains a bonus DSP object named *PeakLimiter*—use it in place of the nonlinear processor (waveshaper) and compare your results.

12.9 Bibliography

Antošová, V. and Davídek, V. 2001. "Design and Implementation of Wave Digital Filters." *Radioengineering*, Vol. 10, No. 3.

De Pavia, P. and Välimäki, T. 2011. "Real-Time Transformer Emulation for Virtual Tube Amplifiers." *EURASIP Journal on Advances in Signal Processing*, Vol. 2011, No. 537645.

Elsie Filter Design Tool, http://tonnesoftware.com/elsie.html, Accessed August 1, 2018.

Fettweis, A. 1986. "Wave Digital Filters: Theory and Practice." *Proceedings of the IEEE.*, Vol. 74, No. 2. https://ieeexplore.ieee.org/document/1457726

Gonzalez, G. 1984. *Microwave Transistor Amplifiers*, Chap. 1, 2. Englewood Cliffs: Prentice-Hall.

Kammeyer, K. and Kroschel, K. 1997. *Digitale Signalverarbeitung*. Stuttgart: B.G. Teubner.

Karjalainen, M. 2008. "Efficient Realization of Wave Digital Components for Physical Modeling and Sound Synthesis." *IEEE Transactions on Audio, Speech and Language Processing*, Vol. 16, No. 5.

Lindquist, C. 1989. *Adaptive and Digital Signal Processing*, Chap. 12. Miami: Steward and Sons.

Psenicka, B., Ugalde, F. and Romero, M. *Design of the Wave Digital Filters*, www.intechopen.com

Schlichthärle, D. 2011. *Digital Filters Basics and Design, 2nd Ed.*, Chap. 5. Heidelberg: Springer.

Yeh, D. and Smith, J. *Discretization of the '59 Fender Bassman Tone Stack*. Proc. 9th Intl. Conference on Digital Audio Effects, Vol. DAFx-06, No. DAFX-1.

Zavalishin, V. 2012. *The Art of VA Filter Design*, www.native-instruments.com/fileadmin/ni_media/downloads/pdf/VAFilterDesign_1.0.3.pdf, Accessed June 2014.

CHAPTER 13

Modulators: LFOs and Envelope Detectors

The basic definition of "modulate" is to change something. In our plugins, we will use modulators to alter plugin parameters in real time on a constant update basis. This means that our plugin parameters will undergo a constant updating of their values and we will generally make updates on every sample interval. Modulators will take static plugins like filters and delays and turn them into dynamic time-varying processors such as envelope followers and chorus/flanger/vibrato devices. There are two fundamental kinds of modulators we will use in our FX plugins: LFOs and envelope detectors (or simply detectors). LFOs generate time-varying waveforms such as sine, sawtooth, and triangle with relatively low oscillation frequencies of about 0.02 Hz to 20 Hz. These waveforms are not designed to create a pitched oscillator that we might use in a synthesizer. Instead, they are designed to slowly change plugin parameters. Envelope detectors output a value that is related to the envelope of an input signal. The envelope output value may be in a raw form, usually from 0.0 to 1.0, or in dB form. The input to the envelope detector might be the audio input signal into the plugin, or it might be an external audio signal that is supplied in a *side-chain*.

13.1 LFO Algorithms

LFOs are sometimes called *trivial oscillators* and are simple to design and implement. The LFOs will use a modulo counter as the time base for the oscillator. A modulo counter is a counter that rolls over and back to zero at some specified event. The modulo counter is unipolar and counts from zero up to 1.0. Once the counter hits or exceeds 1.0, it rolls over back to zero plus the distance it exceeded 1.0 prior to the rollover. The counter is reset to zero when the oscillator is triggered and counts upwards with a constant increment called *inc*. The *inc* value represents the current phase of the oscillator and is also called the *phase increment*. Thus the *inc* value also determines the frequency of the oscillator. The output of the modulo counter is a unipolar ramp or sawtooth waveform. Converting this modulo counter output to bipolar creates an LFO sawtooth waveform. We can also convert it to a triangle wave using this formula.

$$tri = 2|saw| - 1 \tag{13.1}$$

If we multiply the modulo counter by 2π, we can use it to create a sinusoid using several different algorithms. This allows us to easily create three LFO waveforms from one time-base: sawtooth, triangle, and sine. Figure 13.1 shows our LFO block diagram.

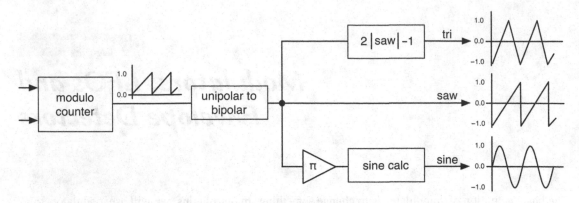

Figure 13.1: The LFO uses a modulo counter and some trivial math to generate useful waveforms for modulation.

The modulo counter has two variables, f_o (frequency of oscillation) and f_s (sampling frequency). The value of *inc* is calculated simply as follows.

$$inc = \frac{f_o}{f_s} \tag{13.2}$$

We will be using a sinusoidal approximation to avoid calling the *sin* function for the sine wave output. The other math is simple and trivial, so our *LFO* object will be very efficient right from the start. A sinusoid may be approximated using two modified parabolas: one for the top half of the waveform and the other for the bottom half. This concept has been around since the seventh century A.D. and originated with the mathematician Bhaskra I. A more recent refinement generates a sinusoid that is nearly pure—certainly close enough for us to use as a modulator. The *parabolicSine* function is declared in your *fxobjects.h* file. It takes an argument between −π and +π. The function is shown here; you can find the original source at the stackoverflow.com link in the bibliography as the original site has been deleted.

```
const double B = 4.0 / kPI;
const double C = -4.0 / ( kPI* kPI);
const double P = 0.225;

// --- input is -pi to +pi
double parabolicSine(double x)
{
      double y = B * x + C * x * fabs(x);
      y = P * (y * fabs(y) -y) + y;
      return y;
}
```

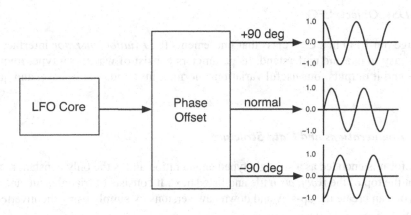

Figure 13.2: The LFO features three output waveforms in three different phases; here the sinusoidal output is shown

Our *LFO* object will feature a tri-phase output. It will generate three outputs that are shifted by 90 degrees: normal (in phase), +90 degrees (called *quadrature-phase* or *quad-phase*), and −90 degrees (*inverted quad-phase*) as depicted in Figure 13.2. Having the multiple phases instantly available will allow you to create some amazing stereo effects. The quad-phase output is easily accomplished with a second modulo counter that is offset to start at a value of 0.25 rather than 0.0—and everything else remains the same so the calculations remain simple. Notice that the LFO outputs a bipolar waveform. If we need a unipolar modulator instead, we can simply convert it using a built-in function *bipolarToUnipolar* in the *fxobjects.h* file. We also have a *unipolarToBipolar* function so that we may freely change back and forth as needed.

13.1.1 *The* IAudioSignalGenerator *Interface*

Our C++ FX building blocks so far have implemented the *IAudioSignalProcessor* interface to simplify programming as well as aggregating *IAudioSignalProcessor* objects together to create more complex devices. The LFO is a type of signal generator as it has no actual audio input signal. The needs and requirements for signal generators are a bit different that those for a signal processor, so we introduce a second interface for use in our plugins: *IAudioSignalGenerator*. To further simplify coding, we'll also introduce the *SignalGenData* structure that will hold the generator output samples. These are discussed in detail in Section 13.1.2.

The *IAudioSignalGenerator* interface is also simple and straightforward; it consists of only pure abstract virtual functions, so any object that implements the interface must override and implement each of the functions. The functions are as follows.

- *reset*: reset any internal counters, phase offsets or other variables for a fresh run
- *setGeneratorFrequency*: set a new oscillator frequency in Hz
- *setGeneratorWaveform*: set the current output waveform for the LFO
- *renderAudioOutput*: render the next oscillator output sample

13.1.2 C++ DSP Object: LFO

We will package our LFO in a C++ class that implements the *IAudioGenerator* interface. The *LFO* object is easy to use and understand. Its parameters consist of waveform type, frequency, and amplitude and it outputs four useful variations: normal, inverted, quad-phase, and quad-phase inverted.

13.1.3 LFO: *Enumerations and Data Structure*

The LFO waveform is encoded as a strongly typed enumeration and is the only constant required. It is declared at the top of the *fxobjects.h* file and listed next. It consists of triangle, sin, and saw waveforms—you can create the up-saw and down-saw versions by simply using the inverted output member variable.

```
enum class generatorWaveform { kTriangle, kSin, kSaw };
```

The object is updated via the custom structure *OscillatorParameters* that contains members for the fundamental controls: waveform and frequency.

```
struct OscillatorParameters
{
    OscillatorParameters() {}
    OscillatorParameters& operator=(. . .){. . .}

    // --- individual parameters
    generatorWaveform waveform = generatorWaveform::kTriangle;
    double frequency_Hz = 0.0;
};
```

13.1.4 LFO: *Members*

Tables 13.1 and 13.2 list the *LFO* member variables and member functions.

Table 13.1: *LFO* member variables

LFO Member Variables		
Type	**Name**	**Description**
OscillatorParameters	*lfoParameters*	The object parameters
double	*modCounter*	Modulo counter for normal output
double	*modCounterQP*	Modulo counter for quad-phase outputs
double	*phaseInc*	The current phase increment value
double	*sampleRate*	Sample rate

Table 13.2: *LFO* member functions

LFO Member Functions		
Returns	**Name**	**Description**
OscillatorParameters	*getParameters*	Get all parameters at once
void	*setParameters* Parameters: —*OscillatorParameters _parameters*	Set all parameters at once
bool	*reset* Parameters: —*double sampleRate*	Reset counters, reset indexes
SignalGenData	*renderAudioOutput*	Render the LFO synthesized waveform
void	*setSampleRate* Parameters: —*double _sampleRate*	Set sample rate
bool	*checkAndWrapModulo* Parameters: —*double& moduloCounter* —*double phaseInc*	Check a modulo counter to see if it needs to be wrapped; if so, do the wrapping
bool	*advanceAndcheckAndWrapModulo* Parameters: —*double& moduloCounter* —*double phaseInc*	Advance modulo counter, then check to see if it needs to be wrapped; if so, do the wrapping
void	*advanceModulo* Parameters: —*double& moduloCounter* —*double phaseInc*	Advance modulo counter, do not check for wrap
double	*parabolicSine* Parameters: —*double angle*	Do the parabolic sine calculation at input angle in radians

13.1.5 LFO: *Programming Notes*

The majority of the work happens in the *IAudioSignalGenerator* overrides so let's look at those. Using the *LFO* object is straightforward:

1. Reset the object to prepare it for rendering.
2. Set the waveform and frequency in the *OscillatorParameters* custom data structure.
3. Call *setParameters* to update the calculation type.
4. Call *renderAudioOutput* to generate the output data on each sample interval.

Function: *reset*
Details:

- Store the new sample rate.
- Recalculate the *phaseInc* based on new sample rate and Equation 13.2.
- Reset the *modCounter* to 0.0 and the quad-phase *modCounterQP* to 0.25.

```
virtual bool reset(double _sampleRate)
{
        // --- store and recalc
        sampleRate = _sampleRate;
        phaseInc = oscFrequency / sampleRate;

        // --- timebase variables
        modCounter = 0.0; // --- reset modCounter
        modCounterQP = 0.25; // --- reset quad-phase modCounter
        return true;
}
```

Function: *renderAudioOutput*
Details:

- Check and wrap the modulo counter if needed.
- Advance the quad-phase modulo counter by 0.25; wrap if needed.
- Decode the waveform type and render the various waveforms.
- Invert the quad-phase output sample.
- Advance the *modCounter* for the next sample period by adding the *phaseInc* to it; the counter will be checked and wrapped at the start of the next call to *renderAudioOutput*.

The *renderAudioOutput* is the only complicated function in the object, but it follows the details provided in a step-by-step manner. The various outputs are rendered according to the block diagram in Figure 13.1, and helper functions do most of the work involving wrapping the modulo counter and approximating the sinusoid.

```
virtual const SignalGenData renderAudioOutput()
{
        // --- always first!
        checkAndWrapModulo(modCounter, phaseInc);

        // --- QP output always follows location of current modulo
        modCounterQP = modCounter;

        // --- advance QP modulo by 0.25 = 90 degrees; wrap if needed
        advanceAndCheckWrapModulo(modCounterQP, 0.25);

        SignalGenData output;

        // --- calculate the oscillator value
        if (waveform == generatorWaveform::kSin)
        {
                // --- calculate normal angle
                double angle = modCounter*2.0*kPI - kPI;

                // --- norm output with parabolicSine approximation
                output.normalOutput = parabolicSine(-angle);
```

```
        // --- calculate QP angle
        angle = modCounterQP*2.0*kPI-kPI;

        // --- calc QP output
        output.quadPhaseOutput_pos = parabolicSine(-angle);
}
    else if (waveform == generatorWaveform::kTriangle)
    {
        // --- triv saw
        output.normalOutput = unipolarToBipolar(modCounter);

        // --- bipolar triangle
        output.normalOutput = 2.0*fabs(output.normalOutput) - 1.0;

        // --- quad phase
        output.quadPhaseOutput_pos =
                            unipolarToBipolar(modCounterQP);

        // --- bipolar triangle
        output.quadPhaseOutput_pos =
                    2.0*fabs(output.quadPhaseOutput_pos) - 1.0;
    }
    else if (waveform == generatorWaveform::kSaw)
    {
        output.normalOutput = unipolarToBipolar(modCounter);

        output.quadPhaseOutput_pos =
                            unipolarToBipolar(modCounterQP);
    }
    // --- invert two main outputs to make the opposite versions
    output.quadPhaseOutput_neg = -output.quadPhaseOutput_pos;
    output.invertedOutput = -output.normalOutput;

    // --- setup for next sample period
    advanceModulo(modCounter, phaseInc);

    return output;
}
```

13.2 Envelope Detection

The second type of modulator we'll use is the envelope detector. It is based on a simple analog circuit found in compressors, signal meters, and other devices where tracking the envelope of the signal is important. An envelope detector may track the peak, mean-square (MS) or root-mean-square (RMS) value of an audio signal. The detector may track only the positive or negative portion of the audio input (called *half-wave*) or both positive and negative portions (called *full-wave*); a rectifier circuit generates this signal from the audio input. The detector may produce a *log* output in dB or a linear output. The analog detector uses a capacitor and two potentiometers (variable resistors) to implement the detection function on the rectified signal. One potentiometer controls how fast the rectified signal charges a capacitor and thus the detector's attack time, and the other controls how fast the charge is bled off of the capacitor as the signal drops. The two resistors and capacitor in combination create a RC charge and discharge curve that creates the detector's output. Figure 13.3a shows a simple positive peak detector while Figure 13.3b shows the textbook RC charge and discharge characteristics. The analog definition of the RC rise-time is the time it takes for the capacitor to charge up to 63.2% of the total during the charging phase, while the RC fall-time is the time it takes for the capacitor to discharge down to 36.8% of the total. We will rename the rise- and fall-times to the *attack* and *release* times of the detector.

As it turns out, there are numerous digital versions of the envelope detector, each containing multiple variations. If you are interested, you can find many more versions, along with their response characteristics in Reiss (2011), which includes details of this envelope detector as well as the others. Our envelope detector is based on an old design but with a few new features. Figure 13.4 shows the block diagram of the *AudioDetector* object that you can find in your *fxobjects.h* file. It has the following features:

- uses a full-wave rectifier
- detects peak, MS, or RMS values
- includes attack and release time controls in milliseconds (mSec)
- RC simulator uses analog charge/discharge values
- can output true *log* (dB) values or linear, unipolar values
- can be set to clamp the detection output range to unity [0.0, 1.0] or allow the envelope to exceed unity values in linear or dB format

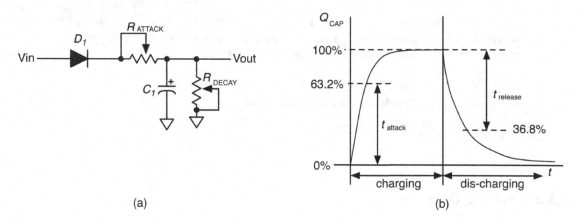

(a) (b)

Figure 13.3: (a) A simple analog positive peak detector. (b) The capacitor charging and discharging times are controlled with the attack and decay potentiometers.

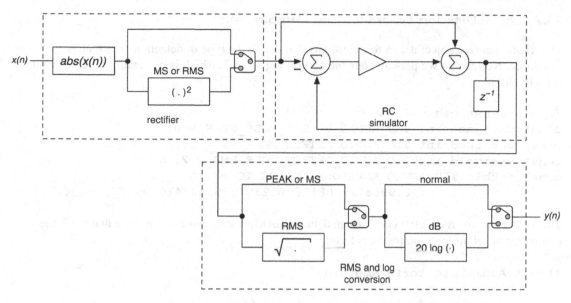

Figure 13.4: Our envelope detector consists of a rectifier stage followed by a RC simulator that feeds a RMS and/or *log* converter; as shown, the switches are set up to implement a peak detector with a linear output.

13.2.1 C++ DSP Object: AudioDetector

The *AudioDetector* object implements the block diagram in Figure 13.4. It implements the *IAudioSignalProcessor* interface. It processes an input sample *x(n)* and produces a value *y(n)* that is the detected envelope of the input. The range of output values depends on the mode of operation and the *clamping* flag. When set, the clamping flag will limit the RC simulator block's output to a maximum value of 1.0. The variable output range of *y(n)* is shown here.

Clamping enabled

- normal: [0.0, 1.0]
- dB: [−96.0, 0.0]

Clamping disabled

- normal: [0.0, numerical max value]
- dB: [−96.0, numerical max value]

With clamping disabled, the envelope maximum is gigantic, technically the most positive *double* data-type value. We will want to disable the clamping when we create compressors, limiters, and expanders so we can detect signals above unity gain, which may happen in some designs.

13.2.2 AudioDetector: *Enumerations and Data Structure*

The *AudioDetector* object uses a few constants that define the type of detection. The names are easy to decode. Notice that we pre-calculate the natural log of 36.7, which is the analog time constant for a RC filter.

```
// --- constants
const unsigned int TLD_AUDIO_DETECT_MODE_PEAK = 0;
const unsigned int TLD_AUDIO_DETECT_MODE_MS = 1;
const unsigned int TLD_AUDIO_DETECT_MODE_RMS = 2;
const double TLD_AUDIO_ENVELOPE_ANALOG_TC =
                -0.99967234081320612357829304641019; // ln(36.7%)
```

The *AudioDetector* is updated via a custom data *AudioDetectorParameters* structure that holds the pertinent control information for the detector.

```
struct AudioDetectorParameters
{
      AudioDetectorParameters() {}
      AudioDetectorParameters& operator=(. . .){. . .}

      // --- individual parameters
      double attackTime_mSec = 0.0;
      double releaseTime_mSec = 0.0;
      unsigned int detectMode = TLD_AUDIO_DETECT_MODE_PEAK;
      bool detect_dB = false; /* DEFAULT = false (linear NOT log) */
      bool clampToUnityMax = true;
};
```

13.2.3 AudioDetector: *Members*

Tables 13.3 and 13.4 list the *AudioDetector* member variables and member functions. This object follows the same design pattern as the other DSP objects. It includes an additional function to change the sample rate because this is a CPU intensive operation.

Table 13.3: *AudioDetector* **member variables**

AudioDetector Member Variables		
Type	**Name**	**Description**
AudioDetectorParameters	*parameters*	The object parameters
double	*attackTime*	The calculated attack coefficient for the given attack time in mSec
double	*releaseTime*	The calculated release coefficient for the given release time in mSec
double	*lastEnvelope*	The last envelope value that was generated; used to calculate next value
double	*sampleRate*	current sample rate

Table 13.4: *AudioDetector* **member functions**

AudioDetector Member Functions		
Returns	**Name**	**Description**
AudioDetectorParameters	*getParameters*	Get all parameters at once
void	*setParameters* Parameters: —*AudioDetectorParameters* _parameters	Set all parameters at once
bool	*reset* Parameters: —*double* sampleRate	Reset last envelope output to 0.0
double	*processAudioSample* Parameters: —*double* xn	Generate an output envelope value from the current input xn
bool	*setSampleRate* Parameters: —*double* _sampleRate	Set sample rate
void	*setAttackTime* Parameters: —*double* attack_in_ms	Calculate the attack coefficient with an attack time in mSec
void	*setReleaseTime* Parameters: —*double* attack_in_ms	Calculate the release coefficient with a release time in mSec

13.2.4 AudioDetector: *Programming Notes*

Using the *AudioDetector* object is straightforward:

1. Reset the object to prepare it for streaming.
2. Set the attack, release, and detection type in the *AudioDetectorParameters* custom data structure.
3. Call *setParameters* to update the information.
4. Call *processAudioFrame*, passing the input value in and receiving the output envelope value as the return variable.

Function: *setAttackTime* and *setReleaseTime*
Details: Before looking at the detection code, first examine the functions that calculate the attack and release time coefficients. These follow Equation 13.3.

$$coeff = e^{\frac{TC}{(f_s)(time\_ms)(0.001)}}$$
$$TC = \ln(0.368)$$

(13.3)

Both functions are short and simple; notice that we first check to see if the stored user time in *mSec* has changed before recalculating the coefficients. This is an expensive calculation because it uses the *exp* function, so we want to recalculate only if necessary. The *setSampleRate* function also issues calls to these coefficient calculation functions, but only if the sample rate has changed.

```
void AudioDetector::setAttackTime(double attack_in_ms)
{
        if (attackTime_mSec == attack_in_ms)
            return;

    attackTime_mSec = attack_in_ms;
    attackTime = exp(TLD_AUDIO_ENVELOPE_ANALOG_TC / (attack_in_ms *
                                            sampleRate * 0.001));
}

void AudioDetector::setReleaseTime(double release_in_ms)
{
        if (releaseTime_mSec == release_in_ms)
            return;

    releaseTime_mSec = release_in_ms;
    releaseTime = exp(TLD_AUDIO_ENVELOPE_ANALOG_TC / (release_in_ms *
                                            sampleRate * 0.001));
}
```

Function: *reset*
Details:

- Clear the *lastEnvelope* value.
- store the new sample rate and update the coefficients only if the sample rate changed; this is done with the *setSampleRate* function.

```
virtual bool reset(double _sampleRate)
{
    setSampleRate(_sampleRate);
    lastEnvelope = 0.0;
    return true;
}
```

Function: *processAudioSample*
Details:

- Full-wave rectify the signal by taking the absolute value of the input
- Square the rectified value for MS and RMS calculations.
- Detect whether the current input value is above or below the last envelope value; if it is rising, apply the attack time coefficient and if falling, the release time coefficient to the RC simulator in Figure 13.4 to generate the current envelope value.
- Clamp the RC simulator output to 1.0 if the clamping flag is set.
- Store the current envelope value as the last envelope value for next sample interval.
- For RMS, take the square root of the current envelope value.
- If detecting dB, run the *log* calculation, or just return the envelope value if not.

You should be able to match the code that follows with the bullet points for this function, so examine it closely.

```
virtual double processAudioSample(double xn)
{
    // --- all modes do Full Wave Rectification
    double input = fabs(xn);

    // --- square it for MS and RMS
    if (detectMode == TLD_AUDIO_DETECT_MODE_MS ||
        detectMode == TLD_AUDIO_DETECT_MODE_RMS)
            input *= input;

    // --- to store current
    double currEnvelope = 0.0;

    // --- do the RC simulator detection with attack or release
    if (input > lastEnvelope)
        currEnvelope = attackTime*(lastEnvelope-input) + input;
    else
        currEnvelope = releaseTime*(lastEnvelope-input) + input;

    // --- we are recursive so need to check underflow
    checkFloatUnderflow(currEnvelope);

    // --- bound them if desired
    if (clampToUnityMax)
        currEnvelope = fmin(currEnvelope, 1.0);

    // --- can not be (-)
    currEnvelope = fmax(currEnvelope, 0.0);

    // --- store envelope prior to sqrt for RMS version
    lastEnvelope = currEnvelope;

    // --- if RMS, do the SQRT
    if (detectMode == TLD_AUDIO_DETECT_MODE_RMS)
        currEnvelope = pow(currEnvelope, 0.5);

    // --- if not dB, we are done
    if (!detect_dB)
        return currEnvelope;

    // --- setup for log( )
    if (currEnvelope <= 0) return -96.0;

    // --- true log output in dB, can go above 0dBFS!
    return 20.0*log10(currEnvelope);
}
```

13.3 Modulating Plugin Parameters

Now that we have two C++ objects that can generate modulation values, we need to understand how to use their outputs. Ultimately this is going to depend on their role in the different plugins, but we can make some generalizations about parameter modulation first to make the code easier to understand later.

13.3.1 Modulation Range, Polarity, and Depth

The first thing you need to establish when modulating a parameter is the range of modulation that is needed or available. This means setting a maximum and minimum value that the parameter may take after the modulation calculation. In some cases, we might limit this range for musical sensibility reasons. In other cases, there may be practical reasons that the limits must be observed, such as filter stability requirements. The two limits set our *modulation range* and will be included in the final modulation calculation. For our audio filters, we will usually use the filter's stable minimum and maximum values over a 10-octave range for f_s = 44.1 kHz, or 20 Hz to 20,480 Hz. Let's use the filter f_c as the modulation target parameter for these next few examples.

The plugin algorithm usually dictates the *modulation polarity*, which may be either *unipolar* or *bipolar* about a center point. Suppose we are modulating the f_c of a filter in a unipolar manner using a sinusoidal LFO. When the LFO is at a minimum value, we want the filter f_c to be 20 Hz (minimum) and when the LFO is at a maximum value, f_c should be 20,480 Hz. In this case, it makes sense to convert the LFO's normal bipolar [−1, +1] output to a unipolar one [0, +1] before applying the calculation. Then, we may easily find the modulated value using this C++ pseudo-code, where the unipolar LFO value is named *unipolarModValue*:

```
// --- modulate from minimum value upwards
modValue = unipolarModValue*(maxValue - minValue) + minValue;
```

We may also modulate from the maximum value downwards (we will use this for the tremolo homework plugin):

```
// --- modulate from maximum value downwards
modValue = maxValue - (1.0 - unipolarModValue)*(maxValue - minValue);
```

We might also do bipolar modulation about a center point in the modulation range using the bipolar output of the LFO. For our example here we would find the arithmetic center value between 20 Hz and 20,480 Hz as the half range. During the positive half of the LFO output, we will modulate from the center up this half-range amount, and during the negative half we will modulate down from the center.

```
// --- calculate half-range and midpoint
double halfRange= (maxValue - minValue) / 2.0;
double midpoint = halfRange + minValue;

// --- modulate about center
modValue = bipolarModValue*(halfRange) + midpoint;
```

Figure 13.5 shows (a) unipolar and (b) bipolar modulation with a sinusoidal LFO sweeping across the modulation range.

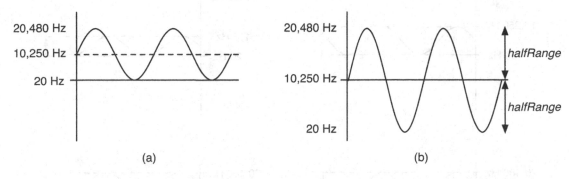

Figure 13.5: (a) Unipolar modulation from the minimum value and up. Compared this with (b) bipolar modulation about the center point. In both cases the LFO depth is 100% and the modulation result is identical.

If we manipulate the amplitude of the unipolar or bipolar modulator value prior to using the modulation calculation, we can affect the *modulation depth*. The depth of modulation will have a different effect depending the modulation polarity, and in the case of unipolar modulation, on the direction as well (*min-up* or *max-down*). Figure 13.6 shows the modulation scaling effect for unipolar (a) 50% and (b) 25% depths while Figure 13.7 shows the effect for bipolar (a) 50% and (b) 25% modulation depths. The unipolar version produces very different results depending on the direction of modulation and center of modulation changes accordingly.

The block diagram for a filter whose f_c is modulated with a LFO is shown in Figure 13.8 in two different forms. Figures 13.8a and b are equivalent. Notice that neither diagram shows the polarity of modulation. The depth control is implemented as a simple coefficient multiplier whose value is on the range of [0.0, +1.0]. The upward arrow is the standard symbol for a modulation path, with the parameter being modulated noted next to it.

When we prescribe a modulator device and a modulation destination parameter in our plugin algorithms, we will always specify the modulation range and polarity. The depth control may or may not be used depending on the plugin architecture. If it is omitted, we assume the depth is 100% or a coefficient of 1.0.

13.3.2 Modulation With the Envelope Detector

Whereas modulation with the LFO is fairly straightforward—the LFO generates values that we apply to plugin parameters across a fixed range—algorithms that use the envelope detector as the modulator are usually a bit more complicated. The reason is that these algorithms usually incorporate a *threshold* parameter. The envelope detector tracks an audio signal, and its output is compared with the threshold value. Then, a decision is made on how to process the audio input signal depending on whether the detected envelope is above or below the threshold. This usually involves a calculation that depends on the distance the detector's output lies above or below the threshold. In some cases the threshold and detector outputs may be compared in dB form rather than a linear raw value. The algorithms may also require that the detector can output values above unity when the internal size of the plugin audio signal goes above it.

This chapter features two plugin projects, one each that use the *LFO* and *AudioDetector* objects. The *ModFilter* plugin block diagram shown in Figure 13.9a uses the envelope detector to move the

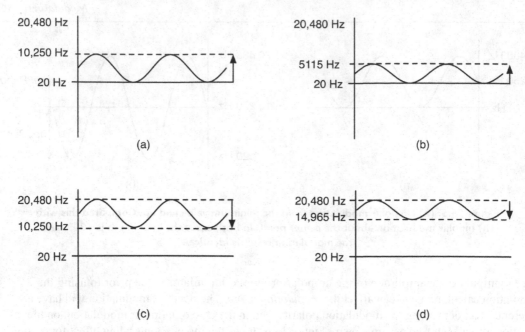

Figure 13.6: Unipolar modulation from the minimum up with (a) 50% and (b) 25% depth; and unipolar modulation from the maximum down with (c) 50% and (d) 25% depth. The center of modulation changes in all cases.

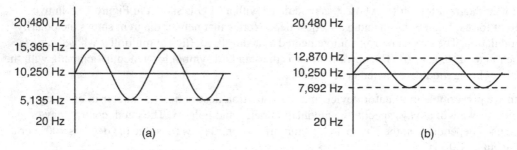

Figure 13.7: Bipolar modulation with (a) 50% depth and (b) 25% depth of modulation. The center of modulation does not change.

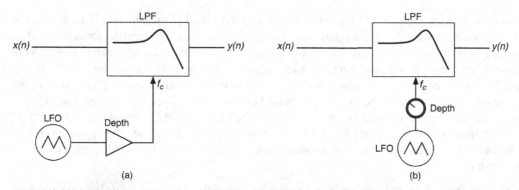

Figure 13.8: Two block diagrams that depict modulating a filter's f_c with a LFO and depth control (a) shows the depth control as a coefficient multiplier while (b) shows it as a GUI knob control.

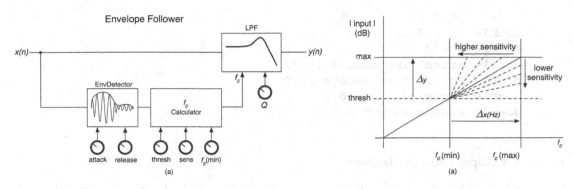

Figure 13.9: The envelope detector (a) block diagram and (b) envelope-to-f_c tracking graph.

location of the cutoff frequency of a filter. The filter's cutoff frequency travels across some range that the user (or manufacturer) sets as $[f_c(\text{min}), f_c(\text{max})]$. In our *ModFilter* plugin we will track the input envelope level. When the input envelope is below a user-established threshold, the low-pass filter will use the GUI control f_c setting as its cutoff frequency. When the signal travels above the threshold, we measure the distance and move the cutoff frequency up, starting from the GUI control f_c setting. The f_c value will move up and down linearly as the envelope changes above the threshold, as shown in Figure 13.9b. Increasing the sensitivity control makes the cutoff frequency move more quickly through its excursion, while decreasing it does the opposite. The sensitivity is implemented as a simple post-detector gain control. In the *EnvelopeFollower* object that follows, we do the dB conversion in the audio detector (this is only one variation) so the detector's attack and release slopes are in dB as well. This is easily changed in the object so that you may experiment with your own versions.

13.4 *C++ Effect Object:* **EnvelopeFollower**

The *EnvelopeFollower* object contains two DSP objects: *AudioDetector* and *ZVAFilter*, which implement the detector and filter respectively. The *AudioDetector* is set for RMS detection, converted to dB, with the ability to detect values above 0 dB Full Scale (dBFS). The *ZVAFilter* is configured as a 2nd order SVF LPF with gain compensation (reduced gain with increased Q), nonlinear processing (a waveshaper from Chapter 19), and matched analog Nyquist frequency (there is some error at high Q and high f_c values).

13.4.1 EnvelopeFollower: *Enumerations and Data Structure*

The *EnvelopeFollower* object does not require any special enumerations other than those needed for its member detector. The object is updated via the custom structure *EnvelopeFollowerParameters,* which contains a combination of members for both the filter and detector.

```
struct EnvelopeFollowerParameters
{
    EnvelopeFollowerParameters() {}
    EnvelopeFollowerParameters& operator=(. . .){. . .}
```

```
        // --- individual parameters
        double fc = 0.0;
        double Q = 0.707;
        double attackTime_mSec = 10.0;
        double releaseTime_mSec = 10.0;
        double threshold_dB = 0.0;
        double sensitivity = 1.0; // 0.25 to 10 is range
};
```

13.4.2 EnvelopeFollower: *Members*

Tables 13.5 and 13.6 list the *EnvelopeFollower* member variables and member functions. The filter and detector member objects do the majority of the work. The *EnvelopeFollower* applies the modulation calculation logic (this is the place to modify the object for the homework). Notice how compact this object is: three member variables and the *IAudioSignalProcessor* functions plus the *get/set* parameter functions.

13.4.3 EnvelopeFollower: *Programming Notes*

The *EnvelopeFollower* object combines the filter and detector objects that do the bulk of the low-level work. Using the *EnvelopeFollower* object is straightforward.

- Reset the object to prepare it for streaming.
- Set the object parameters in the *EnvelopeFollower Parameters* custom data structure.

Table 13.5: *EnvelopeFollower* member variables

EnvelopeFollower Member Variables		
Type	Name	Description
EnvelopeFollowerParameters	*parameters*	Filter parameters, f_c, Q, boost/cut, and algorithm
ZVAFilter	*filter*	The 2nd order LPF filter object
AudioDetector	*detector*	The RMS dB detector

Table 13.6: *AudioFilter* member functions

EnvelopeFollower Member Functions		
Returns	Name	Description
EnvelopeFollowerParameters	*getParameters*	Get all parameters at once
void	*setParameters* Parameters: —*EnvelopeFollowerParameters* _parameters	Set all parameters at once
bool	*reset* Parameters: —*double* sampleRate	Reset member objects
double	*processAudioSample* Parameters: —*double* xn	Process input xn; detect and modulate filter

- Call *setParameters* to update the calculation type.
- Call *processAudioSample*, passing the input value in and receiving the output value as the return variable.

Function: *Constructor*

Details: The object constructor sets up the filter and detector object's hardwired components. The user may only adjust the attack and release times. This is where you can make modifications for the homework and change the detector's characteristics—the detect mode and whether the envelope is converted to dB or not and the filter's features may be changed. Notice the stock manner of initializing the objects.

```
EnvelopeFollower() {
        // --- setup the filter
        ZVAFilterParameters filterParams;
        filterParams.filterAlgorithm = vaFilterAlgorithm::kSVF_LP;
        filterParams.fc = 1000.0;
        filterParams.enableGainComp = true;
        filterParams.enableNLP = true;
        filterParams.matchAnalogNyquistLPF = true;
        filter.setParameters(filterParams);

        // --- setup the detector
        AudioDetectorParameters adParams;
        adParams.attackTime_mSec = -1.0;
        adParams.releaseTime_mSec = -1.0;
        adParams.detectMode = TLD_AUDIO_DETECT_MODE_RMS;
        adParams.detect_dB = true;
        adParams.clampToUnityMax = false;
        detector.setParameters(adParams);

}                    /* C-TOR */
```

Function: *reset*

Details: resets sub-components

```
virtual bool reset(double _sampleRate){
        filter.reset(_sampleRate);
        detector.reset(_sampleRate);
        return true;
}
```

Function: *setParameters*

Details: The *setParameters* function behaves identically to the others. It checks the incoming parameters to see if either the filter or detector parameters have changed; if so, it fills out the appropriate data structure and calls one or both object's *setParameters* function to update it.

Function: *processAudioSample*

Details: The *processAudioSample* function follows the flowchart for the modulated filter plugin in Figure 13.9. It detects the input envelope in dB, compares the value with the threshold, and then modulates the filter using the helper function *doUnipolarModulationFromMin*. This is where you can make serious tweaks to the algorithm.

```
virtual double processAudioSample(double xn)
{
        // --- calc threshold
        double threshValue = pow(10.0, parameters.threshold_dB / 20.0);

        // --- detect the signal
        double detect_dB = detector.processAudioSample(xn);
        double detectValue = pow(10.0, detect_dB / 20.0);
        double deltaValue = detectValue - threshValue;

        ZVAFilterParameters filterParams = filter.getParameters();
        filterParams.fc = parameters.fc;

        // --- if above the threshold, modulate the filter fc
        if (deltaValue > 0.0)
        {
                // --- fc Computer
                double modulatorValue = 0.0;

                // --- best results-linear when detector is in dB mode
                modulatorValue = (deltaValue * parameters.sensitivity);

                // --- calculate modulated frequency
                filterParams.fc =
                        doUnipolarModulationFromMin(modulatorValue,
                        parameters.fc, kMaxFilterFrequency);
        }

        // --- update with new modulated frequency
        filter.setParameters(filterParams);

        // --- perform the filtering operation
        return filter.processAudioSample(xn);
}
```

13.5 *Chapter Plugin 1:* ModFilter

For the first plugin example, we will implement an envelope follower using the 2nd order virtual analog SVF filter that we've implemented in the *ZVAFilter* object. This plugin uses the block diagram in Figure 13.10c.

13.5.1 ModFilter: *GUI Parameters*

The GUI parameter table is shown in Table 13.7—use it to declare your GUI parameter interface for the plugin framework you are using. If you are using ASPiK, remember to first create your

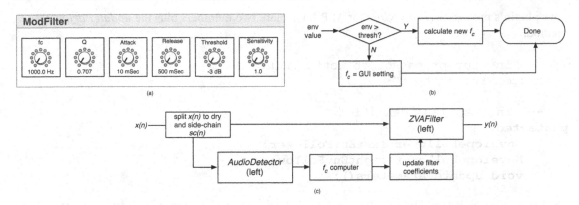

Figure 13.10: The *ModFilter* plugin's (a) GUI and (b) modulation calculation flowchart.
(c) The conceptual block diagram for the left channel; the right channel is identical.

Table 13.7: The *ModFilter* plugin's GUI parameter and linked-variable list

Control Name	Units	Min/max/default or string-list	Taper	Linked Variable	Linked Variable Type
f_c	Hz	20 / 10,000 / 1,000	1V/oct	*filterFc_Hz*	*double*
Q	-	0.707 / 20 / 0.707	linear	*filterQ*	*double*
Attack	mSec	1 / 250 / 20	linear	*attackTime_mSec*	*double*
Release	mSec	1 / 2,000 / 500	linear	*releaseTime_mSec*	*double*
Threshold	dB	−20 / 0 / −6	linear	*threshold_dB*	*double*
Sensitivity	-	0.25 / 5/ 1	anti-log	*sensitivity*	*double*

enumeration of control ID values at the top of the *plugincore.h* file, then use automatic variable binding to connect the linked variables to the GUI parameters.

ASPiK: top of *plugincore.h* file

```
enum controlID {
    filterFc_Hz = 0,
    filterQ = 1,
    attackTime_mSec = 2,
    releaseTime_mSec = 3,
    threshold_dB = 4,
    sensitivity = 5
};
```

13.5.2 ModFilter: *Object Declarations and Reset*

The *ModFilter* plugin uses two *EnvelopeFollower* members to provide the processing for the left and right channels of a stereo plugin. To support more channels, just add more member objects. In your plugin's processor (*PluginCore* for ASPiK) declare two members plus the *updateParameters* method.

```
// --- in your processor object's .h file
#include "fxobjects.h"

// --- in the class definition
protected:
    EnvelopeFollower leftEnvFollower;
    EnvelopeFollower rightEnvFollower;
    void updateParameters();
```

The components require the current DAW sample rate to properly calculate their coefficients. Call the *reset* function and pass in the current sample rate in your plugin framework's prepare-for-audio-streaming function. In this example, *resetInfo.sampleRate* holds the current DAW sample rate.

ASPiK: *reset( )*

```
    // --- reset the filters
    leftEnvFollower.reset(resetInfo.sampleRate);
    rightEnvFollower.reset(resetInfo.sampleRate);
```

13.5.3 ModFilter: *GUI Parameter Update*

Update the left and right channel objects at the top of the buffer processing loop. Gather the values from the GUI parameters and apply them to the custom data structure for each of the envelope follower objects using the *getParameters* method to get the current parameter set, and then overwrite only the parameters your plugin actually implements. Do this for all channels you support. Notice that since we want both envelope followers to share identical parameters, we just reuse the parameter structure and set the right channel with it.

ASPiK: *updateParameters( )* function

```
EnvelopeFollowerParameters params = leftEnvFollower.getParameters();
params.fc = filterFc_Hz;
params.Q = filterQ;
params.attackTime_mSec = attackTime_mSec;
params.releaseTime_mSec = releaseTime_mSec;
params.threshold_dB = threshold_dB;
params.sensitivity = sensitivity;

// --- set objects identically
leftEnvFollower.setParameters(params);
rightEnvFollower.setParameters(params);
```

13.5.4 ModFilter: *Process Audio*

Add the code to process the plugin object's input audio samples through each filter and write the output samples. The ASPiK code is shown here for grabbing the input samples and writing the output samples; use the method that your plugin framework requires to read and write these values.

ASPiK: *processAudioFrame( )* function

```
//--- read input (adapt to your framework)
double xnL = processFrameInfo.audioInputFrame[0]; //< framework input sample L
double xnR = processFrameInfo.audioInputFrame[1]; //< framework input sample R

// --- process the audio to produce output
double ynL = leftEnvFollower.processAudioSample(xnL);
double ynR = rightEnvFollower.processAudioSample(xnR);

// --- write output (adapt to your framework)
processFrameInfo.audioOutputFrame[0] = ynL; //< framework output sample L
processFrameInfo.audioOutputFrame[1] = ynR; //< framework output sample R
```

Rebuild and test the plugin. Notice the interplay between the threshold and sensitivity controls. Test the plugin at extremes of the ranges. Also notice how the level drops when the Q is increased, which we implemented by design. You can change this in the *EnvelopeFollower* constructor where you initialize the filter.

13.6 The Phaser Effect

We'll use the *LFO* object to implement the phaser effect. The phaser effect was an attempt at creating a flanging effect without needing to create a delay line. A flanger uses a modulated delay that is blended with the input signal. The phaser uses all-pass filters in an attempt to do the same. All-pass filters have a delay-response that is flat across the first one-third or so of the DC-to-Nyquist band. In fact, all-pass filters can be designed to perform fractional delay as an alternative to interpolation. The phaser didn't sound like a flanger at all. The reason is that in a flanger, the notches in the inverse-comb filtering are mathematically related as simple multiples of each other. This is not the case for the phaser, whose peaks and notches are related in a complex fashion. The block diagram for an analog phaser is shown in Figure 13.11.

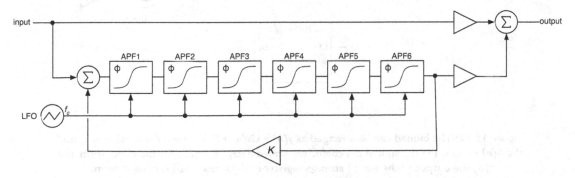

Figure 13.11: The analog phaser block diagram.

The phaser splits the signal into a dry path and an effected path. The effected path is run through a set of 1st order all-pass filters in series; each filter is called a *stage*. The number of stages depends on the manufacturer's design. In the classic designs, each of the all-pass filters operates on a band of frequencies that is different, but highly overlapped for each stage. A LFO modulates the center frequency of the all-pass filters together, where each APF f_c value modulates between a minimum and maximum value that is established for that stage. The output of the phase-shifters is fed back into its input to create a resonant feedback loop, which is noted as "intensity" or "resonance" on some models. The dry and effected outputs are summed together in equal proportions.

In order to implement the phaser design, we need to address the feedback loop present in the analog version. This feedback loop is delay-free—and that presents a problem for discrete designs. Zavalishin published an algebraic method for resolving the delay-free loops that specifically use his filter topologies. I derived an alternate system to resolve the delay-free loop that extended the work of an existing algorithm by Härmä that may be applied to the direct form or transposed canonical form of the biquad structure in addition to Zavalishin's structures (Pirkle, 2013). This method requires re-factoring the biquad structures in a form of $y(n) = Gx(n) + S(n)$ where $S(n)$ refers to the delayed component of the biquad; this is shown in Figure 13.12.

In order to use the *Biquad* object in the phaser, we've added two hidden functions to allow an outer container object to access the G and $S(n)$ values. These are named *getG_value( )* and *getS_value( )*. The proof of the resolution of the delay-free loop may be found at www.willpirkle.com/forum/; it will require understanding either Zavalishin's method or our modified Härmä method. Both methods are

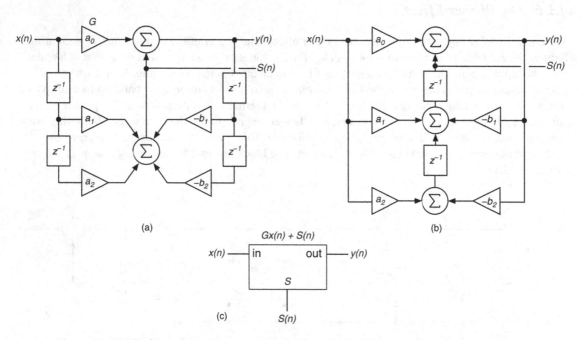

Figure 13.12: The biquad can be arranged as $y(n) = Gx(n) + S(n)$ where the G value is a_0 and the $S(n)$ value is (a) the sum of the coefficient-scaled delayed signals for the direct form and (b) the output of the top z^{-1} storage register in the transposed canonical form. (c) The brief schematic version of the biquad with input, output and $S(n)$ output port.

described in extreme detail in *Designing Software Synthesizers in C++,* so I won't repeat that theory here. The phaser design here was derived from the analog circuit in the *National Semiconductor (NSC) Audio/Radio Handbook,* a 1970s source of old App Notes from National Semiconductor (it also includes numerous filtering, distortion, EQ, and power amp designs). The NSC design used six 1st order all-pass stages that were modulated from a common *LFO.* The ranges of each all-pass unit are shown in Table 13.8.

The block diagram of the phaser design with the delay-free loop resolution is shown in Figure 13.13. The feedback $-K$ value varies between 0% and 100%, with the upper value producing sinusoidal oscillations. This phaser design sounds best with feedback (intensity) values between 75% and 95%.

The design equations for the gamma and alpha coefficients are based on the delay-free loop resolution method. Note that for each all-pass stage shown here, the G value is simply the a_0 coefficient for that filter.

$$\gamma_1 = G_6 \qquad \gamma_2 = G_5\gamma_1 \quad \gamma_3 = G_4\gamma_2 \quad \gamma_4 = G_3\gamma_3 \quad \gamma_5 = G_2\gamma_4 \quad \gamma_6 = G_1\gamma_5$$

$$\alpha_0 = \frac{1}{1 + K\gamma_6} \tag{13.4}$$

13.6.1 C++ Effect Object: PhaseShifter

The *PhaseShifter* object implements the six-stage phaser effect described in Section 13.6. It includes six *AudioFilter* objects and one *LFO* object. These do the bulk of the low-level work. The PhaseShifter's

Table 13.8: Minimum and maximum phase rotation
frequencies for the NSC phaser APFs

APF	Minimum f_c	Maximum f_c
1	16 Hz	1.6 kHz
2	33 Hz	3.3 kHz
3	48 Hz	4.8 kHz
4	98 Hz	9.8 kHz
5	160 Hz	16 kHz
6	260 Hz	26 kHz*

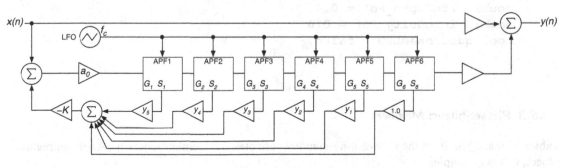

Figure 13.13: The six-stage phaser design.

main responsibility is in connecting the all-pass filters in the delay-free loop, calculating the γ and α values, propagating the signal through the series all-pass stages, and mixing the dry and effected signals together.

13.6.2 PhaseShifter: *Enumerations and Data Structure*

The *PhaseShifter* object uses some constant declarations to set the ranges of the various APF stages. Feel free to modify them as you like to experiment with different versions. These follow the values in Table 13.8.

```
const unsigned int PHASER_STAGES = 6;
const double apf0_minF = 16.0;
const double apf0_maxF = 1600.0;
const double apf1_minF = 33.0;
const double apf1_maxF = 3300.0;
const double apf2_minF = 48.0;
const double apf2_maxF = 4800.0;
const double apf3_minF = 98.0;
const double apf3_maxF = 9800.0;
const double apf4_minF = 160.0;
const double apf4_maxF = 16000.0;
const double apf5_minF = 260.0;
const double apf5_maxF = 20480.0;
```

The object is updated via the custom structure *PhaseShifterParameters* that contains members for the LFO and the phaser settings. Notice the addition of the Boolean flag to set the quadrature phase—this is useful in stereo phasers and we'll implement it in our second chapter plugin.

```
struct PhaseShifterParameters
{
        PhaseShifterParameters() {}
        PhaseShifterParameters& operator=(. . .){. . .}

        // --- individual parameters
        double lfoRate_Hz = 0.0;
        double lfoDepth_Pct = 0.0;
        double intensity_Pct = 0.0;
        bool quadPhaseLFO = false;

};
```

13.6.3 PhaseShifter: *Members*

Tables 13.9 and 13.10 list the *PhaseShifter* member variables and member functions. As an aggregate object, it is very simple.

Table 13.9: *PhaseShifter* **member variables**

PhaseShifter Member Variables		
Type	**Name**	**Description**
PhaseShifterParameters	*parameters*	Custom data structure
AudioFilter	*apf[6]*	Six all-pass filter objects
LFO	*lfo*	The LFO

Table 13.10: *PhaseShifter* **member functions**

PhaseShifter Member Functions		
Returns	**Name**	**Description**
PhaseShifterParameters	*getParameters*	Get all parameters at once
void	*setParameters* Parameters: —*PhaseShifterParameters* _parameters	Set all parameters at once
bool	*reset* Parameters: —*double* sampleRate	Resets all filters and LFO
double	*processAudioSample* Parameters: —*double* xn	Process input xn through the phaser

13.6.4 PhaseShifter: *Programming Notes*

The *PhaseShifter* object combines six filter and the LFO objects that do the bulk of the low-level work. Using the *PhaseShifter* object is straightforward.

- Reset the object to prepare it for streaming.
- Set the object parameters in the *PhaseShifterParameters* custom data structure.
- Dall *setParameters* to update the calculation type.
- Call *processAudioSample*, passing the input value in and receiving the output value as the return variable.

Function: *Constructor*

Details: The object constructor sets up the LFO as a sinusoidal oscillator, then it sets all of the audio filters to the APF1 algorithm.

```
PhaseShifter(void) {
    OscillatorParameters lfoparams = lfo.getParameters();
    lfoparams.waveform = generatorWaveform::kSin;// sine LFO
    lfo.setParameters(lfoparams);

    AudioFilterParameters params = apf[0].getParameters();

    params.algorithm = filterAlgorithm::kAPF1;
```

```
        for (int i = 0; i < PHASER_STAGES; i++)
        {
               apf[i].setParameters(params);
        }
}    /* C-TOR */
```

Function: *reset*

Details: The *reset* function simply resets all of the member objects with the current sample rate and implements no other functionality.

Function: *setParameters*

Details: The *setParameters* function checks to see if LFO parameters have changed, then applies them; the bulk of the updating happens on each sample interval where the six different modulated cutoff frequency values are calculated.

```
void setParameters(const PhaseShifterParameters& params)
{
       // --- update LFO rate
       if (params.lfoRate_Hz != parameters.lfoRate_Hz)
       {
              OscillatorParameters lfoparams = lfo.getParameters();
              lfoparams.frequency_Hz = params.lfoRate_Hz;
              lfo.setParameters(lfoparams);
       }


       // --- save new
       parameters = params;
}
```

Function: *processAudioSample*

Details: The *processAudioSample* function completes the phaser design. You should study this function and relate it to the block diagram in Figure 13.13. Notice the use of the special *Biquad* functions to obtain the *G* and *S* values.

```
virtual double processAudioSample(double xn)
{
       SignalGenData lfoData = lfo.renderAudioOutput();

       // --- create the bipolar modulator value
       double lfoValue = lfoData.normalOutput;
       if (parameters.quadPhaseLFO)
              lfoValue = lfoData.quadPhaseOutput_pos;

       double depth = parameters.lfoDepth_Pct / 100.0;
       double modulatorValue = lfoValue*depth;

       // --- calculate modulated values for each APF
       AudioFilterParameters params = apf[0].getParameters();
```

```
params.fc = doBipolarModulation(modulatorValue,
                                apf0_minF, apf0_maxF);
apf[0].setParameters(params);

params = apf[1].getParameters();
params.fc = doBipolarModulation(modulatorValue,
                                apf1_minF, apf1_maxF);
[apf[1].setParameters(params);

// --- NOTE: follows same pattern for the rest of the APF stages
//     <SNIP SNIP SNIP>

// --- calculate gamma values
double gamma1 = apf[5].getG_value();
double gamma2 = apf[4].getG_value() * gamma1;
double gamma3 = apf[3].getG_value() * gamma2;
double gamma4 = apf[2].getG_value() * gamma3;
double gamma5 = apf[1].getG_value() * gamma4;
double gamma6 = apf[0].getG_value() * gamma5;

// --- set the alpha0 value
double K = parameters.intensity_Pct / 100.0;
double alpha0 = 1.0 / (1.0 + K*gamma6);

// --- create combined feedback
double Sn = gamma5*apf[0].getS_value() +
            gamma4*apf[1].getS_value() +
            gamma3*apf[2].getS_value() +
            gamma2*apf[3].getS_value() +
            gamma1*apf[4].getS_value() +
            apf[5].getS_value();

// --- form input to first APF
double u = alpha0*(xn - K*Sn);

// --- cascade of APFs (could also nest these)
double APF1 = apf[0].processAudioSample(u);
double APF2 = apf[1].processAudioSample(APF1);
double APF3 = apf[2].processAudioSample(APF2);
double APF4 = apf[3].processAudioSample(APF3);
double APF5 = apf[4].processAudioSample(APF4);
double APF6 = apf[5].processAudioSample(APF5);

// --- sum with -3dB coefficients
double output = 0.707*xn + 0.707*APF6;
return output;
}
```

13.7 Chapter Plugin 2: Phaser

The GUI and overall block diagram for the *Phaser* plugin is shown in Figure 13.14. For this design, we will throw the right channel's *LFO* into quadrature phase, producing an interesting stereo effect.

13.7.1 Phaser: *GUI Parameters*

The GUI parameter table is shown in Table 13.11—use it to declare your GUI parameter interface for the plugin framework you are using. If you are using ASPiK, remember to first create your enumeration of control ID values at the top of the *plugincore.h* file, and then use automatic variable binding to connect the linked variables to the GUI parameters.

ASPiK: top of *plugincore.h* file

```
enum controlID {
    lfoRate_Hz = 0,
    lfoDepth_Pct = 1,
    intensity_Pct = 2
};
```

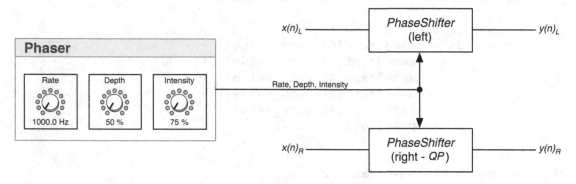

Figure 13.14: The *Phaser* GUI and block diagram.

Table 13.11: The *Phaser* plugin's GUI parameter and linked-variable list

Control Name	Units	Min/max/default or string-list	Taper	Linked Variable	Linked Variable Type
Rate	Hz	0.02 / 20 / 0.2	1V/oct	*lfoRate_Hz*	*double*
Depth	%	0 / 100 / 50	linear	*lfoDepth_Pct*	*double*
Intensity	%	0 / 100 / 75	linear	*boostCut_dB*	*double*

13.7.2 Phaser: *Object Declarations and Reset*

The *Phaser* plugin uses two *PhaseShifter* members to perform the audio processing, and it is very simple in design. In your plugin's processor (*PluginCore* for ASPiK) declare two members plus the *updateParameters* method.

```
// --- in your processor object's .h file
#include "fxobjects.h"

// --- in the class definition
protected:
    PhaseShifter leftPhaseShifter;
    PhaseShifter rightPhaseShifter;
    void updateParameters();
```

The APF and LFO objects require the current DAW sample rate to properly calculate their coefficients. Call the *reset* function and pass in the current sample rate in your plugin framework's prepare-for-audio-streaming function. In this example, *resetInfo.sampleRate* holds the current DAW sample rate.

ASPiK: *reset( )*

```
    // --- reset the filters
    leftPhaseShifter.reset(resetInfo.sampleRate);
    rightPhaseShifter.reset(resetInfo.sampleRate);
```

13.7.3 Phaser: *GUI Parameter Update*

Update the left and right channel objects at the top of the buffer processing loop. Gather the values from the GUI parameters and apply them to the custom data structure for each of the filter objects by using the *getParameters* method to get the current parameter set, and then overwrite only the parameters your plugin actually implements. Do this for all channels you support.

ASPiK: *updateParameters( )* function

```
// --- update with GUI parameters
PhaseShifterParameters params = leftPhaseShifter.getParameters();
params.lfoRate_Hz = lfoRate_Hz;
params.lfoDepth_Pct = lfoDepth_Pct;
params.intensity_Pct = intensity_Pct;

// --- update
leftPhaseShifter.setParameters(params);
rightPhaseShifter.setParameters(params);
```

13.7.4 Phaser: *Process Audio*

Add the code to process the plugin object's input audio samples through each phase shifter object and write the output samples. The ASPiK code is shown here for grabbing the input samples and writing the output samples; use the method that your plugin framework requires to read and write these values.

ASPiK: *processAudioFrame( )* function

```
// --- read input (adapt to your framework)
double xnL = processFrameInfo.audioInputFrame[0]; //< input sample L
double xnR = processFrameInfo.audioInputFrame[1]; //< input sample R

// --- process the audio to produce output
double ynL = leftPhaseShifter.processAudioSample(xnL);
double ynR = rightPhaseShifter.processAudioSample(xnR);

// --- write output (adapt to your framework)
processFrameInfo.audioOutputFrame[0] = ynL; //< output sample L
processFrameInfo.audioOutputFrame[1] = ynR; //< output sample R
```

Rebuild and test the plugin. Notice the effect of the quadrature phase right channel. How would you modify the phaser for your own design?

13.8 Homework

1. Add an amplitude parameter to the *OscillatorParameters* structure in Section 13.1.3 and use this value to scale the output of the LFO in the object's rendering function. You will need to connect a GUI control to adjust and test it in the plugin of your choice.
2. Modify the *LFO* object to add a special max-down variation. In this case, increasing the amplitude parameter of the *LFO* creates a signal that is stuck at the top of the range—as shown in Figure 13.6c and d—that gets progressively larger in the downward direction as the amplitude is increased.

    Hint: first create a unipolar LFO and scale it by the amplitude value. Then, add an offset to the LFO that pushes it up by the exact amount required. Use the amplitude parameter to figure out how much to shift the LFO.
3. Design the tremolo plugin using this new LFO output to modulate the gain coefficient multiplier as shown in Figure 13.15. Use the built-in function for max-down modulation so that when the depth control (aka the LFO amplitude) is at a minimum there is no effect: as the depth control value increases, the signal is attenuated more. Try both linear and dB-based attenuation calculations. Design a stereo version that uses the inverted LFO for one of the outputs.

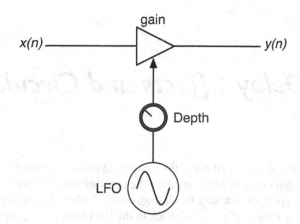

Figure 13.15: Tremolo block diagram.

13.9 Bibliography

Anderton, C. 1981. *Electronic Projects for Musicians*. London: Wise Publications.

Dodge, C. and Jerse, T. 1997. *Computer Music Synthesis, Composition and Performance*, Chap. 2. New York: Schirmer.

Giles, M., ed. 1980. *The Audio/Radio Handbook*. Santa Clara: National Semiconductor Corp.

musicdsp.org. *Envelope Follower with Different Attack and Release*, www.musicdsp.org/archive.php?classid=0#205, Accessed August 1, 2018.

Pirkle, W. 2013. *Resolving Delay-Free Loops in Recursive Filters using the Modified Härmä Method*. Presented at the 137th Audio Engineering Society Convention, Los Angeles.

Pirkle, W. 2015. *Designing Software Synthesizers in C++*, Chap. 7. New York: Focal Press.

Reiss, J. 2011. *Under the Hood of a Dynamic Range Compressor*, 131st Audio Engineering Society Convention (Tutorials).

stackoverflow.com, *Fastest Implementation fo Sine and Cosine*, https://stackoverflow.com/questions/18662261/fastest-implementation-of-sine-cosine-and-square-root-in-c-doesnt-need-to-b, Accessed January 23, 2019.

CHAPTER 14

Delay Effects and Circular Buffers

Before we can start looking at some FIR algorithms, we need to deal with the concept of long delay lines and circular buffers. Not only are they used for the delay effects, but also they are needed to make long FIR filters. In this chapter we'll take a break from the DSP filter algorithms and develop some digital delays for our delay plugin. If you think back to the IIR filters you've worked on so far, you will remember that after implementing the difference equation you need to shuffle the z^{-1} delay element values. You do this by overwriting the delays backwards in the *Biquad* object's *processAudioSample* function:

```
// --- update states
stateArray[x_z2] = stateArray[x_z1];
stateArray[x_z1] = xn;

stateArray[y_z2] = stateArray[y_z1];
stateArray[y_z1] = yn;
```

But what happens when the delay line needs to be very long, such as z^{-1024} or (for a one-second digital delay with f_s = 44.1 kHz) $z^{-441000}$? It's going to be difficult to implement the delay shuffling this way. Not only would it be tedious to code, it would also be very inefficient to have to implement all those read/ write operations each sample period. The answer to the problem of long delay lines is called *circular buffering*. A circular buffer is also called a *ring buffer*. We are going to use the term *ring buffer* to mean a fixed circular buffer used in convolution, while *delay buffer* will refer to a variable circular buffer that is designed to implement time delay.

We can create an ordinary buffer of doubles with length 1024 like this:

```
double buffer[1024];
```

We access the first slot (or location or cell) in the array at *buffer[0]* and the last slot at *buffer[1023]*. The first slot is sometimes called the "top" of the buffer and the last slot is sometimes called the "bottom" of the buffer.

Suppose we set up a long processing loop, accessing every fifth element in the array, using a read index *n*:

```
for (uint32_t n=0; n < 10000; n+=5)
{
      double data = buffer[n];
}
```

It is easy to see that this code is going to crash when the buffer access goes outside the buffer boundary, as shown in Figure 14.1. In this case the buffer access location *buffer[1025]* is actually two locations past the end of the buffer at *buffer[1023]*.

In a circular buffer, the index value is automatically wrapped back around to the top of the buffer plus any offset as needed, allowing a circular access pattern as shown in Figure 14.2. Here the attempt to use the index 1025 is mitigated as the index is automatically wrapped back around the buffer and offset by two slots to *buffer[1]*.

You might be thinking that this buffer loop example is obvious since the end of loop counter is 10,000 or far outside the last legal array index. But in the context of an audio plugin, the access "loop" is going to start when the user begins streaming data through the plugin, and end only when the user has stopped the audio.

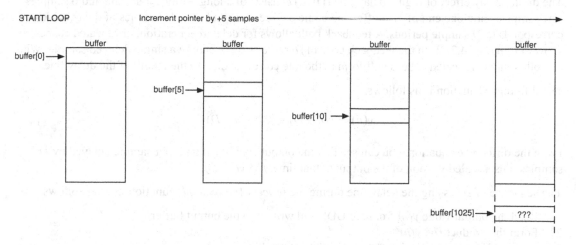

Figure 14.1: After some accesses, the read index goes outside the buffer, usually resulting in a crash.

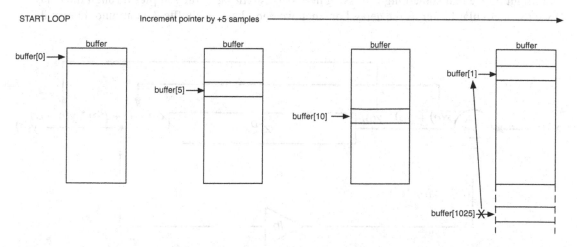

Figure 14.2: In a circular buffer, the pointer is automatically wrapped back to the top and offset by the proper amount to continue the access-by-five-samples loop.

Digital signal processor ICs typically allow you to declare a circular buffer and use a pointer to access it. You increment its address to move through the buffer. When you try to add an offset to the address that would cause it to move outside the buffer, it automatically wraps the address to the proper circular value. In those DSPs, a restriction is that the starting address of the circular buffer must be a power of two. In our efficient circular buffer implementation, we will set the buffer length to a power of two as well, and we will use a similar mechanism for wrapping the buffer index access values automatically. Circular buffers are useful in audio signal processing. You can create circular buffers of audio samples or buffers of coefficients and loop through and access them automatically. Let's start with the most obvious use and make a digital delay effect.

14.1　The Basic Digital Delay

The digital delay effect or digital delay line (DDL) consists of a long buffer for storing audio samples. The samples enter one end of the buffer and come out the other end after D samples of delay, which corresponds to D sample periods. A feedback path allows for delay regeneration, or repeated echoes, as shown in Figure 14.3. With the feedback control fb set to 0, there is only a single delayed sample. With any other value, repeated echoes will form at the rate corresponding to the length of the delay line.

The difference equation is as follows.

$$y(n) = x(n-D) + fb * y(n-D) \qquad (14.1)$$

From the difference equation, you can see that the output consists of an input sample delayed by D samples plus a scaled version of the output at that time, $fb * y(n - D)$.

The sequence of accessing the delay line during the *processAudioSample* function goes as follows.

1.　Read the output value $y(n)$ from the DDL and write it to the output buffer.
2.　Form the product $fb * y(n)$.
3.　Write the input value $x(n) + fb * y(n)$ into the delay line.

You might notice that something is missing here—the shuffling of the samples through the delay. If we use a circular buffer as the delay line, then we don't have to shuffle data around, but we do

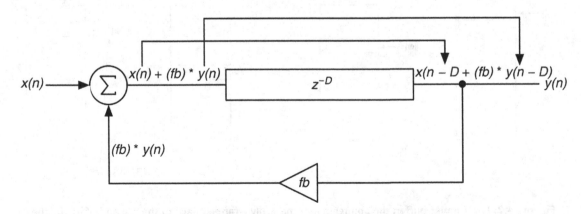

Figure 14.3: The basic DDL consists of a delay line and feedback path.

have to keep track of the read and write access locations in the buffer and wrap the index values as needed.

In order to understand how the buffer operates to make a time delay, consider a circular buffer version of the *buffer* array of size 1024 in the previous section. Suppose that we've been writing samples into the buffer sequentially on each sample period, incrementing and automatically wrapping the index back to the top of the buffer as needed. In this case, we will always be incrementing the write index by exactly one. For the code, we might write something like this:

```
// --- write to the buffer
buffer[writeIndex] = audioSample;

// --- increment index
writeIndex++;

// --- check: if we go outside the buffer
if(writeIndex == 1024)
    writeIndex = 0;    // <- then wrap the pointer back to the top
```

Suppose this has been going on for some time, and we are left with the buffer looking like Figure 14.4a just before we write in the current value, *x(n)*. In this state, we can see the value we wrote on the previous sample period is now *x(n − 1)* and it is located one slot *behind* our write location. Two slots back is *x(n − 2)* and so on. If we keep moving backwards through the array, accessing increasingly older data, we will eventually hit the top and then wrap backwards. Eventually, we find that the oldest sample in the delay is right where our write index value is pointing. After we do the write operation in Figure 14.4b, we can see that we've overwritten the oldest sample with the newest sample, *x(n)*. This leads to our first rule when accessing delay lines in our plugins: **we will always read the buffer before writing to it so that we have access the oldest sample in the delay.**

Now that we understand how the write mechanism operates, let's consider the read operation. We know that we can access any delayed sample we wish from one sample of delay up to the maximum

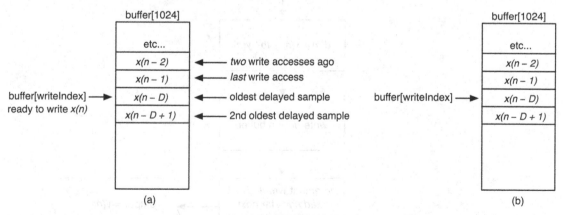

Figure 14.4: (a) Just before we write the current input into the delay line, we can see how the samples are oriented from youngest to oldest. (b) Writing the current value into the delay line destroys the oldest value.

length of the delay line or *D* samples. When we design our delay effect plugin, we will first decide on the maximum delay time that the plugin will allow. A good starting point is 2 seconds—but of course you can set this however you wish. We can set up a write index to handle the writing of the delay line, incrementing and wrapping it as needed. To create a time delay, we will base our read index location such that it is always lagging behind the write index location by the number of samples needed to create the desired delay time. When the user adjusts the time delay control on the GUI, they are actually manipulating the distance between the write index and read index. As the read index moves farther behind the write index, the delay time is increased.

A simple flowchart for the buffer accesses is shown in Figure 14.5. Figure 14.6 shows each step of the operation to read and write the *buffer* array for a fixed delay time of 100 samples. Notice that at the end of the operation we are set up for the next sample period and the read index is lagging exactly 100 samples behind the write index.

The steps are as follows.

1. Figure 14.6a: read the output sample *y(n)* 100 samples behind the current write location.
2. Figure 14.6b: form the input to the delay line *d(n) = x(n) + fb * y(n)* and write it into the buffer at the write location.
3. Figure 14.6c: increment the write index by +1 and wrap if needed.
4. Figure 14.6d: increment the read index by +1 and wrap if needed. This read index location remains 100 samples behind the current write location.

When the user changes the delay time, we must recalculate the offset between the read and write indexes. During this operation, we will need to subtract the delay in samples from the write index, and then check to see if we need to wrap this new value around from the top to the bottom of the buffer.

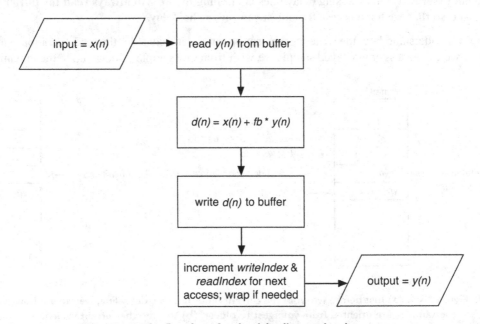

Figure 14.5: The flowchart for the delay line read/write accesses.

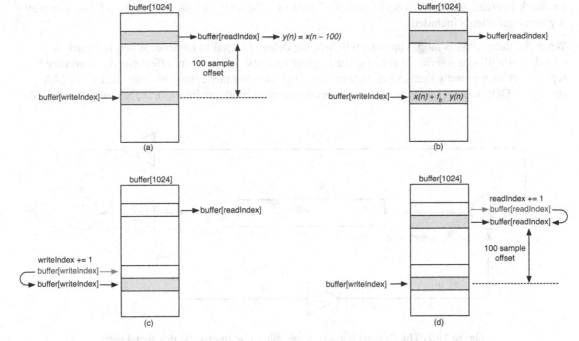

Figure 14.6: (a) Read the delay line and (b) form the delay line input and write it, (c) then increment the write index and (d) increment the read index.

We could do this by adding the buffer length value to the negative *readIndex* value. Some simple code to implement this operation might look like this (make sure you understand the math in the wrapping statement):

```
// --- user selects numSamples delay
// --- move readIndex behind writeIndex by numSamples
readIndex = writeIndex-numSamples;

// --- check and wrap BACKWARDS if the index is negative
if (readIndex < 0)
        readIndex += bufferLength;   // amount of wrap is -Read + Length
```

We'll create a more efficient circular buffer for use in our plugins. Let's begin by looking at the frequency and impulse responses of the DDL.

14.2 Digital Delay With Wet/Dry Mix

A more useful version of the basic DDL combines the output of the delay line with the current input. The delay line output is called the *wet* value and the unprocessed audio input is the *dry* value. The two audio streams are combined together in some ratio that we might call the *wet/dry mix*. In this kind of effect, we hear the current input plus a delayed version of itself. The delayed version may include feedback to create multiple layers of echoes. Figure 14.7 shows the block diagram of a DDL with wet/dry mix coefficients included.

When the delay time is long, your ears will hear the delayed signal as an echo. When feedback is added, you will hear a series of echoes. The impulse response of the delay effect clearly shows the repeated echoes as individual events in time and this makes the most sense in analysis. Figure 14.8 shows the DDL with 1:1 wet/dry mix and several settings of both positive (a), (b) and negative (c), (d)

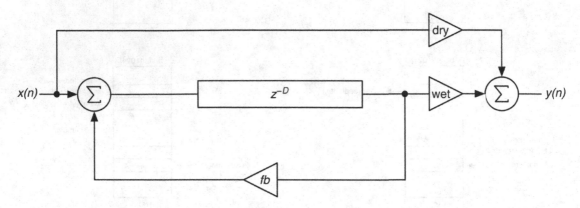

Figure 14.7: The DDL with wet/dry mix allows us to mix the dry signal with the delayed version in any ratio.

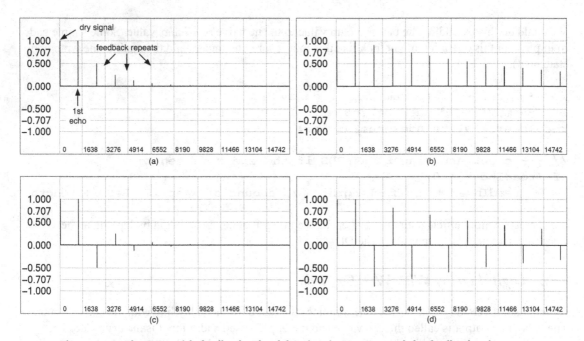

Figure 14.8: The DDL with feedback. The delay time is 30 mSec and the feedback values are (a) 50%, (b) 90%, (c) −50%, and (d) −90%.

feedback. In Figure 14.8a the feedback is 50%. After the first echo, each repeating echo is 50% of the amplitude of the preceding echo.

Notice that the negative feedback inverts every other echo. In a normal delay plugin, you won't notice that the even-numbered echoes are negative (phase inverted). We will take advantage of this when we use the DDL module as a building block in reverb algorithms.

When the delay time becomes shorter than about 30 mSec, your ears cannot resolve the echoes as independent events in time; instead they smear together with the dry signal and this produces a noticeably "filtered" output rather than echoes. MIDI musicians are familiar with this effect as it sometimes happens when a MIDI loop accidentally re-triggers the sending device, producing a second audio signal that is only slightly delayed from the original and their blend sounds hollow or tinny. In this case, the frequency response will tell us more about what is happening than the impulse response. To understand how the DDL plus wet/dry mix produces filtering, first simplify the algorithm by setting the wet and dry coefficients to 1.0 each and setting the feedback amount to 0%. This will produce a single echo combined with the original dry signal at equal amplitudes. Removing the feedback path and setting the wet/dry mix with a 1:1 ratio produces the following difference equation for the output of the system.

$$y(n) = x(n) + x(n - D) \tag{14.2}$$

This difference equation would read "the output *y(n)* equals the input *x(n)* plus a version of the input delayed by D samples." Taking the z Transform of the difference equation yields the transfer function of the algorithm.

$$
\begin{aligned}
y(n) &= x(n) + x(n - D) \\
Y(z) &= X(z) + X(z)z^{-D} \\
&= X(z)(1 + z^{-D}) \\
H(z) &= \frac{Y(z)}{X(z)} = 1 + z^{-D}
\end{aligned}
\tag{14.3}
$$

The filter is a pure feed-forward topology having only zeros in its transfer function. The zero locations must satisfy Equation 14.4.

$$0 = 1 + z^{-D}$$

or

$$z^D = -1$$

let $z = e^{j\omega}$

$$e^{jD\omega} = -1 \tag{14.4}$$

apply Euler

$$\cos(D\omega) + j\sin(D\omega) = -1$$

We will get a zero whenever $D\omega$ causes the equation to evaluate to $-1 + j0$, which will happen for each odd multiple of π. This can be stated as shown in Equation 14.5.

$$H(z) = 1 + z^{-D}$$

$$\text{produces zeros at: } \omega = \pm\frac{k\pi}{D} \quad k = 1, 3, 5, ..., D \tag{14.5}$$

This means that there are D zeros spread equally around the unit circle. This makes sense: the fundamental theorem of algebra predicts D roots for a polynomial of order D. Now consider the simple case of $D = 2$ samples, then Equation 14.6.

$$\cos(2\omega) + j\sin(2\omega) = -1$$

$$\text{if}$$

$$\omega = \pm\frac{k\pi}{D} \quad k = 1, 3, 5, ..., D \tag{14.6}$$

$$\omega = \pm\frac{\pi}{2}$$

There are two zeros, one at $+\pi/2$ and the other at $-\pi/2$. Plot those on the unit circle in the z-plane and you can see that the frequency response has a notch at $+\pi$ and another at $-\pi/2$. If we increase the delay length to four samples and then again to five samples, we find the zeros in a similar manner.

$$D = 4$$

$$\text{zeros at } \omega = \pm\frac{\pi}{4}, \pm\frac{3\pi}{4}$$

$$\tag{14.7}$$

$$D = 5$$

$$\text{zeros at } \omega = \pm\frac{\pi}{5}, \pm\frac{3\pi}{5}, \pm\pi$$

Figure 14.9 shows these pole/zero and frequency response plots for 2, 4, 5, and 32 samples of delay in our DDL with wet/dry mix. We observe a series of notches in the frequency response at equal intervals across the spectrum. This response is called an *inverse comb filter* response. A true *comb filter* response shows the teeth of the comb pointing upwards.

When we add the feedback path, the difference equation and transfer function change to the following.

$$y(n) = x(n) + x(n - D) - fb * x(n - d) + fb * y(n - D)$$

$$H(z) = \frac{1 + (1 - fb)z^{-D}}{1 - fbz^{-D}}$$

$$\text{if } fb = 1.0, \text{ then}$$

$$H(z) = \frac{1}{1 - z^{-D}}$$

$$\tag{14.8}$$

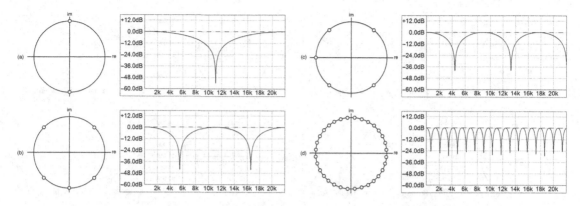

Figure 14.9: The pole/zero and frequency response plots for the DDL with wet/dry mix and (a) two samples, (b) four samples, (c) five samples, and (d) 32 samples of delay.

Deriving this new difference equation and transfer function is not overly complex, but it does involve some algebra. You can find a line-by-line derivation of it at http://www.willpirkle.com/derivations/. Examining the transfer function in Equation 14.7, we can see that there are both zeros (from the numerator) and poles (from the denominator). We've already figured out where the zeros must lay, so let's look at the pole locations. In the same mathematical manner as before, we can deduce that the poles of the system must occur when the denominator of the transfer function is zero. This is easy to find if we let the feedback value *fb* = +1.0, as shown in Equation 14.9.

$$0 = 1 - z^{-D}$$
$$z^{-D} = 1$$

let $z = e^{j\omega}$
$$e^{jD\omega} = 1$$

apply Euler
$$\cos(D\omega) + j\sin(D\omega) = 1 + j0$$
if
$$D\omega = 0, \ \pm 2\pi, \ \pm 4\pi, \ \pm 6\pi, \ etc...$$
$$H(z) = \frac{1}{1 - z^{-D}}$$

produces poles at: $\omega = \pm \dfrac{k\pi}{D} \quad k = 0, 2, 4, ..., D$

(14.9)

So, the poles are going to be spaced evenly *between* the zeros in the system. The distance from the poles or zeros to the unit circle will depend on the feedback amount. As the feedback is increased, the poles move closer to the unit circle while the zeros move in the opposite direction. Figure 14.10a and

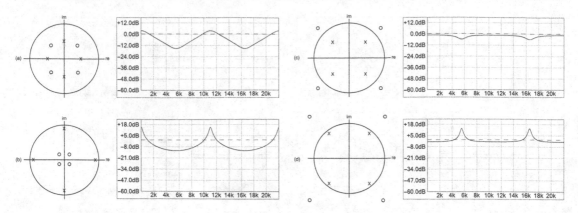

Figure 14.10: The four-sample delay line with wet/dry mix and feedback values of (a) +50%, (b) +90%, (c) –50%, (d) –90%.

b show the frequency response for $D = 4$ and feedback values of +50% and +90%. As the feedback increases, the peaks in the frequency response become sharper as the poles move toward the unit circle.

If we invert the feedback and make it negative (–) then two interesting things happen. The first is that the feedback coefficient in the numerator $(1 - fb)$ becomes a value greater than 1.0, effectively moving the zeros outside the unit circle. In the denominator, the fb value's sign will change the overall denominator to $1 + fbz^{-1}$. This will place the poles at the same frequency (angle) as the zeros. The effect can be seen in Figure 14.10c and d. If the feedback value is –62%, the zeros will lie at radii that are exactly inverted from the pole radii. As with the all-pass filter, this arrangement produces a flat frequency response.

14.3 An Efficient Circular Buffer Object

Your *fxobjects.h* and *fxobjecgts.cpp* files contain a C++ implementation of a circular buffer named *CircularBuffer*. This object may be used to create both time delay and ring buffers. This object will be used as a member object of higher-level processing objects including *AudioDelay* and *AudioConvolver*, therefore it does not implement the *IAudioSignalProcessor* interface. Two goals of the *CircularBuffer* are as follows.

- can create a circular buffer of *float* or *double* data types
- highly efficient code for wrapping write and read indexes

We can use a class template to handle the chore of creating both *float* and *double* circular buffers, so that part is straightforward. Reading and writing to the buffer is already efficient; the only recurring mathematical operation has to do with manipulating the read and write index values to wrap around the buffer as needed. There are a couple of types of wrapping to address.

- incrementing the write index value to wrap around from the bottom back up to the top of the buffer when needed
- finding the delay offset by subtracting the desired samples of delay from the current write index location, then wrapping this value backwards from the top around to the bottom when needed

Some might suggest indexing the array with the modulo (or remainder) operator as a way of automatically wrapping values. But the modulo (%) operator is not cheap; it requires a combination of subtraction, multiplication and division operations (three in total, see Jones, n.d.). However, we can take a clue from the hardware DSPs that implement circular buffers but require that the buffer's starting address is a power of two. In our case, we can't easily control the actual address of the buffer, but we can control its length and we can force it to be the nearest power of two larger than the maximum desired size. This strategy will allow us to perform the buffer boundary checking and wrapping—in both the forward and reverse direction—with one simple math operation called *wire-AND-ing* on the index we are manipulating. In this case, the length of the buffer must be a power of two. Remember that the delay time is based on the distance between the write and read index values and is *not* strictly equal to the length of the buffer. The buffer length only sets the maximum delay time that we may achieve. Let's look at the first problem of incrementing and wrapping the write index value.

Suppose the circular buffer stores eight samples; we declare it like this:

```
double buffer[8];
```

Then the top of the buffer would be at location *buffer[0]* and the bottom (end) of the buffer would be at location *buffer[7]*. If we were sequentially pushing values into the array and updating our write index location, we would want it to follow a pattern, starting at zero, incrementing by one on each sample period, then wrapping around from value 7 back to value 0 at the end-of-buffer boundary.

```
0 - 1 - 2 - 3 - 4 - 5 - 6 - 7 - 0 - 1 - 2 - 3 - etc . . .
```

The binary value of any number that is a positive power of two is going to consist of a single binary 1 value in the power-of-two binary position. For example, the number eight in binary is 1000 (or 000 . . . 1000 if we append the leading 0s) while the number 256 is 0001 0000 0000; both are powers of two. For our efficient wrapping mechanism, we actually want to observe the binary value that is exactly one less than the array length. In our case, that would be the binary value for seven.

    8:    1000
    7:    0111

For a buffer of length 256, we would look at the number 255:

    256:   1 0000 0000
    255:   0 1111 1111

We see that the (buffer length -1) value is always a series of trailing 1s in binary. We can use this value as our *wrapping mask*. In order to implement the masking operation, you *wire-and* the current write index with the masking value. The self *wire-anding* operator in C++ is *&=* so our code would look like this:

```
// --- set mask
int mask = 7;

// --- increment writeIndex
writeIndex++;

// --- wire-and it with the mask to wrap forward
writeIndex &= mask;
```

Table 14.1: *Wire-anding* with the *mask* value creates the roll-over wrapping mechanism for the *writeIndex* value

writeIndex	mask	writeIndex &= *mask*
0000	0111	0000
0001	0111	0001
0010	0111	0010
0011	0111	0011
0100	0111	0100
0101	0111	0101
0110	0111	0110
0111	0111	0111
1000	0111	0000

The *wire-anding* operation will return the same value back to the *writeIndex* variable when the index value is less than or equal to the mask. When the index equals the buffer length, the *wire-anding* will wrap it around back to the top location (0). You can see this operation in Table 14.1. This works because the *and* operation produces a 1 only when *both* operands are 1, and a 0 when *either* operand is 0.

Interestingly, the mechanism for wrapping a negative index value around in the opposite direction is to *wire-and* it with the same mask—it's the identical operation, but the theory is slightly more complicated. We are going to use this method like this:

```
// --- set mask
int mask = 7;

// --- calculate read index by subtracting delay-in-samples
int readIndex = writeIndex - delaySamples;

// --- wire-and it with the mask to wrap backwards
readIndex &= mask;
```

If the delay in samples is less than the current *writeIndex* value, then the *readIndex* will be a value that is somewhere between zero and *writeIndex*. When you *wire-and* it with the mask, its value does not change, which is exactly as the previous example shows in Table 14.1.

However, if the *readIndex* takes on a negative value, then its most significant bits will be set to 1s— recall how the values are encoded as two's complement for the bipolar (or signed) integer format from Section 1.4. Let's suppose that after the subtract operation the *readIndex* value is −3. We would like this to wrap around backwards to a value that is three less than the end of the buffer, which would be at location *buffer[5]*. The 32-bit binary values for the *readIndex* and *mask* are shown here:

readIndex (−3):	1111 1111 1111 1111 1111 1111 1111 1101
mask (7):	0000 0000 0000 0000 0000 0000 0000 0111

When we *wire-and* the bits together we get:

readIndex & mask:	0000 0000 0000 0000 0000 0000 0000 0101

We see that the *readIndex* final value is 0101 or five—exactly what we need.

The *CircularBuffer* object is clean and simple. It uses a smart pointer to maintain the circular buffer's lifecycle, automatically deleting it when it is resized or when the object is destroyed. It uses a class template to allow the creation of *float* and *double* circular buffers. It has one function to write data into the buffer, and two functions for reading delayed values out. One function specifies the delay in samples as an *int* value and the other as a *double*. But why do we need the second function?

14.3.1 C++ DSP Object: CircularBuffer With Fractional Delay

When we use the circular buffer as a simple ring buffer for convolution, we will always access values from locations that are indexed with exact integers. However, when we use the circular buffer as a time-delay buffer, we allow the user to specify the delay time in mSec and we allow them to dial-in very precise values. The reason for allowing this follows.

- The user may need very accurate delay settings for some kind of precision audio operation.
- The user may want to synchronize the delay times to the host BPM value so that echoes land precisely on musical timing intervals.

So, we will need to create delays that may implement *fractional delay times*, or non-integer delay in samples. For example, suppose the sample rate is f_s = 44.1 kHz and the user needs 247 mSec of delay. The fractional delay in samples is calculated as:

$$delay_{samples} = \left(delay_{m\,Sec}\right)\left(\frac{f_s}{1000}\right) \tag{14.10}$$

In this case, the delay in samples is 10.8927, so we need to interpolate the circular buffer and find the value that is 0.8927 between delayed sample 10 and delayed sample 11. We may use any interpolation scheme we like—linear interpolation will do a good job of creating the fractional delay value. Note that there are several approaches to the best interpolation choice, and it may get complicated depending on the interpolation method chosen.

For linear interpolation between two adjacent *x* values (10 and 11 here), the algorithm is simple and becomes a weighted-sum method, not unlike the 1st order feed-forward filter we used throughout Chapters 9 and 10. Knowing the *y1* and *y2* values (the values of the samples located at *buffer[10]* and *buffer[11]*) and the fractional amount (0.8927), we can use the simple function *doLinearInterpolation*, defined in your *fxobjects.h* file. The operation is shown graphically in Figure 14.11.

```
double doLinearInterpolation(double y1, double y2, double fractional_X)
{
        // --- check invalid condition
        if (fractional_X >= 1.0) return y2;

        // --- use weighted sum method of interpolating
        return fractional_X*y2 + (1.0 - fractional_X)*y1;
}
```

For this example, the interpolated value is:

```
value = 0.8927*y2 + 0.1073*y1
```

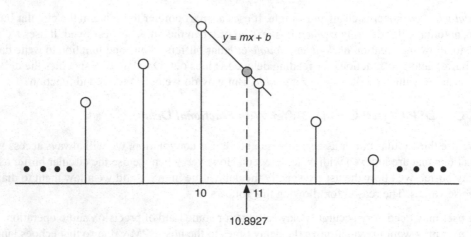

$$y = mx + b$$

10   11

10.8927

Figure 14.11: The linear interpolation operation finds a fractional delay value by drawing an imaginary line through the two endpoints and using a weighted sum to calculate the interpolated value.

14.3.2 CircularBuffer: *Enumerations and Data Structure*

This object does not require any enumerations or a custom data structure, as it is a low-level object that we won't manipulate from a GUI. It is designed for use in aggregate DSP and effect objects. However, this object uses a class template in its definition so that it may be used to implement buffers of any data type; typically these will consist of *float* and *double* types.

14.3.3 CircularBuffer: *Members*

Tables 14.2 and 14.3 list the *CircularBuffer* members. First, note that the object is defined using a class template. This manifests itself in the declaration of the object, but also as function argument and return types.

```
template <typename T>
class CircularBuffer
```

The smart pointer to the buffer is declared as a private member variable that points to an array of *T*-type values (*T[]*):

```
std::unique_ptr<T[]> buffer = nullptr;
```

So, you must specify the template data type when declaring or creating the object.

Static Declaration of *double* buffer:

```
CircularBuffer<double> delayBuffer;
```

Dynamic Declaration of *double* buffer:

```
CircularBuffer<double>* delayBuffer = new CircularBuffer<double>;
```

Table 14.2: *CircularBuffer* member variables

CircularBuffer Member Variables		
Type	**Name**	**Description**
std::unique_ptr<T[]>	*buffer*	Smart pointer to a buffer
unsigned int	*writeIndex*	Index for writing to array
unsigned int	*bufferLength*	Total length of buffer as a nearest power of two
unsigned int	*wrapMask*	*(bufferLength −1)* for mask

Table 14.3: *CircularBuffer* member functions

CircularBuffer Member Functions		
Returns	**Name**	**Description**
void	*createCircularBuffer* Parameters: —*unsigned int* _bufferLength	Create or resize the circular buffer
void	*flushBuffer*	Clear the circular buffer values and set to 0.0
void	*writeBuffer* Parameters: —*T* input	Write the *input* value of type *T* into the buffer
T	*readBuffer* Parameters: —*int* delayInSamples	Read a specific location from buffer that is *delayInSamples* old
T	*readBuffer* Parameters: —*double* delayInFractionalSamples —*bool* interpolate	Read a fractional delay location from buffer; interpolate if desired

14.3.4 CircularBuffer: *Programming Notes*

The two functions in our *CircularBuffer* object that read the buffer at a desired delay time are named the same (*readBuffer),* but the second function takes a fractional delay time as an argument and interpolates the value according to the *interpolate* argument (notice that it defaults to *true,* so leaving it out will always perform interpolation). The return value is the template type *T,* either a *float* or *double* depending on how the object was declared.

```
// --- read an arbitrary location that is delayInSamples old
T readBuffer(int delayInSamples)
{
    // --- subtract to make read index
    int readIndex = writeIndex - delayInSamples;

    // --- autowrap index
    readIndex &= wrapMask;

    // --- read it
    return buffer[readIndex];
}
```

Notice how we cast the *double* to an *int* and subtract it to create the fractional component for the second function—this truncates the value rather than rounding it, which is what we need.

```
// --- read an arbitrary location that includes a fractional sample
T readBuffer(double delayInFractionalSamples, bool interpolate = true)
{
        // --- truncate delayInFractionalSamples and read the int part
        T y1 = readBuffer((int)delayInFractionalSamples);

        // --- if no interpolation, just return value
        if (!interpolate) return y1;

        // --- else do interpolation
        //
        // --- read the sample at n+1 (one sample OLDER)
        T y2 = readBuffer((int)delayInFractionalSamples + 1);

        // --- get fractional part
        double fraction = delayInFractionalSamples-
                                    (int)delayInFractionalSamples;

        // --- do the interpolation (you could try different types here)
        return doLinearInterpolation(y1, y2, fraction);
}
```

Tables 14.2 and 14.3 list the *CircularBuffer* member variables and member functions. We will show an example of using *CircularBuffer* in the *AudioDelay* object in Section 14.4. You can check out the code to see all the details of operation using the wrapping mask and interpolation mechanisms. In addition to the two *readBuffer* functions, the highlights include the function to create or resize the buffer given a desired *_bufferLength* in size.

```
void createCircularBuffer(unsigned int _bufferLength)
{
        // --- reset to top
        writeIndex = 0;

        // --- find nearest power of 2 for buffer, save as bufferLength
        bufferLength = (unsigned int)(pow(2, ceil(log(_bufferLength) /
        log(2))));

        // --- save (bufferLength -1) for use as wrapping mask
        wrapMask = bufferLength-1;

        // --- create new buffer
        buffer.reset(new T[bufferLength]);
```

```
      // --- flush buffer
      flushBuffer();
}
```

They also include this function to write data to the buffer:

```
void writeBuffer(T input)
{
      // --- write and increment index counter
      buffer[writeIndex++] = input;

      // --- wrap if index > bufferlength-1
      writeIndex &= wrapMask;
}
```

14.4 Basic Delay Algorithms

You can find dozens of delay algorithms from numerous sources and you can also easily invent your own schemes. For our delay plugin, we will implement a normal stereo delay with feedback and a ping-pong delay with feedback. We'll specify the maximum delay time at 2 seconds and allow the user to adjust the delay time, feedback, and wet/dry mix. For the ping-pong delay, we will allow adjustment of the stereo delay ratio. Both effects will operate in our three standard channel formats: mono in/ mono out, mono in/stereo out, and stereo in/stereo out. Both of the algorithms will feature/allow the following.

- The left and right delay times may be different but will be limited to a maximum of 2 seconds (easily changed)
- Feedback values for each channel are identical and set with one GUI control.
- The dry and wet controls are likewise shared across channels; there is one GUI control for each, *dry* and *wet*, and the values are adjusted in dB.

14.4.1 Stereo Delay With Feedback

Our stereo delay with feedback simply combines two DDLs with wet/dry mix capabilities along with a single feedback control that governs both left and right feedback values. The block diagram is shown in Figure 14.12.

14.4.2 Stereo Ping-Pong Delay

The ping-pong delay crosses both the input and feedback paths to the opposing channel. It produces a circular kind of delay sound that starts in the center then moves to one side, then the other side, then back to the center where it starts all over. A key ingredient is the ability to set the left and right delays as ratios of one another. When the ratios are perfect (1:2, 3:5, 8:7 etc.) the effect becomes quite rhythmic in nature. If the delay times are identical across channels, then the ping-pong effect is usually lost. Figure 14.13 shows the block diagram of the ping-pong delay.

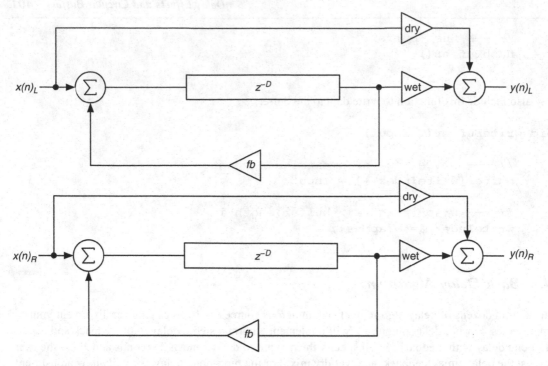

Figure 14.12: The stereo delay consists of two independent delay lines, one for each channel.

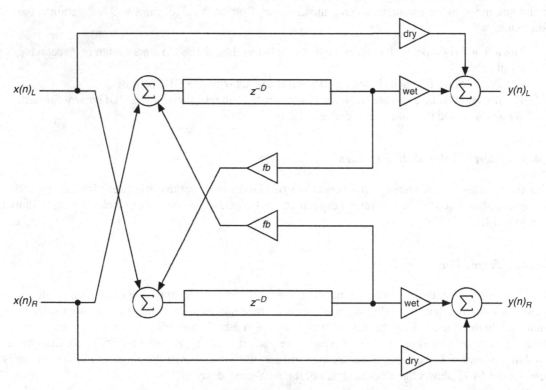

Figure 14.13: The ping-pong delay has crossed inputs and feedback paths, creating a figure-of-eight pattern that the delayed signal follows

14.5 C++ *Effect Object:* AudioDelay

To implement our two-delay algorithms, and extend the possibility for many more delay effect designs, we have created an *AudioDelay* object that includes two *CircularBuffer* member objects to handle mono or stereo operation. This object will implement the familiar *IAudioSignalProcessor* interface; however, due to the stereo nature of the ping-pong delay, it will implement the optional function *processAudioFrame*. This object is more complex than previous plugin objects for several reasons.

- It processes stereo.
- It requires an additional function call to set the sizes of the circular buffers.
- The left and right delay times may be set in two different ways, depending on your plugin architecture.

14.5.1 AudioDelay: *Enumerations and Data Structure*

The *AudioDelay* object requires two strongly typed enumerations: one for the algorithm (normal or ping-pong) and the other for the method of updating the left and right delay times (individually, or as left delay and right-ratio):

```
enum class delayAlgorithm { kNormal, kPingPong };
enum class delayUpdateType { kLeftAndRight, kLeftPlusRatio };
```

The *delayUpdateType* explains how to update the left and right delay times using the information in the *AudioDelayParameters* structure.

```
struct AudioDelayParameters
{
        AudioDelayParameters() {}
        AudioDelayParameters& operator=(. . .){. . .}

        // --- individual parameters
        delayAlgorithm algorithm = delayAlgorithm::kNormal;
        double wetLevel_dB = -3.0;
        double dryLevel_dB = -3.0;
        double feedback_Pct = 0.0;

        delayUpdateType updateType = delayUpdateType::kLeftAndRight;
        double leftDelay_mSec = 0.0;
        double rightDelay_mSec = 0.0;
        double delayRatio_Pct = 100.0; // rDly = (ratio)*(left dly)
};
```

There are two variables to set the left and right delay times in mSec. There is also a ratio control that allows you to specify the left delay time and a ratio that is used to calculate the right delay time. This is useful for the ping-pong algorithm where the effect is usually inaudible when the left and right delay times are identical and the left and right channels contain correlated signals. The *updateType* variable specifies the type of calculation you need.

14.5.2 AudioDelay: *Members*

Tables 14.4 and 14.5 list the *AudioDelay* member variables and member functions. Notice that this object can process frames, which may be mono, stereo, or multi-channel. The *processAudioSample* method still works if you only need to process a mono channel.

Table 14.4: *AudioDelay* member variables

AudioDelay Member Variables		
Type	**Name**	**Description**
AudioDelayParameters	parameters	Fractional delay in samples for left channel
CircularBuffer<double>	delayBuffer_L	Circular buffer for left channel
CircularBuffer<double>	delayBuffer_R	Circular buffer for right channel
double	wetMix	Wet mix as scalar value (not dB)
double	dryMix	Dry mix as scalar value (not dB)
double	delayInSamples_L	Delay time in samples for L CH
double	delayInSamples_R	Delay time in samples for R CH
double	bufferLength_mSec	Maximum buffer size in mSec that user requests
unsigned int	bufferLength	Maximum buffer size in samples that user requests
double	sampleRate	current sample rate
double	samplesPerMSec	Variable for easily calculating fractional delay in samples from user delay time in mSec

Table 14.5: *AudioDelay* member functions

AudioDelay Member Functions		
Returns	**Name**	**Description**
AudioDelayParameters	getParameters	Get all parameters at once
void	setParameters Parameters: —AudioDelayParameters _parameters	Set all parameters at once
bool	reset Parameters: —double sampleRate	Reset circ buffers; create new if sample rate changes
double	processAudioSample Parameters: —double xn	Process input xn through left delay
bool	processAudioFrame Parameters: —const float* inputFrame —float* outputFrame —unsigned int inputChannels —unsigned int outputChannels	Process a frame through L and R delays

14.5.3 AudioDelay: *Programming Notes*

The *AudioDelay* object will implement the majority of the plugin's functionality, so it is worthwhile to examine some of its functions. For each function, make sure you understand how the function implements the details. Using the *AudioDelay* object is straightforward:

1. Reset the object to prepare it for streaming.
2. Call *createDelayBuffers* to set up the delay lines; pass the maximum delay time in mSec your plugin supports (usually around 2,000 mSec = 2 seconds).
3. Set the object parameters in the *AudioDelayParameters* custom data structure.
4. Set the *updateType* variable to specify the left and right delay time calculation.
5. Call *setParameters* to update the calculation type.
6. Call *processAudioFrame* passing the input frame and output frame pointers, and the channel counts; the function processes both left and right channels.
7. Call *processAudioSample* to process only one channel (for mono plugins).

Function: *reset*
Details: if the sample rate has not changed, you just flush the delay buffers and return. Otherwise, call the *createDelayBuffers* function to set up new buffers. After reset, you can call *createDelayBuffers* whenever you want to resize the maximum buffer length (do not do this during audio streaming).

```
virtual bool reset(double _sampleRate)
{
    // --- if sample rate did not change
    if (sampleRate == _sampleRate)
    {
        // --- just flush buffer and return
        delayBuffer_L.flushBuffer();
        delayBuffer_R.flushBuffer();
        return true;
    }

    // --- create new buffer, will store sample rate and length(mSec)
    createDelayBuffers(_sampleRate, bufferLength_mSec);

    return true;
}
```

Function: *createDelayBuffers*
Details:

- Convert input buffer length in mSec to a buffer size in samples.
- Add one to the buffer length to accommodate the fractional delay.
- Store the sample rate and a samples-per-mSec value for later.
- Create the circular buffers.

```
void createDelayBuffers(double _sampleRate, double _bufferLength_mSec)
{
    // --- store for math
    bufferLength_mSec = _bufferLength_mSec;
    sampleRate = _sampleRate;
    samplesPerMSec = sampleRate / 1000.0;

    // --- total buffer length including fractional part
    bufferLength = (unsigned int)(bufferLength_mSec*(samplesPerMSec))
                    + 1; // +1 for fractional part

    // --- create new buffers
    delayBuffer_L.createCircularBuffer(bufferLength);
    delayBuffer_R.createCircularBuffer(bufferLength);
}
```

Function: *processAudioFrame*
Details:

- Check for supported channel I/O counts.
- Check for supported delay algorithm.
- If the channel format is mono-in to stereo-out, feed the mono input into *both* delay lines.
- Decode the algorithm and do the audio processing.

Let's observe the two block diagrams in Figures 14.12 and 14.13 and then implement the function. We can see from the block diagrams that:

- The outputs for both delay algorithms are the same. For example, the output for the left channel is the sum of the dry left input and processed left delay line output scaled by the wet/dry mix amounts; the same is true for the right output.
- An input is formed for each delay line consisting of a dry input plus a delayed signal that is scaled with the feedback value.
- The only real difference is that in the ping-pong algorithm, the input to each delay line is swapped from the normal delay algorithm.

```
virtual bool processAudioFrame(float* inputFrame,
                               float* outputFrame,
                               uint32_t inputChannels,
                               uint32_t outputChannels)
{
    // --- make sure we have input and outputs
    if (inputChannels == 0 || outputChannels == 0)
        return false;

    // --- make sure we support this delay algorithm
    if (parameters.algorithm != delayAlgorithm::kNormal &&
        parameters.algorithm != delayAlgorithm::kPingPong)
        return false;
```

```cpp
    // --- LEFT channel
    double xnL = inputFrame[0];

    // --- RIGHT channel (duplicate left input if mono-in)
    double xnR = inputChannels > 1 ? inputFrame[1] : xnL;

    // --- read delay LEFT
    double ynL = delayBuffer_L.readBuffer(delayInSamples_L);

    // --- read delay RIGHT
    double ynR = delayBuffer_R.readBuffer(delayInSamples_R);

    // --- create input for delay buffer with LEFT channel info
    double dnL = xnL + (parameters.feedback_Pct / 100.0) * ynL;

    // --- create input for delay buffer with RIGHT channel info
    double dnR = xnR + (parameters.feedback_Pct / 100.0) * ynR;

    // --- decode
    if (parameters.algorithm == delayAlgorithm::kNormal)
    {
        // --- write to LEFT delay buffer with LEFT channel info
        delayBuffer_L.writeBuffer(dnL);

        // --- write to RIGHT delay buffer with RIGHT channel info
        delayBuffer_R.writeBuffer(dnR);
    }
    else if (parameters.algorithm == delayAlgorithm::kPingPong)
    {
        // --- write to LEFT delay buffer with RIGHT channel info
        delayBuffer_L.writeBuffer(dnR);

        // --- write to RIGHT delay buffer with LEFT channel info
        delayBuffer_R.writeBuffer(dnL);
    }

    // --- form mixture out = dry*xn + wet*yn
    double outputL = dryMix*xnL + wetMix*ynL;

    // --- form mixture out = dry*xn + wet*yn
    double outputR = dryMix*xnR + wetMix*ynR;

    // --- set left channel
    outputFrame[0] = outputL;

    // --- set right channel
    outputFrame[1] = outputR;

    return true;
}
```

14.6 *Chapter Plugin:* StereoDelay

Our delay FX plugin will include the normal stereo and ping-pong delay algorithms. The GUI will include controls for adjusting the delay parameters as well as a switch to choose the algorithm. All of the real audio signal processing work will be done in the *AudioDelay* object. Figure 14.14 shows the stereo delay GUI and block diagram.

14.6.1 StereoDelay: *GUI Parameters*

The GUI parameter table is shown in Table 14.6—use it to declare your GUI parameter interface for the plugin framework you are using. If you are using ASPiK, remember to first create your enumeration of control ID values at the top of the *plugincore.h* file and use automatic variable binding to connect the linked variables to the GUI parameters.

ASPiK: top of *plugincore.h* file

```
enum controlID {
      delayTime_mSec = 0,
      delayFeedback_Pct = 1,
      delayRatio_Pct = 2,
      delayType = 3,
      wetLevel_dB = 4,
      dryLevel_dB = 5
};
```

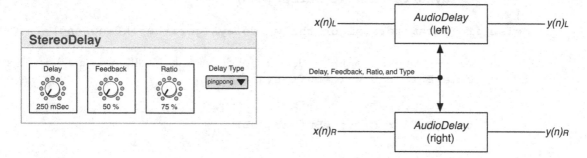

Figure 14.14: *StereoDelay* GUI and block diagram.

Table 14.6: The *StereoDelay* plugin's GUI parameter and linked-variable list (*) for experimental purposes; you usually want to lower this limit in the final release

Control Name	Units	Min/max/default or string-list	Taper	Linked Variable	Linked Variable Type
Delay Time	mSec	0 / 2000 / 250	linear	*delayTime_mSec*	*double*
Feedback	%	0 / 100* / 50	linear	*delayFeedback_Pct*	*double*
Ratio	%	1 / 100 / 50	linear	*delayRatio_Pct*	*double*
Wet Level	dB	−60 / 12 / −3	linear	*wetLevel_dB*	*double*
Dry Level	dB	−60 / 12 / −3	linear	*dryLevel_dB*	*double*
Delay Type	-	"normal, pingpong"		*delayType*	*int*

14.6.2 StereoDelay: *Object Declarations and Reset*

The *StereoDelay* plugin needs only one *AudioDelay* object to implement its stereo operation. We declare one *AudioDelay* object and the usual *updateParameters* function.

```
// --- in your processor object's .h file
#include "fxobjects.h"

// --- in the class definition
protected:
    AudioDelay stereoDelay;
    void updateParameters();
```

Call the *reset* function and pass in the current sample rate in your plugin framework's prepare-for-audio-streaming function. In this example, *resetInfo.sampleRate* holds the current DAW sample rate. After the reset operation, create the delay buffers—this plugin has a maximum delay time of 2 seconds or 2,000 mSec.

ASPiK: *reset( )*

```
    // --- reset the delay
    stereoDelay.reset(resetInfo.sampleRate);

    // --- create 2 second delay buffers
    stereoDelay.createDelayBuffers(resetInfo.sampleRate, 2000.0);
```

14.6.3 StereoDelay: *GUI Parameter Update*

Update the single stereo object at the top of the buffer processing loop. Gather the values from the GUI parameters and apply them to the custom data structure. Do this for all channels you support. For ASPiK, we use a helper function to convert the GUI integer data into the strongly typed enumeration.

ASPiK: *updateParameters( )* function

```
AudioDelayParameters params = stereoDelay.getParameters();
params.leftDelay_mSec = delayTime_mSec;
params.feedback_Pct = delayFeedback_Pct;
params.delayRatio_Pct = delayRatio_Pct;
params.updateType = delayUpdateType::kLeftPlusRatio;

params.dryLevel_dB = dryLevel_dB;
params.wetLevel_dB = wetLevel_dB;

// --- use helper
params.algorithm = convertIntToEnum(delayType, delayAlgorithm);

// --- set them
stereoDelay.setParameters(params);
```

14.6.4 StereoDelay: *Process Audio*

Add the code to process the plugin object's input audio samples. If you are using ASPiK, you can pass it the frame pointers and count variables directly, then return.

ASPiK: *processAudioFrame( )* function

```
// --- object does all the work on frames
stereoDelay.processAudioFrame(processFrameInfo.audioInputFrame,
                             processFrameInfo.audioOutputFrame,
                             processFrameInfo.numAudioInChannels,
                             processFrameInfo.numAudioOutChannels);
```

Non-ASPiK: you gather your frame input and output samples into small arrays (you can statically declare them to minimize cost). For example:

```
double xnL = //< get from your framework: input sample L
double xnR = //< get from your framework: input sample R

float inputs[2] = { xnL, xnR };
float outputs[2] = { 0.0, 0.0 };

// --- process (in, out, 2 inputs, 2 outputs)
stereoDelay.processAudioFrame(inputs, outputs, 2, 2);

//< framework Left output sample = outputs[0];
//< framework Right output sample = outputs[1];
```

14.6.5 Synchronizing the Delay Time to BPM

All APIs allow the plugin to retrieve information about the current session's musical timing, including BPM, time signature numerator, and time signature denominator. You can use this information to synchronize plugin parameters such as delay time or LFO rate to the musical tempo. The equations for converting BPM to delay times are readily available from numerous websites. Some of the more common calculations for converting BPM to musical note duration are in Equation 14.11.

$$T_{quarter} = \frac{60}{BPM} \quad T_{eighth} = \frac{T_{quarter}}{2} \quad T_{triplet\_eighth} = \frac{T_{quarter}}{3} \tag{14.11}$$

14.6.5.1 ASPiK Users: BPM and Time Signature Information

This information is delivered to your plugin in the *HostInfo* structure that is passed into the *processAudioFrame* or *processAudioBuffer* argument. The information is valid for the duration of the frame or buffer. A *HostInfo* pointer is delivered in both the *ProcessFrameInfo* and *ProcessBufferInfo* structures for the *processAudioFrame* and *processAudioBuffer* functions respectively. An example of using this information to calculate eighth note delay times synchronized to the BPM in the *processAudioFrame* function is as follows.

```
if (processFrameInfo.hostInfo)
{
        double quarterNoteDelay = 60.0 / processFrameInfo.hostInfo->dBPM;
        double eighthNoteDelay = quarterNoteDelay / 2.0;

        etc . . .
}
```

You may wish to play with the time signature as well. The numerator and denominator are found in:

```
processFrameInfo.hostInfo->fTimeSigNumerator
processFrameInfo.hostInfo->uTimeSigDenomintor
```

Notice that the numerator is a floating point value, so it may be fractional (such as 3.5) while the denominator is an unsigned integer and will always be a whole number.

14.7 More Delay Algorithms

Sections 14.7.1 through 14.7.4 demonstrate more delay algorithms and their block diagrams. You can find a plethora of delay algorithms in the effects documentation for the Korg Triton® or Korg Wavestation® synthesizers. The Korg documentation is typically excellent with a high level of detail, including the parameter minimum and maximum values.

14.7.1 Analog Modeling Delay

Analog delays using the Bucket Brigade Delay ICs (BBDs) are popular for their organic sounds. Contrary to the name, analog delays include a sample-and-hold along with an anti-aliasing low-pass filter, and they sample the input, chopping it into pieces. The "analog" part is that instead of digital numeric values, analog charges are stored and delayed using a system of switches and capacitors, which pass the analog voltage slices along in the same way as a bucket brigade passes buckets of water for fire-fighting. Most of these analog devices also include a *compander* (complimentary compression and expansion) around the BBD since these ICs are limited in dynamic range: the signal is compressed going into the BBD, and then expanded coming out (note that this is true upward expansion). The filtering, companding and generally noisy operation of the BBD ICs contributes to this organic sound. The most obvious audible effect is that as the echoes are fed back into the BBD, they come back out slightly noisier, and with slightly less high frequency content. To mimic the loss of high frequencies with each lap around the feedback path, we place a low-pass filter in it, either before or after the feedback attenuation coefficient. Make sure you do not use a resonant low-pass filter or one with any gain above 0.0 dB as this will cause the feedback loop to become unstable. Figure 14.15 shows what a basic analog modeling delay algorithm might look like. Improvements would include the addition of a compander around the delay line and an injection of noise (perhaps EQ'd to model a specific tape noise) into the delay line.

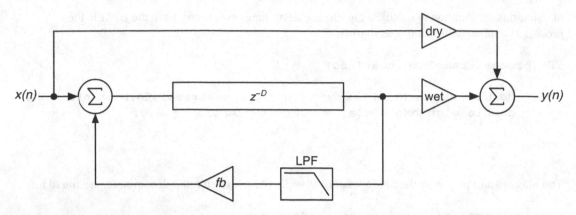

Figure 14.15: A simple analog modeling delay algorithm includes a LPF in the feedback path.

14.7.2 Multi-Tap Delay

A multi-tapped delay reads multiple delayed values from one delay line. The Roland Space Echo® line of effects is a great example. This kind of delay is often used in music synthesizer patches as well. There are many variations on this effect usually involving the number of taps, the timing relationship between the taps, delay tap amplitude controls, and which taps are fed back into the input. Figure 14.16 shows a basic multi-tap delay that produces four different echoes that are blended with the dry input signal. Only the primary (longest) delay signal is fed back to the input. You may use any scheme you like for setting the relationship between the different delay taps, including the use of musical timing relationships that are synchronized with the host BPM value.

14.7.3 LCR Delay

The Left-Center-Right (LCR) delay shown in Figure 14.17 is based on the Korg Triton delay effect. There are three delay lines and only the center includes a feedback path. The delay time for each channel (L, C, and R) is independently adjustable. The filters in the feedback path have bypass switches so that they may be used alone or together in series.

14.7.4 TC Electronics TC-2290 Dynamic Delay

Introduced in 1985, the TC-2290 Dynamic Delay helped define a decade of electric guitar solos. The use of delays on electric guitar was already in vogue, but a problem was that the delays tended to muddy the virtuosic guitar work, especially with high delay feedback values. Some guitarists such as Steve Morse used a volume pedal on the wet channel of the delay to fade the echoes in and out as needed. One popular approach was to remove the delays while the guitar signal's amplitude was high, and then fade in the echoes as the guitar signal diminished in amplitude. The TC-2290 accomplished the same technique (and many others) by using an envelope detector in a side chain. When the input signal's envelope is above the threshold, the echo amplitude is lowered significantly, or even reduced to nothing. When the signal falls below the threshold, the amplitude of the echoes is increased, making the notes swell or "bloom" as the echo signals grow. This is known as a "ducking delay" because the echo signal is reduced or "ducked" in response to

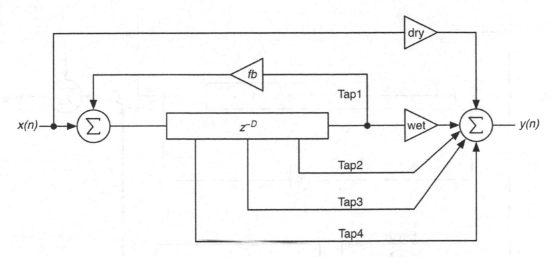

Figure 14.16: A multi-tap delay with four taps (delay times).

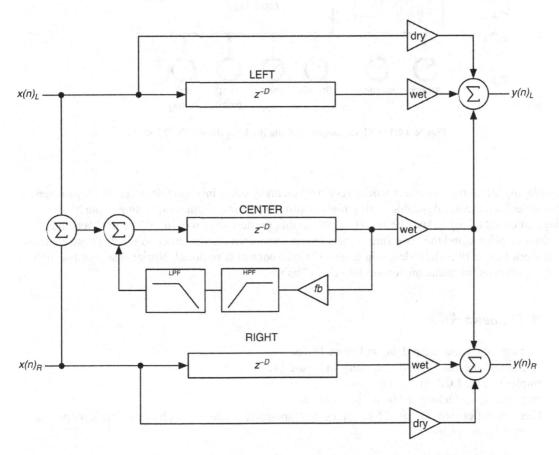

Figure 14.17: The LCR delay includes three independent delay lines and filters in the center channel's feedback path.

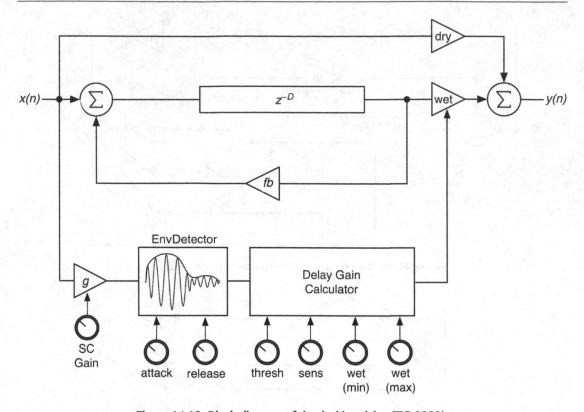

Figure 14.18: Block diagram of the ducking delay (TC-2290).

the playing strength. The effect works very well on many other instruments or audio loops where intense echoes are not desirable during high amplitude periods. Figure 14.18 shows the block diagram of a ducking delay like the TC-2290. In this version, the "wet (min)" control sets the ducked amplitude and the "wet (max)" sets the un-attenuated maximum echo level. These controls will work best in dB. The side-chain gain (SC Gain) control is optional. Notice also that the logic may be reversed to create an "expanding delay" as well.

14.8 Homework

1. Implement the multi-tap delay in Figure 14.16.
2. Implement the analog modeling delay in Figure 14.15.
3. Implement the LCR delay in Figure 14.17.
4. Implement the ducking delay in Figure 14.18.
5. Circular buffers are also used for looping and tape-reverse effects. Implement a basic looper and reverse-looper plugin.

14.9 Bibliography

Coulter, D. 2000. *Digital Audio Processing*, Chap. 11. Lawrence: R&D Books.

Jones, N. n.d. *Efficient C Tip #13: Use the Modulus (%) Operator with Caution*, https://embeddedgurus.com/stack-overflow/2011/02/efficient-c-tip-13-use-the-modulus-operator-with-caution/, Accessed August 1, 2018.

Korg Inc. 2000. *Triton Rack Parameter Guide*. Tokyo: Korg Inc.

Motorola, Inc. 1992. *DSP56000 Digital Signal Processor Family Manual. DSP56KFAM/AD*, Chap. 4. Schaumburg: Motorola Inc.

Roads, C. 1996. *The Computer Music Tutorial*, Chap. 3. Cambridge: The MIT Press.

T.C. Electronics. 1986. *TC 2290 Owners Manual*, https://toneprints.com/media/687091/tcelectronic-tc2290-manual-english.pdf, Accessed August 1, 2018.

CHAPTER 15

Modulated Delay Effects

Perhaps the most interesting kinds of audio effects, from both listening and engineering standpoints, are the modulated delay effects. They include the chorus, flanger, and vibrato; additionally, modulated delays are also found in some reverb algorithms. These time-varying filters are actually quite sophisticated and fun to implement. We already have C++ objects designed for audio delays with circular buffers, and LFOs. These are key ingredients for modulated delay effects. Figure 15.1a shows the standard delay line with feedback. The delay is constant until the user changes it to another fixed value. In the modulated delay line in Figure 15.1b the amount of delay is constantly changing over time. The system uses only a portion of the total available delay time. In the modulated delay, the read location is recalculated on every sample period.

The modulation *depth* relates to the size of the section of the delay line that is being read. The *modulation rate* is the speed at which the read index moves back and forth over the modulated delay section. An LFO is used to modulate the delay amount—and can be just about any kind of waveform. The typical rate for a LFO is between 0.02 and 20 Hz, and variations in the rate can produce dramatic changes to the resulting effect. The most common waveforms are triangle and sinusoid. The LFO output value is used to calculate the new read index on each sample interval. In our discussion, we will describe the modulated delay time with the following.

- min-delay: the minimum delay time
- max-delay: the maximum delay time
- mod-depth: the difference between the max-delay and min-delay

Figure 15.1b shows the relationship between the min-delay, max-delay, and mod-depth values. It is easy to see that the max-delay = min-delay + mod-depth.

15.1 The Flanger/Vibrato Effect

The flanger effect gets its name from the original analog method of running two tape machines slightly out of sync with each other by placing a thumb on the flange (the ring that holds the tape) of one machine's reel and applying pressure to slow down the tape ever so slightly. Other variations include rubbing the flange circularly. The effect creates a time-varying delay that falls out of sync, then back in sync with the first tape machine. The effect has been compared to the whoosh of a jet engine, or the sound of an automatic wah-wah effect. No matter how you define it, the flanger has a unique sound and the effect is heard on numerous recordings. In the flanger effect, the amount of delay time varies between 0 and about 2 to 7 mSec, though this is very subjective; higher values produce a deeper effect, but the addition of high levels of feedback may produce audible pinging or zipper noise. This means that the min-delay is 0.0 and the max-depth is around 2 to 7 mSec.

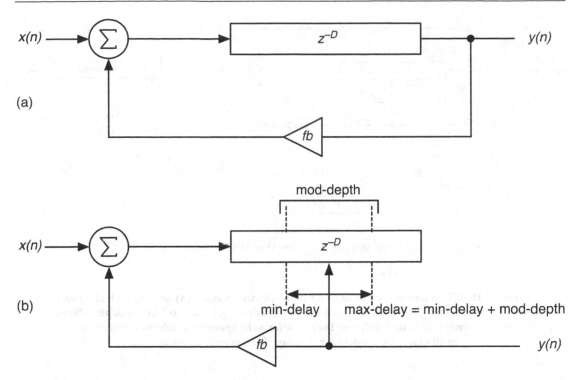

Figure 15.1: (a) A static delay and (b) modulated delay effect.

The flanger uses unipolar modulation from the minimum delay time of zero up to the maximum value, and back down to zero again; we called this *min-up* modulation. We use the same equations and code from Chapter 13.3.1 applied to the delay time. When the read index moves increasingly further away from the write index on each sample interval, the pitch is modulated down. When the index turns around and moves increasingly closer to the write index, the pitch is modulated up—this is the Doppler effect in practice.

When the output of the modulated delay is sent out 100% wet, with no dry signal added and no feedback, the effect is that of vibrato: the pitch shifts up and down as the read index moves back and forth along the delay range. The speed at which the read index moves controls the amount of pitch shifting; faster rates yield more pitch shift, which is most audible in the vibrato effect. The LFO waveform is usually triangular for the flanger and sinusoidal for the vibrato. The max-depth for the vibrato is slightly higher than the flanger, but again this is very subjective and open to experimentation. Feedback is usually not employed with vibrato.

When the dry signal is mixed with the wet signal in equal ratios, the pitch shifting changes into the flanging effect. In the flanger, feedback is usually employed to increase the effect by creating resonances or anti-resonances depending on whether the feedback is normal or inverted. This control is often labeled "resonance" on the GUI and is usually in phase (no inversion) but of course we will allow our plugins to invert the phase for experimentation. Figures 15.2a and b show two versions of the flanger/vibrato module, with the LFO connection to the delay symbolized in an abstract manner.

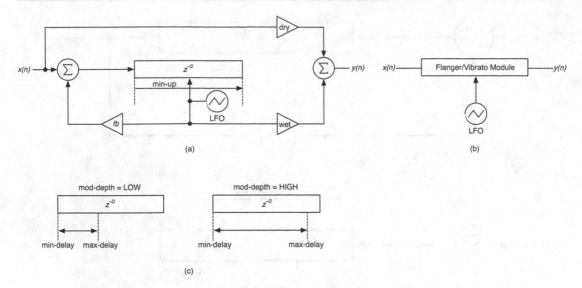

Figure 15.2: Two different versions of a flanger/vibrato module, (a) and (b) each showing the LFO modulating the delay time in an abstract fashion. (c) The modulation depth affects the maximum delay time only and the rate affects the speed at which the index moves. In all cases, the modulation is always from the minimum upwards.

The feedback path is taken from the same location as the moving delay tap (read index). Because the only major difference between the flanger and vibrato effects is the wet/dry mix ratio, these effects are easy to package together. Since the modulation is from the minimum delay time upwards, the depth control changes the maximum delay value, as shown in Figure 15.2c.

The flanger/vibrato GUI controls usually consists of the following.

- **Depth:** how much of the available delay time range is used
- **Rate:** the rate of the LFO driving the modulation
- **LFO type:** usually sinusoidal or triangular, but may be anything
- **Feedback or resonance:** the amount of feedback in the system, may be a positive or negative (inverted feedback) value
- **Flanger/vibrato control:** affects wet/dry ratio −100% wet = vibrato, 50/50 = flanger

15.1.1 Stereo Flanger

A stereo version of the flanger/vibrato usually consists of two identical flanger/vibrato modules that are modulated by a common LFO as shown in Figure 15.3a. One modification specific to the flanger is to phase shift the LFO of one channel with respect to the other, as found in many Korg synth effects. When the phase shift amount is 90 degrees, it is called *quadrature phase* and it produces a stereo quad-phase flanger shown in Figure 15.3b. When the phase is shifted more than 90 degrees, the flanger begins to lose its effect as the two channels are mixed acoustically in the air, where phase cancellations reduce the comb filtering effect (Dattorro, 1997, p. 776).

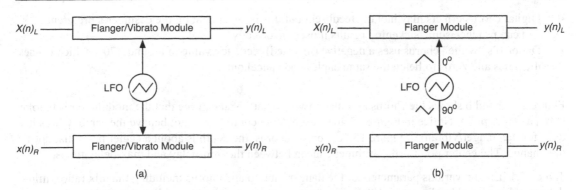

Figure 15.3: (a) A stereo flanger/vibrato uses two identical modules, each modulated by a common LFO. (b) The stereo quad-phase flanger uses a 90 degree phase-shifted LFO for one channel.

The flanger technically modulates the delay right down to its min-delay value of 0 samples. For analog tape flanging, this occurs when the two tape machines come back into perfect synchronization. This is called *Through Zero Flanging* (TZF). This means that for an instant, the output is double the input value. For low frequencies and/or high feedback values, this can cause a problem when the flanger delay time is between 0.0 and about 1.0 mSec and the waveform is at a peak location. In this case, large-valued samples near each other on the waveform sum together producing an increased and sometimes distorted bass response.

15.2 The Chorus Effect

The chorus effect appeared in the late 1970s. It was used on countless recordings during the 1980s and it became a signature sound effect of that decade. The chorusing effect can produce a range of effects from a subtle moving sheen to a swirling, seasick, de-tuned madness. Like the flanger it is also based on a modulated delay line, driven by a LFO. Although different manufacturers adopted different designs, the basic algorithm for a single chorusing unit is the same. The chorus effect is somewhat more loosely specified than the flanger effect, and different manufacturers developed their own interpretations, often involving sets of chorusing structures whose outputs are blended together. Whereas the flanger/vibrato's max-depth value is subjective, both the min-delay and max-depth for the chorus are extremely subjective, and this leads to countless variations from the different manufacturers. The main differences between the flanger and chorus effects are listed here.

- In the chorus, delay time modulation is bipolar about a center point; in the flanger it is unipolar from the minimum value upwards. This specifically affects the algorithm's depth control (compare Figures 15.2 and 15.4).
- The chorus never modulates the time delay all the way down to zero as the flanger does.
- The overall range of delay time modulation is significantly greater for the chorus as compared to the flanger.
- Analog chorus effects typically do not include feedback.

- Digital chorus effects may include feedback, but this is dependent on the designer's preference and can produce some sickening results if used incorrectly.
- Dattorro's "white" chorus uses a negative (inverted) feedback value of around −70% which causes the poles and zeros to lie on the same angles and cancel out.

Figure 15.4a and b shows the chorus module in two formats. You can see that the modulation is bipolar about a center point. Notice in Figure 15.4c that the depth control does not behave the same way as it does for the flanger/vibrato in Figure 15.2c. For the chorus, the depth is about a center point that does not change. The center point is the arithmetic mean between the min-delay and max-delay times.

Table 15.1 lists the various parameters of the flanger/vibrato and chorus including the mix ratios, min-delays and mod-depths (Dattorro, 1997). The max-delay is calculated by adding the mod-depth to the min-delay time.

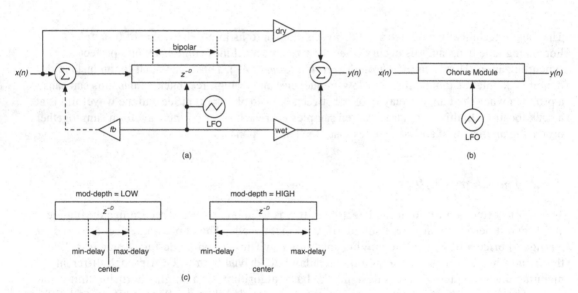

Figure 15.4: Two different versions of a chorus module, (a) and (b), each showing the LFO modulating the delay time in an abstract fashion; in (a) the feedback is dashed as it is optional. (c) The modulation depth affects both the minimum and maximum delay times; the rate affects the speed at which the index moves. In all cases, the modulation is bipolar about the center point.

Table 15.1: Parameter values for the modulated delay effects; these are subjective and you may certainly play around with different values as you wish

	min-delay (mSec)	mod-depth (mSec)	wet-mix (dB)	dry-mix (dB)	feedback (%)	LFO waveform
Flanger	1	2–7	−3	−3	−99 to +99	triangle
Vibrato	0	3–7	0	−96 (off)	0	sine
Chorus	1–30	1–30	−3	0	0	triangle
White Chorus	1–30	1–30	−3	0	−70	triangle

15.2.1 Stereo Chorus

The chorus effect is usually designed as a stereo effect. Its intended use is that of simulating multiple musicians or of simulating recording the same musical part many times. The modulated delay produces the timing shifts that occur in both situations. Manufacturers tend to really go out of their way to differentiate the chorus effect in the marketplace. This usually involves the combination of multiple chorus modules. Unlike the stereo flanger, these stereo chorus effects often capitalize on running multiple LFOs out of phase, at mathematically related rates, or a combination of both specifically to eliminate the cyclic nature of the LFO and try to make the effect more natural and almost transparent. The delay times of the modules may also be manipulated mathematically to try to reduce the obvious nature of the periodic LFO. There are some bonus chorus algorithms later in the chapter, but consider the Korg LCR chorus from the Wavestation synthesizer shown in Figure 15.5.

The LCR chorus here consists of three independent chorus modules, each with its own LFO with rate, depth, and phase control. It produces a wide range of chorus effects from subtle to almost disorienting. Altering it to modify the min-delay and mod-depth of each module would produce even more variety. The center channel module receives a mono version of the stereo input and then blends it with the left and right outputs. Numerous variations are available and the Sony Deca chorus in Section 15.5.5 may win the prize for most complex and thickest sounding of them all.

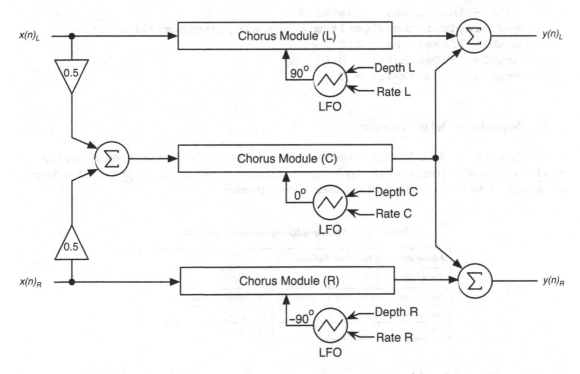

Figure 15.5: The Korg LCR chorus diagrammed here produces a dense chorus effect.

15.3 C++ *Effect Object:* ModulatedDelay

The *ModulatedDelay* object implements the three effect types: flanger, chorus, and vibrato. It uses Table 15.1 to configure the delay modulation wet/dry levels and feedback (disables it for vibrato and chorus). This effect object aggregates the *AudioDelay* and *LFO* objects together, and they do the low-level work. The only real work for the *ModulatedDelay* object is in calculating the modulated delay values for the left and right channels—everything else is handled in the other objects.

15.3.1 ModulatedDelay: *Enumerations and Data Structure*

The *ModulatedDelay* object requires one strongly typed enumeration for the effect algorithm (flanger, chorus, vibrato):

```
enum class modDelaylgorithm { kFlanger, kChorus, kVibrato };
```

The object's parameters are updated via the custom *ModulatedDelayParameters* structure that contains members for the ordinary GUI controls for this type of plugin effect.

```
struct ModulatedDelayParameters
{
      ModulatedDelayParameters() {}
      ModulatedDelayParameters& operator=(. . .){. . .}

      // --- individual parameters
      modDelaylgorithm algorithm = modDelaylgorithm::kFlanger;
      double lfoRate_Hz = 0.0;
      double lfoDepth_Pct = 0.0;
      double feedback_Pct = 0.0;
};
```

15.3.2 ModulatedDelay: *Members*

Tables 15.2 and 15.3 list the *ModulatedDelay* member variables and member functions. Notice that this object can process frames, which may be mono, stereo, or multi-channel. The *processAudioSample* method still works if you simply need to process a mono channel.

Table 15.2: *ModulatedDelay* member variables

ModulatedDelay Member Variables		
Type	Name	Description
ModulatedDelayParameters	*parameters*	Object parameter structure
AudioDelay	*delay*	Delay object
LFO	*lfo*	LFO object

Table 15.3: *ModulatedDelay* member functions

ModulatedDelay **Member Functions**		
Returns	**Name**	**Description**
ModulatedDelayParameters	*getParameters*	Get all parameters at once
void	*setParameters* Parameters: —*ModulatedDelayParameters* _parameters	Set all parameters at once
bool	*reset* Parameters: —*double* sampleRate	Reset sub-components
double	*processAudioSample* Parameters: —*double* xn	Process mono input through modulated delay
bool	*processAudioFrame* Parameters: —*const float** inputFrame —*float** outputFrame —*unsigned int* inputChannels —*unsigned int* outputChannels	Process a frame through L and R delays

15.3.3 ModulatedDelay: *Programming Notes*

The *ModulatedDelay* object uses the functionality in the AudioDelay and LFO for the majority of its processing work. Most of the GUI parameters are simply forwarded to the appropriate member objects.

Using the *ModulatedDelay* object is straightforward.

1. Reset the object to prepare it for streaming.
2. Set the effect parameters in the *ModulatedDelayParameters* custom data structure.
3. Call *setParameters* to update the object.
4. For mono devices, call *processAudioSample*, passing the input value in and receiving the output value as the return variable.
5. For stereo devices, call the *processAudioFrame* function passing the frame pointers and channel counts.

Function: *reset*
Details: reset the sub-components; set up the delay for a maximum 100.0 mSec delay (we'll use only a max of 30 mSec).

```
virtual bool reset(double _sampleRate)
{
    // --- create new buffer, 100mSec long
    delay.reset(_sampleRate);
    delay.createDelayBuffers(_sampleRate, 100.0);

    // --- lfo
    lfo.reset(_sampleRate);

    return true;
}
```

Function: *setParameters*

Details: We simply forward the incoming GUI parameter changes to the sub-component objects; note that we decode the algorithm type and set the LFO waveform according to Table 15.1 (of course, you should experiment with this).

```
void setParameters(ModulatedDelayParameters _parameters)
{
     // --- save
     parameters = _parameters;

     // --- update LFO
     OscillatorParameters lfoParams = lfo.getParameters();
     lfoParams.frequency_Hz = parameters.lfoRate_Hz;
     if (parameters.algorithm == modDelaylgorithm::kVibrato)
          lfoParams.waveform = generatorWaveform::kSin;
     else
          lfoParams.waveform = generatorWaveform::kTriangle;

     lfo.setParameters(lfoParams);

     // --- update delay
     AudioDelayParameters adParams = delay.getParameters();
     adParams.feedback_pct = parameters.feedback_pct;
     delay.setParameters(adParams);
}
```

Function: *processAudioFrame*

Details:

- Render LFO output sample.
- Check the algorithm: set the min-delay, max-modulation, and wet/dry mix values based on Table 15.1.
- Disable the feedback (by setting it to 0.0) for chorus and vibrato.
- Modulate the delay value: for flanger, use unipolar modulation from the min delay out to the max and use bipolar modulation about the center of the min/max range for chorus and vibrato.
- Set the left and right channels with the same modulated delay value (homework).
- Set the new delay, wet/dry mix, and feedback values on the delay object—it will handle the rest of the work.

```
virtual bool processAudioFrame(. . .)
{
     // --- make sure we have input and outputs
     if (inputChannels == 0 || outputChannels == 0)
          return false;

     // --- render LFO
     SignalGenData lfoOutput = lfo.renderAudioOutput();
```

```
// --- setup delay modulation
AudioDelayParameters params = delay.getParameters();
double minDelay_mSec = 0.0;
double maxDepth_mSec = 0.0;

// --- set delay times, wet/dry and feedback
if (parameters.algorithm == modDelaylgorithm::kFlanger)
{
    minDelay_mSec = 0.1; maxDepth_mSec = 7.0;
    params.wetLevel_dB = -3.0; params.dryLevel_dB = -3.0;
}
else if (parameters.algorithm == modDelaylgorithm::kChorus)
{
    minDelay_mSec = 10.0; maxDepth_mSec = 30.0;
    params.wetLevel_dB = -3.0; params.dryLevel_dB = 0.0;
    params.feedback_pct = 0.0;
}
else if (parameters.algorithm == modDelaylgorithm::kVibrato)
{
    minDelay_mSec = 0.0; maxDepth_mSec = 7.0;
    params.wetLevel_dB = 0.0; params.dryLevel_dB = -96.0;
    params.feedback_pct = 0.0;
}

// --- calc modulated delay times
double depth = parameters.lfoDepth_Pct / 100.0;
double modulationMin = minDelay_mSec;
double modulationMax = minDelay_mSec + maxDepth_mSec;

// --- flanger—unipolar
if (parameters.algorithm == modDelaylgorithm::kFlanger)
{
    double modulationValue = bipolarToUnipolar(depth *
                                lfoOutput.normalOutput);
    params.leftDelay_mSec =
        doUnipolarModulationFromMin(modulationValue,
                            modulationMin, modulationMax);
}
else
{
    double modulationValue = depth * lfoOutput.normalOutput;
    params.leftDelay_mSec =
        doBipolarModulation(modulationValue,
                            modulationMin, modulationMax);
}
```

```
        // --- set right delay to match (*Hint Homework!)
        params.rightDelay_mSec = params.leftDelay_mSec;

        // --- modulate the delay
        delay.setParameters(params);

        // --- just call the function and pass our info in/out
        return delay.processAudioFrame(inputFrame, outputFrame,
                                       inputChannels, outputChannels);
}
```

15.4 Chapter Plugin: ModDelay

Our delay FX plugin will include the normal stereo and ping-pong delay algorithms. The GUI will include controls for adjusting the delay parameters as well as a switch to choose the algorithm. All of the real audio signal processing work will be done in the *AudioDelay* object. Figure 15.6 shows the stereo delay GUI and block diagram.

15.4.1 ModDelay: *GUI Parameters*

The GUI parameter table is shown in Table 15.4—use it to declare your GUI parameter interface for the plugin framework you are using. If you are using ASPiK, remember to first create your enumeration of control ID values at the top of the *plugincore.h* file, then use automatic variable binding to connect the linked variables to the GUI parameters.

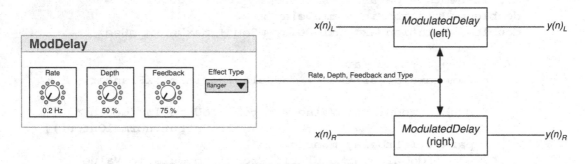

Figure 15.6: The GUI and block diagram for the *ModDelay* plugin.

Table 15.4: The *ModDelay* plugin's GUI parameter and linked-variable list (*) for experimental purposes; you usually want to lower this limit in the final release

Control Name	Units	Min/max/default or string-list	Taper	Linked Variable	Linked Variable Type
Rate	Hz	0.02 / 20 / 0.2	linear	*lfoRate_Hz*	*double*
Depth	%	0 / 100 / 50	linear	*lfoDepth_Pct*	*double*
Feedback	%	0 / 100(*) / 50	linear	*feedback_Pct*	*double*
Effect Type	-	"flanger,chorus,vibrato"		*effectType*	*int*

ASPiK: top of *plugincore.h* file

```
enum controlID {
     lfoRate_Hz = 0,
     lfoDepth_Pct = 1,
     feedback_Pct = 2,
     modDelayEffect = 3
};
```

15.4.2 ModDelay: *Object Declarations and Reset*

The *ModDelay* plugin uses one *AudioDelay* and one *LFO* member object to provide the stereo processing via *processAudioFrame*. In your plugin's processor (*PluginCore* for ASPiK), declare the member plus the *updateParameters* method.

```
// --- in your processor object's .h file
#include "fxobjects.h"

// --- in the class definition
protected:
     ModulatedDelay modDelay;
     void updateParameters();
```

You need only reset the object with the current sampling rate—it will reset its sub-components and so on.

ASPiK: *reset( )*

```
     // --- reset
     modDelay.reset(resetInfo.sampleRate);
```

15.4.3 ModDelay: *GUI Parameter Update*

The *ModulatedDelay* object requires only four parameters for updating, which are shown on the GUI. Gather the values from the GUI parameters and apply them to the custom data structure using the *getParameters* method to get the current parameter set, and then overwrite only the parameters your plugin actually implements. For ASPiK we use the helper *convertIntToEnum* to convert the user's selected value (a zero-indexed integer) to the corresponding strongly typed enumeration as an integer.

ASPiK: *updateParameters( )* function

```
ModulatedDelayParameters params = modDelay.getParameters();
params.lfoDepth_Pct = lfoDepth_Pct;
params.lfoRate_Hz = lfoRate_Hz;
params.feedback_Pct = feedback_Pct;
params.algorithm = convertIntToEnum(modDelayEffect,
modDelaylgorithm);

modDelay.setParameters(params);
```

15.4.4 ModDelay: *Process Audio*

The *ModulationDelay* object does all of the heavy lifting. All we need to do is update the parameters, then process the audio frame.

ASPiK: *processAudioFrame( )* function

```
// --- object does all the work on frames
modDelay.processAudioFrame(processFrameInfo.audioInputFrame,
                           processFrameInfo.audioOutputFrame,
                           processFrameInfo.numAudioInChannels,
                           processFrameInfo.numAudioOutChannels);
```

Non-ASPiK: You gather your frame input and output samples into small arrays (you can statically declare them to minimize cost). See the following for an example.

```
double xnL = //< get from your framework: input sample L
double xnR = //< get from your framework: input sample R

float inputs[2] = { xnL, xnR };
float outputs[2] = { 0.0, 0.0 };

// --- process (in, out, 2 inputs, 2 outputs)
modDelay.processAudioFrame(inputs, outputs, 2, 2);

//< framework left output sample = outputs[0];
//< framework Right output sample = outputs[1];
```

Rebuild and test the plugin. All of the effect algorithms, especially the chorus, are very subjective in quality and nature—the minimum delay and maximum modulation depth controls are very sensitive. You should definitely play with the min and max settings in the *processAudioFrame* function to experiment with your own values. The maximum delay length is 100 mSec, so you have much more leeway in the upper limit values. As an experiment, set the flanger maximum depth to 90 mSec. With a slow LFO rate, you will hear the effect morph between flanging, chorus and then doubling/echoes, then back again.

15.5 More Modulated Delay Algorithms

15.5.1 Korg Stereo Cross-Feedback Flanger/Chorus

Figure 15.7 shows the block diagram for a crossed-feedback configuration for the flanger/chorus. The left and right channels run on normal and quadrature phase LFO outputs. The feedback path is crossed between channels. Try experimenting with different ranges of min depth and max modulation across the two channels so that the crossed feedback goes through different modulations on each trip around the figure-8 pattern.

15.5.2 Sony DPS-M7 Multi-Flanger

The DPS M7 has some intensely thick modulation algorithms. The multi-flanger shown in Figure 15.8 includes two flanger circuits that can be combined in parallel or series on either channel. The channels are also cross-mixable. Each module marked "Flanger" contains a complete flanger module: depth,

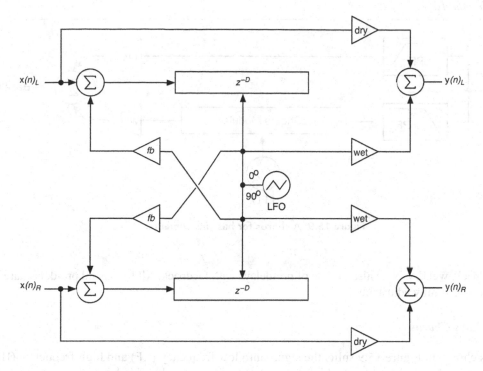

Figure 15.7: A stereo cross-feedback flanger/chorus from the Korg Wavestation synth.

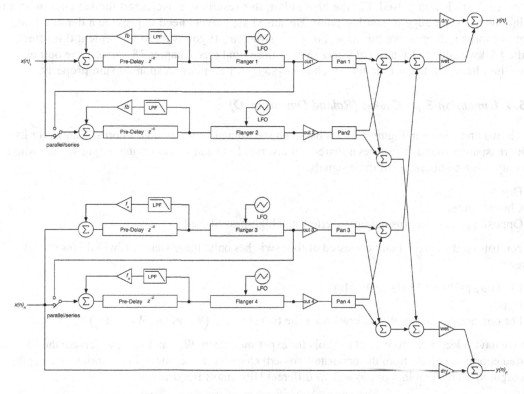

Figure 15.8: The Sony DPS-M7 Multi-Flanger.

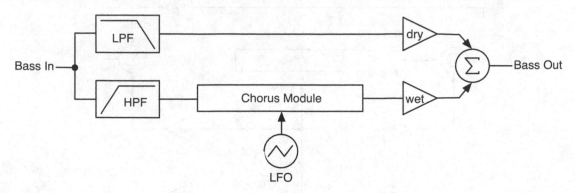

Figure 15.9: A chorus for bass instruments.

rate, feedback, wet/dry. Note also the use of pre-delays with feedback. All LFOs and pre-delays are independent and fully adjustable.

15.5.3 Bass Chorus

The bass chorus in Figure 15.9 splits the signal into low frequency (LF) and high frequency (HF) components and then processes the high frequency component only. This leaves the instrument's fundamental frequencies intact. The comb filtering of the chorus effect smears time according to how much delay is being used. For the bass guitar, this results in a decreased fundamental with an ambiguous transient edge or starting point. Because bass players need to provide a defined beat, the bass chorus will preserve this aspect of the instrument. If you want to implement this effect, use the Linkwitz-Riley LP and HP filters to split the signal (see Chapter 18). Invert the output of one of the filters—it doesn't matter which one—so that their phase responses sum properly.

15.5.4 Dimension-Style Chorus (Roland Dimension D)

This chorus unit shown in Figure 15.10 is based on the Roland Dimension D® chorus. Known for its subtle transparent sound, it features a shared but inverted LFO and an interesting output section where each output is a combination of three signals:

*   Dry
*   Chorus output
*   Opposite channel chorus output, inverted, and high-pass filtered

The controls on the original unit consisted of four switches only; these were hardwired presets and changed:

*   LFO rate (either 0.25 Hz or 0.5 Hz)
*   Depth
*   For one preset, the wet/dry mix ratios for the two channels (W_L vs D_L, W_R vs D_R)

Here we have added two more level controls for experimentation, W_{CL} and W_{CR}, which are the filtered, inverted chorused signals from the opposite (crossed) channels. Experiment with various rate, depth, and output mixture combinations as well as different HPF cutoff frequencies.

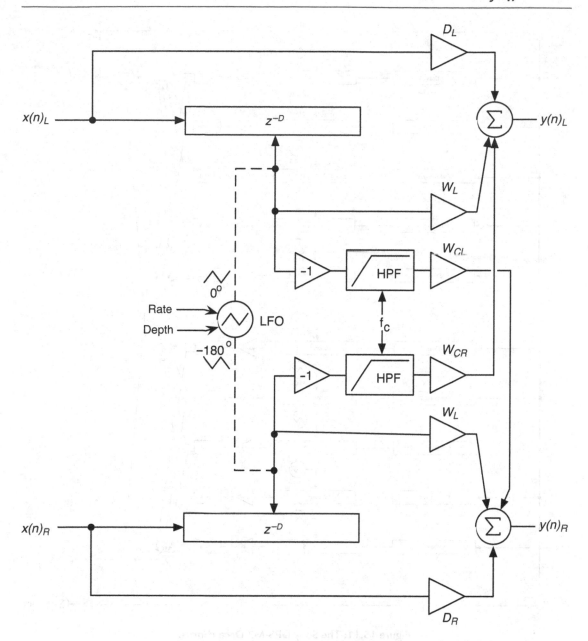

Figure 15.10: A Roland Dimension–style chorus.

15.5.5 *Sony DPS-M7 Deca Chorus*

The Deca chorus shown in Figure 15.11 includes 10 (deca) chorus units, five per channel. Each chorus has its own pre-delay, LFO, gain, and pan control. It can run also run in a mono mode (10 chorus units in parallel) as well.

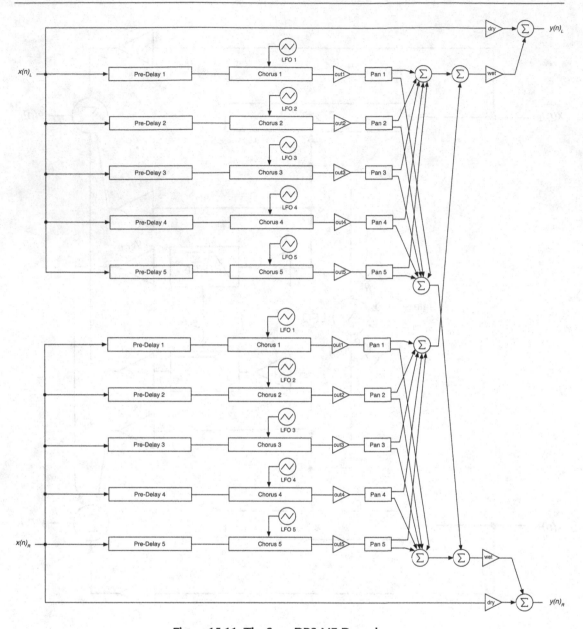

Figure 15.11: The Sony DPS-M7 Deca chorus.

15.6 Homework

1. Design your own chorus plugin using multiple chorusing modules in a variety of connections.
2. The Dyna-Flanger® used an envelope detector in a side chain to allow the modulation of effect parameters with the strength of the audio signal. Implement your own dynamic-flanger plugin.

15.7 Bibliography

Cole, M. *Roland Dimension C Clone for Eventide 7000, Orville and H8000*, http://web.eventide.com, Accessed January 2011.

Coulter, D. 2000. *Digital Audio Processing*, Chap. 11. Lawrence: R&D Books.

Dattorro, J. 1997. "Effect Design Part 2: Delay Line Modulation and Chorus." *Journal of the Audio Engineering Society*, Vol. 45, No. 10.

Korg Inc. 1991. *Wavestation SR Reference Guide*. Tokyo: Korg Inc.

Roads, C. 1996. *The Computer Music Tutorial*, Chap. 3. Cambridge: The MIT Press.

Roland Corporation. 1990. *Boss CH-1 Schematics*. Tokyo: Roland Corp.

White, P. 2001. *Understanding & Emulating Vintage Effects*. Sound on Sound Magazine, www.soundonsound.com/sos/jan01/articles/vintage.asp, Accessed August 1, 2018.

CHAPTER 16

Audio Filter Designs: FIR Filters

IIR filters have several attractive properties:

- They require only a few delay elements and math operations.
- You can design them directly in the *z*-plane.
- You can use existing analog designs and convert them to digital with the BZT.
- You can get extreme resonance/gain by placing poles very near the unit circle.
- You can make filters, EQs, etc. with controls that link to the coefficients directly or indirectly for real-time manipulation of the plugin.

A drawback is that IIR filters may become unstable and blow up, or oscillate. Their impulse responses can be infinite. FIR filters have all zeros and a finite impulse response. They are unconditionally stable so their designs can never blow up: they may output all zeros, a constant value, series of clicks or pulses, and other erroneous data, but they don't actually go unstable.

You can also make a linear phase filter with an FIR, just like the simple feed-forward (FF) topology we analyzed in Chapters 2 and 3; a real-time linear phase filter is technically impossible to make with an IIR topology although you can approximate it by adding phase compensation filters. But the one thing that separates FIR filters from all other kinds, including analog counterpart filters, is that their coefficients $a_0, a_1, \ldots, a_N$ are the impulse response of the filter. We proved that when we manually pushed an impulse through the simple FF filter in Chapter 2, and then again when taking the *z* Transform of the impulse response of the same filter in Chapter 3.

16.1 The Impulse Response Revisited: Convolution

In Chapter 1 you saw how the digitized signal was reconstructed into its analog counterpart by filtering through an ideal low-pass filter. When the series of impulses is filtered, the resulting set of *sin(x)/x* pulses overlap with each other and their responses all add up linearly. The addition of all the smaller curves and damped oscillations reconstructs the inter-sample curves and damped oscillations shown in Figure 16.1.

In the time domain, you can see how the impulse response of each sample is overlaid on the others and that the summing together of the peaks and valleys of the *sin(x)/x* shape ultimately creates the portions in between the samples that appeared to have been lost during the sampling process. The overlapping-summation idea is also central to fast convolution, which we'll study in Chapter 20.

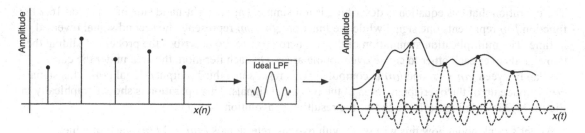

Figure 16.1: The *sin(x)/x* outputs of the LPF are overlapped and summed together to reconstruct the original band-limited input waveform, as shown here.

If you know how a system affects one single impulse, you can exactly predict how it will affect a stream of impulses (i.e. a signal) by doing the time domain overlay.

If you capture the *impulse response* of a system, you have captured the state of the system and you have the *algorithm* for the system coded in a single function. For a dynamic system, whose impulse response changes over time, you will need a set of impulse responses that encode each state.

The process of overlaying the impulse response on top of the input stream of impulses *x(n)* and adding up the results to get the final time domain output *y(n)* is called *convolution*. Convolution is a mathematical operation used in many fields of science; digital audio signal processing is just one of them. The mathematical symbol for convolution is a star with a circle around it, but in most cases we simply use the asterisk *, which can be confusing because this is used to represent the multiplication operation in C/C++. In this section, it refers to convolution. In Figure 16.1, you convolved the input signal *x(n)* with the impulse response *h(n)* by overlaying the *h(n)* signal on top of each input impulse, then summing up signals to create the final output *y(n)*. Mathematically, you would write this as *y(n) = x(n) * h(n)*. Visually, it's easy to understand the concept of overlaying the signals and adding them to get the final result. But how do you describe that mathematically? The answer is that this kind of operation is a special circumstance of a more generalized operation of convolution.

Mathematically, convolution for discrete signals is described as follows.

$$c(n) = f(n) * g(n) = \sum_{m=-\infty}^{+\infty} f(n)g(n-m) \qquad (16.1)$$

In this case, f and g are two generalized signals, and neither of them necessarily needs to be an impulse response. Convolution is commutative so that $f * g = g * f$ or Equation 16.2.

$$c(n) = f(n) * g(n) = \sum_{m=-\infty}^{+\infty} f(n)g(n-m)$$

$$c(n) = g(n) * f(n) = \sum_{m=-\infty}^{+\infty} g(n)f(n-m) \qquad (16.2)$$

The operation that this equation is describing is not simple. On the right-hand side of Equation 16.2 the function $f(n)$ represents one signal while the function $g(n - m)$ represents the second signal reversed in time. The multiplication/summation of the two across $-\infty$ to $+\infty$ describes the process of sliding the two signals over each other to create overlapping areas. On each iteration, the area under the curve of overlap between $g(n - m)$ and $f(n)$ is computed. This results in a third (output) signal $c(n)$. This signal $c(n)$ is made up of the overlapping area of the two input signals. This operation is shown graphically in Figure 16.2a while Figure 16.2b shows the resulting convolution output c(n).

Next, let's think about how this will work with two discrete signals $a(n)$ and $b(n)$, each of which consists of four data points and shown in Figure 16.3. The first signal a has been reversed as $a(n - m)$ while keeping the original indexing in tact. This will allow us to examine the operation on a sample-by sample basis, forming the output signal $y(n)$ in the process. You can see that on each successive

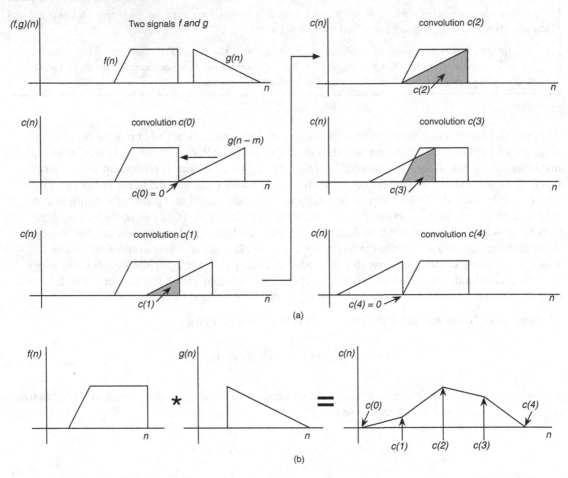

Figure 16.2: Two signals $f(n)$ and $g(n)$ are convolved. These are discrete signals, but the sample symbols have been removed to make it easier to see; instead they are shown as continuous.
(a) In the first step $c(0)$, one of the signals is reversed and the two are pushed up next to each other. As the convolution progresses through each step $c(1)$ to $c(4)$, the overlapping areas are calculated and stored. (b) Another way to show the convolution along with the resulting signal.

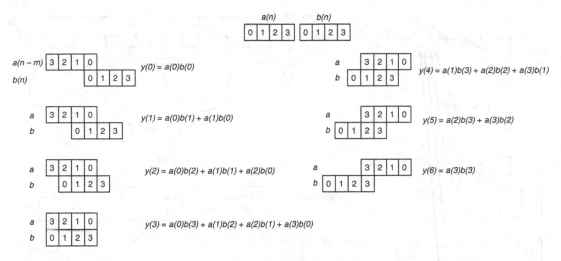

Figure 16.3: The operation of linear convolution, sample by sample, showing the algorithm's operation.

convolution operation, the overlapping samples are multiplied together and then added to produce the result. This operation specifically describes *linear convolution,* and this is the convolution operation we use for audio filtering and convolutional reverb. An important observation of Figure 16.3 is that we started with two signals, each containing four samples, but we wound up with an output $y(n)$ consisting of seven samples. In a linear convolution of two sequences of the same length N, the output sequence will have length $2N - 1$.

If you know the impulse response of a system $h(n)$, you can convolve it with the input signal $x(n)$ to produce the output signal $y(n)$. This is the equivalent of multiplying the transfer function $H(z)$ with the input signal $X(z)$ to produce the output $Y(z)$. Thus convolution in the time domain is multiplication in the frequency (z) domain (digital) or the $j\omega$ domain (analog).

$$y(t) = x(t) * h(t) \leftrightarrow Y(j\omega) = X(j\omega)H(j\omega)$$
$$y(n) = x(n) * h(n) \leftrightarrow Y(z) = X(z)H(z)$$

(16.3)

Now for the fine print: the relationships in Equation 16.3 hold true only for a special type of convolution called *circular convolution*, which we will investigate in Chapter 20. We can force circular convolution to be equivalent with linear convolution, so the notion of multiplication in the frequency domain is something you should get comfortable thinking about. We'll create filters that multiply in the frequency domain as well in Chapter 20.

16.2 FIR Filter Structures

To understand how a feed-forward filter implements convolution first rearrange the block diagram. Let's consider a long FIR filter with N + 1 coefficients shown in Figure 16.4a. You can mentally rotate the structure so that it resembles Figure 16.4b, which is called a *transversal structure*. In Figure 16.4 you can see that at any given time a portion of the input signal $x(n)$ is trapped in the delay line. On each sample period, the input signal slides to the right and a summation is formed with the product of

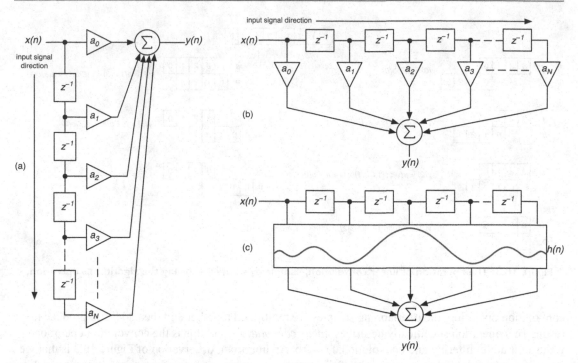

Figure 16.4: Two structures that implement the FIR filter are shown in (a) and (b). Notice that the filter has only feed-forward branches. (c) You can also imagine the impulse response living inside of the coefficient array.

the coefficients with the samples $x(n - d)$. The words *sliding*, *summation*, and *product* are key here—they're the same words used to describe convolution. The input signal $x(n)$ moves through the delay line being convolved with the impulse response on each sample period. Since each sample in the delay line is an impulse, and each impulse is symmetrical when reversed, this is the same as conceptually overlapping the impulse response on top of each sample and scaling by the sample value. The result is the final summation of all the peaks and valleys of the impulse response with the delayed signal $x(n)$. So, a FIR filter exactly implements discrete linear convolution in the time domain. This ultimately gives us a whole new way to filter a signal—by convolving it against an impulse. So, if an ideal LPF has an impulse response in the shape $sin(x)/x$ and we sample the impulse response, we get a discrete impulse response $h(n)$. The more samples we take, the more accurate our version of the impulse response becomes. This introduces the first way to design a FIR filter: find an impulse response you like, sample it, and convolve with the impulse response (IR) to produce the filtered output.

A fundamental issue exists with FIR filters that involves the mathematical complexity of convolution. The more finely detailed our impulse response is—in other words, the more samples $h(n)$ has—the more accurate the resulting filter's output will be. Although the process of multiplying samples with coefficients, then summing the results is fairly simple, the sheer number of operations that must occur becomes an obstacle; most of our filters will require hundreds or thousands of $h(n)$ samples and therefore a_N coefficients. Unfortunately, linear convolution is CPU intensive and limits our ability to convolve with long impulse responses. We will take advantage of the multiplication in the frequency domain property we observe in Equation 16.3 to perform fast convolution in Chapter 20.

16.3 Generating Impulse Responses

The FIR filter operates on impulse responses, so we need methods of generating these impulse responses. The impulse responses are then stored in arrays. One typical way of storing and transmitting impulse responses is to encode them in audio files such as the *.wav* file. These files are never meant to render audio, but rather hold the sampled impulse response in a predetermined and well-specified way that includes channel count, sample encoding, and other information.

16.3.1 Impulse Responses of Acoustic Environments

One way to obtain an impulse response is to generate a stimulus signal inside of an acoustic environment, and then use a microphone to capture the resulting impulse response. In this case, we've captured the impulse response of an acoustic space. The methods of generating the impulse stimulus range from primitive devices—shooting cap pistols, popping balloons, or smacking wood or metal rods together—to very complex methods of deterministic noise and swept sinusoids. In any event, the result is a *.wav* file that captures the impulse response. Convolution with the impulse response should produce an exact perfect replica of the reverberation that occurs in this space. However, the exact replica is only really valid for the location of the microphones used to make the measurement.

16.3.2 Impulse Responses of Speakers and Cabinets

Another popular use of IR convolution is to emulate loudspeakers and their enclosures for musical instrument amplifier-modeling plugins. The impulse response is usually measured with pseudo-random noise, or more recently using the swept-sine de-convolution. The IR files are relatively short in length—usually 1024 to 4096 points in length.

16.3.3 Impulse Responses by Frequency Sampling

With the exception of the impulse invariant filters in Chapter 11, we typically specify our audio filters using a frequency response, rather than an impulse response. We can design FIR filters by first specifying the desired frequency response, then using the inverse Fourier transform in Section 20.4 to convert the frequency domain information into an impulse response. We will wait until Chapter 20 to get involved in the Fourier math, but it is convenient to know that we can very easily convert any frequency domain specifications we can imagine—from exact replicas of analog filters to crazy and bizarre zig-zag shapes and everything in between—to impulse responses. That frequency domain specification is really a set of discrete points from a desired frequency response plot. This is called the Frequency Sampling Method (FSM).

This method is demonstrated in Figure 16.5. If you are using RackAFX, you have a built-in and powerful frequency sampling plugin that allows you to specify your desired filter frequency response with a series of line segments shown in Figure 16.5—and don't worry if you aren't using RackAFX. We have several functions already coded for you to extract frequency domain samples from desired curves that you supply.

16.3.4 Sampling Arbitrary Frequency Responses

The first step in the frequency sampling method is to divide the frequency domain from 0 Hz to Nyquist by half of the order of the final filter. As an example, we will create a $N = 16$-point FSM

FIR filter and divide the frequency into M = eight segments. The sampling locations will be at linear equal intervals of Nyquist/M, so our eight frequency sample points will be spaced at ~2.7 kHz intervals. Note that this also sets the accuracy of the final filter. Figure 16.5a shows our frequency domain divided into 2.7 kHz slices. In Figure 16.5b we've created an arbitrary frequency response curve using a set of line segments. Figure 16.5c shows the resulting FIR filter. It is a fairly poor rendition of the desired response; however, it is guaranteed that the actual frequency response will exactly match the desired response at the sample locations. At the places in between the frequency samples, the curve is allowed to do whatever it wants. If we keep the original set of line segments, but increase the value of M to perform the frequency sampling at smaller intervals, the resulting filter gets closer and closer to the desired response as shown in Figure 16.5d–f. For 1024 points, the actual response matches the desired response so closely that they are nearly indistinguishable. The ability to create filters with completely arbitrary magnitude responses makes the frequency sampling method a powerful tool for plugin designers. The fact that the resulting filters are always linear phase is icing on the cake.

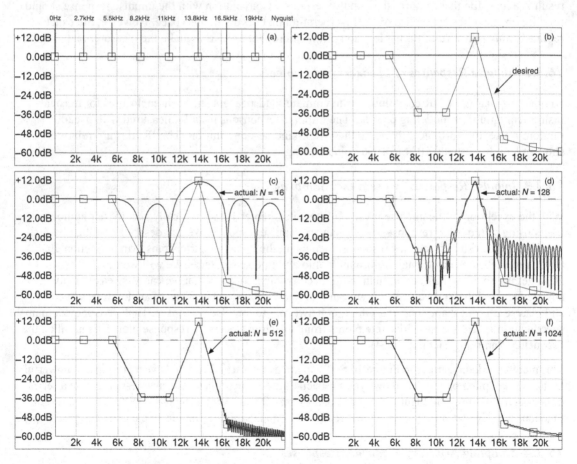

Figure 16.5: (a) A desired frequency response drawn as a series of line segments and the frequency responses of the resulting frequency sampling method filters with lengths (a) N = 16, (b) N = 128, and (c) N = 512, and the impulse responses for (e) N = 16 (f) N = 128, (g) N = 512.

The frequency sampling method involves these steps:

1. Decide on a desired frequency response and plot it in the frequency domain.
2. Sample the frequency response at evenly spaced intervals determined by the filter order you choose.
3. Take the inverse Discrete Cosine Transform (Chapter 20) of the sampled frequency response to get the sampled impulse response.

16.3.4.1 Linear Phase FIR Using the Frequency Sampling Method

Choose:

- N = number of coefficients

Calculate for N = odd:

- $(N + 1)/2$ = number of samples in frequency domain, starting at 0 Hz

Calculate for N = even:

- $N/2$ = number of samples in frequency domain, starting at 0 Hz

$\Delta f = fs/N$ = frequency spacing, starting at 0 Hz

Calculate the filter coefficients a_0 to $a_{N/2}$

For N = odd:

$$a_n = \frac{1}{N}\left[H(0) + 2\sum_{i=1}^{(N-1)/2} |H(i)|\cos|[2\pi i\left[n - \frac{N-1}{2}\right]/N]|\right]$$

For N = even:

$$a_n = \frac{1}{N}\left[H(0) + 2\sum_{i=1}^{N/2-1} |H(i)|\cos|[2\pi i\left[n - \frac{N-1}{2}\right]/N]|\right] \qquad (16.4)$$

Note: this produces half of the coefficients; the other half are a mirror image because the impulse response is guaranteed to be symmetrical about its center, producing a linear phase FIR filter every time. Equation 16.4 represents the real part of the discrete inverse Fourier transform; it takes its name "Inverse Discrete Cosine Transform" from the cosine term in the equation. The great news is that Equation 16.4 is already encoded for you as a C++ function named *frequencySample* and is available in your *fxobjects.h* file. You need only supply an input array of frequency sampled points and the function returns the sampled impulse response.

16.3.5 Sampling Analog Filter Frequency Responses

The magnitude responses for the classical analog filters are well known and may be calculated knowing the order, type, cutoff frequency f_c, and Q of the desired analog filter. This means that we can calculate the magnitudes of the analog filter at some number of equally spaced points, and then use the frequency sampling method to generate the sampled impulse response to use in our FIR filters. The resulting filters, like all frequency sampling method filters, are guaranteed to match the magnitude perfectly at the frequency sample locations. **We are not advocating this as an efficient method of**

designing digital versions of analog filters, but it is an excellent demonstration of the design method. In addition, there is no frequency domain warping from the BZT as we have with many IIR filters, and as FSM filters, these analog-matched-magnitude filters are also linear phase.

Your *fxobjects.h* file contains a function *calculateAnalogMagArray* that generates an array of sampled analog filter magnitude responses for the following types.

- 1st and 2nd order analog LPF and HPF
- 2nd order analog BPF and BSF

The function takes the f_c and Q values for the filter and generates a sampled set of points from 0 Hz to Nyquist that are the magnitudes of the analog filter at the sampled frequencies. It uses the same equations from Chapter 11 for calculating these magnitude values. You can then use the built-in function *freqSample* to generate an impulse response array that is twice the length of the original sampled frequency response, as per Equation 16.4. Figure 16.6 shows the output of a 1024-point FIR filter using the frequency sampling method on sampled (a) 1st and (b) 2nd order analog low-pass filters. If you are absolutely determined to match an analog filter's response as closely as possible, these filters may be useful even if you only use them for comparison purposes.

16.3.6 Sampling Ideal Filter Frequency Responses

In addition to drawing arbitrary waveforms and sampling analog magnitude responses, we may also sample ideal filter responses in an attempt to implement them digitally. These filters have perfect or infinitely sharp filter edges, and are also known as *brick-wall filters*. The good news is that we can certainly sample these ideal responses and implement them as FIR filters. The bad news is that the sharp discontinuities are problematic for the inverse Fourier transform step. While we are guaranteed to have perfect matches at the sampled frequency points, the response can theoretically do anything it wants in between those points, and for the sharp filter edges this results in rippling. Figures 16.7a–d show the ideal brick-wall filter frequency responses for LPF, HPF, BPF and BSF respectively.

Your *fxobjects.h* file contains a function *calculateBrickwallMagArray* that generates an array of sampled ideal filter magnitude responses for the following types.

- LPF and HPF
- BPF and BSF

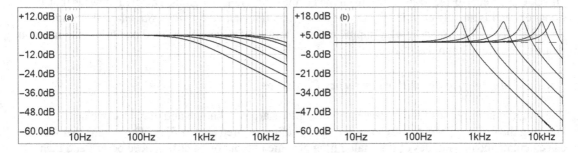

Figure 16.6: Actual frequency responses of 1024-point FIR filters using the frequency sampling method on analog magnitude responses for (a) 1st order LPF and (b) 2nd order LPF. Here the f_c is varied between 500 Hz, 1 kHz, 2.5 kHz, 5 kHz, 10 kHz and 15 kHz and for the 2nd order filters Q = 5.

The function takes the f_c and Q values for the filter and generates a sampled set of points from 0 Hz to Nyquist that are the magnitudes of the ideal filter at the sampled frequencies. You can then use the built-in function *freqSample* to generate an impulse response array that is twice the length of the original sampled frequency response, as per Equation 16.4. Figure 16.8a shows the result of applying the ideal LPF response to a 512-point frequency sampling method FIR filter. There is severe rippling

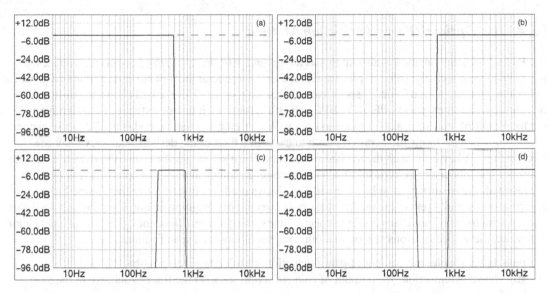

Figure 16.7: Brick-wall filter responses for (a) LPF, (b) HPF, (c) BPF, and (d) BSF filters with f_c = 500 Hz; for the LPF and HPF, Q = 0.707 while for the BPF and BSF, Q = 1.0.

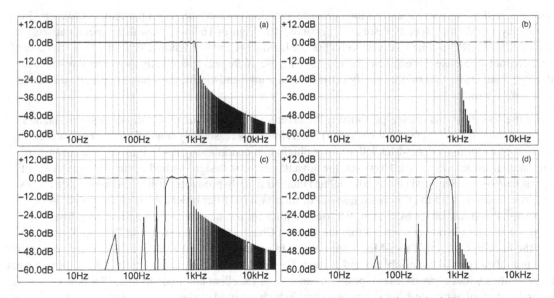

Figure 16.8: These 512-point FIR filters use the frequency sampling method on ideal filter responses for (a) ideal brick-wall LPF, (b) brick-wall LPF with relaxed edge specification, (c) ideal brick-wall BPF, and (d) brick-wall BPF with relaxed edge specifications.

in the stop band near the discontinuity (edge). Much better results can be achieved if we relax the specifications of the filter by changing the slope of the brick-wall edge. Figure 16.8b shows the same filter but with the edge specifications relaxed by only two frequency samples so that the edge is only slightly less steep. Figure 16.8c shows the brick-wall BPF while Figure 16.8d shows the same filter with the edge slope relaxed by two frequency samples.

16.4 The Optimal/Parks-McClellan Method

The Parks-McClellan method is also known as the *optimal FIR design method*. In this technique, we use an iterative search algorithm to place and move zeros around in the z-plane in an attempt to generate a filter that matches a set of "loose" specifications that we provide. The method allows the design of the four classical filter types: LPF, HPF, BPF, and BSF. However, it can be extended to design filters that have multiple sets of peaks (or plateaus) and valleys. Central to the method is the Remez exchange algorithm, which performs the iterative search to try to fit a filter to the desired frequency response specifications. A key issue with the method is that for a given set of specifications, the Remez exchange algorithm may not converge on a solution and therefore the filter cannot be designed. This is in sharp contrast with the frequency sampling method that always produces filter coefficients. These loose specifications actually reflect the preferred textbook method of designing analog filters, which seems to contrast with the way musicians prefer to specify filters using a cutoff frequency f_c and resonance factor Q. Instead, these specifications rely on the following set of loosened specifications.

- **pass-band ripple:** the amount of gain fluctuation we tolerate in the pass-band
- **stop-band attenuation:** the amount of gain reduction we wish the filter to achieve in the stop band of frequencies
- **transition-band width:** the steepness of the filter edge(s)

We refer to these as loose specifications because they don't exactly specify a particular cutoff or -3 dB frequency, nor do they specify resonance or bandwidth. Figures 16.9a–d show these specifications for the LPF, HPF, BPF and BSF respectively. For the band-pass and band-stop filters, there are two sets of specifications because there are two band edges. Figure 16.9c was specifically drawn to show that for the BPF (and BSF), you might specify different transition bandwidths, pass-band ripple, or stop-band attenuation values.

Once the specifications are set, performing the iterative Remez exchange algorithm is not trivial and can take time to converge (or not converge). Therefore, it is reasonably impossible to create a real-time version of a Optimal (Parks-McClellan) filter plugin that continuously re-designs itself according to user specifications via a GUI. However, the advantage to this design method is that for a given set of specifications, the Optimal (Parks-McClellan) method usually produces a filter that meets or exceeds the specifications with the minimum number of filter taps (or filter order). As with the FSM filters, these are also linear phase by design. We use this filter for the steep anti-aliasing low-pass filters required for sample rate conversion that we study in Chapter 22. If you are using RackAFX, you can use the built-in Optimal (Parks-McClellan) plugin to generate LPF, HPF, BPF, or BSF filter coefficients you may then copy to the clipboard and paste as C++ code. If you are not using RackAFX, or you want to generate a wider array of filter types, you can use the excellent and free Optimal (Parks-McClellan) filter design tool at http://t-filter.engineerjs.com/, which we used exclusively for designing the filters in Chapter 22. As with the RackAFX FIR designer, you may copy C++ code to your clipboard and paste it directly into your code.

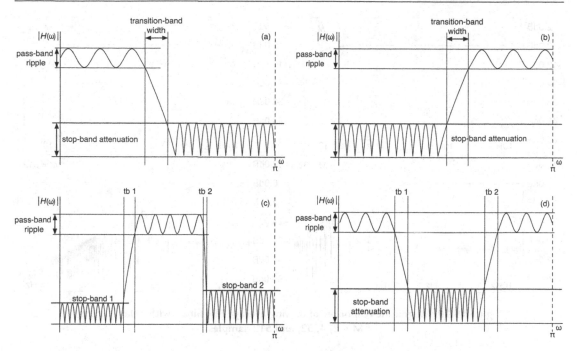

Figure 16.9: The "loose" specification bands for (a) LPF, (b) HPF, (c) BPF, and (d) BSF. The latter two include two transition bands labeled "tb 1" and "tb 2." The BSF may also specify two different pass-band ripple values.

16.5 Other FIR Design Methods

There are several other FIR design methods you may want to check out. One of these is the windowing method, in which we use the curved, pulse shape of the FFT windows in Chapter 20 to create filter magnitude response shapes. This method is highly limited in the variations on these designs and tends to be used sparingly for audio filtering purposes. A very simple but useful FIR filter is the moving average (MA) filter. In this case, the coefficients are simple to calculate and are each identical to one another. We simply average the values of samples inside of a moving window. In most designs, the window of samples includes the current input $x(n)$ plus some number of past inputs $x(n - M)$ that are all averaged together with equal weighting. The coefficients a_N are all equal to $1/(M + 1)$. Averaging is an inherently low-pass smoothing operation so the MA filter is always going to reduce high frequency content. Figure 16.10 shows the MA filter magnitude responses for windows lengths of 3, 5, 33, and 513 points corresponding to delay lengths of $M = 2, 4, 32,$ and 512 samples. You can see that as the number of averaged samples increases, the stop-band attenuation gets better, but the apparent LPF cutoff frequency becomes lower. Because of these issues, this type of filter is usually used only when cheap and inexpensive data point smoothing is required.

There are many more FIR design methods you may wish to investigate, including:

- numerical differentiation filters
- numerical integration filters
- median filters
- adaptive filters

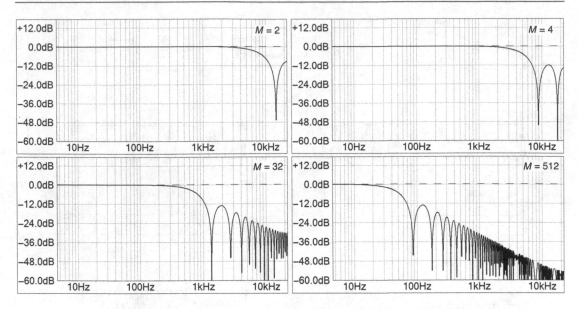

Figure 16.10: Magnitude responses of moving average FIR filters with delay lengths of M = 2, 4, 32, and 512 samples.

Some of these designs are actually quite interesting, especially the methods that rely on numerical analysis techniques, so don't necessarily discount them as being a useful addition to your arsenal of filter design techniques. You can find numerous sources in the Bibliography.

16.6 C++ DSP Function: freqSample

The *freqSample* function directly implements the frequency sampling calculations of Equation 16.4. It is derived from a long established and very old C-function. The function accepts an array of sampled frequency response points and finds the impulse response for the resulting FIR filter. The function has been left fundamentally intact. We will be sampling only frequency response curves with positive symmetry. You supply the IR length (N) the array of sampled frequency points (A), the output array to hold the impulse response (H), and the symmetry ($POSITIVE$).

```
#define NEGATIVE 0
#define POSITIVE 1

inline void freqSample(int N, double A[], double h[], int symm)
```

16.7 C++ DSP Function: calculateAnalogMagArray

Our chapter plugin requires an additional global function that calculates an array full of analog filter frequency response points. You supply the array and information about the analog filter, and it fills the array with sampled magnitudes of the basic analog filters. There is also a bonus function *calculateBrickwallMagArray* that constructs ideal filters using simple line segments, then frequency samples those responses.

16.7.1 calculateAnalogMagArray: *Enumerations and Data Structure*

This function requires a strongly typed enumeration to encode the analog filter type. It defines the common 1st and 2nd order filters:

```
enum class analogFilter {kLPF1, kHPF1, kLPF2, kHPF2, kBPF2, kBSF2};
```

A custom data structure is used for the function argument, allowing us to pass data into and out of the simple function easily. Note the *mirrorMag* Boolean flag. With this flag set, the function will generate a mirror image in the array, suitable for manipulating FFT magnitude data (be careful about that—see Chapter 20 for more information).

```
struct AnalogMagData
{
    AnalogMagData() {}

    analogFilter filterType = analogFilter::kLPF2;
    double* magArray = nullptr;
    unsigned int dftArrayLen = 0;
    double sampleRate = 44100.0;

    // --- for LPF, HPF fc = corner frequency
    //      for BPF, BSF fc = center frequency
    double fc = 1000.0;
    double Q = 0.707;
    bool mirrorMag = true;
};
```

16.7.2 calculateAnalogMagArray: *Calculations*

This function implements the very well-known and documented functions for exactly calculating typical analog frequency responses of filters. These equations are shown here.

$$f_o = \frac{f}{f_c} \qquad \zeta = \frac{1}{2Q}$$

1st order filters:

$$\left|H(s)_{LPF}\right| = \frac{1}{\sqrt{\left(1+f_o^2\right)^2}} \qquad \left|H(s)_{HPF}\right| = \frac{f_o}{\sqrt{\left(1+f_o^2\right)^2}}$$

(16.5)

2nd order filters:

$$\left|H(s)_{LPF}\right| = \frac{1}{\sqrt{\left(1-f_o^2\right)^2 + 4\zeta^2 f_o^2}} \qquad \left|H(s)_{HPF}\right| = \frac{f_o^2}{\sqrt{\left(1-f_o^2\right)^2 + 4\zeta^2 f_o^2}}$$

$$\left|H(s)_{BPF}\right| = \frac{2f_o\zeta}{\sqrt{\left(1-f_o^2\right)^2 + 4\zeta^2 f_o^2}} \qquad \left|H(s)_{BSF}\right| = \frac{\sqrt{\left(1-f_o^2\right)^2}}{\sqrt{\left(1-f_o^2\right)^2 + 4\zeta^2 f_o^2}}$$

16.8 *C++ DSP Object:* **LinearBuffer**

We provide a very simple *LinearBuffer* object that uses a smart pointer to maintain its own dynamically declared buffer array. You never need to bother with allocating or de-allocating memory, and the object will delete its own array during its destruction. Note that this is identical to the *CircularBuffer* object. We will store impulse responses in a linear buffer for convolving with our audio input signal, which will flow through a *CircularBuffer* object in order to easily implement the transversal convolution structure in Figure 16.4.

16.8.1 LinearBuffer: *Enumerations and Data Structure*

This object does not require any enumerations or a custom data structure, as it is a low-level object that we won't manipulate from a GUI. It is designed for use in aggregate DSP and effect objects. However, this object uses a class template in its definition so that it may be used to implement buffers of any data type; typically these will consist of *float* and *double* types.

16.8.2 LinearBuffer: *Members*

Tables 16.1 and 16.2 list the **LinearBuffer** members. First, note that the object is defined using a class template. This manifests itself in the declaration of the object, but also as function argument and return types.

```
template <typename T>
class LinearBuffer
```

Table 16.1: *LinearBuffer* **member variables**

LinearBuffer **Member Variables**		
Type	**Name**	**Description**
std::unique_ptr<T[]>	*buffer*	Smart pointer to a buffer
unsigned int	*bufferLength*	Total length of buffer as a nearest power of 2

Table 16.2: *LinearBuffer* **member functions**

LinearBuffer **Member Functions**		
Returns	**Name**	**Description**
void	*createLinearBuffer* Parameters: —*unsigned int* _bufferLength	Create or resize the linear buffer
void	*flushBuffer*	Clear the linear buffer values and set to 0.0
void	*writeBuffer* Parameters: —*T* input	Write the *input* value of type *T* into the buffer
T	*readBuffer* Parameters: —*int* index	Read the *buffer[index]* location

The smart pointer to the buffer is declared as a private member variable that points to an array of *T*-type values (*T[]*):

```
std::unique_ptr<T[]> buffer = nullptr;
```

So, you must specify the template data type when declaring or creating the object.

Static declaration of *double* buffer:
```
        LinearBuffer <double> irBuffer;
```

Dynamic declaration of *double* buffer:
```
        LinearBuffer <double>* irBuffer = new LinearBuffer <double>;
```

16.8.3 LinearBuffer: *Programming Notes*

The *LinearBuffer* object simply wraps a smart pointer buffer. The read and write functions just access locations within the smart buffer. Its implementation is simple and doesn't warrant reprinting here.

16.9 C++ DSP Object: ImpulseConvolver

The *ImpulseConvolver* object convolves the incoming audio with a pre-stored impulse response. The impulse response is set via a simple function that allows you to easily create a dynamic convolver whose impulse response changes over time—there is little CPU overhead as long as the IR length does not change as well.

16.9.1 ImpulseConvolver: *Enumerations and Data Structure*

The *ImpulseConvolver* object does not require special enumerations, constant declarations, or data structures. It is primarily targeted for use as a member object of a larger class that selects the impulse response and interacts with the user.

16.9.2 ImpulseConvolver: *Members*

Tables 16.3 and 16.4 list the *ImpulseConvolver* member variables and member functions.

Table 16.3: *ImpulseConvolver* **member variables**

ImpulseConvolver Member Variables		
Type	**Name**	**Description**
CircularBuffer<double>	*signalBuffer*	Circular buffer for signal
LinearBuffer<double>	*irBuffer*	Linear buffer for IR
unsigned int	*irLength*	Length of the IR—also length of convolution

Table 16.4: *ImpulseConvolver* member functions

ImpulseConvolver Member Functions		
Returns	Name	Description
bool	reset Parameters: —double sampleRate	Resets buffers
double	processAudioSample Parameters: —double xn	Convolves input xn with stored IR
void	init Parameters: —unsigned int lengthPowerOfTwo	Initializes both buffers to the length (must be a power of two)
void	setImpulseResponse Parameters: —double* irArray —unsigned int lengthPowerOfTwo	Sets the new IR for the convolution; IR is passed in array of double

16.9.3 ImpulseConvolver: *Programming Notes*

The *ImpulseConvolver* object implements textbook linear convolution using a simple for-loop. The signal (circular) buffer maintains the transversal flow through the structure and the linear buffer stores the IR.

Function: *Constructor*
Details: The constructor needs only to initialize itself to set up the circular and linear buffers and prepare them to receive input.

```
ImpulseConvolver() { init(512); }          /* C-TOR */
```

Function: *reset*
Details: The reset function simply flushes out the circular buffer to prepare for new data.

```
virtual bool reset(double _sampleRate)
{
    // --- flush signal buffer; IR buffer is static
    signalBuffer.flushBuffer();
    return true;
}
```

Function: *setImpulseResponse*
Details: This function stores the incoming impulse response, first testing the length to see if it needs to be changed. If so, it calls the functions on its buffers to allocate the memory.

```
void setImpulseResponse(double* irArray, unsigned int lengthPowerOfTwo)
{
    if (lengthPowerOfTwo != irLength)
    {
```

```
            irLength = lengthPowerOfTwo;
            signalBuffer.createCircularBufferPowerOfTwo(
                                          lengthPowerOfTwo);
            irBuffer.createLinearBuffer(lengthPowerOfTwo);
    }

    // --- load up the IR buffer
    for (unsigned int i = 0; i <irLength; i++)
            irBuffer.writeBuffer(i, irArray[i]);
}
```

Function: *processAudioSample*

Details: The processing function simply performs direct linear convolution on the input signal in a simple for-loop.

```
virtual double processAudioSample(double xn)
{
        double output = 0.0;

        // --- write buffer; x(n) overwrites oldest value
        //      this is the only time we do not read before write!
        signalBuffer.writeBuffer(xn);

        // --- do the convolution
        for (unsigned int i = 0; i < irLength; i++)
        {
                // --- y(n) += x(n)h(n)
                //      for signalBuffer.readBuffer(0) -> x(n)
                //              signalBuffer.readBuffer(n-D)-> x(n-D)
                double signal = signalBuffer.readBuffer((int)i);
                double ir = irBuffer.readBuffer((int)i);
                output += signal*ir;
        }

        return output;
}
```

16.10 C++ Effect Object: AnalogFIRFilter

As a fun FIR project that is quite over-engineered, I have created a FIR filter object that uses the frequency sampling method on sampled analog magnitude responses. This project is to show how the frequency sampling method works and **is not intended to be a useful analog-modeling filter**. However, due to the fact that its frequency response exactly matches that of the analog filter (at the sampling locations), you might use it as a measuring stick to examine and test other analog modeling filters. The idea is for you to generate your own magnitude arrays using whatever calculations you like—piecewise linear, polynomial approximations, etc.

16.10.1 AnalogFIRFilter: *Enumerations and Data Structure*

This object requires the strongly typed enumeration to define the analog filter for sending to the analog magnitude sampling function and is the same as in Section 16.7.1. The object is updated with a *AnalogFIRFilterParameters* custom data structure that holds the required parameters for operation. We define one constant that sets the length of the convolution (512 points).

```
const unsigned int IR_LEN = 512;

struct AnalogFIRFilterParameters
{
        AnalogFIRFilterParameters() {}
        AnalogFIRFilterParameters& operator=(...){...}

        // --- individual parameters
        analogFilter filterType = analogFilter::kLPF1;
        double fc = 0.0;
        double Q = 0.0;
};
```

16.10.2 AnalogFIRFilter: *Members*

Tables 16.5 and 16.6 list the *AnalogFIRFilter* member variables and member functions.

Table 16.5: *AnalogFIRFilter* **member variables**

AnalogFIRFilter Member Variables		
Type	Name	Description
AnalogFIRFilterParameters	*parameters*	Filter parameters, f_c, Q, boost/cut and algorithm
ImpulseConvolver	*convolver*	The convolver object
double	*analogMagArray[IR_LEN]*	Array to store analog magnitude samples
double	*irArray[IR_LEN]*	Array to store impulse response
double	*sampleRate*	Current sample rate

Table 16.6: *AnalogFIRFilter* **member functions**

AnalogFIRFilter Member Functions		
Returns	Name	Description
AnalogFIRFilterParameters	*getParameters*	Get all parameters at once
void	*setParameters* Parameters: —*AnalogFIRFilterParameters* _parameters	Set all parameters at once
bool	*reset* Parameters: —*double* sampleRate	Flush the signal buffer
double	*processAudioSample* Parameters: —*double* xn	Convolve one input sample with the IR

16.10.3 AnalogFIRFilter: *Programming Notes*

The *AnalogFIRFilter* object is super simple. Its *ImpulseConvolver* does the work, but this object is responsible for performing the frequency sampling operation: you will want to modify the *setParameters* method to insert your own code and frequency sample any magnitude array you like. Using the *AnalogFIRFilter* object is straightforward:

1. Reset the object to prepare it for streaming.
2. Set the parameters in the *AnalogFIRFilterParameters* custom data structure.
3. Call *setParameters* to update the calculation type.
4. Call *processAudioSample*, passing the input value in and receiving the output value as the return variable.

Function: *reset*
Details: Initializes the impulse convolver and stores the sample rate, which is needed to perform the analog magnitude sampling.

```
virtual bool reset(double _sampleRate)
{
    sampleRate = _sampleRate;
    convolver.reset(_sampleRate);
    convolver.init(IR_LEN);
    return true;
}
```

Function: *setParameters*
Details: The *setParameters* function checks to see if the analog filter parameters have changed; if so, it recalculates the matched analog magnitude response, and then calls the *freqSamp* method to perform the frequency sampling operation. Due to the CPU intensive nature, this function should be called only once per buffer cycle, and not during audio frame processing.

```
void setParameters(AnalogFIRFilterParameters _parameters)
{
    if (_parameters.fc != parameters.fc ||
        _parameters.Q != parameters.Q ||
        _parameters.filterType != parameters.filterType)
    {
        // --- set the filter IR for the convolver
        AnalogMagData analogFilterData;
        analogFilterData.sampleRate = sampleRate;
        analogFilterData.magArray = &analogMagArray[0];
        analogFilterData.dftArrayLen = IR_LEN;
        analogFilterData.mirrorMag = false;

        analogFilterData.filterType = _parameters.filterType;
        analogFilterData.fc = _parameters.fc; // 1000.0;
        analogFilterData.Q = _parameters.Q;
```

```
                  // --- calculate the analog mag array
                  calculateAnalogMagArray(analogFilterData);

                  // --- frequency sample the mag array
                  freqSample(IR_LEN, analogMagArray, irArray, POSITIVE);

                  // --- update new frequency response
                  convolver.setImpulseResponse(irArray, IR_LEN);
         }

         parameters = _parameters;
}
```

Function: *processAudioSample*
Details: The *processAudioSample* method simply calls the impulse convolver's method and lets it handle all of the hard work.

```
virtual double processAudioSample(double xn)
{
         // --- do the linear convolution
         return convolver.processAudioSample(xn);
}
```

16.11 Chapter Plugin: AnalogFIR

We will use the *AnalogFIRFilter* object directly for this chapter plugin. We will re-investigate the Optimal/Parks-McClellan method for sample rate conversion use later, in Chapter 21. Due to the heavy CPU use that is incurred through both the frequency sampling and the linear convolution, we will make a few changes to this plugin.

- We will build the project in the *Release* configuration to optimize speed; building for the *Debug* configuration will cause glitches and CPU overloads (unless you have a blazing-fast CPU).
- The plugin will be monaural only—feel free to add another convolver and make it work in stereo if you would like (and you have a fast CPU).
- We will update the GUI parameters only once per buffer processing cycle, and not on every sample period (even with parameter smoothing engaged) as the CPU overhead is extreme.

We will re-investigate this plugin again when we use fast convolution in Chapter 20 to implement sample rate conversion, in which case we may run longer filters in stereo without incurring too much CPU overhead. The GUI and block diagrams are shown in Figure 16.11.

16.11.1 AnalogFIR: *GUI Parameters*

The GUI parameter table is shown in Table 16.7—use it to declare your GUI parameter interface for the plugin framework you are using. If you are using ASPiK, remember to first create your enumeration of control ID values at the top of the *plugincore.h* file and use automatic variable binding to connect the linked variables to the GUI parameters.

Table 16.7: AnalogFIR plugin's GUI parameter and linked-variable list

Control Name	Units	Min/max/default or string-list	Taper	Linked Variable	Linked Variable Type
fc	Hz	20 / 20480 / 1000	1V/oct	filterFc_Hz	double
Q	-	0.707 / 20 / 0.707	linear	filterQ	double
Filter Type		string-list follows this table	-	filterType	int

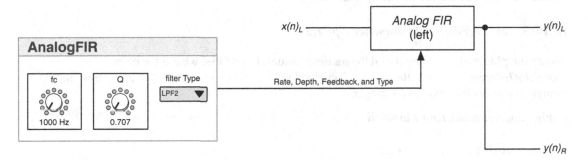

Figure 16.11: The *AnalogFIR* plugin processes only the left channel and splits the output to the right to conserve CPU cycles; the plugin uses the magnitude responses of basic analog filters to apply to the frequency sampling method.

ASPiK: top of *plugincore.h* file

```
enum controlID {
    filterFc_Hz = 0,
    filterQ = 1,
    filterType = 2
};
```

The string-list parameter for the filter type exactly matches the ordering of the *analogFilter* strongly typed enumeration:

```
"LPF1, HPF1, LPF2, HPF2, BPF2, BSF2"
```

16.11.2 AnalogFIR: *Object Declarations and Reset*

The *AnalogFIR*plugin is monaural, to conserve CPU time, and needs only one member plus the *updateParameters* method:

```
// --- in your processor object's .h file
#include "fxobjects.h"

// --- in the class definition
protected:
      AnalogFIRFilter filter;
      void updateParameters();
```

The analog magnitude function requires the proper sample rate for calculation. Call the *reset* function and pass in the current sample rate in your plugin framework's prepare-for-audio-streaming function. In this example, *resetInfo.sampleRate* holds the current DAW sample rate.

ASPiK: *reset( )*

```
// --- reset the filters
filter.reset(resetInfo.sampleRate);
```

16.11.3 AnalogFIR: GUI Parameter Update

Update the plugin object using the GUI parameter values. For ASPiK we use the helper *convertIntToEnum* to convert the user's selected value (a zero-indexed integer) to the corresponding strongly typed enumeration as an integer.

ASPiK: *updateParameters( )* function

```
// --- update the object
AnalogFIRFilterParameters params = filter.getParameters();
params.fc = filterFc_Hz;
params.Q = filterQ;
params.filterType = convertIntToEnum(filterType, analogFilter);
filter.setParameters(params);
```

16.11.4 AnalogFIR: Process Audio

Add the code to process the plugin object's input audio samples through the left channel object—then copy it to the right output. The ASPiK code is shown here for grabbing the input samples and writing the output samples; use the method that your plugin framework requires to read and write these values.

ASPiK: *processAudioFrame( )* function

```
// --- read input (adapt to your framework)
double xnL = processFrameInfo.audioInputFrame[0]; //< input sample L

// --- process the audio to produce output
double ynL = filter.processAudioSample(xnL);

// --- write output (adapt to your framework)
processFrameInfo.audioOutputFrame[0] = ynL; //< output sample L
processFrameInfo.audioOutputFrame[1] = ynL; //< output sample R = L
```

Rebuild and test the plugin making sure to use the *Release* configuration. While these filters do have the same magnitude as their analog counterparts, they are also linear phase. Compare the outputs of this plugin with a fixed f_c and Q to the other analog modeling filters. Can you perceive a difference it the outputs?

16.12 Homework

1. Design the MA filter and allow the user to control the width of the averaging window.
2. Research *integration* and *differentiation* FIR filters and write code to implement them.

16.13 Bibliography

Farina, A. 2007. *Advancements in Impulse Response Measurements by Sine Sweeps*, 122nd Audio Engineering Society Convention.

Ifeachor, E. C. and Jervis, B. W. 1993. *Digital Signal Processing: A Practical Approach*, Chap. 4, 6. Menlo Park: Addison Wesley.

Kwakernaak, H. and Sivan, R. 1991. *Modern Signals and Systems*, Chap. 3, 9. Eaglewood Cliffs: Prentice-Hall.

Lindquist, C. 1977. *Active Network Design*, Chap. 2–6. Miami: Steward and Sons.

Lindquist, C. 1989. *Adaptive and Digital Signal Processing*, Chap. 10. Miami: Steward and Sons.

Oppenheim, A. V. and Schafer, R. W. 1999. *Discrete-Time Signal Processing, 2nd Ed.*, Chap. 7. Eaglewood Cliffs: Prentice-Hall.

TFilter Design Tool, http://t-filter.engineerjs.com/, Accessed August 1, 2018.

CHAPTER 17

Reverb Effects

Reverb effects have an appeal that seems universal, perhaps because we live in a reverberant world. Our ears are time-integrating devices that use time domain cues and transients for information, so we are sensitive to anything that manipulates these cues. In this chapter, we discuss reverb algorithms. There are two general ways to create the reverberation effect:

- reverb by direct convolution—the physical approach
- reverb by simulation—the perceptual approach

In the physical approach, the impulse response of a room is convolved with the input signal in a long FIR filter. For large rooms, these impulse responses could be 5 to 10 seconds in length. In the mid 1990s, Gardner developed a hybrid system for fast convolution that combined direct convolution with block FFT processing (see Section 20.7). Around the same time, Reilley and McGrath described a new commercially available system that could process 262,144-tap FIR filters for convolving impulses more than 5 seconds in length. Even using the most efficient fast convolution methods, the required processing power is often too large to be practical in a plugin. Aside from the computing expense, another drawback to this approach is that an impulse response is frozen in time and measures the room at one location only, under a certain set of conditions. A reverb unit (or *reverberator*) that is general enough to provide many different reverbs of different spaces would require a large library of IR files. In addition, its parameters can't be easily adjusted.

The perceptual approach aims to simulate the reverberation with enough quality to fool the ears and give the same perception of real reverb, but with a much lower signal processing cost. The advantages are numerous from the minimal processing required to the ability to vary many parameters in real time. There are several key engineers who developed much of the theory still in use today. These include Schroeder's initial work in the early 1960s with continued research and contributions from Moorer, Griesinger, Gerzon, Gardner, Jot, Chaigne, Smith, Roscchesso, and others across the decades. Much of this chapter is owed to their work in the field. We will focus on the perceptual approach and try to find computationally efficient algorithms for interesting reverberation effects. Griesinger states that it is impossible to perfectly emulate the natural reverberation of a real room and thus the algorithms will always be approximations. It seems that the area of reverberation design has the most empirically derived or trial-and-error research of just about any audio signal processing field. There is no single "correct" way to implement a reverb algorithm, so this chapter focuses on giving you many different reverberator modules to experiment with.

17.1 Anatomy of a Room Impulse Response

The first place to start is by examining impulse responses of actual rooms. There are several popular methods for capturing the impulse response, from cap pistols and balloons to de-convolution of chirp signals and pseudo-random sequences. The resulting time domain plot is useful for investigating the properties of reverberation. For a purely digital plugin, the impulse response may be taken by feeding a single sample value of +1.0 followed by a stream of zeros.

Figure 17.1 shows the impulse response plots for two very different spaces: a large concert hall and a cathedral. The initial impulse is followed by a brief delay called the *pre-delay*. As the impulse pressure wave expands, it comes into contact with the nearby structures—walls, floor, and ceiling—and the first echoes appear. These initial echoes called *early reflections* are important to the simulation of reverb because of the auditory cues we get from them. The pressure wave continues to expand and more reflections occur, with reflected signal upon reflected signal piling on top of each other while decaying in energy. The resulting reverb "tail" is called *late reverberation* or *subsequent reverberation*.

The top impulse response in Figure 17.1 is a modern concert hall designed for a pleasing reverberant quality. The initial reflections are from irregularly shaped back and side walls and they pile up in amplitude; they are also piling up in density shown as the impulse gets "fatter" in the middle section. The dense reverberant tail follows, which decays in an exponential fashion as the seats and acoustical room treatments absorb the sound pressure energy. The cathedral is a different story. After a short pre-delay, a few large early reflections arrive from the nearby structures but they don't increase in amplitude or density in the same way as the hall. There is a blurry line between when the reflections end and the reverb tail starts. The reverb's decay is also much longer in time as there is little in the cathedral to absorb the sound pressure wave energy. Figure 17.2's block diagram shows the three components we observe in the impulse responses from Figure 17.1 and represents one possible way to

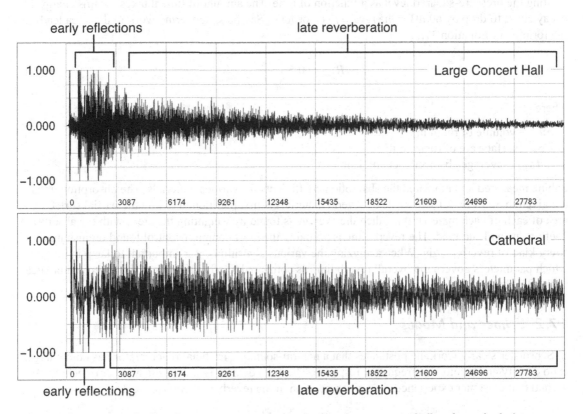

Figure 17.1: The impulse response measurements for a large concert hall and a cathedral.

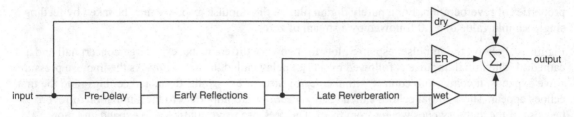

Figure 17.2: A possible reverb algorithm block diagram that simulates each component in the final room impulse response.

model the reverberant space. However, there is substantial debate as to whether or not this is a good way to approach reverb design. Some feel that the reverb algorithm should not need to separate out the early reflections from the late reverberation. In other words, a good algorithm will create all of the reverberation aspects at once.

17.1.1 RT$_{60}$

The most common measurement for reverb is the RT_{60} reverb time. Reverb time is measured by first stimulating the room into a reverberant state using and audio source, then turning off the source and plotting the pressure-squared level as a function of time. The amount of time it takes for this energy decay curve to drop by 60 dB is the *reverb time,* or RT_{60}. Sabine's pioneering work in this area leads to the formula in Equation 17.1.

$$RT_{60} = 0.5 \frac{V_R}{S_R A_{RAve}} \tag{17.1}$$

where

V_R = volume of room in ft^3
S_R = surface are of room in ft^2
A_{RAve} = average absorption coefficient

Sabine measured and tabulated the absorption coefficients for various materials. The absorption units are given in *Sabines*. A room made of several materials is first partitioned to find the partial surface area of each of each material type, then the average is found by weighting the areas with the absorption coefficients and summed. The reverb time is probably the most common control found on just about every kind of reverb plugin. When analyzing the various algorithms, it is important to keep in mind which parameters govern the overall reverb time as compared to other aspects of the reverberator such as the density of echoes or the frequency response.

17.2 Echoes and Modes

In Schroeder's early work, he postulates that a natural-sounding artificial reverberator has both a large echo density and a colorless frequency response. The echo density is simply the number of echoes per second that a listener experiences in a given position in the reverberant environment.

$$E_D = \frac{\text{echoes}}{\text{second}} \tag{17.2}$$

If the echo density is too low, the ear discerns the individual echoes and a fluttering sound is the result. As the echo density increases, the echoes fuse together to produce a wash of sound. Schroeder postulated that the echo density should be at least 1,000 echoes/second for a natural sounding reverb. Greisinger recommends 10,000 echoes/second instead. Statistically, the Echo Density is shown in Equation 17.3, which reveals that the echoes build up as the square of time.

$$\text{Echo Density} = \frac{4\pi c^3}{V_R} t^2 \tag{17.3}$$

where

    c = speed of sound
    t = time
    V_R = volume of room

Reverb is often modeled statistically as decaying white noise, which implies that ideal reverberant rooms have flat frequency responses. A room's geometry can be used to predict its frequency response. The dimensions of the room can be used to calculate the frequencies that will naturally resonate as they bounce back and forth between parallel surfaces. An untreated rectangular room will have multiple resonances and anti-resonances. Given a rectangular room with length, width, and height of l, w, and h, the resonances are found with Equation 17.4, a well-known and useful equation for predicting room resonances, also called *modes* (Beranek, 1986).

$$f_e = \frac{c}{2} \sqrt{\left(\frac{n_x}{l}\right)^2 + \left(\frac{n_y}{w}\right)^2 + \left(\frac{n_z}{h}\right)^2} \tag{17.4}$$

where

    n_x, n_y, n_z = half wave numbers 0, 1, 2, 3. . .

    l, w, h = length(x), width(y) and height(z) of the room

Above a certain frequency, the resonances overlap and fuse together. Each resonant frequency has its own envelope whose curve is bell shaped (like a band-pass filter), meaning that it has a peak resonance at the mode frequency, but is still excitable at frequencies around it as well and thus the Quality factor (Q) of the curve relates to this excitability. The number of resonant peaks increases with frequency. A room with a good reverberation response will have many resonant peaks which all add together to create that wash of white noise sound.

The *modal density* is the number of resonant peaks per Hz in the room's frequency response. Physicists call these resonant frequencies or modes *eigenfrequencies* (note this is not an acoustics-specific term; an eigenfrequency is the resonant frequency of any system). Schroeder's second postulation is that for a colorless frequency response, the modal density should be 0.15 eigenfrequencies/Hz. or 1 eigenfrequency every 6.67 Hz. or approximately 3,000 resonances spread across the audio spectrum. So, it makes sense that good reverberant environments have interesting geometries with many non-parallel walls. The many paths an impulse can take from the source to the listener create the multitude of resonances. Kuttruff derived the approximation for the modal density as it relates to the volume of the room and modal frequency. Equation 17.5 shows that the resonances build up as the square of frequency.

$$D_m = \frac{4\pi V_R}{c^3} f_m^2 \tag{17.5}$$

where

    V_R is the volume of the room
    f_m is the modal frequency in question

Schroeder's rules for natural reverb:

- **echo density:** at least 1,000 echoes/sec (Greisinger: 10,000 echoes/sec)
- **modal density:** at least 0.15 eigenfrequencies/Hz

In physical rooms we know that:

- echo density increases with the square of time.
- modal density increases with the square of frequency.

The *Energy Decay Relief* plot (EDR) shows how the energy decays over both frequency and time for a given impulse response of a room. Figure 17.3 shows a very simple, fictitious EDR of a room with a poor frequency response; notice that the frequency axis comes out of the page; low frequencies are in the back. It also shows that this room has a resonant peak, which forms almost right away in time. In an EDR, the modal density is shown across the *z*-axis (frequency) while the echo density appears across the *x*-axis (time). This simplified plot shows just one resonance and no echo density build up for ease of viewing.

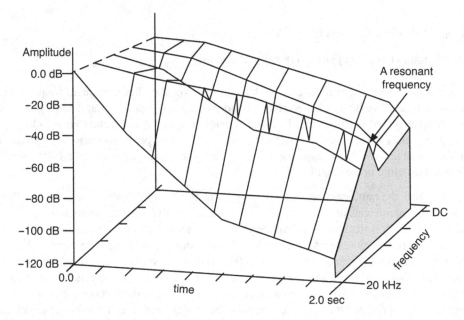

Figure 17.3: The EDR shows time (*x*-axis), frequency (*z*-axis), and amplitude (*y*-axis) of the energy decay of a room; here we can see a resonant frequency ridge that forms early on and persists throughout the duration of the measurement.

Figure 17.4 shows the EDR of an actual room, in this case a theater. The eigenfrequencies are visible as the ridges that run perpendicularly to the frequency axis.

Figure 17.5 shows the EDR for a cathedral. In this case, the echo pile-ups are clearly visible running perpendicular to the time axis. Comparing the EDRs shows that both rooms have high echo and modal

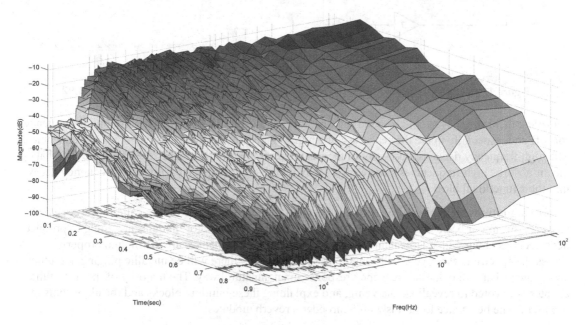

Figure 17.4: The EDR for a theater.

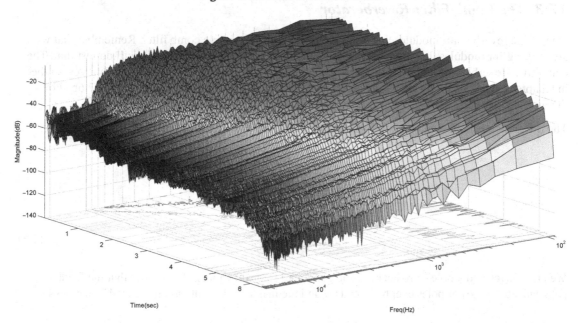

Figure 17.5: EDR of a cathedral.

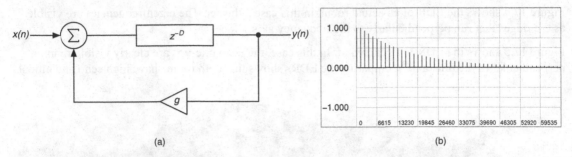

Figure 17.6: (a) A basic comb filter. (b) Its impulse response with $D = 30$ mSec and $g = 0.92$.

densities; therefore they should be good reverberant spaces. Both EDRs show an initial high frequency roll-off and, especially in the theater's case, the high frequencies decay faster than the low frequencies. The high frequency decay is a property of the treatment of the room surfaces along with the fact that the air absorbs high frequencies more easily. Many reverb algorithms factor this high frequency decay into their functional blocks.

Reverb algorithms are typically made of arrangement of these functional blocks called *reverberator modules*. From our basic observations we can tell that good reverberator modules are going to produce dense echoes along with a large number of resonances. If the resonances are distributed properly across the spectrum, the reverb will sound flat. If they are not, there will be metallic pinging and other annoyances that will color the frequency response in an unnatural way. The majority of the rest of the chapter is devoted to revealing, analyzing, and explaining these building blocks and the algorithms that use them. The best place to start is with Schroeder's reverb modules.

17.3 The Comb Filter Reverberator

One of the reverberator modules that Schroeder proposed is a simple comb filter. Remember that we are looking for modules that produce echoes—and the comb filter with feedback will do just that. The comb filter reverberator looks familiar because it's also the delay with feedback algorithm you studied in Chapter 14. Figure 17.6a shows the comb filter with feedback control g while Figure 17.6b shows the module's impulse response with a delay time of 30 mSec and feedback value $g = 0.92$.

The difference equation and transfer function for the comb filter are as follows.

$$y(n) = x(n-D) + gy(n-D)$$

and

$$H(z) = \frac{z^{-D}}{1 - gz^{-D}} \tag{17.6}$$

We also performed a pole–zero analysis and generated frequency plots. We showed that the feedback path caused a series of poles evenly spaced around the inside of the unit circle. Figure 17.7a shows

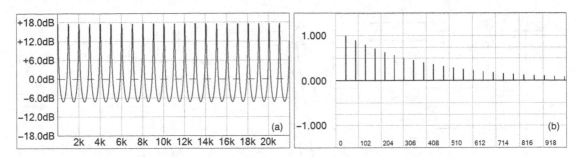

Figure 17.7: (a) The frequency response and (b) impulse response of a comb filter with D = 1 mSec, f_s = 44.1 kHz and g = 0.98.

the frequency and impulse responses for a comb filter with D = 1 mSec delay time and f_s = 44.1 kHz. This results in 44 poles and 44 zeros spread around the unit circle and 22 evenly spaced peaks in the frequency response. The resulting impulse response in Figure 17.7b is easy to predict as the echo re-circulates through the feedback path, being attenuated by multiplied by g each time through the loop.

While Figure 17.7 might look simple, the results certainly trend in the right direction. The frequency response plot shows a set of resonant peaks that could correspond to a room's eigenfrequencies and the impulse response shows a set of decaying impulses, mimicking the energy loss as the echoes bounce off of surfaces. Jot defines the modal density equation for a comb filter as:

$$M_d = \frac{D}{f_s}$$

$$\Delta f = \frac{f_s}{D}$$

(17.7)

where

D = the delay length
f_s = the sample rate

So, the comb filter produces D resonances across DC to f_s (or $D/2$ resonances from DC to Nyquist) with each resonance separated by $\Delta f = f_s/D$. This is exactly what we found when we analyzed the comb filter in Chapter 14. In order to get Schroeder's desired 0.15 eigenfrequencies/hz from DC to Nyquist at f_s = 44.1 kHz, you would need D = 6,622. This also makes sense because we need about 3,000 ($D/2$) resonances to cover the range from DC to Nyquist. The problem is that while we will get the desired number of resonances, they will be linearly spaced rather than piling up as the square of frequency. Additionally, it will take 6,622 samples (~150 mSec with f_s = 44.1 kHz) before the reverberator begins outputting samples. Schroeder was able to achieve high density by placing comb filters in parallel, then summing the results. This would then cause the comb filter responses to overlap. However, care must be taken when choosing the delay lengths: if they are mathematically related, then resonances or anti-resonances will color the sound.

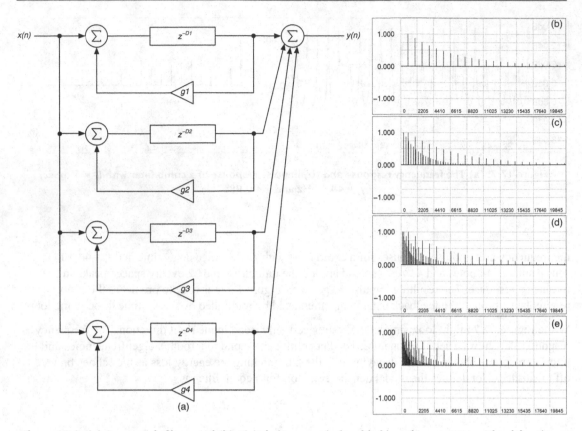

Figure 17.8: (a) Four-comb filters and (b)—(e) their successively added impulse responses. The delay times range from 1.7 mSec to 16 mSec; all feedback values g1–g4 are set identically at 0.86.

Figure 17.8a shows an example of four-comb filters in parallel. Each has its own delay time ($D1$–$D4$) and gain ($g1$–$g4$). The impulse responses in Figure 17.8b–e represent the accumulative outputs of each comb filter in succession. The delay times are $D1$ = 16 mSec, $D2$ = 6.5 mSec, $D3$ = 3.0 mSec, and $D4$ = 1.7 mSec. The feedback values $g1$–$g4$ are all identically set at 0.86. The combination of the impulse responses reveals a fundamental problem with setting the feedback values identically. Each set of impulses decays with the same feedback percentage, which means that the shorter comb filters will decay down to zero more quickly than the longer filters. You can see that in Figure 17.8e where four separate decay envelopes can be identified. This attempt at a reverberator will fail miserably as the resulting sound will be metallic and unnatural. We might also observe that the impulse response is one-sided. All of the impulses have positive values, whereas the real room impulse responses in Figure 17.1 include both positive and negative going pulses.

In order to produce a more realistic reverberator, we need to set the comb filter feedback g values so that the impulse envelopes decay at the same rate across the different comb filter delay lengths. Kahrs derived the feedback g value equation given the comb filter delay length D and the desired RT_{60} time shown in Figure 17.8.

$$RT_{60} = \frac{3DT_s}{\log(1/g)}$$

or

$$1/g = 10^{\frac{3DT_s}{RT_{60}}} \qquad (17.8)$$

or

$$g = 10^{\frac{-3DT_s}{RT_{60}}}$$

where T_s is the sample rate.

Equation 17.8 shows that we can control the reverb time by using the feedback gain factor *or* the comb filter delay length D. The tradeoff is that if we increase g to increase reverb time, the poles get very near the unit circle causing resonances. If we increase D, then the echoes become distinctly audible rather than smearing together. This means that we must make a tradeoff of the modal density vs. the echo density for the design. Figure 17.9a shows the effect of applying Equation 17.8 to each comb filter for a RT_{60} time of 500 mSec while Figure 17.9b shows the effect of alternating the sign of every *other* comb filter output (multiplying by −1.0, which flips the phase by 180 degrees). Now we are starting to get closer to those real room impulse responses we are trying to simulate, but it is clear that the echo density needs to increase as we can see visual gaps in the impulse responses in Figure 17.9.

The Jot defines the modal density (M_d) for a parallel bank of N comb filters as:

$$M_d = \sum_{i=0}^{N} D_i T_s$$

where

D_i = the delay length of the ith comb filter
T_s = the sample period

or

$$M_d = N\overline{D}T_s \qquad (17.9)$$

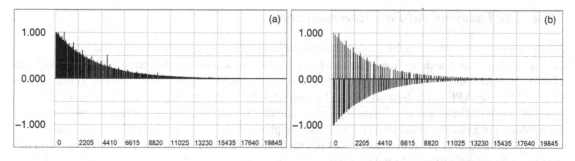

Figure 17.9: (a) The impulse response for the structure in Figure 17.8a with the comb filter feedback g values adjusted for 500 mSec of reverb time. (b) Flipping the phase of the output of every other comb filter results in a more symmetric impulse response.

where

$\bar{D}$ = the mean delay length of all filters averaged

From Equation 17.9 you can see that the modal density is not a function of frequency—which is what happens in real rooms where the modal density increases with the square of the frequency.

The Echo Density (E_d) for the parallel combs is given as:

$$E_d = \sum_{i=0}^{N} \frac{1}{D_i T_s}$$

where

D_i = the delay length of the ith comb filter

or

$$E_d = \frac{N}{\overline{D} T_s} \tag{17.10}$$

where

$\bar{D}$ = the mean delay length of all filters averaged

Knowing the desired M_d and E_d, you can then calculate the number of parallel comb filters N with average delay time DT_s as:

$$N = \sqrt{E_d M_d}$$
$$\overline{DT}_s = \sqrt{M_d / E_d} \tag{17.11}$$

Plugging Schroeder's values of M_d = 0.15 eigenfrequencies/Hz and E_d = 1,000 echoes/sec into Equation 17.11 yields N = 12 and the average delay time DT_s = 12 mSec. Then, the final values for the feedback gains g may be calculated for each module using Equation 17.8. Schroeder's reverb design includes only four parallel comb filters rather than the 12 that would be predicted here. He was able to increase the M_d and E_d by further smearing out the comb filter echoes with another structure: a delaying all-pass filter.

17.4 The Delaying All-Pass Reverberator

A Dth order all-pass filter may be implemented using the block diagram in Figure 17.10a. In this design, the length of the delay line is the order of the filter and a single coefficient g controls all aspects of the filter operation. For a 1st order filter, g controls the 90-degree phase shift frequency f_o just like the 1st order APF did in Section 12.4.1. For higher order filters, g controls the steepness of the phase shift of $90*D$ degrees at a frequency f_o = Nyquist/D. For audio designs that require independent manipulation of both steepness of phase shift and the phase shift frequency, the bilinear transform version in Section 11.3.2.7 is better suited. It is straightforward to prove that the difference equation of the Dth order all-pass filter is found as follows.

$$y(n) = -gx(n) + x(n-D) + gy(n-D) \tag{17.12}$$

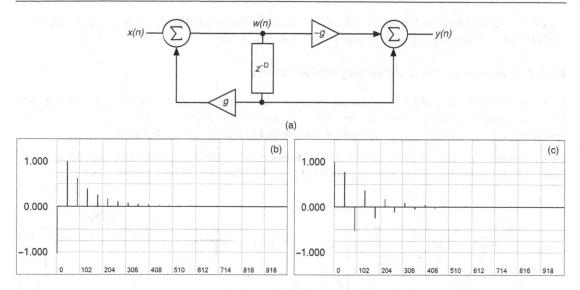

Figure 17.10: (a) a *D*th order all-pass filter structure. The impulse response for *N* = 45 with (b) *g* = +0.7 and (c) *g* = −0.7.

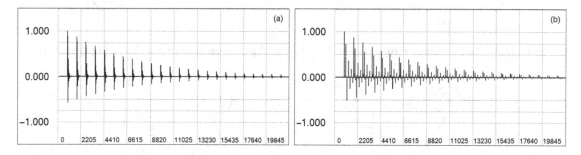

Figure 17.11: A comb filter followed by a delaying all-pass filter with (a) a 1 mSec and (b) 3 mSec all-pass delay lengths. The all-pass feedback *g* value is −0.7 in both cases.

You can download this proof along with the others from this chapter at http://www.willpirkle.com/derivations/.

For reverb designs, we are interested in the impulse response of this structure. By observation of the structure we can see that for positive values of *g*, the impulse response will consist of a downward pulse of strength −*g*, followed by a series of decaying pulses that fall off at an exponential rate of g^M where *M* is the number of trips through the feedback loop, shown in Figure 17.10b. For negative values of *g*, the polarity of the pulses alternates between positive and negative as shown in Figure 17.10c. The reverb version of this filter is called the *delaying all-pass* reverberator.

Figure 17.11 shows the series combination of a comb filter followed by a delaying all-pass structure. The comb filter length is 20 mSec with a feedback value *g* that corresponds to a 1-second RT_{60} time; the all-pass delay feedback value *g* is −0.7. In Figure 17.11a the all-pass delay length is 1 mSec, and in

Figure 17.11b the length is 3 mSec. We observe that each of the original comb filter pulses is further smeared out and the echo density increases dramatically as a result.

17.4.1 *Alternate and Nested Delaying APF Structures*

There are two more delaying APF structures that yield identical results as the one in Figure 17.11a and these derivations are also available at http://www.willpirkle.com/derivations/ . Figure 17.12 shows the two structures with an interesting location for the *x(n)* and *y(n)* nodes on the left side.

These structures are easy to nest. Nesting all-pass filters involves placing a second all-pass filter in series with the delay line of the first, as shown in Figure 17.13a. Each all-pass section has its own

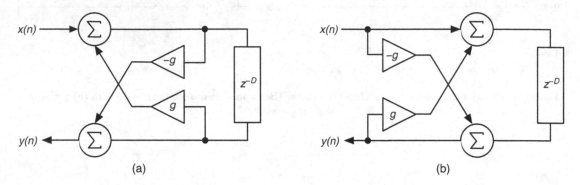

Figure 17.12: Two alternate structures for the delaying all-pass reverberator are shown in (a) and (b).

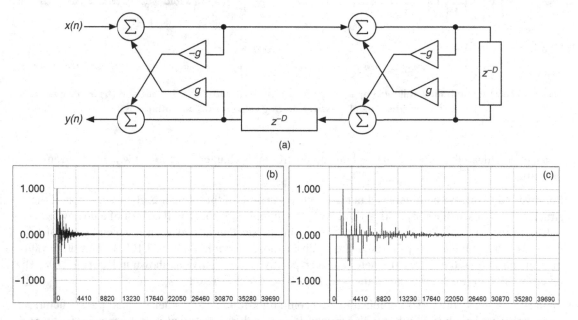

Figure 17.13: (a) A nested all-pass reverberator structure with impulse responses for delay lengths as follows: (b) 7 mSec and 4 mSec; and (c) 29 mSec and 22 mSec. Both feedback g values are 0.6.

delay length and feedback g value. The nested version produces an impulse response that is different from either the ordinary series or parallel combination of the two structures. Figure 17.13b shows the impulse response with relatively short delay lengths of 4 mSec to 7 mSec while Figure 17.13c shows the impulse response with longer delays of 22 mSec to 29 mSec. In both cases we get more smearing and echo pile-ups from a simple structure that uses very little memory.

17.5 Schroeder's Reverberator

Schroeder combined a parallel four-comb filter bank with two all-pass filters to create his first design. The comb filters produce the long series of echoes and the all-pass filters multiply the echoes, overlaying their own decaying impulse response on top of each comb echo. The resulting reverberation unit in Figure 17.14 sounds fairly good, considering that it is so simple to implement. Many consider it to be poor compared to current designs but we can certainly glean important information about its structure and performance.

Figure 17.15 traces the impulse response through each stage of Schroder's reverb. You can see that the output of every other comb filter has its phase flipped. It is also interesting to notice just how much more the echo density increases as the signal passes through the two all-pass structures on the output. Schroeder suggests that the ratio of smallest to largest delay in the comb filters should be about 1:1.5 and originally specified a range of 30 mSec to 45 mSec in total. The 1:1.5 ratio rule turns out to be useful for just about any set of parallel comb filters.

The comb filters have the following properties.

*   Choose the delays to have the 1:1.5 ratio.
*   Choose delay times that have no common factors or divisors.
*   Set the comb feedback g values according to Equation 17.8.

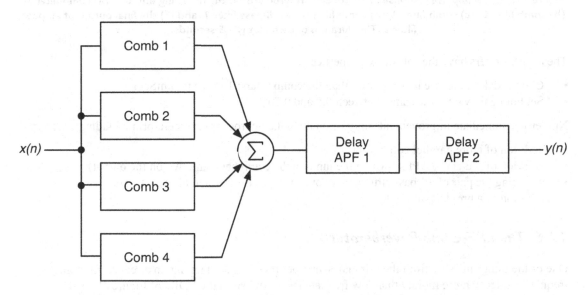

Figure 17.14: Schroeder's reverberator.

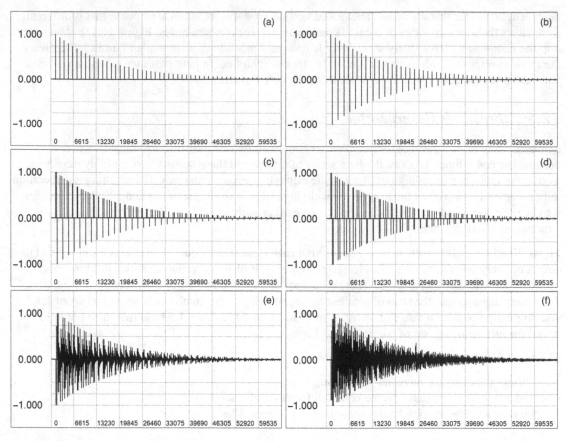

Figure 17.15: The impulse response of Schroeder's reverberator at the following outputs: (a) comb filter 1, (b) comb filter 2, (c) comb filter 3, (d) comb filter 4, (e) all-pass filter 1, and (f) the final output of all-pass filter 2. The duration of each plot is 1.5 seconds.

The all-pass filters have the following properties.

- Choose delays that are much shorter than the comb filters: 1 mSec to 5 mSec.
- Set both gain values the same: between 0.5 and 0.707.

Numerous immediate improvements may be made to the Schroeder reverberatorm including:

- addition of more parallel comb filters
- addition of pre-comb APFs (two on the input to the comb bank and two on the output)
- changing the pair of all-pass structures to one or two sets of nested all-pass filters
- addition of a pre-delay module

17.6 The LPF-Comb Reverberator

One of the things missing from the original Schroeder reverb is the fact that in a real room, high frequencies decay more rapidly than low frequencies, as shown in the EDRs of Figures 17.4–17.5. Placing a low-pass filter in the comb filter's feedback path will roll off the HF content of successive

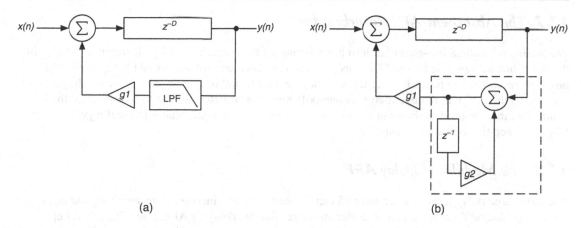

(a)   (b)

Figure 17.16: (a) A comb filter with a low-pass filter embedded in the feedback path. (b) A simple realization of the structure; the dotted line shows the one-pole low-pass filter.

echoes exponentially, which is closer to what we want. The LPF-Comb reverberator block diagram is shown in Figure 17.16a and a simple implementation of it using a one-pole low-pass filter is shown in Figure 17.16b.

In this realization, a second feedback value *g2* is specified for the low-pass filter while the comb filter's feedback is called *g1*. Finding the difference equation of the resulting structure in Equation 17.13 is somewhat tricky, but you can find the complete line-by-line derivation at http://www.willpirkle.com/derivations/.

$$y(n) = x(n-D) - g_2 x(n-D-1) + g_2 y(n-1) + g_1 y(n-D) \qquad (17.13)$$

In order for the filter combination to remain stable (after all, it is a massive feedback system) g_1 and g_2 can be related as:

$$g = \frac{g_2}{1-g_1} \qquad (17.14)$$

where

$g < 1.0$

Because the RT_{60} determines the pole radii, and therefore the value of g_1, you can rearrange Equation 17.14 as follows.

$$g_2 = g(1-g_1) \qquad (17.15)$$

where

$g < 1.0$

This means that we may give the user control over the low-pass filter with a GUI knob or slider that adjusts from 0.0 to 0.9999, and we might label it "damping"—there is no need to let the user attempt to adjust the filter by its cutoff frequency as this will surely be done by ear during use.

17.7 The Absorbent-APF Reverberator

We can likewise add a low-pass filter into the delaying all-pass reverberator by placing it in series with the delay line as shown in Figure 17.17a as prescribed by Dahl and Jot and named the *Absorbent All-pass reverberator*. The post LPF coefficient *a* may be used to add further absorption or to add gain into the loop as the LPF will remove energy. A one-pole low-pass structure similar to the one from the LPF-comb filter may be used as shown in Figure 17.17b, noting that it requires an input scaling coefficient (*1-g2*) to keep the loop gain at unity.

17.8 The Modulated Delay APF

The modulated delay lines from Chapter 15 can be used to further increase time smearing and echo density by placing them within reverberator structures like the delaying APF. The LFO rate is kept low (< 1 Hz) and the depth also very low (10–20 samples) to ensure that the detuning and chorusing effect is not overly obvious. Frenette showed that the use of modulated delay lines could reduce the computational requirements without a perceptual reduction in quality of reverb. Frenette used modulated comb filters (chorus modules) in addition to modulated all-pass filters to further reduce the overall complexity and memory requirements. The modulated APF is shown in Figure 17.18; only the very end of the delay line is modulated, producing a delay described by Equation 17.16.

$$delay = z^{(-D+u(nT))} \tag{17.16}$$

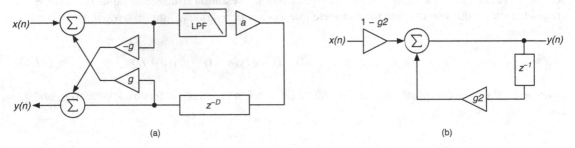

(a)

(b)

Figure 17.17: (a) A delaying all-pass structure with embedded low-pass filter. (b) A simple one-pole realization of the LPF.

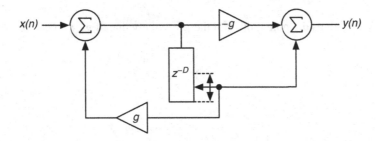

Figure 17.18: The modulated delaying APF uses a LFO to slowly modulate the delay time in the structure.

where

$u(nT)$ is the discrete time modulation signal

17.9 Moorer's Reverberator

Moorer proposed a design similar to Schroeder's reverberator that uses a parallel bank of LPF comb filters. Because low-pass filters remove energy from the system, more units are needed in parallel for a given reverb time. Some feel that Moorer's reverb sounds better because it mimics the HF losses of a real room. The block diagram for Moorer's reverberator is shown in Figure 17.19. As with Schroeder's reverb, numerous changes and enhancements are possible, including the addition of more modules, the modulated APF, and the use of nested all-pass diffusors instead of, or in addition to, the single output all-pass filter.

17.10 Dattorro's Plate Reverb

In 1997, Dattorro presented a plate reverb algorithm "in the style of Greisinger," revealing the use of a modulated all-pass filter. He notes that the modulated filter actually mimics a room whose walls are slowly moving back and forth. Dattorro's plate reverb algorithm in Figure 17.20 has a block diagram that reveals its figure-8 recirculating "tank" circuit. The tank circuit traps the signal, which loses energy

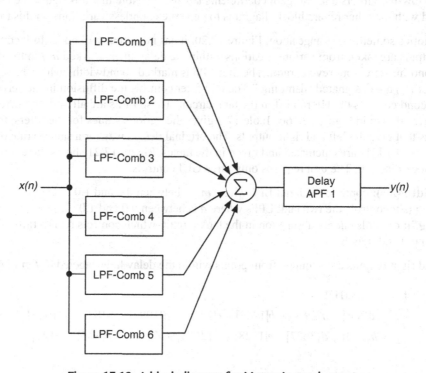

Figure 17.19: A block diagram for Moorer's reverberator.

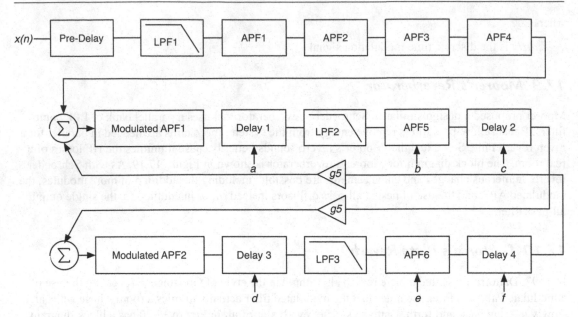

Figure 17.20: A block diagram of Dattorro's plate reverb algorithm.

through two low-pass filters and two gain coefficients labeled *g5*. Note that this figure-8 circuit could be combined with the other reverb block diagrams to generate countless variations on this theme.

You might notice something strange about Figure 17.20: it has no output *y(n)* node. In fact, the left and right outputs are taken from various locations within the delay lines, marked *a–f* in the diagram. This is a mono-in, stereo-out reverberator. The first LPF is marked "bandwidth" while the second pair (LPF2 and LPF3) are designated "damping." The first filter controls the diffusion in the series APFs while the second controls the HF roll off in the tank circuit. These LPFs are all dc-normalized one-pole feedback filters shown in Figure 17.16b. Table 17.1 gives the various values for the filters, followed by the equations that give the left and right outputs. The original design was for a sample rate of 29.8 kHz. Values for *fs* = 44.1 kHz are calculated and given in the tables. Figure 17.21 shows the entire reverb algorithm block diagram. The unit requires only three GUI controls:

- **bandwidth (*bw*):** controls the input LPF where *bw* is between 0.0 and 1.0
- **damping (*d*):** controls the two tank LPFs where *d* is between 0.0 and 1.0
- **decay (*g5*):** controls energy absorption in the tank circuit, which controls reverb time where *g5* is between 0.0 and 0.9999

The left and right outputs are summed from points within the delay lines labeled *a–f* in Figure 17.18:

$$f_s = 29.8\text{kHz}:$$
$$y_L = a[266] + a[2974] - b[1913] + c[1996] - d[1990] - e[187] - f[1066]$$
$$y_R = d[353] + d[3627] - e[1228] + f[2673] - a[2111] - b[335] - c[121]$$

$$(17.17)$$

$$f_s = 44.1\text{kHz}:$$
$$y_L = a[394] + a[4401] - b[2831] + c[2954] - d[2945] - e[277] - f[1578]$$
$$y_R = d[522] + d[5368] - e[1817] + f[3956] - a[3124] - b[496] - c[179]$$

Table 17.1: Gain and delay values for Dattorro's plate reverb

APF *Dx*	Delay (samples) Fs = 29.8k	Delay (samples) Fs = 44.1k	g (index)	g (value)
1	142	210	1	0.75
2	107	158	1	0.75
3	379	561	2	0.625
4	277	410	2	0.625
9	2,656	3,931	3	0.5
10	1,800	2,664	3	0.5

Fixed Delay Dx	Delay (samples) Fs = 29.8k	Delay (samples) Fs = 44.1k	–	–
7	4,217	6,241	–	–
8	4,453	6,590	–	–
11	3,136	4,641	–	–
12	3,720	5,505	–	–

Mod APF Dx	Delay (samples) Fs = 29.8k	Delay (samples) Fs = 44.1k	g (index)	g (value)
5	908 ±8	1,343 ±12	4	0.7
6	672 ±8	995 ±12	4	0.7

17.11 The Spin Semiconductor® Reverb Tank

Around 2007, Spin Semiconductor released an application note (App Note) that described a reverb tank circuit that was clearly influenced by Dattorro's plate reverb; the author refers to modulated APF units, the addition of low-pass filters, and other hints. The App Note is light on details, but it describes a recirculating tank circuit made of four similar branches, each consisting of two delaying all-pass units followed by a fixed delay line. The output of each branch is attenuated with a coefficient marked *kRT* (k-Reverb Time) and the raw input value is fed into each branch, along with a *kRT* scaled output of the previous branch. A global loop is formed between the output of the last branch and the input to the tank. This reverberator is shown in Figure 17.22.

The only relevant information the App Note discloses is that the feedback *g* values for the all-pass modules are all set to 0.5. No other information, including the delay lengths for any of the modules, is listed. The App Note states that the left and right outputs are taken from various locations within the four fixed delays. The various *y(n)* outputs in Figure 17.22 are added here as they relate to the reverb plugin design in Section 17.15. This tank circuit does not make a super convincing reverb on its own, but we will combine it with other modules for this chapter's reverb plugin. We will discuss the details about our chosen delay lengths and output locations in that section. As with the previous designs, there are numerous variations on this theme including the addition of filters, altering the number of APF units in each branch, and taking multiple output values from numerous locations within the structure. There is plenty of experimentation possible here.

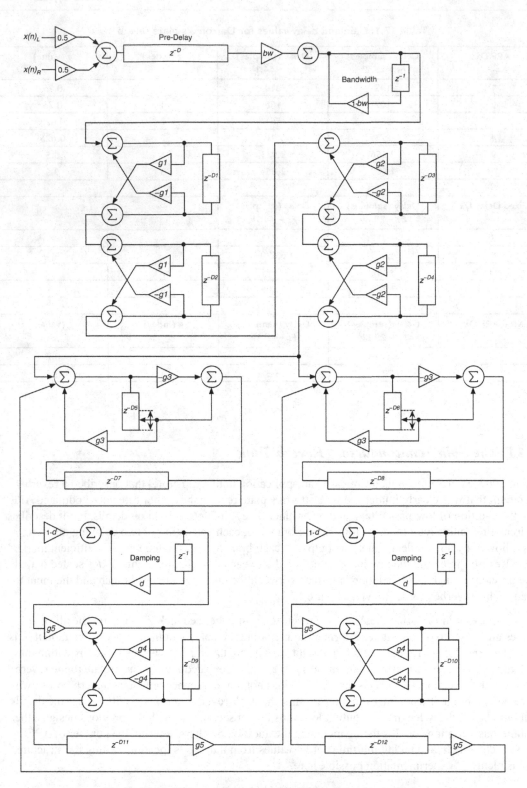

Figure 17.21: The complete design for Dattorro's plate reverb.

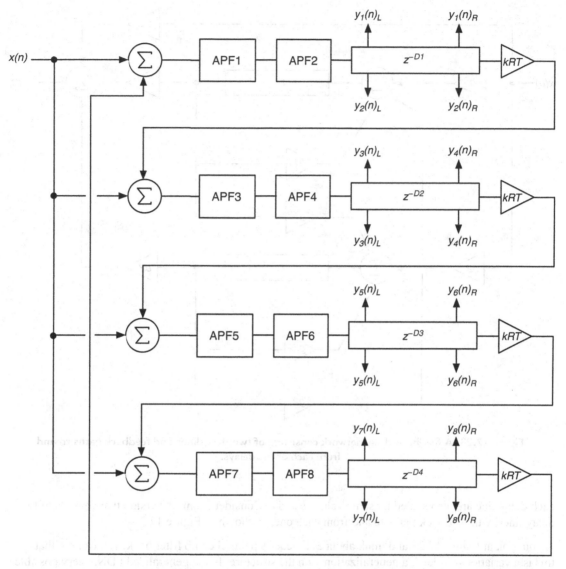

Figure 17.22: The Spin Semiconductor reverb tank. Note that the outputs $y_M(n)_L$ and $y_M(n)_R$ were not included in the original App Note; where $M = [1, 8]$.

17.12 Generalized Feedback Delay Network Reverbs

Another way to look at reverberation is to realize that the listener is experiencing dense echoes caused by multiple reflections off of surfaces along with potentially constructive or destructive interference. The generalized *Feedback Delay Network* (FDN) approach began with Gerzon's work in 1972 on preserving energy in multi-channel reverberators. It was continued by Stautner and Puckette as well as Jot and Chaigne. Generally speaking, the idea is to model the room with some number of delay lines with potential feedback paths to and from every delay line in the system. The inputs and outputs of

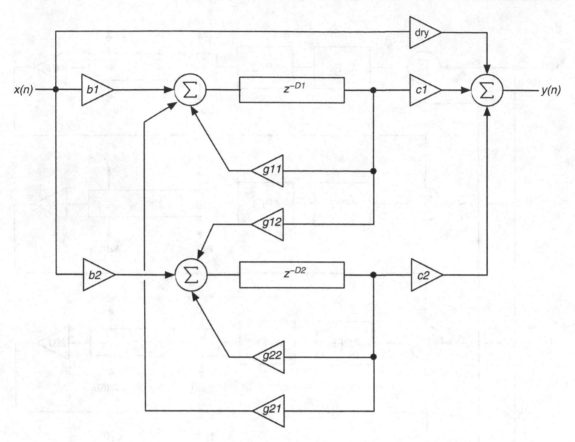

Figure 17.23: A feedback delay network consisting of two delay lines and feedback paths to and from each of the delays.

each delay line are also scaled by some values b and c. Consider a simple version that consists of two delay lines with feedback paths to and from each one, as shown in Figure 17.23.

If you look at Figure 17.23 and think about Schroeder's parallel comb filter bank, you can see that this is a variation—indeed, a generalization—on the structure. In the generalized FDN, every possible feedback path is accounted for. Additionally, each delay line has input and output amplitude controls. If you let $b1$, $b2$, $c1$, and $c2$ all equal 1.0, and set $g12$ and $g22$ to 0.0, you get Schroeder's parallel comb filter network exactly.

Notice how the feedback coefficients are labeled.

   $g11$: feedback from delay-line 1 into delay-line 1
   $g12$: feedback from delay-line 1 into delay-line 2
   $g21$: feedback from delay-line 2 into delay-line 1
   $g22$: feedback from delay-line 2 into delay-line 2

An equivalent version of Figure 17.23 is shown in Figure 17.24. You can see that the feedback coefficients have been grouped together in what looks like a linear algebra matrix. Gerzon found that the total energy is

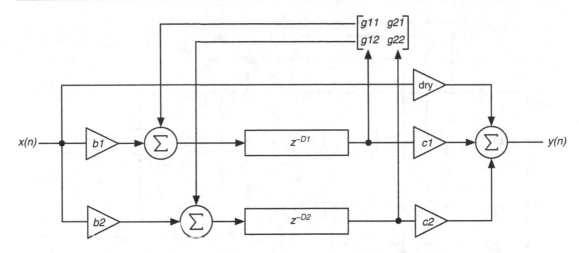

Figure 17.24: An equivalent version of a two-delay line FDN.

preserved if the matrix of coefficients is *unitary*. A unitary matrix multiplied by its *transpose matrix* results in the *identity matrix*. To form the transpose matrix, you swap the rows and columns. These matrices are square. The identity matrix is a square matrix whose diagonal values are all 1. A FDN that uses a unitary matrix for its coefficients is called a *Unitary Feedback Delay Network* (UFDN).

Examine the gain matrix G:

$$G = \begin{bmatrix} 0 & 1 \\ -1 & 0 \end{bmatrix} \tag{17.18}$$

Equation 17.18 is a unitary matrix. Multiplying by the transpose matrix yields the identity matrix.

$$GG^T = \begin{bmatrix} 0 & 1 \\ -1 & 0 \end{bmatrix}\begin{bmatrix} 0 & -1 \\ 1 & 0 \end{bmatrix} = \begin{bmatrix} 1 & 0 \\ 0 & 1 \end{bmatrix} \tag{17.19}$$

This reverberator would ring forever because of the 1.0 gain values. Jot proposed adding an absorptive coefficient k to each delay line. For a colorless reverb, the value for k in Equation 17.20 is given by Jot:

$$k = \gamma^M \tag{17.20}$$

where

M = the length of the delay
γ = the delay factor, set by the user

Addition of these post-delay coefficients will set the proper attenuation for each delay line to keep the reverb colorless. However, the structure still does not include the frequency-dependent absorptive losses we noted from the EDR diagrams. To accomplish this, Jot suggested inserting low-pass filters instead of the static attenuators. The low-pass filter magnitudes are cleverly chosen in relation to the frequency-dependent reverb time, $T_{60}(\omega)$.

$$20\log(|h|) = \frac{-60T_s}{RT_{60}(\omega)} M \tag{17.21}$$

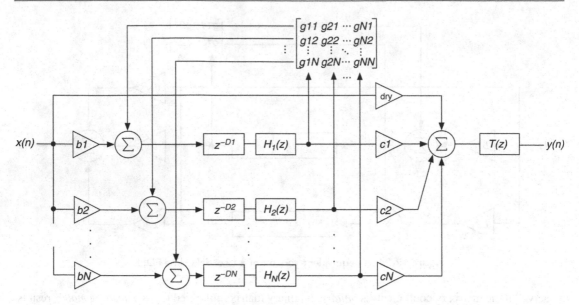

Figure 17.25: The FDN with delay filter-attenuators and a final correction filter *T(z)* on the wet output.

where

M = delay length
h = the magnitude of the filter at some frequency ω

The problem with this setup is that the pole radii are no longer the same (circular in the z-plane) and we know that this produces coloration in the signal. Jot's solution was to add a correction filter at the end of the whole system, $T(z)$, whose magnitude is proportional to the inverse square root of the frequency-dependent reverb time $RT_{60}(\omega)$. This is shown in Figure 17.25. Notice that we now need to specify the frequency-dependent reverb time in order to design the reverberator.

Finally, Figure 17.26 shows a generalized version of the FDN. For clarity, it does not include the absorptive loss components (either static k values or H_{LPFs}). The feedback matrix is always square $N \times N$ where N is the number of delay lines. Because it is square, it can be made to be unitary.

As an experiment, you can try to implement a four delay line version using the unitary matrix in Equation 17.22 and gain coefficients g_{NN} proposed by Stautner and Puckette, who were researching multi-channel reverb algorithms.

$$G = \frac{g}{\sqrt{2}} \begin{bmatrix} 0 & 1 & 1 & 0 \\ -1 & 0 & 0 & -1 \\ 1 & 0 & 0 & -1 \\ 0 & 1 & -1 & 0 \end{bmatrix} \tag{17.22}$$

where

$g < 1.0$

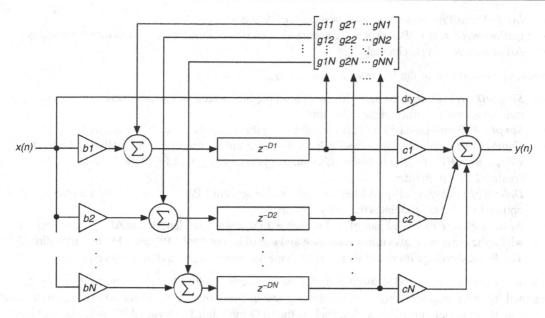

Figure 17.26: A generalized FDN with *N* delay branches.

17.12.1 Searching for FDN Coefficients

An issue with the FDN reverberators is the sheer number of coefficients that must be calculated and set. For an $N \times N$ FDN with only the simple matrix coefficients and a and b branch coefficients, there are potentially $N^2 + 2N$ coefficients to find. Even with unitary matrices that tend to be sparse, there are still many variables involved. Chemistruck, Marcolini, and Pirkle experimented with FDNs for reverb algorithms. They used the Genetic Algorithm (GA) to generate coefficients for feedback delay blocks in their proposed system. Coggin and Pirkle took that research further with an advanced hybrid reverberator that used a more robust search methodology and inclusion of a short convolution with the first few milliseconds of the target room impulse response as part of the design. In addition, the GA was used to search for the FDN branch filter coefficients and final correction filter. In the first project, a random unitary matrix generator was coded to seed the GA so that the starting point was a UFDN, while the second used a Hadamard matrix. An impulse response was taken for the target listening environment—this IR could be the response of an actual acoustic listening space or a digital reverb. The GA fitness function used information from the time domain envelope of the target impulse response and its frequency content as the matching criteria on the output of the system. More details can be found in the original references (Chemistruck et all., 2012).

17.13 C++ DSP Objects: Reverb Objects

The *fxobjects.h* and *fxobjects.cpp* files contain a set of objects specially designed to help in our reverb algorithms. We will use most of them in our chapter project, a long reverb plugin. Due to the number of objects, we will list only the most important attributes and details here. These objects are tiny and extremely specialized. Consult the documentation for more detailed information. Each of these objects implements identical sets of functions.

- *IAudioSignalProcessor*: *reset* and *processAudioSample*
- *getParameter* and *setParameter* are used with each object's custom data structure for updates.
- All objects are single channel.

These objects consist of the following.

- **SimpleDelay**: single fixed delay line with multiple buffer read and write functions to allow random access to values in the delay line
- **SimpleLPF**: one-pole LPF with one coefficient *g* that controls the cutoff frequency
- **CombFilter**: comb filter with feedback coefficient *g* and the ability to optionally set the feedback value with a RT_{60} time in milliseconds with an optional one-pole LPF in the feedback path to create a LPF-comb filter
- **DelayAPF**: delaying all-pass filter with optional absorbent LPF in the feed-forward branch, and optional LFO that modulates the delay lookup location
- **NestedDelayAPF**: nested pair of APFs, each with its own delay length and APF *g* coefficient in which the outer APF also allows enabling and use of its optional LPF and LFO modules directly
- **TwoBandShelvingFilter**: simple filters to shape the reverb signal's frequency response

These objects have been coded specifically for use in reverb plugins, though of course they may be adapted for other projects as well. These objects contain numerous functions for setting parameters and enabling the optional internal modules, such as the LFO-modulated delaying APF. With these objects and a bit of code, you can easily assemble every reverberator algorithm in the chapter, plus many more. Note that not every function is documented here to save space, so please take some time to poke around the objects as you like.

17.13.1 C++ DSP Object: SimpleDelay

The *SimpleDelay* object implements a simple delay without feedback or wet/dry mix controls. Figure 17.27 shows a block diagram of this simple object.

17.13.2 SimpleDelay: Custom Data Structure

The object is updated via the custom structure *SimpleDelayParameters*. The structure also holds a delay-in-samples version of the delay time for easy access to the delay's circular buffer during the processing. The interpolation flag is used to signify interpolation during circular buffer accesses.

```
struct SimpleDelayParameters
{
    SimpleDelayParameters() {}
    SimpleDelayParameters& operator=(. . .){ . . . .}

    // --- individual parameters
    double delayTime_mSec = 0.0;
    bool interpolate = false;
    double delay_Samples = 0.0;
};
```

17.13.3 SimpleDelay: Members

Tables 17.2 and 17.3 list the *SimpleDelay* member variables and member functions.

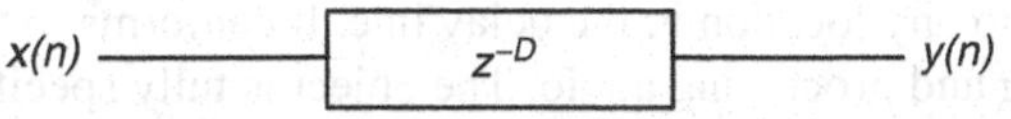

$$x(n) \longrightarrow \boxed{z^{-D}} \longrightarrow y(n)$$

Figure 17.27: The *SimpleDelay* block diagram.

Table 17.2: *SimpleDelay* member variables

SimpleDelay Member Variables		
Type	**Name**	**Description**
SimpleDelayParameters	*parameters*	Object parameters
CircularBuffer<double>	*delayBuffer*	circular buffer delay line
double	*samplesPerMSec*	Stored variable that converts delay time in *mSec* to samples
double	*bufferLength_mSec*	The maximum buffer length in *mSec*
double	*bufferLength*	The maximum buffer length in samples
double	*sampleRate*	Current sample rate

Table 17.3: *SimpleDelay* member functions

SimpleDelay Member Functions		
Returns	**Name**	**Description**
SimpleDelayParameters	*getParameters*	Get all parameters at once
void	*setParameters* Parameters: —SimpleDelayParameters _parameters	Set all parameters at once
bool	*reset* Parameters: —double sampleRate	Flush delay line
double	*processAudioSample* Parameters: —double xn	Process input xn through the audio filter
void	*createDelayBuffer* Parameters: —double _sampleRate —double _bufferLength_mSec	Create the delay buffer dynamically
double	*readDelay*	Return the delayed value specified in the *SimpleDelayParameters*
double	*readDelayAtTime_mSec* Parameters: —double _delay_mSec	Read arbitrary delay value at _delay_mSec
double	*readDelayAtPercentage* Parameters: —double delayPercent	Read arbitrary delay value at a percentage of the delay length
double	*writeDelay* Parameters: —double xn	Write the value xn into the delay line

17.13.4 SimpleDelay: *Programming Notes*

This is a simplified version of the *AudioDelay* object, containing only the single delay line. The only parameter that may be adjusted is the delay time. The object features multiple *read* functions to allow easy access to any location in the delay line. It conforms to the *IAudioSignalProcessor* interface for resetting and processing audio. The object is fully specified in the *PluginKernel* Doxygen documentation; please refer to it for details.

Initialization:

Call the object's *reset* function, then create the buffer.

```
myDelay.reset(resetInfo.sampleRate);
myDelay.createDelayBuffer(resetInfo.sampleRate, 1000.0);
```

Update From GUI:

You update the object with the *SimpleDelayParameters* structure, passing the delay time in milliseconds; usually this comes from a GUI control bound variable.

Processing Audio:

This object processes single samples only, so use the *processAudioSample( )* function to process an input completely through to the output.

```
double yn = myDelay.processAudioSample(xn);
```

After calling the *processAudioSample( )* function, which returns the sample at the currently selected delay time, you may optionally use either of the *read* functions to access any location within the delay line. You may read the delay as a percentage of the total length; here we read out the delay at 23.7% of the total length:

```
double yn = myDelay.readDelayAtPercentage(23.7);
```

You may also read the delay at a specific delay time in milliseconds; here we read out the delay at 100.234 mSec behind the current write location:

```
double yn = myDelay.readDelayAtTime_mSec(100.234);
```

17.13.5 C++ DSP Object: SimpleLPF

The *SimpleLPF* object implements the one-pole LPF with a single coefficient g; an input scaling factor $a_0 = (1 - g)$ forces the filter to have unity gain at DC. The filter is shown in Figure 17.28a while its frequency responses are shown in Figure 17.28b for a variety of settings of the lone coefficient g.

17.13.6 SimpleLPF: Custom Data Structure

The object is updated via the custom structure *SimpleLPFParameters*.

```
struct SimpleLPFParameters
{
      SimpleLPFParameters() {}
      SimpleLPFParameters& operator=(. . .){ . . . }

      // --- individual parameters
      double g = 0.0;
};
```

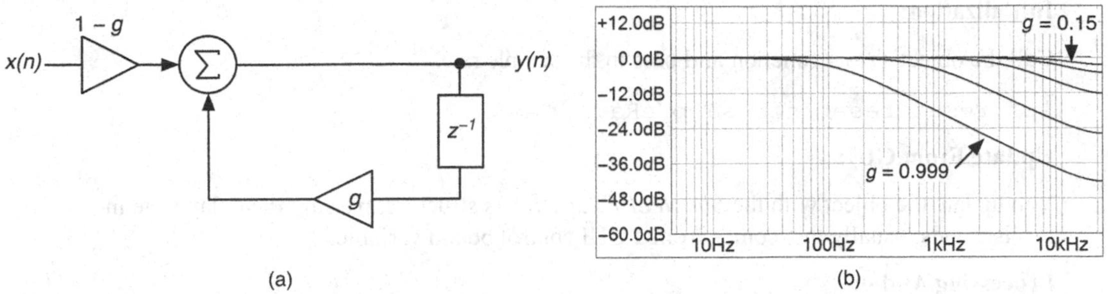

(a)   (b)

Figure 17.28: (a) The one-pole *SimpleLPF* block diagram. (b) Its frequency responses with *g* = 0.999 (most HF reduction), *g* = 0.9, 0.6, 0.3; *g* = 0.15 (least HF reduction).

Table 17.4: *SimpleLPF* member variables

SimpleLPF Member Variables		
Type	Name	Description
SimpleLPFParameters	*parameters*	Object parameters
double	*state*	The single state variable register

Table 17.5: *SimpleLPF* member functions

SimpleLPF Member Functions		
Returns	Name	Description
SimpleLPFParameters	*getParameters*	Get all parameters at once
void	*setParameters* Parameters: —*SimpleLPFParameters* _parameters	Set all parameters at once
bool	*reset* Parameters: —*double* sampleRate	Flush storage register
double	*processAudioSample* Parameters: —*double* xn	Process input *xn* through the LPF

17.13.7 SimpleLPF: Members

Tables 17.4 and 17.5 list the *SimpleDelay* member variables and member functions.

17.13.8 SimpleLPF: Programming Notes

This code implements the same one-pole filter from Chapter 10. The signal processing simply implements the single feedback branch of a direct form biquad using a simple *double* as the state storage register. It conforms to the *IAudioSignalProcessor* interface for resetting and processing audio.

Initialization:

Call the object's *reset* function and pass in the sample rate.

```
lpf.reset(resetInfo.sampleRate);
```

Update From GUI:

You update the object with the *SimpleLPFParameters* structure, passing the delay time in milliseconds; usually this comes from a GUI control bound variable.

Processing Audio:

This object processes single samples only, so use the *processAudioSample( )* function.

```
double yn = lpf.processAudioSample(xn);
```

17.13.9 C++ DSP Object: CombFilter

The *CombFilter* object implements a comb filter with feedback and an optional one-pole LPF that may be engaged in the feedback path as shown in Figure 17.14b. There are two feedback *g* coefficients, one for the comb filter (*g1*) and the other for the LPF (*g2*). The coefficients are set using Equation 17.14.

17.13.10 CombFilter: *Custom Data Structure*

The object is updated via the custom structure *CombFilterParameters*.

```
struct CombFilterParameters
{
        CombFilterParameters() {}
        CombFilterParameters& operator=(. . .){ . . . }

        // --- individual parameters
        double delayTime_mSec = 0.0;
        double RT60Time_mSec = 0.0;
        bool enableLPF = false;
        double lpf_g = 0.0;
        bool interpolate = false;
};
```

17.13.11 CombFilter: *Members*

Tables 17.6 and 17.7 list the *CombFilter* member variables and member functions.

17.13.12 CombFilter: *Programming Notes*

This code implements the comb filter with LPF. It conforms to the *IAudioSignalProcessor* interface for resetting and processing audio.

Initialization:

Call the object's *reset* function and pass in the sample rate, then create the buffer.

Table 17.6: *CombFilter* member variables

CombFilter Member Variables		
Type	**Name**	**Description**
CombFilterParameters	*parameters*	Object parameters
double	*comb_g*	The feedback value g
double	*lpf_g*	The g (feedback) value for the optional LPF
double	*lpf_state*	State storage register for optional LPF
double	*bufferLength_mSec*	Maximum buffer
double	*sampleRate*	Current sample rate

Table 17.7: *CombFilter* member functions

CombFilter Member Functions		
Returns	**Name**	**Description**
CombFilterParameters	*getParameters*	Get all parameters at once
void	*setParameters* Parameters: —*CombFilterParameters* _parameters	Set all parameters at once
bool	*reset* Parameters: —*double* sampleRate	Flush storage register
double	*processAudioSample* Parameters: —*double* xn	Process input *xn* through the LPF
void	*createDelayBuffer* Parameters: —*double* _sampleRate —*double* _bufferLength_mSec	Create the delay buffer dynamically

```
combFilter.reset(resetInfo.sampleRate);
combFilter.createDelayBuffer(resetInfo.sampleRate, 100.0);
```

Update From GUI:

You update the object with the *SimpleLPFParameters* structure, passing the parameters; usually this comes from a GUI control bound variable. To enable the LPF, set the *enableLPF* flag in the custom data structure. Then, you may adjust the filter cutoff frequency with the *lpf_g* variable with the argument for *g* on the range of [0.0, 0.99999].

Processing Audio:

This object processes single samples only, so use the *processAudioSample( )* function to process an input completely through to the output.

```
double yn = combFilter.processAudioSample(xn);
```

17.13.13 C++ DSP Object: DelayAPF

The *DelayAPF* object implements a delaying APF with optional absorbent LPF and optional LFO-based modulated delay shown in Figure 17.29. The delay modulation is from the maximum delay time

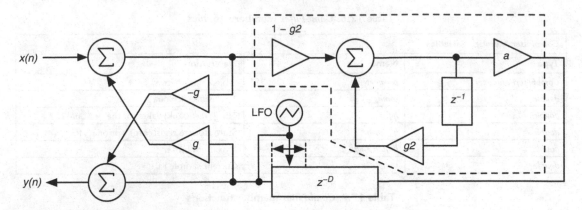

Figure 17.29: The *DelayAPF* structure includes an optional absorbent LPF (in dotted line) and an optional LFO-based modulated delay lookup.

downward, so if the delay length is 100 mSec and you specify a maximum modulation depth of 10%, the delay time would modulate between 90 mSec and 100 mSec. A separate LFO depth control further refines this value.

17.13.14 DelayAPF: *Custom Data Structure*

The object is updated via the custom structure *DelayAPFParameters*. Note that there are variables for all of the internal components, including the optional LPF and LFO. The member names and their functions should be evident.

```
struct DelayAPFParameters
{
        DelayAPFParameters() {}
        DelayAPFParameters& operator=(. . .){ . . . }

        // --- individual parameters
        double delayTime_mSec = 0.0;
        double apf_g = 0.0;
        bool enableLPF = false;
        double lpf_g = 0.0;
        bool interpolate = false;

        bool enableLFO = false;
        double lfoRate_Hz = 0.0;
        double lfoDepth = 0.0;
        double lfoMaxModulation_mSec = 0.0;
};
```

17.13.15 DelayAPF: *Members*

Tables 17.8 and 17.9 list the *DelayAPF* member variables and member functions.

Table 17.8: *DelayAPF* member variables

DelayAPF Member Variables		
Type	Name	Description
DelayAPFParameters	parameters	Object parameters
SimpleDelay	delay	The delay object
LPF	modLFO	The optional modulating LFO
double	lpf_state	State storage register for optional LPF
double	bufferLength_mSec	Maximum buffer
double	sampleRate	Current sample rate

Table 17.9: *DelayAPF* member functions

DelayAPF Member Functions		
Returns	Name	Description
DelayAPFParameters	getParameters	Get all parameters at once
void	setParameters Parameters: —DelayAPFParameters _parameters	Set all parameters at once
bool	reset Parameters: —double sampleRate	Flush storage register
double	processAudioSample Parameters: —double xn	Process input xn through the LPF
void	createDelayBuffer Parameters: —double _sampleRate —double _bufferLength_mSec	Create the delay buffer dynamically

17.13.16 DelayAPFParameters: *Programming Notes*

This code implements the delaying APF with LPF and modulated delay lookup. It conforms to the *IAudioSignalProcessor* interface for resetting and processing audio.

Initialization:

Call the object's *reset* function, then create the delay line. The delay line must be re-created each time *reset* is called since the delay line length is dependent on the sample rate. In this example, we create (or re-create) a 100.0 mSec delay after resetting the object with the current sample rate.

```
delayAPF.reset(resetInfo.sampleRate);
delayAPF.createDelayBuffer(resetInfo.sampleRate, 100.0);
```

Update From GUI:

You update the object with the *DelayAPFParameters* structure, passing the parameters; usually this comes from a GUI control bound variable. To enable the LPF, set the *enableLPF* flag in the custom data structure. Then, you may adjust the filter cutoff frequency with the *lpf_g* variable with the argument for *g* on the range of [0.0, 0.99999]. To enable the LFO and modulate the delay lookup, set the *enableLFO* flag in the custom data structure. Next, set the maximum LFO modulation depth in either milliseconds or a percentage of the overall delay length.

Processing Audio:

This object processes single samples only, so use the *processAudioSample( )* function to process an input completely through to the output.

```
double yn = delayAPF.processAudioSample(xn);
```

17.13.17 C++ DSP Object: NestedDelayAPF

The *NestedDelayAPF* object adds a second "inner" or "nested" all-pass filter in series with the "outer" APF's delay line as shown in Figure 17.30. You can think of this as a kind of "double object" because there are two required values to be set: the delay length of each APF (and they need to be different) and the coefficient *g* for each structure. The outer APF may also be directly set to use the optional LPF and LFO as described in the previous section. You may want to modify this object to access the inner APF's LPF and LFO operations as an exercise.

17.13.18 NestedDelayAPF: *Custom Data Structure*

The object is updated via the custom structure *NestedDelayAPFParameters*. Note that there duplicated variables to handle the pair of embedded APF structures. There are also flags and other members to handle the optional LFO.

```
struct NestedDelayAPFParameters
{
    NestedDelayAPFParameters() {}
    NestedDelayAPFParameters& operator=(. . .){ . . .]

    // --- individual parameters
    double outerAPFdelayTime_mSec = 0.0;
    double innerAPFdelayTime_mSec = 0.0;
    double outerAPF_g = 0.0;
    double innerAPF_g = 0.0;
};
```

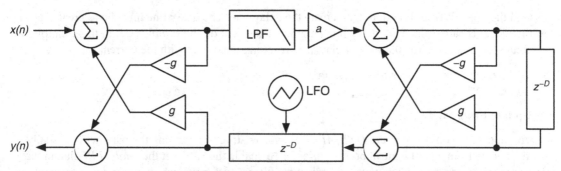

Figure 17.30: The *NestedDelayAPF* structure includes an inner or nested APF structure in addition to the fully featured outer APF.

17.13.19 NestedDelayAPF: *Members*

Tables 17.10 and 17.11 list the *NestedDelayAPF* member variables and member functions.

17.13.20 NestedDelayAPF: *Programming Notes*

Initialization:

Call the object's *reset* function, then create the delay lines (one for each APF); the delay lines must be re-created each time *reset* is called since the delay line length is dependent on the sample rate. In this example, we create (or re-create) two 100.0 mSec delay lines after resetting the object with the current sample rate.

```
nestedAPF.reset(resetInfo.sampleRate);
nestedAPF.createDelayBuffers(resetInfo.sampleRate, 100.0, 100.0);
```

Update From GUI:

You update the object with the *NestedAPFParameters* structure, passing the parameters; usually this comes from a GUI control bound variable. The optional LPF and LFO parameters are set using the same functions as the *DelayAPF*.

Table 17.10: *NestedDelayAPF* **member variables**

NestedDelayAPF Member Variables		
Type	Name	Description
NestedDelayAPFParameters	parameters	Nested object
DelayAPF	nestedAPF	Delay object

Table 17.11: *NestedDelayAPF* **member functions**

NestedDelayAPF Member Functions		
Returns	Name	Description
NestedDelayAPFParameters	getParameters	Get all parameters at once
void	setParameters Parameters: —NestedDelayAPFParameters _parameters	Set all parameters at once
bool	reset Parameters: —double sampleRate	Flush storage register
double	processAudioSample Parameters: —double xn	Process input *xn* through the LPF
void	createDelayBuffer Parameters: —double _sampleRate —double _bufferLength_mSec	Create the delay buffer dynamically

Processing Audio:

This object processes single samples only, so use the *processAudioSample( )* function to process an input completely through to the output.

```
double yn = nestedAPF.processAudioSample(xn);
```

17.13.21 C++ DSP Object: TwoBandShelvingFilter

The *TwoBandShelvingFilter* is a simple filtering object that combines a 1st order low shelving filter and a 1st order high shelf filter in series and in the same object. The user may adjust the shelving frequency (Hz) and the shelf gain/attenuation (dB). In this object, the two filters are hardwired in series and both are always active (on). The block diagram and typical frequency responses are shown in Figure 17.31.

17.13.22 TwoBandShelvingFilter: *Custom Data Structure*

The object is updated via the custom structure *TwoBandShelvingFilterParameters*.

```
struct TwoBandShelvingFilterParameters
{
        TwoBandShelvingFilterParameters() {}
        TwoBandShelvingFilterParameters& operator=(. . .){ . . . }

        // --- individual parameters
        double lowShelf_fc = 0.0;
        double lowShelfBoostCut_dB = 0.0;
        double highShelf_fc = 0.0;
        double highShelfBoostCut_dB = 0.0;
};
```

17.13.23 TwoBandShelvingFilter: *Members*

Tables 17.12 and 17.13 list the *TwoBandShelvingFilter* member variables and member functions.

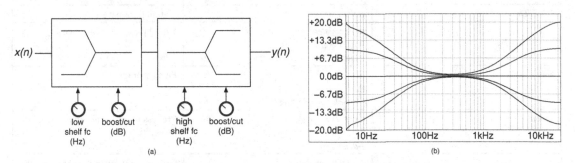

Figure 17.31: (a) The *TwoBandShelvingFilter* functional block diagram and (b) example frequency response plot; here the f_L = 100 Hz, f_H = 5 kHz, and the boost/cut values range from −20.0 to +20.0 dB.

Table 17.12: *TwoBandShelvingFilter* member variables

TwoBandShelvingFilter Member Variables		
Type	**Name**	**Description**
TwoBandShelvingFilterParameters	*parameters*	The nested object
AudioFilter	*lowShelfFilter*	Low shelf object
AudioFilter	*highShelfFilter*	High shelf object

Table 17.13: *TwoBandShelvingFilter* member functions

TwoBandShelvingFilter Member Functions		
Returns	**Name**	**Description**
TwoBandShelvingFilterParameters	*getParameters*	Get all parameters at once
void	*setParameters* Parameters: —*TwoBandShelvingFilterParameters* _parameters	Set all parameters at once
bool	*reset* Parameters: —*double* sampleRate	Reset filters
double	*processAudioSample* Parameters: —*double* xn	Process input *xn* through filters

17.13.24 TwoBandShelvingFilter: *Programming Notes*

This code implements a pair of shelving filters. It conforms to the *IAudioSignalProcessor* interface for resetting and processing. It uses a pair of inner *AudioFilter* objects to implement the shelving filters.

Initialization:

Call the object's *reset* function. For stereo operation you will use two objects, one for the left channel and the other for the right.

```
leftOutputFilter.reset(resetInfo.sampleRate);
rightOutputFilter.reset(resetInfo.sampleRate);
```

Update From GUI:

You update the object with the *TwoBandShelvingFilterParameters* structure, passing the parameters; usually this comes from a GUI control bound variable. There are two sets of f_c and boost/cut parameters, one for each filter.

Processing Audio:

This object processes single samples only, so use the *processAudioSample( )* function to process an input completely through to the output. To process the left and right input samples (*xnL* and *xnR*) you would use this code. The object processes the low shelf filter first, then applies its output to the high shelf component, from which the output is taken.

```
double ynL = leftOutputFilter.processAudioSample(xnL);
double ynR = rightOutputFilter.processAudioSample(xnR);
```

17.14 C++ *Effect Object:* ReverbTank

Our reverb plugin will make use of some of the more specialized reverb objects along with the Spin Semiconductor reverb tank. In testing, I was never able to get exceptional results from the tank circuit without adding parallel comb filters, which while thickening the reverb signal, often introduced too much of a latency, as if a pre-delay were in use (in effect, the parallel comb filters act as a pre-delay unit). However, by making a few changes to the original circuit, we can generate a beautiful reverb unit that can implement both short and long reverbs. Referring back to the Spin Semiconductor tank in Section 17.11 you can see that the outputs are tapped from each of the four fixed delay lines at various and different locations. Meanwhile, the input is applied to each of the four branches in parallel, so that tapping the delay lines generates the early reflections for us. The last branch output is fed back into the input, so as the impulse decays away, and the multiple output taps generate the echo density as well. For our reverb, we will keep the overall architecture of the tank but will make the following changes.

- The structure is composed of four fundamentally identical *branches*.
- Each branch features two APFs in series;. We will swap these out for *nested pairs* of APFs instead, greatly multiplying the echo density—this creates *2 * branches* = eight total APF units.
- We will use the LFO option in the first (outer) APF of each nested pair and slowly modulate the delay by a small percentage of the total length; the LFO modulation rates and depths are hard-coded, and you may experiment with them as you like.
- We will insert a one-pole low-pass filter in each branch, after the nested APF pair, and before the fixed delay line; the LPFs will all be controlled with one GUI knob that sets the LPF *g* coefficient on each object simultaneously.
- We will keep the fixed delay lines as-is (no feedback) and tap either one or two outputs per channel from each delay line; this will give us either four or eight samples to sum together for each output channel—the user can choose "sparse" (four) or "dense" (eight) output taps.
- We will invert the phase of every *other* output tap prior to summing the values.
- We will use prime numbers for tapping the output locations in the fixed delays in mSec.
- The nested APFs will use *g* coefficients that are inverted from one another.
- We will add a two-band shelving filter to the reverb signal only, to further shape its response.

The block diagram for our reverb is shown in Figure 17.32 where (a) shows the input summation to mono, pre-delay, and tank circuit modules while (b) shows the fixed delay output tap summations, output filtering, and blending with the dry signal to create the final design. The *ReverbTank* object will implement the entire reverb effect in one object. The *ReverbTank* is used directly in the chapter *Reverb* plugin but with a secret back-door interface for voicing the reverb plugin. During the voicing phase, you adjust the fixed and all-pass delay lengths. We will do this with a pair of controls for each bank of fixed and nested all-pass modules. We expose four knobs (which we name *tweaker controls*) to use during the voicing phase, which you may want to hide from the end user, or provide a limited range of controls, or provide a series of presets (e.g. short, medium, long, or hall, plate, cathedral, etc.). Our strategy for setting up the eight all-pass and four branch delays with a bunch of different and uncorrelated delay lengths will work as follows.

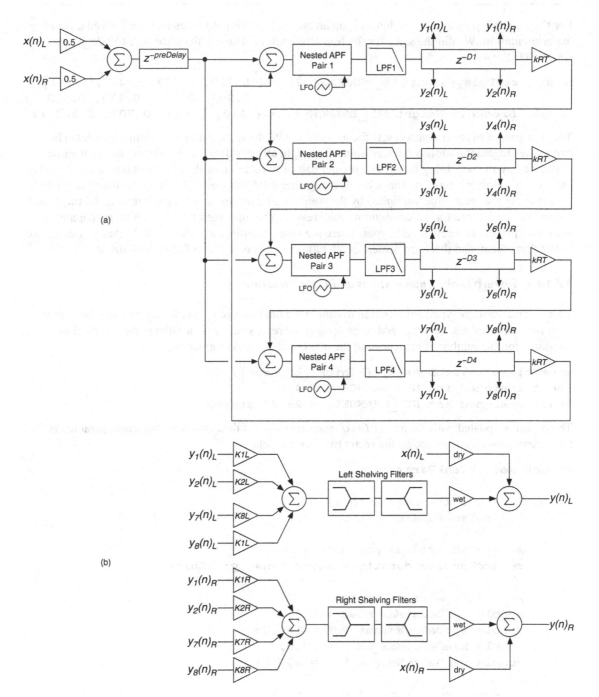

Figure 17.32: The *Reverb* plugin's (a) input and tank modules and (b) output taps, filters, and wet/dry blend.

For the eight all-pass delays, we have set up an array of weighting values that we derived through experimentation. We did the same for the four fixed delays. Those values are here (*NUM_BRANCHES* = 4):

```
double apfDelayWeight[NUM_BRANCHES*2] = { 0.317, 0.873, 0.477, 0.291,
                                          0.993, 0.757, 0.179, 0.575 };
double fixedDelayWeight[NUM_BRANCHES] = { 1.0, 0.873, 0.707, 0.667 };
```

The first pair of tweaker controls sets the maximum APF delay time and a weighting percent. The two values together establish the seed value. The eight all-pass delays are set using the seed value multiplied by the weighting factor. The second pair of tweaker controls sets the maximum fixed delay time and a weighting percent. The two values together establish the seed value. The four fixed delays are set using the seed value multiplied by the weighting factor. In the GUI parameter table, the default values we suggest are for a nice-sounding long reverb. You must then take it from there to generate your own presets and new reverb types. There are many beautiful sounding reverbs that you can coax out of this object, and there are numerous alterations you may make to fully customize it.

17.14.1 ReverbTank: *Enumerations and Data Structure*

There is one strongly typed enumeration to handle the thick/sparse option for tapping the four delay lines for output values. You may add more options to create your own variations. We also declare constants for the number of branches and the maximum channels supported.

```
enum class reverbDensity {kThick, kSparse};
const unsigned int NUM_BRANCHES = 4;
const unsigned int NUM_CHANNELS = 2; // stereo
```

The object is updated with the *ReverbTankParameters* custom data structure that stores parameters for the reverb sub-objects, including the secret tweaker controls.

```
struct ReverbTankParameters
{
      ReverbTankParameters() {}
      ReverbTankParameters& operator=(. . .){ . . . }

      // --- individual parameters
      reverbDensity density = reverbDensity::kThick;

      // --- tweaker parameters
      double apfDelayMax_mSec = 5.0;
      double apfDelayWeight_Pct = 100.0;
      double fixeDelayMax_mSec = 50.0;
      double fixeDelayWeight_Pct = 100.0;

      // --- direct control parameters
      double preDelayTime_mSec = 0.0;
      double lpf_g = 0.0;       // damping, 0 to 1
      double kRT = 0.0;         // reverb time, 0 to 1

      // --- shelving filters
      double lowShelf_fc = 0.0;
```

```
        double lowShelfBoostCut_dB = 0.0;
        double highShelf_fc = 0.0;
        double highShelfBoostCut_dB = 0.0;

        // --- output controls
        double wetLevel_dB = -3.0;
        double dryLevel_dB = -3.0;
};
```

17.14.2 ReverbTank: *Members*

Tables 17.14 and 17.15 list the *ReverbTank* member variables and member functions. Notice that this object can process frames, which may be mono, stereo, or multi-channel. The *processAudioSample* method still works if you only need to process a mono channel.

17.14.3 ReverbTank: *Programming Notes*

The reverb tank implements the block diagram in Figure 17.30 and is an advanced effect object. This is a great object for you to scrutinize to match up the C++ code with the block diagram. It also has numerous simple alterations, enhancements, and tweaks that you may add to customize it even further.

Using the *ReverbTank* object is straightforward.

1. Reset the object to prepare it for streaming.
2. Set the effect parameters in the *ReverbTankParameters* custom data structure.
3. Call *setParameters* to update the object.
4. For mono devices, call *processAudioSample*, passing the input value in and receiving the output value as the return variable.
5. For stereo devices, call the *processAudioFrame* function, passing the frame pointers and channel counts.
6. For voicing, expose the four tweaker controls: use them to either generate presets, or allow the user some subset of control values (e.g. allow the user to alter these parameters, but give them a severely narrowed range [limits] rather than the full ranges, which can generate some awful-sounding reverbs).

Table 17.14: *ReverbTank* **member variables**

ReverbTank Member Variables		
Type	**Name**	**Description**
ReverbTankParameters	*parameters*	Custom param struct
SimpleDelay	*preDelay*	Fixed pre-delay object
SimpleDelay	*branchDelays[NUM_BRANCHES]*	Branch delays
NestedDelayAPF	*bestedNestedAPFs[NUM_BRANCHES]*	APFs
SimpleLPF	*branchLPFs[NUM_BRANCHES]*	LPFs
TwoBandShelvingFilter	*shelvingFilters[NUM_CHANNELS]*	Shelving filters
double	*apfDelayWeight[NUM_BRANCHES*2]*	Weighting values
double	*fixedDelayWeight[NUM_BRANCHES]*	Weighting values
double	*sampleRate*	Current sample rate

Table 17.15: *ReverbTank* member functions

ReverbTank Member Functions		
Returns	**Name**	**Description**
ReverbTankParameters	*getParameters*	Get all parameters at once
void	*setParameters* Parameters: —*ReverbTankParameters* _parameters	Set all parameters at once
bool	*reset* Parameters: —*double* sampleRate	Reset sub-components
double	*processAudioSample* Parameters: —*double* xn	Process mono input through reverb
bool	*processAudioFrame* Parameters: —*const float** inputFrame —*float** outputFrame —*unsigned int* inputChannels —*unsigned int* outputChannels	Process a frame through reverb

Function: *reset*

Details: reset the sub-components; set up all of the delay lines (we default to 100.0 mSec maximums here, but you may alter these if you like).

```
virtual bool reset(double _sampleRate)
{
      // ---store
      sampleRate = _sampleRate;

      // ---set up preDelay
      preDelay.reset(_sampleRate);
      preDelay.createDelayBuffer(_sampleRate, 100.0);

      for (int i = 0; i < NUM_BRANCHES; i++)
      {
            branchDelays[i].reset(_sampleRate);
            branchDelays[i].createDelayBuffer(_sampleRate, 100.0);

            branchNestedAPFs[i].reset(_sampleRate);
            branchNestedAPFs[i].createDelayBuffers(_sampleRate,
                                                  100.0, 100.0);

            branchLPFs[i].reset(_sampleRate);
      }
      for (int i = 0; i < NUM_CHANNELS; i++)
            shelvingFilters[i].reset(_sampleRate);

      return true;
}
```

Function: *setParameters*

Details: We simply forward the incoming GUI parameter changes to the sub-component objects, which will update themselves only if their parameters have changed. This is also where we hard-wire the LFOs in the nested APFs (you may experiment with those as well). The updates here are in no particular order.

```
void setParameters(const ReverbTankParameters& params)
{
        // --- 1) shelving filters
        TwoBandShelvingFilterParameters filterParams =
                        shelvingFilters[0].getParameters();
        filterParams.highShelf_fc = params.highShelf_fc;
        filterParams.highShelfBoostCut_dB = params.highShelfBoostCut_dB;
        filterParams.lowShelf_fc = params.lowShelf_fc;
        filterParams.lowShelfBoostCut_dB = params.lowShelfBoostCut_dB;

        // --- copy to both channels
        shelvingFilters[0].setParameters(filterParams);
        shelvingFilters[1].setParameters(filterParams);

        // --- 2) simple LPFs
        SimpleLPFParameters lpfParams = branchLPFs[0].getParameters();
        lpfParams.g = params.lpf_g;

        for (int i = 0; i < NUM_BRANCHES; i++)
                branchLPFs[i].setParameters(lpfParams);

        // --- 3) pre delay
        SimpleDelayParameters delayParams = preDelay.getParameters();
        delayParams.delayTime_mSec = params.preDelayTime_mSec;
        preDelay.setParameters(delayParams);

        // --- set nested apf and delay parameters
        int m = 0;
        NestedDelayAPFParameters apfParams =
                        branchNestedAPFs[0].getParameters();
        delayParams = branchDelays[0].getParameters();

        // --- global max Delay times (from tweakers - you can hardcode)
        double globalAPFMaxDelay = (parameters.apfDelayWeight_Pct /
                                100.0)*parameters.apfDelayMax_mSec;
        double globalFixedMaxDelay = (parameters.fixeDelayWeight_Pct /
                                100.0)*parameters.fixeDelayMax_mSec;
        // --- LFO option
        apfParams.enableLFO = true;
        apfParams.lfoMaxModulation_mSec = 0.3;
        apfParams.lfoDepth = 1.0;
```

```
for (int i = 0; i < NUM_BRANCHES; i++)
{
    // --- setup APFs
    apfParams.outerAPFdelayTime_mSec =
                    globalAPFMaxDelay*apfDelayWeight[m++];
    apfParams.innerAPFdelayTime_mSec =
                    globalAPFMaxDelay*apfDelayWeight[m++];
    apfParams.innerAPF_g = -0.5;
    apfParams.outerAPF_g = 0.5;
    if (i == 0)
            apfParams.lfoRate_Hz = 0.15;
    else if (i == 1)
            apfParams.lfoRate_Hz = 0.33;
    else if (i == 2)
            apfParams.lfoRate_Hz = 0.57;
    else if (i == 3)
            apfParams.lfoRate_Hz = 0.73;

    branchNestedAPFs[i].setParameters(apfParams);

    // --- fixedDelayWeight
    delayParams.delayTime_mSec =
                    globalFixedMaxDelay*fixedDelayWeight[i];
    branchDelays[i].setParameters(delayParams);
}

// --- save our copy
parameters = params;
}
```

Function: *processAudioFrame*

Details: This follows the block diagram of Figure 17.30; refer to it while comparing the code. Note that the input is mono-ized before sending into the tank.

There are 25 prime numbers between 1 and 100: { 2, 3, 5, 7, 11, 13, 17, 19, 23, 29, 31, 37, 41, 43, 47, 53, 59, 61, 67, 71, 73, 79, 83, 89, and 97 } and we choose 16 of them for the maximum eight taps per channel (density = thick): { 23, 29, 31, 37, 41, 43, 47, 53, 59, 61, 67, 71, 73, 79, 83, and 89}. We use these values and the fixed delay helper function *readDelayAtPercentage( )* to grab various outputs at prime-number related locations within the four branches.

We will use a for-loop to iterate over the branches, noting that they are all identical except the output tap locations, which are located outside of the loop. For the thick density, we double the number of taps, but still use prime numbers to locate them.

```
virtual bool processAudioFrame(const float* inputFrame,
    float* outputFrame,
    uint32_t inputChannels,
    uint32_t outputChannels)
```

```
{
    // --- global feedback from delay in last branch
    double globFB = branchDelays[NUM_BRANCHES-1].readDelay();

    // --- feedback value
    double fb = parameters.kRT*(globFB);

    // --- mono-ized input signal
    double xnL = inputFrame[0];
    double xnR = inputChannels > 1 ? inputFrame[1] : 0.0;
    double monoXn = double(1.0 / inputChannels)*xnL +
                    double(1.0 / inputChannels)*xnR;

    // --- pre delay output
    double preDelayOut = preDelay.processAudioSample(monoXn);

    // --- input to first branch = preDelay + globFB
    double input = preDelayOut + fb;
    for (int i = 0; i < NUM_BRANCHES; i++)
    {
        double apfOut =
                branchNestedAPFs[i].processAudioSample(input);

        double lpfOut = branchLPFs[i].processAudioSample(apfOut);

        double delayOut =
        parameters.kRT*branchDelays[i].processAudioSample(lpfOut);
        input = delayOut + preDelayOut;
    }

    double weight = 0.707;
    double outL= 0.0;
    outL += weight*branchDelays[0].readDelayAtPercentage(23.0);
    outL -= weight*branchDelays[1].readDelayAtPercentage(41.0);
    outL += weight*branchDelays[2].readDelayAtPercentage(59.0);
    outL -= weight*branchDelays[3].readDelayAtPercentage(73.0);

    double outR = 0.0;
    outR -= weight*branchDelays[0].readDelayAtPercentage(29.0);
    outR += weight*branchDelays[1].readDelayAtPercentage(43.0);
    outR -= weight*branchDelays[2].readDelayAtPercentage(61.0);
    outR += weight*branchDelays[3].readDelayAtPercentage(79.0);

    if (parameters.density == reverbDensity::kThick)
    {
        outL += weight*branchDelays[0].readDelayAtPercentage(31.0);
```

```
        outL -= weight*branchDelays[1].readDelayAtPercentage(47.0);
        outL += weight*branchDelays[2].readDelayAtPercentage(67.0);
        outL -= weight*branchDelays[3].readDelayAtPercentage(83.0);

        outR -= weight*branchDelays[0].readDelayAtPercentage(37.0);
        outR += weight*branchDelays[1].readDelayAtPercentage(53.0);
        outR -= weight*branchDelays[2].readDelayAtPercentage(71.0);
        outR += weight*branchDelays[3].readDelayAtPercentage(89.0);
    }

    // --- filter
    double tankOutL = shelvingFilters[0].processAudioSample(outL);
    double tankOutR = shelvingFilters[1].processAudioSample(outR);

    // --- sum with dry
    double dry = pow(10.0, parameters.dryLevel_dB / 20.0);
    double wet = pow(10.0, parameters.wetLevel_dB / 20.0);

    if (outputChannels == 1)
        outputFrame[0] = dry*xnL +
                            wet*(0.5*tankOutL + 0.5*tankOutR);
    else
    {
        outputFrame[0] = dry*xnL + wet*tankOutL;
        outputFrame[1] = dry*xnR + wet*tankOutR;
    }
    return true;
}
```

17.15 Chapter Plugin: Reverb

Our reverb FX plugin will connect directly to the *ReverbTank* object as shown in Figure 17.33, which processes mono or stereo signals. The effect object does all of the work, and the plugin shell simply updates the parameters and calls the processing function.

17.15.1 Reverb: *GUI Parameters*

The GUI parameter table is shown in Table 17.16—use it to declare your GUI parameter interface for the plugin framework you are using. If you are using ASPiK, remember to first create your enumeration of control ID values at the top of the *plugincore.h* file, and then use automatic variable binding to connect the linked variables to the GUI parameters. The delay time variable name *kRT* is chosen to match the original App Note. The reverb time control on the final GUI should be labeled ambiguously as it is not in milliseconds or other absolute time. The same applies to the damping control, which governs the *g* coefficient of the branch LPFs.

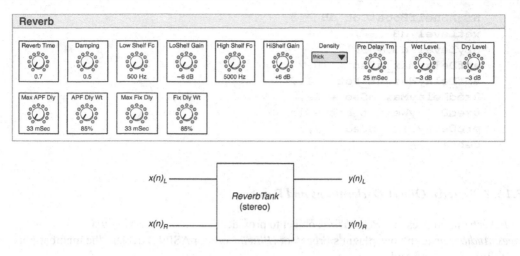

Figure 17.33: The GUI and block diagram for the *Reverb* plugin; note the tweaker controls on the second row.

Table 17.16: The *Reverb* plugin's GUI parameter and linked-variable list; the default values produce a fine-sounding long reverb whose impulse response is shown in Figure 17.31

Control Name	Units	Min/max/default or string-list	Taper	Linked Variable	Linked Variable Type
Reverb Time	–	0.0 / +1.0 / +0.9	a-log	kRT	double
Damping	–	0.0 / 0.5 / 0.3	linear	lpf_g	double
Low Shelf Fc	Hz	20 / 2,000 / 150	1V/oct	lowShelf_fc	double
LoShelf Gain	dB	−20 / 20 / −20	linear	lowShelfBoostCut_dB	double
High Shelf Fc	Hz	1,000 / 5,000 / 4,000	1V/oct	highShelf_fc	double
HiShelf Gain	dB	−20 / 20 / −6	linear	highShelfBoostCut_dB	double
Density	–	{ thick, thin }	–	density	int
Pre Delay	mSec	0 / 100 / 25	linear	preDelay_mSec	double
Wet Level	dB	−60.0 / 12.0 / −12.0	linear	wetLevel_dB	double
Dry Level	dB	−60.0 / 12.0 / 0.0	linear	dryLevel_dB	double
Max APF Dly	mSec	0 / 100 / 33	linear	apfDelayMax_mSec	double
APF Dly Wt	%	1 / 100 / 85	linear	apfDelayWeight_Pct	double
Max Fix Dly	mSec	0 / 100 / 81	linear	fixedDelayMax_mSec	double
APF Fix Wt	%	1 / 100 / 100	linear	fixedDelayWeight_Pct	double

ASPiK: top of *plugincore.h* file

```
enum controlID {
    kRT = 0,
    lpf_g = 1,
    lowShelf_fc = 2,
    lowShelfBoostCut_dB = 3,
    highShelf_fc = 4,
```

```
        highShelfBoostCut_dB = 5,
        wetLevel_dB = 8,
        dryLevel_dB = 9,
        apfDelayMax_mSec = 10,
        apfDelayWeight_Pct = 11,
        fixedDelayMax_mSec = 12,
        fixedDelayWeight_Pct = 13,
        preDelayTime_mSec = 7,
        density = 6
};
```

17.15.2 Reverb: *Object Declarations and Reset*

The *Reverb* plugin uses one *ReverbTank* object to provide the stereo processing via *processAudioFrame*. In your plugin's processor (*PluginCore* for ASPiK) declare the member plus the *updateParameters* method.

```
// --- in your processor object's .h file
#include "fxobjects.h"

// --- in the class definition
protected:
        ReverbTank reverb;
        void updateParameters();
```

You need only reset the object with the current sampling rate—it will reset its sub-components and so on.

ASPiK: *reset( )*

```
// --- reset the filters
reverb.reset(resetInfo.sampleRate);
```

17.15.3 Reverb: *GUI Parameter Update*

The *Reverb* object requires numerous parameters for updating that are shown on the GUI, including the secret tweaker controls. Gather the values from the GUI parameters and apply them to the custom data structure, using the *getParameters* method to get the current parameter set, and then overwrite only the parameters your plugin actually implements. For ASPiK we use the helper *convertIntToEnum* to convert the user's selected value (a zero-indexed integer) to the corresponding strongly typed enumeration as an integer.

ASPiK: *updateParameters( )* function

```
ReverbTankParameters params = reverb.getParameters();
params.kRT = kRT;          // "Reverb Time"
params.lpf_g = lpf_g;      // "Damping"

params.lowShelf_fc = lowShelf_fc;
params.lowShelfBoostCut_dB = lowShelfBoostCut_dB;
```

```
params.highShelf_fc = highShelf_fc;
params.highShelfBoostCut_dB = highShelfBoostCut_dB;

params.preDelayTime_mSec = preDelayTime_mSec;
params.wetLevel_dB = wetLevel_dB;
params.dryLevel_dB = dryLevel_dB;

// --- tweakers
params.apfDelayMax_mSec = apfDelayMax_mSec;
params.apfDelayWeight_Pct = apfDelayWeight_Pct;
params.fixeDelayMax_mSec = fixeDelayMax_mSec;
params.fixeDelayWeight_Pct = fixeDelayWeight_Pct;

params.density = convertIntToEnum(density, reverbDensity);
reverb.setParameters(params);
```

17.15.4 Reverb: *Process Audio*

The *ReverbTank* object does all of the heavy lifting. All we need to do is update the parameters, then process the audio frame.

ASPiK: *processAudioFrame( )* function

```
// --- object does all the work on frames
reverb.processAudioFrame(processFrameInfo.audioInputFrame,
        processFrameInfo.audioOutputFrame,
        processFrameInfo.numAudioInChannels,
        processFrameInfo.numAudioOutChannels);
```

Non-ASPiK

You gather your frame input and output samples into small arrays (you can statically declare them to minimize cost). For example:

```
double xnL = //< get from your framework: input sample L
double xnR = //< get from your framework: input sample R

float inputs[2] = { xnL, xnR };
float outputs[2] = { 0.0, 0.0 };

// --- process (in, out, 2 inputs, 2 outputs)
reverb.processAudioFrame(inputs, outputs, 2, 2);

//< framework left output sample = outputs[0];
//< framework Right output sample = outputs[1];
```

Rebuild and test the plugin. Play around with the tweaker controls and listen to the wide range of reverbs you can get from this structure. The default values in Table 17.16 are used to establish a nice-sounding long reverb. You can go from there and generate your own presets or other algorithms. Using the values in Table 17.16, we measured the long impulse responses in RackAFX for the reverb with thick and thin density settings (these are shown in Figure 17.34).

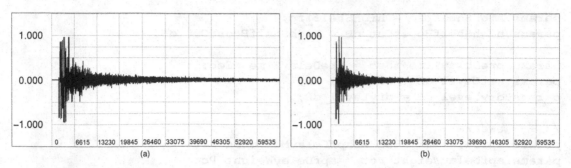

Figure 17.34: Long impulse responses for the *Reverb* plugin with the default values in Table 17.16 and (a) thick and (b) thin density settings.

17.16 Homework

1.   Design a Schroeder reverb plugin.
2.   Design a Moorer reverb plugin.
3.   Implement Datorro's plate reverb algorithm.
4.   Design a reverb plugin that uses solely all-pass and nested all-pass modules.
5.   Modify the *ReverbTank* object with the addition of more branches and/or multiple feedback paths from one branch to another.

17.17 Bibliography

Beranek, L. 1986. *Acoustics*. New York: American Institute of Physics.

Browne, S. 2001. *Hybrid Reverberation Algorithm Using Truncated Impulse Response Convolution and Recursive Filtering*. MS diss., University of Miami., Florida.

Chemistruck, M., Marcolini, K. and Pirkle, W. 2012. *Generating Matrix Coefficients for Feedback Delay Networks Using Genetic Algorithms*. Proc. 133rd Audio Engineering Society Convention, San Francisco.

Dahl, L. and Jot, J.-M. 2000. *A Reverberator Based On Absorbent All-Pass Filters*. Proceedings of the COST G-6 Convention on Digital Audio Effects, Preprint DAFX-1.

Dattorro, J. 1997. "Effect Design Part 1: Reverberators and Other Filters." *Journal of the Audio Engineering Society*, Vol. 45, No. 9.

Frenette, J. 2000. *Reducing Artificial Reverberation Requirements using Time-Variant Feedback Delay Networks*. MS diss., University of Miami., Florida.

Gardner, W. G. 1992. *The Virtual Acoustic Room*. MS diss., Massachusetts Institute of Technology.Massachusetts.

Gardner, W. G. 1995. "Efficient Convolution without Input-Output Delay." *Journal of the Audio Engineering Society*, Vol. 43, No. 3.

Gerzon, M. A. 1972. "Synthetic Stereo Reverberation." *Studio Sound Magazine*, No. 14.

Greisinger, D. 1989. *Practical Processors and Programs for Digital Reverberation*. 7th International Conference of the Audio Engineering Society. Toronto.

Greisinger, D. 1995. "How Loud Is My Reverberation?" *Journal of the Audio Engineering Society*, Preprint 3943.

Jot, J.-M. and Chaigne, A. 1995. "Digital Delay Networks for Designing Artificial Reverberators." *Journal of the Audio Engineering Society*, Preprint 3930.

Kahrs, M. and Brandenberg, K. 1998. *Applications of Digital Signal Processing to Audio and Acoustics*, Chap. 3. Boston: Klower Academic Publishers.

Kuttruff, H. 1991. *Room Acoustics*. New York: Elsevier Science Publishing Company.

Moorer, J. A. 1979. "About This Reverberation Business." *Computer Music Journal*, Vol. 3, No. 2.

Reilly, A. and McGrath, D. S. 1995. "Convolution Processing for Realistic Reverb." *Journal of the Audio Engineering Society*, Preprint 3977.

Roads, C. 1996. *The Computer Music Tutorial*, Chap. 3. Cambridge: The MIT Press.

Sabine, W. 1972. "Reverberation." In *Acoustics: Historical and Philosophical Development*, ed. by R. D. Lindsay, Dowden, Hutchinson, and Ross. Stroudsburg, Pennsylvania.

Schroeder, M. 1962. "Natural Sounding Artificial Reverberation." *Journal of the Audio Engineering Society*, Vol. 10, No. 3.

Schroeder, M. 1984. "Progress in Architectural Acoustics and Artificial Reverberation: Concert Hall Acoustics and Number Theory." *Journal of the Audio Engineering Society*, Vol. 32, No. 4.

Smith, J. O. 1985. *A New Approach to Digital Reverberation Using Closed Waveguide Networks*. Proceedings of the 1985 International Computer Music Conference, Computer Music Association. Burnaby, British Columbia, Cananda.

Smith, J. O. and Rocchesso, D. 1994. *Connections between Feedback Delay Networks and Waveguide Networks for Digital Reverberation*. Proceedings of the 1994 International Computer Music Conference, Computer Music Association. Aarhus, Denmark.

Spin Semiconductor, Corp. *Audio Effects*, www.spinsemi.com/knowledge_base/effects.html#Reverberation, Accessed August 1, 2018.

Stautner, J. and Puckette, M. 1982. "Designing Multi-Channel Reverberators." *Computer Music Journal*, Vol. 6, No. 1.

CHAPTER 18

Dynamics Processing

Dynamics processors are designed to automatically control the amplitude, or gain, of an audio signal and consist of two families: compressors and expanders. Using the technical definition, compressors and expanders both change the gain of a signal after its level *rises above* a predetermined threshold value. A compressor reduces the gain of the signal once it *goes over* the threshold. An expander raises the gain of a signal after it crosses above the threshold. With the exception of noise reduction systems, true expanders are rare since they can easily cause instability, runaway gain, and distortion. What is normally called an "expander" today is actually a downward expander. A downward expander reduces the gain of a signal after it *drops below* the threshold. We will be designing a downward expander but will use the common definition and refer to it simply as an "expander." Both compressors and expanders require the user to decide on the threshold of operation as well as a ratio value that tells the device how much gain reduction to implement.

Figure 18.1 shows the input/output transfer functions for the compressor family. The line marked 1:1 is unity gain. A given input signal amplitude x(dB) results in the same output y(dB). Above the threshold, the output level is compressed according to the ratio. For example, above the threshold on the 2:1 line, every increase in 1 dB at the input results in an increase of only 0.5 dB at the output. Above the threshold on the 4:1 line, each 1 dB input increase results in a 0.25 dB output increase. On the ∞:1 line, increasing the input amplitude beyond the threshold results in no output increase at all thus y(dB) is limited to the threshold value. This version of the device is called a *limiter* and represents the most extreme form of compression.

Figure 18.2 shows the input/output transfer functions for the expander family. The ratios are reversed. You can see that as the signal falls below the threshold, it is attenuated by the ratio amount. For example, below the threshold on the 1:2 line, every decrease in 1 dB at the input results in a decrease 2 dB at the output. On the 1: ∞ line, when the input falls below the threshold it receives infinite attenuation—that is, it is muted. This version of the device is called a *gate* (or *noise gate*) and represents the most extreme form of downward expansion. Perhaps the most important aspect of both families is that their gain reduction curves are linear for logarithmic axes. These devices operate in the log domain. So, the two families of dynamics processors yield four combinations: compression and limiting, and expansion and gating. We will implement all four of them.

Figure 18.3 shows the feed-forward and feedback topologies for dynamics processors. There is some debate as to which is best; we'll stay out of that argument and design the feed-forward version first. Then, you can design a feedback version for yourself for homework. Both designs include a detector, gain computer and a DCA (Digitally Controlled Amplifier/Attenuator). The detector analyzes the input signal and the computer calculates a new gain value to be applied to the signal. The detector and computer blocks lie in the side chain of the circuit.

Because the compressor and downward expander are both gain reduction devices, an additional stage of make-up gain is required. In some designs, the make-up gain is calculated in the side chain while in

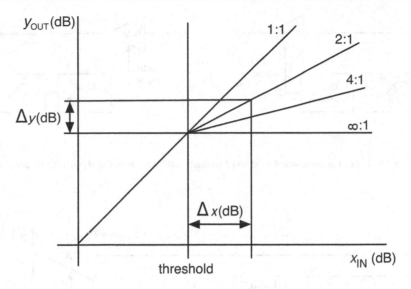

Figure 18.1: The generalized input/output transfer curves for a compressor at various reduction ratios. The ratio is in the form Δx:Δy and is shown for the 2:1 compression curve.

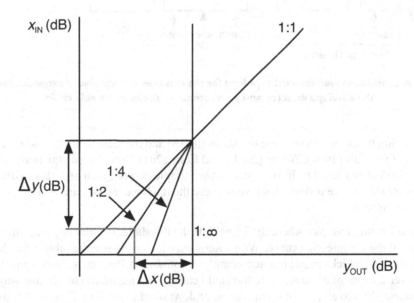

Figure 18.2: The generalized input/output transfer curves for a downward expander at various reduction ratios. The ratio is in the form Δx:Δy and is shown for the 1:2 expander curve.

others it sits outside of the DCA. Both accomplish the same task. Figure 18.4 shows the block diagram of our feed-forward dynamics processor plugin.

The *EnvelopeDetector* object that we used in Chapter 13 has a log output mode, so we can combine the detector and log conversion blocks together as shown in the dotted box in Figure 18.4. These feed

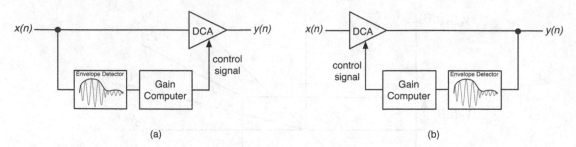

Figure 18.3: (a) Feed-forward and (b) feedback architectures for dynamics processors.

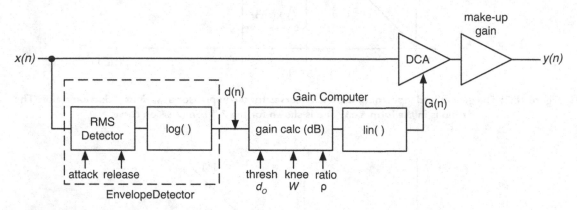

Figure 18.4: A generalized feed-forward topology for the compressor/downward expander families shows the envelope detector and gain computer blocks in the side chain.

the gain calculation block, which uses the threshold (in dB) and the ratio to set the gain for a given detector output. Orfanidis (1996), Zölzer (2011), and Reiss (2011) have given equations that describe the gain reduction calculation. The Reiss equations are simple to work with and also easily create a smooth 2nd order knee curve if desired, so we will use them. However, you are encouraged to research the other sources as well.

The point in the input/output plot where the signal hits the threshold, thereby engaging the device, is called the *knee* of the compression curve. With a *hard-knee* the signal is either above the threshold or below it, and the device kicks in and out accordingly. A *soft-knee* allows the gain control to change more slowly over a width of dB around the threshold called the *knee width* (W). In this way, the device moves more smoothly into and out of compression or downward expansion. Figure 18.5 shows the hard- and soft-knee curves. The dbx Corporation trademarked the name Over Easy® to describe its soft-knee curve setting.

Our *AudioDetector* object will handle the detection and log conversion, and also has the ability to track signals above unity gain (0.0 dB). We will need a gain calculation function to convert the detected value in dB in combination with the threshold, ratio, and knee width controls, into a gain reduction value. We first calculate the desired output $y(n)_{dB}$ that we need for a given detected input value $d(n)_{dB}$. Then, we may convert the necessary gain factor back to a linear value and apply it to the DCA.

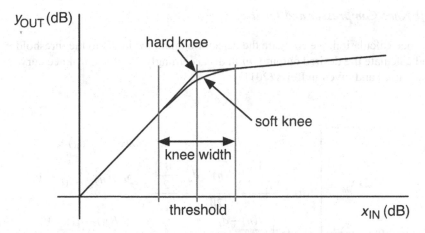

Figure 18.5: Hard and soft-knee compression curves; interpolation is used to calculate the soft-knee portion.

18.1 Compressor Output Calculation

For the compressor, we will use the following output calculations. These follow the designated signal labels in Figure 18.4 and listed here.

- $x(n), y(n)$ = input/output
- $d(n)$ = detector output in dB
- $G(n)$ = gain value for current sample
- d_o = threshold in dB
- ρ = ratio
- W = knee width (dB)

18.1.1 Hard-Knee Compressor and Limiter

For the hard-knee calculation we compare the detector output $d(n)$ in dB to the threshold d_0 and calculate the desired output $y(n)$ in dB accordingly. For a 1:N compression ratio, the variable ρ = N. Note that for a perfect detector, $d(n)_{dB}$ and $x(n)_{dB}$ are interchangeable and we will make that assumption here; you may replace $d(n)$ with $x(n)$ if it helps you understand the theory. We use $d(n)$ here because this is the value we will use in our C++ code.

Compressor

$$y(n)_{dB} = \begin{cases} d(n) & d(n) \le d_0 \\ d_0 + \dfrac{d(n) - \mathrm{d}_0}{\rho} & d(n) > d_0 \end{cases} \qquad (18.1)$$

Limiter

$$y(\mathrm{n})_{dB} = \begin{cases} d(\mathrm{n}) & d(\mathrm{n}) \le d_0 \\ d_0 & d(\mathrm{n}) > d_0 \end{cases}$$

18.1.2 Soft-Knee Compressor and Limiter

For the soft-knee calculation, we compare the detector output $d(n)$ in dB to the threshold d_0, and knee width W, and calculate the desired output $y(n)$ in dB accordingly. For the soft-knee curve, 2nd order interpolation is used and given in Reiss (2011).

Compressor

$$y(n)_{dB} = \begin{cases} d(n) & 2(d(n) - d_0) < -W \\ d(n) + \dfrac{\left(\dfrac{1}{R-1}\right)\left(d(n) - d_0 + \dfrac{W}{2}\right)^2}{2W} & 2(|d(n) - d_0|) \leq W \\ d_0 + \dfrac{d(n) - d_0}{\rho} & 2(d(n) - d_0) > W \end{cases} \tag{18.2}$$

Limiter

$$y(n)_{dB} = \begin{cases} d(n) & 2(d(n) - d_0) < -W \\ d(n) - \dfrac{\left(d(n) - d_0 + \dfrac{W}{2}\right)^2}{2W} & 2(|d(n) - d_0|) \leq W \\ d_0 & 2(d(n) - d_0) > W \end{cases}$$

18.2 Downward Expander Output Calculation

For the downward expander, we will use the following output calculations for $y(n)$ as shown in the next section. These follow the designated signal labels in Figure 18.4.

18.2.1 Hard-Knee Expander and Gate

For the hard-knee calculation we compare the detector output $d(n)$ in dB to the threshold d_0 and calculate the desired output $y(n)$ in dB accordingly. For a $N{:}1$ expansion ratio, the variable $\rho = N$.

Downword Expander

$$y(n)_{dB} = \begin{cases} d(n) & d(n) \geq d_0 \\ d_0 + \rho(d(n) - d_0) & d(n) < d_0 \end{cases} \tag{18.3}$$

Gate

$$y(n)_{dB} = \begin{cases} d(n) & d(n) \geq d_0 \\ -\infty & d(n) < d_0 \end{cases}$$

18.2.2 Soft-Knee Expander

For the soft-knee calculation we compare the detector output $d(n)$ in dB to the threshold d_0, and knee width W, and calculate the desired output $y(n)$ in dB accordingly. For the soft-knee curve, 2nd order interpolation is used (and given in Reiss, 2011). The gate version, with an infinite ratio and with full attenuation to 0.0 (or $-\infty$ dB) cannot truly have a soft knee, as the knee width constraint cannot be met. However, you may approximate a soft-knee gate by using a large expander ratio of greater than 20 or so, and you may also experiment with a gate that does not attenuate all the way to 0.0, which is sometimes preferred as it reduces clicks and distortion (see the Homework section).

Downword Expander

$$
y(n)_{dB} = \begin{cases}
d(n) & 2(d(n)-d_0) > W \\
d(n) - \dfrac{\left(\dfrac{1}{R}\right)\left(d(n)-d_0-\dfrac{W}{2}\right)^2}{2W} & 2(|d(n)-d_0|) > -W \\
d_0 + \rho(d(n)-d_0) & 2(d(n)-d_0) \le -W
\end{cases}
\tag{18.4}
$$

Figure 18.6 shows the results of the output calculations for the (a) compressor, (b) limiter, (c) expander, and (d) quasi-gate. The quasi-gate is an expander with a ratio of 50:1 (notice the y-axis scale), which produces acceptable soft-knee characteristics.

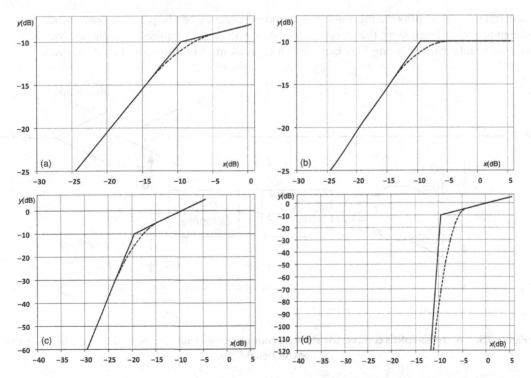

Figure 18.6: The result of input/output calculations for (a) 1:5 compressor, (b) limiter, (c) 5:1 downward expander, and (d) 50:1 downward expander (quasi-gate). In all cases the threshold is −9.5 dB and the soft-knee width is 10 dB. The dotted line shows the soft-knee version.

18.3 Final Gain Calculation

With the desired output level in dB calculated, we can calculate the final linear gain coefficient to apply to the DCA (the DCA is simply a line of code, though we refer to it as an entity). The output/input ratio with dB is performed with a simple subtraction. The gain value is then converted back to linear using the standard formula. For all the processors, the final calculation for the gain factor $G(n)$ is as follows.

$$g(n)_{dB} = y(n)_{dB} - d(n)_{dB}$$

$$G(n) = 10^{\frac{g(n)_{dB}}{20}}$$

(18.5)

18.4 Stereo-Linked Dynamics Processor

A stereo dynamics processor consists of two independent processor components, one each for the left and right channels. The early analog versions such as the dbx 165A® utilize two hardware units that are connected with a stereo-linking cable. The cable effectively connects the side-chain input of one device to the other. The controls are then shared so that one of the unit's attack, release, threshold, etc. controls are inoperable and all modifications are made from the other unit. Figure 18.7 shows the block diagram of a stereo-linked dynamics processor.

18.5 Spectral Dynamics Processing

A spectral processor splits the input into two or more frequency bands, processes each band independently, and then sums the processed signals together. This requires multiple, independent processors and a band-splitting filter bank. The filter bank must be chosen so that it splits the

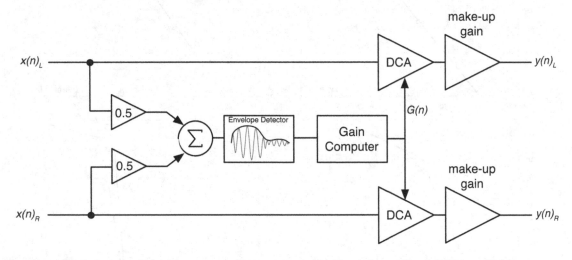

Figure 18.7: A stereo-linked compressor shares a common side-chain, and both DCAs are fed the same control signal $G(n)$.

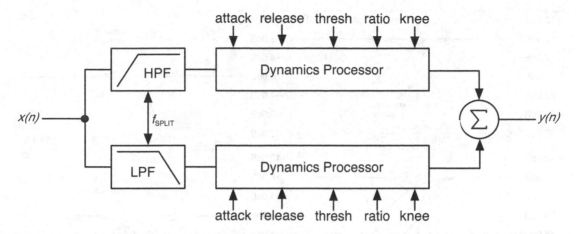

Figure 18.8: A two-band spectral dynamics processor.

frequency bands in a way such that recombining them produces a perfect reconstruction with a flat frequency response. Since we will package a complete dynamics processor in a single C++ object, we may easily add more processors and frequency channels. Laying out the GUI can be cumbersome because there are multiple sets of controls for each parameter, but the underlying plugin will simply use multiple instances of the dynamics processor object to do the grunt-work. Our attention needs to focus on the filter bank more than the actual processing. Figure 18.8 shows a two-band spectral processor.

The filter bank design requires that the combined channels, devoid of any other signal processing, will have the same magnitude characteristics of the original input signal. There are several filter back designs to choose from, but for a simple two-, three-, or four-band spectral compressor, we may use the Linkwitz-Riley filters from Chapter 11. These filters, originally designed for crossovers in loudspeakers, have corner frequencies that are −6 dB down from the maximum, rather than −3 dB (as shown in Figure 11.14). Consider a simple two-way band splitter consisting of a low-pass and high-pass filter in parallel, as seen in the left side of Figure 18.8. The two filters share a common cutoff frequency. If we do an experiment and split, then recombine the parallel outputs without any processing whatsoever, we would like the result to have a flat magnitude response. Figure 18.9a and b show the independent and combined frequency responses of a pair of Butterworth filters, one LPF and the other HPF. In Figure 18.9a the outputs are summed directly, resulting in a massive notch right at the cutoff frequency. In Figure 18.9b, the output of the HPF is inverted, producing a bump in the combined response. Figure 18.9c and d show the same experiment using Linkwitz-Riley filters but in this case, the normal version produces a notch. However, with the phase of the HPF flipped, the resulting magnitude response is flat as shown in Figure 18.9d. In practice, either the LPF or HPF may be inverted with the same results. The *fxobjects.h* file contains a built-in two-band filter bank made with Linkwitz-Riley filters. There is a single adjustment for the shared cutoff frequencies, and the HPF output is inverted.

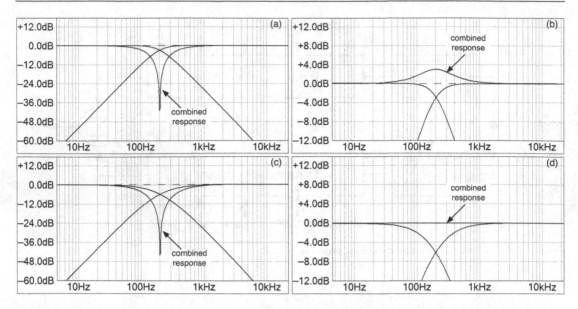

Figure 18.9: Two-band filter bank output for (a) Butterworth filters, normal phase producing a notch (b) Butterworth filters, one output inverted prior to recombination producing a hump (c) Linkwitz-Riley filters, normal phase producing a notch and (d) Linkwitz-Riley filters, one output inverted prior to recombination producing a flat response. In all cases the split frequency is 200 Hz.

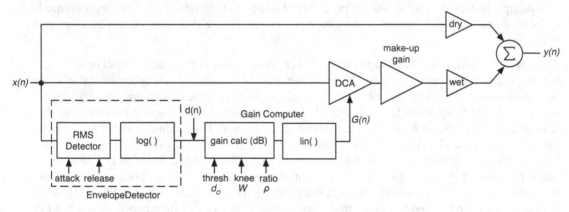

Figure 18.10: A parallel dynamics processor blends the unprocessed dry signal with the processed wet component.

18.6 Parallel Dynamics Processing

Parallel dynamics processing involves simply blending the original unprocessed signal back in with the processed signal in some mix ratio. Typically, the processed signal is severely processed with a large dynamics ratio and/or low threshold. It is especially useful on instruments where heavy processing is overly audible, such as acoustic guitar. For the plugin designer, this is just a matter of allowing the user to alter the output mix ratio with wet and dry controls. We observe that many recording engineers

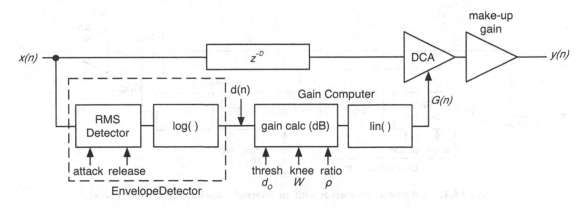

Figure 18.11: A look-ahead dynamics processor uses a delay in the feed-forward path so that the detector may "preview" the signal.

prefer individual wet/dry controls, rather than one shared ratio control, as this allows more freedom of adjustment and mimics the use of the mixing console to implement parallel processing with spare audio channels. Figure 18.10 shows a simple example of a parallel dynamics processor.

18.7 Look-Ahead Processing

A problem with the detector circuit is that for any non-zero attack and release time settings, the detector output lags behind the input in time. Since the detector output lags, the gain reduction actually misses the event that needs to be attenuated. We can accommodate for this detector lag by using a look-ahead technique. We can't look ahead into the future of a signal, but we can delay the present. If we insert a delay line in the forward signal path (not the side chain) we can make up for the detector delay so that the detector begins charging before the signal actually gets to the DCA, as shown in Figure 18.11. The dbx165 manual refers to the look-ahead operation as "previewing" the signal. You may easily use the *DelayLine* object included in *fxobjects.h* to accomplish the previewing operation. The delay line length may be set approximately to match the attack time of the side chain.

Note that the *PluginKernel* allows you to set a latency time for your plugin to accommodate the latency that the delay line introduces. In the *PluginCore* constructor, set the *pluginDescriptor's* latency in samples.

```
pluginDescriptor.latencyInSamples = 1024;
```

If your latency changes during the course of operation, just set this variable to the new latency. Host polling will take care of the rest of the operation.

18.8 External Keying

A popular way to use a dynamics processor is to route an external signal into the side chain and allow that signal's characteristics to alter the gain of the DCA instead of the input signal. The side chain is sometimes called a *keyed* input so this is referred to as "external keying" or creating a "keyed" device (e.g. a "keyed gate"). For musical applications, the side chain may be used as an on-off gating device that adds rhythmic gating to the signal processing. Another application is called a *ducker* where the

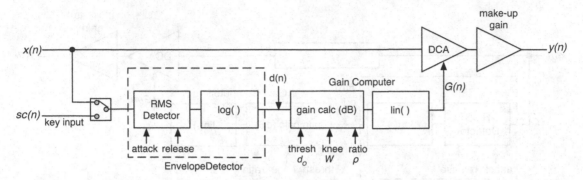

Figure 18.12: A dynamics processor with an external side-chain or key input *sc(n)*.

presence of an audio signal at the side-chain attenuates the feed-forward signal path. This is often heard in radio advertisements, where an announcer's voice causes the background music or audio to "duck" or be attenuated. A keyed dynamics processor block diagram is shown in Figure 18.12.

18.8.1 ASPiK Users: Side Chain Code

The *PluginKernel* is designed to allow side chaining and is simple to set up and use. The first step is to signify that your plugin wants the DAW to expose a side-chain input so that the user may route another audio track into it for processing. In the *PluginCore* constructor, set the *pluginDescriptor's hasSidechain* flag to true (this is done for you during the CMake process, but you can override it manually here):

```
pluginDescriptor.hasSidechain = true;
```

This will notify the DAW to expose the side-chain input on the plugin GUI window header. For an AAX plugin, a small key-shaped icon will appear that allows the user to connect an audio bus to the side chain. For AU plugins, a side chain drop list will appear at the right top of the plugin's GUI window that allows the user to select the track to use. VST plugins vary depending on the host DAW's preference. AU and VST plugins allow both mono and multi-channel side chains. AAX plugins only allow a mono audio signal to be used for the side chain.

In the *processAudioFrame( )* function, the argument is a *ProcessFrameInfo* structure. This structure holds array pointers for the audio input and outputs. The side chain inputs arrive via the *auxInputData* array. There is an accompanying *numAuxInputChannels* variable that lets you know the width of the audio bus connected to the side chain. This variable will almost always be only 1 (mono) or 2 (stereo). For multi-channel side chains, it is up to you as to how to process the data. If the user has the side chain disconnected, the *numAuxInputChannels* variable will be set to 0. A non-zero *numAuxInputChannels* count signifies that the user has enabled and connected a side-chain input. Processing the side chain may be done easily with code similar to the following code chunk. Here we process either a mono or stereo input. The logic here simply picks up the first two channels of any side-chain signal that has more than one input.

```
// --- array to hold side-chain audio samples
double sidechain[2] = { 0.0, 0.0 };
```

```
// --- detect
if (processFrameInfo.numAuxInputChannels == 1)
      sidechain[0] = processFrameInfo.auxInputData[0];
else if (processFrameInfo.numAuxInputChannels > 1)
{
      // --- stereo sidechain
      sidechain[0] = processFrameInfo.auxInputData[0];
      sidechain[1] = processFrameInfo.auxInputData[1];
}
```

18.9 Gain Reduction Metering

Metering is important for our dynamics processor so that the input and output amplitudes may be compared. In addition, we frequently want to show the gain reduction amount on a meter that is usually inverted—either upside down, or indicating from right-to-left as shown in Figure 18.13a and b. For ASPiK users, the *PluginParameter* object that implements the meter may be easily set for inverted operation. This will automatically render the meter as upside down for vertical meters, or right-to-left for horizontal meters. You must supply the gain reduction value to show on the meter on the range of [0.0, 1.0] and the *PluginParameter* and custom VU meter object (found in *customcontrols.h*) will handle the rendering for you.

18.10 Alternate Side-Chain Configurations

As Reiss points out, there are several different configurations for the contents of the side chain. The version we've used so far is a standard linear timing placement configuration shown in Figure 18.14a. One alternate configuration, shown in Figure 18.14b, is called the *biased configuration;* in this

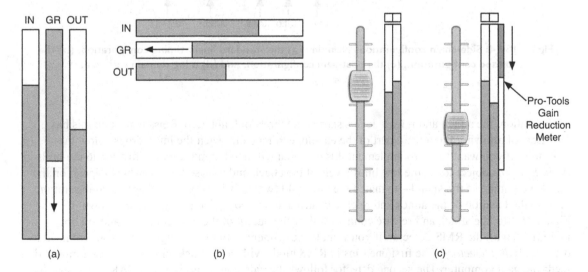

Figure 18.13: (a) vertical and (b) horizontal input, gain reduction (GR) and output meters. (c) Pro Tools meters; the channel on the right features a gain reduction meter and the small arrows denote the direction of the meter's travel.

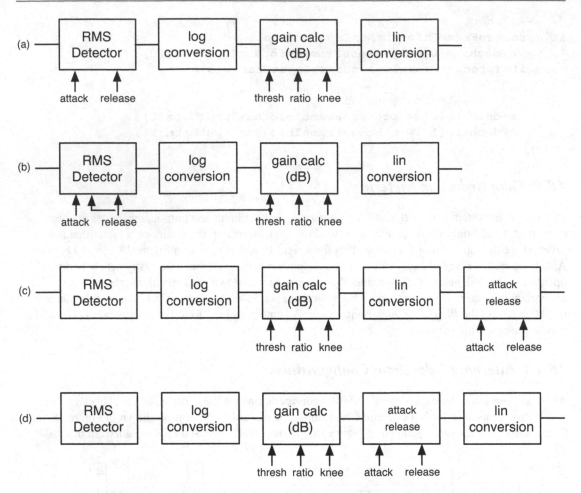

Figure 18.14: Side-chain configurations including (a) the standard linear timing configuration, (b) the biased configuration, (c) the post-gain configuration, and (d) a log domain variation.

configuration, the attack and release levels start at the threshold, not zero. Reiss notes that this has the effect of allowing the release part of the envelope to fade out when the input drops below the threshold. If you want to try to implement this one, you will need to make a modification to the built-in *AudioDetector* object to set the low limit levels of the attack and release times—which depend on the linear magnitude of the signal—to match the threshold (which is in dB). The other two configurations involve the location of the attack and release controls. In the *post-gain configuration*, shown in Figure 18.14c, the attack and release controls follow the output of the gain computer and are not intermixed with the RMS detector. If you want to experiment with this configuration, you will need to use two *AudioDetectors*. The first one runs in RMS mode with zero attack and zero release times and feeds the gain computer. The second detector follows the gain computer, runs in PEAK mode, and has the user-controlled attack and release controls connected to it. A second variation called log domain in Figure 18.14d places the timing controls after the gain computer but before the conversion to the linear domain, so that they operate in the log domain. It could also be implemented using an extra detector placed in the proper location in the side chain.

18.11 C++ DSP Object: **LRFilterBank**

The *LRFilterBank* object implements a pair of Linkwitz-Riley filters that split the input single into two bands: LF and HF. A single f_c control sets the split point between the high pass and low-pass sections. As per the discussion in Section 18.5, the output of the HPF section is inverted. You can also extend this object with your own Linkwitz-Riley band-pass filters: you just cascade a HPF $\rightarrow$ LPF; the HPF f_c control sets the low edge frequency, and the LPF f_c control sets the high edge frequency. You need to invert one of the filters, depending on the filter on the other side of it—only one filter in each overlapping f_c pair is inverted. This object is used in the spectral compressor, or any plugin where you would like to split the signal into separate frequency bands.

18.11.1 LRFilterBank: *Enumerations and Data Structure*

The *LRFilterBank* object requires no custom enumerations. The object's parameters are updated via the custom *LRFilterBankParameters* structure that contains a single member representing the frequency split point between the two bands. If you extend the object to add more bands, you can add more split-frequency members here.

```
struct LRFilterBankParameters
{
        LRFilterBankParameters() {}
        LRFilterBankParameters& operator=(. . .){ . . . }

        // --- individual parameters
        double splitFrequency = 1000.0;
};
```

Another custom structure is declared for the output of the object (one input sample will yield two output samples, one for each frequency band). If you extend the object to add more bands, you can add more split-frequency members.

```
struct FilterBankOutput
{
        FilterBankOutput() {}

        // --- band-split filter output
        double LFOut = 0.0;
        double HFOut = 0.0;

        // --- add more filter channels here; or use an array[]
};
```

18.11.2 LRFilterBank: *Members*

Tables 18.1 and 18.2 list the *LRFilterBank* member variables and member functions.

18.11.3 LRFilterBank: *Programming Notes*

The *LRFilterBank* object is very simple: it has only one parameter, and it is easy to implement as well. Although it shares the *IAudioSignalProcessor* interface, the function *processAudioSample* is inactive. Use *processFilterBank* instead.

Table 18.1: *LRFilterBank* member variables

LRFilterBank Member Variables		
Type	Name	Description
LRFilterBankParameters	parameters	Object parameter structure
AudioFilter	lpFilter	Low-pass filter
AudioFilter	hpFilter	High-pass filter

Table 18.2: *LRFilterBank* member functions

LRFilterBank Member Functions		
Returns	Name	Description
LRFilterBankParameters	getParameters	Get all parameters at once
void	setParameters Parameters: —LRFilterBankParameters _parameters	Set all parameters at once
bool	reset Parameters: —double sampleRate	Resets sub-components
double	processAudioSample Parameters: —double xn	**Function does nothing: use *processFilterBank* instead**
FilterBankOutput	processFilterBank Parameters: —double xn Returns: a FilterBankOutput structure	Split the signal into LF and HF bands, output samples in *FilterBankOutput* structure

Using the *LRFilterBank* object is straightforward.

1. Reset the object to prepare it for streaming.
2. Set the split frequency in the *LRFilterBankParameters* custom data structure.
3. Call *setParameters* to update the object.
4. Call *processFilterBank*, passing the input value in and receiving the two output samples in the output data structure.

Function: *reset*
Details: This function simply resets its two filter objects.

Function: *setParameters*
Details: This function sets the f_c value for each of its filters—they share the same cutoff frequency.

```
void setParameters(const LRFilterBankParameters& _parameters)
{
    // --- update structure
    parameters = _parameters;

    // --- update member objects
    AudioFilterParameters params = lpFilter.getParameters();
```

```
        params.fc = parameters.splitFrequency;
        lpFilter.setParameters(params);
        hpFilter.setParameters(params);
}
```

Function: *processFilterBank*
Details: This function splits the signal into two outputs.

```
FilterBankOutput processFilterBank(double xn)
{
        FilterBankOutput output;

        // --- process the LPF
        output.LFOut = lpFilter.processAudioSample(xn);

        // --- invert the HP filter output so that recombination will
        //     result in the correct phase and magnitude responses
        output.HFOut = -hpFilter.processAudioSample(xn);

        return output;
}
```

18.12 C++ Effect Object: DynamicsProcessor

The *DynamicsProcessor* object implements the compressor and downward expander algorithms with the following parameters:

- threshold (dB)
- knee width (dB)
- ratio
- attack and release times (mSec)
- makeup gain (dB)
- enable hard limiter/gate mode (ratio becomes infinite)
- enable soft knee
- gain reduction value exportable (for metering)
- ability to use external side chain (key) input

18.12.1 DynamicsProcessor: Enumerations and Data Structure

The *DynamicsProcessor* object requires one strongly typed enumeration for the effect type (compressor or downward expander):

```
enum class dynamicsProcessorType { kCompressor, kDownwardExpander };
```

The object's parameters are updated via the custom *DynamicsProcessorParameters* structure that contains members for the ordinary GUI controls for this type of plugin effect. It is longer than most others because there are more GUI controls in this effect. Note also that this is the first custom structure with outbound data members for the gain reduction amount—this is so that the owning object can display a gain reduction meter.

```
struct DynamicsProcessorParameters
{
        DynamicsProcessorParameters() {}
        DynamicsProcessorParameters& operator =(. . .){ . . . }

        // --- individual parameters
        double ratio = 50.0;
        double threshold_dB = -10.0;
        double kneeWidth_dB = 10.0;
        bool hardLimitGate = false; // limit or gate mode
        bool softKnee = true;
        bool enableSidechain = false;
        dynamicsProcessorType calculation =
                                        dynamicsProcessorType::kCompressor
        ;
        double attackTime_mSec = 0.0;
        double releaseTime_mSec = 0.0;
        double outputGain_dB = 0.0;

        // --- outbound values, for owner to use gain-reduction metering
        double gainReduction = 1.0;
        double gainReduction_dB = 0.0;
};
```

18.12.2 DynamicsProcessor: *Members*

Tables 18.3 and 18.4 list the *DynamicsProcessor* member variables and member functions. Notice that this object overrides the *processAuxInputSample* function to accept an external side chain (key) input.

18.12.3 DynamicsProcessor: *Programming Notes*

The *DynamicsProcessor* object uses an *AudioDetector* object to handle the side chain detection. The detector object is set up to detect in dB and to detect values greater than unity gain (0 dBFS).

Using the *DynamicsProcessor* object is straightforward.

1. Reset the object to prepare it for streaming.
2. Set the effect parameters in the *DynamicsProcessorParameters* custom data structure.
3. If you want to enable the external side chain input, set the Boolean flag *enableSidechain*.
4. Call *setParameters* to update the object.
5. Call *processAuxInputAudioSample* to pass the side chain input sample (this will depend on your plugin framework).
6. Call the *processAudioSample* function to process one input sample into a returned output sample.

Function: *reset*
Details: reset the detector and set it for default operation; also zero the side chain input storage register.

Table 18.3: *DynamicsProcessor* member variables

DynamicsProcessor Member Variables		
Type	Name	Description
DynamicsProcessorParameters	parameters	Object parameter structure
AudioDetector	detector	Envelope detector object
double	sidechainInputSample	For external side chain

Table 18.4: *DynamicsProcessor* member functions

DynamicsProcessor Member Functions		
Returns	Name	Description
DynamicsProcessorParameters	getParameters	Get all parameters at once
void	setParameters Parameters: —DynamicsProcessorParameters _parameters	Set all parameters at once
bool	reset Parameters: —double sampleRate	Reset sub-components
double	processAudioSample Parameters: —double xn	Process mono input through modulated delay
double	processAuxInputAudioSample Parameters: —double xn	Process an external side chain input

```
virtual bool reset(double _sampleRate)
{
        sidechainInputSample = 0.0;
        detector.reset(_sampleRate);

        AudioDetectorParameters detectorParams =
            detector.getParameters();
        detectorParams.clampToUnityMax = false;
        detectorParams.detect_dB = true;
        detector.setParameters(detectorParams);
        return true;
}
```

Function: *setParameters*

Details: We simply forward the incoming GUI parameter changes to the detector object—the other parameter changes are handled in our processing function.

```
void setParameters(const DynamicsProcessorParameters& _parameters)
{
      parameters = _parameters;

      AudioDetectorParameters detectorParams =
                                          detector.getParameters();
      detectorParams.attackTime_mSec = parameters.attackTime_mSec;
      detectorParams.releaseTime_mSec = parameters.releaseTime_mSec;
      detector.setParameters(detectorParams);
}
```

Function: *processAuxInputSample*
Details: In this simple function we just store the input sample to be used for the upcoming calculation.

```
virtual double processAuxInputAudioSample(double xn)
{
      sidechainInputSample = xn;
      return sidechainInputSample; // ignoring return — not used
}
```

Function: *computeGain*
Details: This function is the heart of the *DynamicsProcessor* object—it takes all of the parameters and calculates a gain reduction value. The function is quite long, though the calculations are actually very simple and follow exactly those in Equations 18.1–18.5. Please refer to the downloadable project code to review this function's contents.

Function: *processAudioSample*
Details: The processing function implements the feed-forward dynamics processor block diagram in Figure 18.4. Follow that diagram while examining the C++ code.

```
// --- feed-forward dynamics processor flowchart:
/*
      1. detect input signal
      2. calculate gain
      3. apply to input sample
*/
virtual double processAudioSample(double xn)
{
      // --- detect input
      double detect_dB = 0.0;

      // --- if using the sidechain, process the aux input
      if(parameters.enableSidechain)
            detect_dB =
                  detector.processAudioSample(sidechainInputSample);
      else
            detect_dB = detector.processAudioSample(xn);

      // --- compute gain
      double gr = computeGain(detect_dB);
```

```
// --- makeup gain
double makeupGain = pow(10.0, parameters.outputGain_dB / 20.0);

// --- do DCA + makeup gain
return xn * gr * makeupGain;
}
```

18.13 *Chapter Plugin:* Dynamics

Our dynamics processor plugin will implement the feed-forward block diagram in Figure 18.15. The inclusion of an external side chain input will depend on your plugin framework's capabilities and implementation details. Note the use of a single gain reduction meter—it shows a sum of the left and right gain reduction values. For ASPiK users, this meter may also be mapped to the Pro Tools GUI.

18.13.1 Dynamics: *GUI Parameters*

The GUI parameter table is shown in Table 18.5—use it to declare your GUI parameter interface for the plugin framework you are using. If you are using ASPiK, remember to first create your enumeration of control ID values at the top of the *plugincore.h* file, then use automatic variable binding to connect the linked variables to the GUI parameters. The metering information is shown in Table 18.6.

ASPiK: top of *plugincore.h* file

```
enum controlID {
        threshold_dB = 0,
        kneeWidth_dB = 1,
        ratio = 2,
        attackTime_mSec = 3,
        releaseTime_mSec = 4,
        outputGain_dB = 5,
        dynamicsMode = 6,
        hardLimitGate = 7,
        softKnee = 8,
        grMeter = 9
};
```

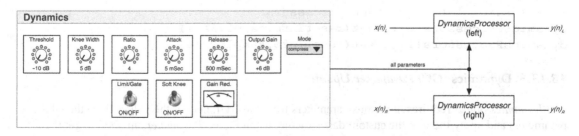

Figure 18.15: The GUI and block diagram for the *Dynamics* plugin.

Table 18.5: The *Dynamics* plugin's GUI parameter and linked-variable list

Control Name	Units	min/max/def or string-list	Taper	Linked Variable	Linked Variable Type
Threshold	dB	−40 / 0 / −10	linear	*threshold_dB*	*double*
Knee Width	dB	0 / 20 / 5	linear	*kneeWidth_dB*	*double*
Ratio	–	1 / 20 / 1	linear	*ratio*	*double*
Attack Time	mSec	1 / 100 / 5	linear	*attackTime_mSec*	*double*
Release Time	mSec	1 / 5000 / 500	linear	*releaseTime_mSec*	*double*
Output Gain	dB	−20 / 20 / 0	linear	*oututGain_dB*	*double*
Dynamics Mode	–	"compress, expand"	–	*dynamicsMode*	*int*
Limiter/Gate	–	"off, on"	–	*hardLimitGate*	*int*
Soft Knee	–	"off, on"	–	*softKnee*	*int*

Table 18.6: The *Dynamics* plugin's gain reduction meter information

Meter Name	Inverted	attack/release	Detect Mode	Calibration	Variable Name
GR	yes	0.0 / 0.0	PEAK	linear	grMeter

18.13.2 Dynamics: *Object Declarations and Reset*

The *Dynamics* plugin uses two *DynamicsProcessor* objects to implement the processing. We use a simple array of two, with one for each of the left and right channels. Of course you would need to declare more members to support more channels. In your plugin's processor (*PluginCore* for ASPiK), declare two members plus the *updateParameters* method.

```
// --- in your processor object's .h file
#include "fxobjects.h"

// --- in the class definition
protected:
    DynamicsProcessor dynamicsProcessors[2];
    void updateParameters();
```

You need only reset the objects with the current sampling rate—it will reset its sub-components and so on.

ASPiK: *reset( )*

```
// --- reset left and right processors
dynamicsProcessors[0].reset(resetInfo.sampleRate);
dynamicsProcessors[1].reset(resetInfo.sampleRate);
```

18.13.3 Dynamics: *GUI Parameter Update*

The *Dynamics* object requires numerous parameters from the GUI. Gather the values from the GUI parameters and apply them to the custom data structure using the *getParameters* method to get the current parameter set, and then overwrite only the parameters your plugin actually implements.

ASPiK: *updateParameters( )* function

```
DynamicsProcessorParameters params =
                           dynamicsProcessors[0].getParameters();
params.threshold_dB = threshold_dB;
params.kneeWidth_dB = kneeWidth_dB;
params.ratio = ratio;
params.attackTime_mSec = attackTime_mSec;
params.releaseTime_mSec = releaseTime_mSec;
params.outputGain_dB = outputGain_dB;
params.calculation = convertIntToEnum(dynamicsMode,
dynamicsProcessorType);
params.hardLimitGate = (hardLimitGate == 1);
params.softKnee = (softKnee == 1);


// --- update objects
dynamicsProcessors[0].setParameters(params);
dynamicsProcessors[1].setParameters(params);
```

18.13.4 Dynamics: *Process Audio and External Keying*

The *Dynamics* objects do the work, but they process only one channel each. If you want to support an external side chain, you need to get the audio sample(s) from your plugin framework and apply them to the object via *processAuxInputSample*. For ASPiK, you use the side chain audio input channel, which is in the *auxAudioInputFrame* array. Notice the logic that detects a mono side chain, in which case it is applied to both dynamics processor objects. This can happen in any DAW, but note that Pro Tools supports only mono side chain inputs. If you are using an external side chain input, first set the dynamics processors to use it (this may be done externally to the process function).

```
DynamicsProcessorParameters params =
                           dynamicsProcessors[0].getParameters();
// --- enable!
params.enableSidechain = true;
dynamicsProcessors[0].setParameters(params);
dynamicsProcessors[1].setParameters(params);
```

ASPiK: *processAudioFrame( )* function

```
// --- read input (adapt to your framework)
double xnL = processFrameInfo.audioInputFrame[0]; //< input sample L
double xnR = processFrameInfo.audioInputFrame[1]; //< input sample R


// --- detect external sidechain: process if it exists
if (processFrameInfo.numAuxAudioInChannels > 0)
     dynamicsProcessors[0].processAuxInputAudioSample(
                           processFrameInfo.auxAudioInputFrame[0]);


// --- process Left channel
double ynL = dynamicsProcessors[0].processAudioSample(xnL);
```

```
// --- detect external sidechain: process if it exists
if (processFrameInfo.numAuxAudioInChannels > 1)
      dynamicsProcessors[1].processAuxInputAudioSample(
                                processFrameInfo.auxAudioInputFrame[1]);
// --- mono sidechain (ProTools)
else if (processFrameInfo.numAuxAudioInChannels > 0)
      dynamicsProcessors[1].processAuxInputAudioSample(
                                processFrameInfo.auxAudioInputFrame[0]);

double ynR = dynamicsProcessors[1].processAudioSample(xnR);

// --- write output (adapt to your framework)
processFrameInfo.audioOutputFrame[0] = ynL; //< output sample L
processFrameInfo.audioOutputFrame[1] = ynR; //< output sample R
```

The gain reduction values may be taken from the dynamics processors after their *processAudioSample* functions have been called. To set our gain reduction meter combined reduction value, and knowing that the meter is inverted (upside down), we can write the following code to implement mono or combined stereo metering.

```
double grLeft = 0.0;
double grRight = 0.0;

DynamicsProcessorParameters params =
                              dynamicsProcessors[0].getParameters();

grLeft = params.gainReduction;
grMeter = 1.0 - grLeft;
```

Then, if the right channel exists, you perform a similar operation and merge the gain reduction values through averaging:

```
DynamicsProcessorParameters params =
                              dynamicsProcessors[1].getParameters();

grRight = params.gainReduction;
grMeter = 1.0-(0.5*grLeft + 0.5*grRight);
```

18.13.5 *Stereo Linking the* DynamicsProcessor *Objects*

To stereo link the two *DynamicsProcessor* objects, you take control of the side chain input and combine the input samples as if their summed value was being used as an externally keyed input. If you want to use both stereo linking *and* an external side chain, then you provide the logic that combines the two external side chain input samples, just like what is shown here where we are using the normal input samples. Assume you have picked up the left and right input samples as *xnL* and *xnR*:

```
DynamicsProcessorParameters params =
                              dynamicsProcessors[0].getParameters();
```

```
// --- combine sidechain input
double sidechain = 0.5*xnL + 0.5*xnR;

// --- enable on object (this can be done outside the process function)
params.enableSidechain = true;
dynamicsProcessors[0].setParameters(params);
dynamicsProcessors[1].setParameters(params);

// --- set the stereo link input
dynamicsProcessors[0].processAuxInputAudioSample(sidechain);
dynamicsProcessors[1].processAuxInputAudioSample(sidechain);
```

18.13.6 ASPiK Users: Enabling the Special Pro Tools Gain Reduction Meter

For AAX plugins, you may also enable the gain reduction meter that shows up in the Pro Tools interface (on the right side of the normal channel meter) easily using the *PluginParameter* object. This single downward meter is narrower and slightly offset (up) from the ordinary channel meters, will reflect whatever information you want to send to it, and is shown in Figure 18.13c on the right side of the normal stereo channel meters. Typically, you will merge all the channel gain reduction values into one combined value for this special meter, just like we did for the chapter plugin. The AAX API specifies the ability to display multiple gain reduction meters, though in practice I have only ever observed single (mono) meters on both the Pro Tools software and Avid hardware interfaces. To set up an ASPiK plugin meter as a Pro Tools gain reduction meter, you just call *setIsProToolsGRMeter( )* when you instantiate the object. The *PluginBase* object will automatically merge all of the meter values for those with the Pro Tools Gain Reduction flag set and send it to the lone meter on the Pro Tools or Avid hardware surface. This is independent of your own GUI metering, which does not assume a monaural metering environment. If you use RackAFX, you check the Pro Tools GR Meter box when you set up the meter on the main UI. To code it by hand for the *Dynamics* plugin, you would write the following.

```
PluginParameter* piParam = new PluginParameter(controlID::grMeter,
                                               "GR", 0.00, 0.00,
                                               ENVELOPE_DETECT_MODE_PEAK,
                                               meterCal::kLinearMeter);
piParam->setInvertedMeter(true);
piParam->setIsProtoolsGRMeter(true);
piParam->setBoundVariable(&grMeter, boundVariableType::kFloat);
addPluginParameter(piParam);
```

18.14 Homework

1. Alter the *DynamicsProcessor* to implement an alternate *feedback* topology instead of the feed-forward version here. Note the difference when using hard limiting.
2. Modify the *DynamicsProcessor calcGain* function's gate mode to attenuate the signal down to some finite −dB rather than 0.0 (−∞ dB).
3. Create a spectral dynamics processor object and/or plugin. Use the built-in *LRFilterBank* object to split the audio signal into HF and LF bands. Process each band through its own independent

dynamics processor, then sum the two outputs to form a single output value. You will need to do this for both left and right channels, requiring a total of four independent dynamics processors.
4. How would you deal with stereo linking the spectral processors in Question 3?
5. Add a fixed input delay to implement a look-ahead compressor. If using ASPiK, remember to report your latency to the host (see Chapter 6).

18.15 Bibliography

Ballou, G. 1987. *Handbook for Sound Engineers*. Carmel: Howard W. Sams & Co., pp. 850–860.
Floru, F. 1998. "Attack and Release Time Constants in RMS-Based Feedback Compressors." *Journal of the Audio Engineering Society*, Preprint 4703.
Orfanidis, S. 1996. *Introduction to Signal Processing*, Chap. 10–11. Eaglewood Cliffs: Prentice-Hall.
Pirkle, W. C. 1995. "Applications of a New High Performance Digitally-Controlled Audio Attenuator in Dynamics Processing." *Journal of the Audio Engineering Society*, Preprint 3970.
Reiss, J. 2011. *Under the Hood of a Dynamic Range Compressor*, 131st Audio Engineering Society Convention (Tutorials).
Zölzer, U. 2011. *DAFX: Digital Audio Effects*, Chap. 4. West Sussex, England: J. Wiley & Sons.

CHAPTER 19

Nonlinear Processing: Distortion, Tube Simulation, and HF Exciters

There are several families of audio processing algorithms that use nonlinear processing as part of their effect. These include the following.

- solid state distortion, overdrive and fuzz emulators
- vacuum tube amplifier simulation
- simulations of analog filters that are designed with transistors or other nonlinear devices (e.g. Moog or Korg35 low-pass filters)
- high frequency "exciters" and enhancers
- ring modulation (linear processing with nonlinear output)

One way to approach nonlinear processing is to look at the input and output time domain transfer function. This is sometimes called the *voltage transfer function* (VTF) and involves the input and output relationship of a circuit as plotted on an *x–y* pair of axes as shown in Figure 19.1a. Here, we have a perfectly linear device whose input and output signal amplitude levels are linearly related. Notice that these graphs only involve the input and output amplitudes and not the signal frequency response. These plots can be confusing at first glance—you are supposed to imagine the input coming up from the bottom of the graph, centered on the *x*-axis, then it is reflected off of the transfer function curve to the right to produce the output signal off of the *y*-axis. Figure 19.1b shows a linear amplifier with gain and hard clipping: when the input signal at the bottom exceeds the linear operating limits, the waveform's top and bottom are chopped off, leaving a flat edge. Figure 19.1c shows a nonlinear transfer relationship that results in soft clipping when the input amplitude just matches the operating range; instead of being chopped off, the top and bottom of the output waveform are rounded. Figure 19.1d shows another soft clipping transfer function that produces even rounder tops and bottoms. All of the transfer functions have been symmetrical up to this point in that the top (+) and bottom (−) portions of the input are amplified in the same manner. Figure 19.1e and f show asymmetrical transfer functions that operate on the positive and negative portions of the waveform in different manners. Figure 19.1f is the extreme example of this, where the bottom portion of the waveform has a much different transfer characteristic than the top portion does. The dotted box that surrounds the transfer functions is there to show the input and output maximum limits—for our digital audio plugins, our inputs and outputs will always be limited to the range of [−1.0, +1.0]. Notice that in all of these cases, the region near the *x–y* origin is linear. The transfer functions only become nonlinear at the extremes.

The time domain transfer functions do a good job at describing the effect of the nonlinear amplification on the waveform. It is easy to see the difference between the hard clipped and soft clipped waveforms insomuch as how they modify the shape of the signal. But the time domain describes only half of the nonlinear processing functionality. When we look at the frequency domain versions, we see what happens to the signal's spectrum as a result of the nonlinear input/output transfer function.

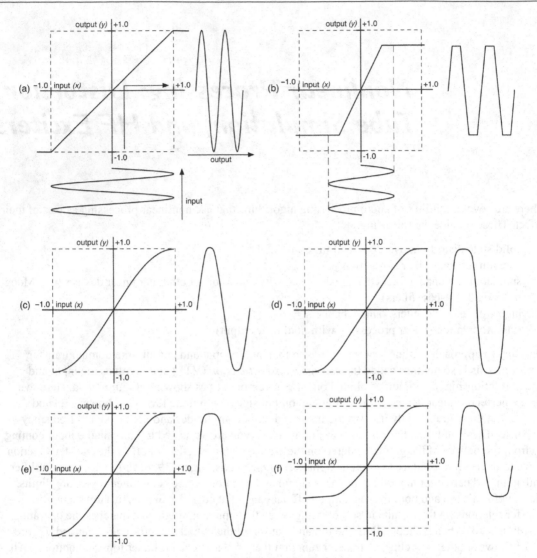

Figure 19.1: Input/output amplitude transfer functions for the following: (a) a perfectly linear unity-gain device, (b) a linear amplifier that has been driven into clipping, (c) symmetrical soft clipping, (d) symmetrical soft clipping with more nonlinearity, (e) asymmetrical soft clipping, and (f) extremely asymmetrical soft clipping. The input signal is omitted in (c)–(f) for brevity.

19.1 Frequency Domain Effects of Nonlinear Processing

A perfect sinusoid has a frequency domain magnitude response that consists of a single spike at the sinusoidal frequency (plus another one in the negative half of the spectrum). When linearly processed (amplified, attenuated, or filtered), the output signal may have a different magnitude and/or phase, but the spectrum still consists of that single spike representing the sinusoid. When nonlinearly processed, harmonics are generated that alter the timbre of the signal. The harmonics are multiples of the original signal frequency and are typically characterized as even or odd depending on the numerical multiplier. Signals that undergo mild nonlinear processing or very soft clipping may generate only a few harmonics that are loud enough to be heard. As the nonlinear amplification increases, more and more harmonics are generated at higher amplitudes. It is these added harmonics that give the distorted sound its characteristics, good or bad.

> All nonlinear processing algorithms will add harmonics to the original signal. The number, size, and amplitude ratios of these harmonics determine how the original sound is altered to change its timbre. These added harmonics will cause aliasing if they exist above the Nyquist frequency. Fortunately, we may use the oversampling techniques in Chapter 22 to eliminate completely, or drastically reduce the aliased components.

Figure 19.2 shows the effect of nonlinear processing on a sinusoid. In Figure 19.2a you can see the time domain (left) and frequency response (right) for a 1 kHz sinusoid. The waveform and spectrum are pure. In Figures 19.2b and c and increasing amount of symmetrical nonlinear processing is applied. You can see the addition of the harmonics above the 1 kHz spike. You can also see that the harmonics

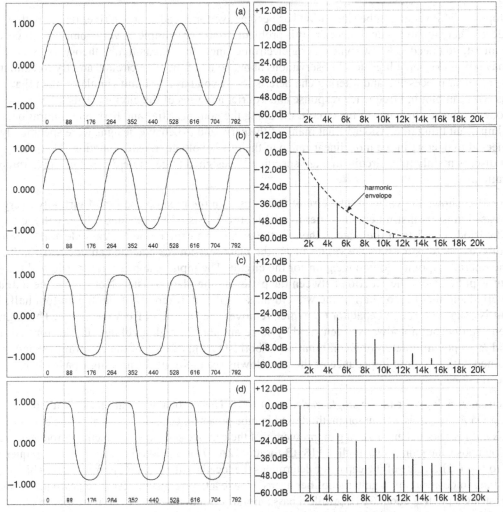

Figure 19.2: The effect of nonlinear processing on a sinusoid with (a) no processing (pure sinusoid), (b) slight nonlinear processing, (c) moderate nonlinear processing, and (d) asymmetrical nonlinear processing. The frequency domain plots are on the right.

are all odd multiples of the fundamental occurring at 3 kHz, 5 kHz, 7 kHz, etc. In Figure 19.2d the nonlinear processing is asymmetrical so that the upper (+) and lower (−) portions of the waveform have different amounts of processing applied. This adds the even harmonics, which are present at 2 kHz, 4 kHz, 6 kHz, etc. So, while the time domain plots show only a distortion to the time domain curvature, the frequency domain plots show the resulting change in the spectrum. Different types of distortion and nonlinear amplifier circuitry yield different harmonic signatures—the presence of even and/or odd harmonics as well as their harmonic amplitude envelopes. The harmonic envelope of the spectrum in Figure 19.2b is shown in the dotted line. Now imagine what the harmonic envelope looks like for the spectrum in Figure 19.2d, where you might also imagine two harmonic envelopes—one for the odd harmonics and one for the even ones. The plots in Figure 19.2 were all made using oversampling techniques from Chapter 22 to eliminate aliasing.

19.2 Vacuum Tubes

The topic of DSP vacuum tube simulation has been active for decades; it is a river that is both wide and deep. There are many approaches to this problem, from very simple to very complex. To go very deep into it, you need to know some analog electronics including vacuum tube theory. However, we may still make some high-level observations about some of the tube circuits and try to emulate their nonlinear processing and filtering characteristics. If you have a circuit simulator, then that is a plus: you can plot the frequency response and Fourier analyses of the circuits. This allows you to model the basic attributes of the circuit, including the nonlinearities and harmonic saturation. One important thing to consider is that many of the early and expensive designs were specifically engineered to operate in the most linear range of the tube's voltage transfer function. Therefore, we'll discuss mainly tube circuits that are specifically designed to generate distortion and harmonic saturation.

Figure 19.3 shows the vacuum tube circuit that is used in the preamp section of virtually every tube-based musical instrument amplifier available. It consists of a triode and a high voltage supply V_p. The triode consists of the plate at the top, the grid in the center, and the cathode at the bottom. The tube characteristics plus the values of V_p, the plate resistor R_p, and the cathode resistor R_k, set the operating point and voltage gain for the circuit. The input is applied at the grid and the output is pulled from the junction between the plate and the plate resistor. Notice that the output is inverted and that the top half of the waveform is amplified differently from the bottom half. The output is coupled through capacitor C_c, which removes the DC offset from the signal. This signal is then usually (but not always) attenuated by some amount through a voltage divider formed with R_1 and R_2, a potentiometer, or a combination of both. The impedance of the coupling capacitor combined with the input impedance of the next downstream circuit forms a high-pass filter between them.

This circuit topology is called a *single-ended* or *Class A*. This particular variation uses cathode self-biasing via R_k, which is a common method used in tube preamps. Capacitor C_k (called the *cathode bypass cap*) is optional and fixes a gain problem that the cathode self-biasing scheme introduces. Its addition boosts the gain substantially (+18 dB for this 12AX7 simulation) but causes a low frequency shelf to form. The value of the capacitor determines the location of the shelving frequency. This is something to be aware of: a guitar amplifier needs to have a flat frequency response down to the open low E string's pitch at about 82 Hz, while a bass amplifier would need to extend its response down to 41 Hz (four-string bass) or 32 Hz (five-string bass).

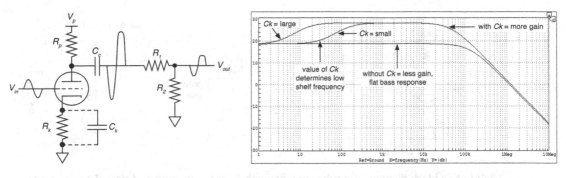

Figure 19.3: A single-ended or Class A tube amplifier (left) and output frequency response (right) of the circuit simulation prior to the coupling capacitor C_c, with and without the cathode bypass capacitor, which produces more gain but a low frequency shelf as well. Two different capacitors were used in two simulations to show the low frequency effect.

The tube circuit in Figure 19.3 is considered to be one complete stage as it fully amplifies its input. Modern and high gain tube preamplifiers consist of a series of cascaded single-ended stages, typically from two to six in total. The input is amplified nonlinearly and inverted, then fed to the next stage, which re-amplifies it, inverts it again, and so on with each stage overdriving the next one. The more stages of amplification, the more harmonically rich the signal becomes at the output of the final tube stage. In addition, some preamps contain cascaded stages that are exact replicas of one another, while others feature stages that incorporate relatively minor changes—a little less gain here, a bit more bass there, etc. The amplifiers' EQ circuit usually sits between one of the multiple stages. These tube stages are usually fixed so that all of the gains and resistor attenuators have been worked out during the design phase. A notable exception is the early Mesa-Boogie line of guitar amplifiers that allowed the user to adjust the gain between stages in the preamp, allowing for more flexibility in adjusting preamp's harmonic quality.

This harmonically rich preamplifier output feeds a power amplifier stage, which may also be single-ended or it may be *complimentary*. In the complimentary version, a phase splitter creates a balanced signal that is fed into two tube amplifier circuits. The outputs of those two circuits are summed together through a transformer that couples the output to the loudspeaker. The output amplifier's pair of tube circuits may each be designed to do the following.

- amplify the full signal, both positive and negative halves (class A)
- amplify half the signal so that one tube amplifies only the positive half and the other tube amplifies the negative half (class B) causing crossover distortion due to the fact that the tubes cannot turn on instantly
- amplify half the signal so that one tube amplifies the positive half plus a bit of the negative half and the other tube amplifies the negative half plus a bit of the positive half (class AB) which removes the crossover distortion

Not including things like efficiency, heat dissipation, and other physical effects, the main thing we can take away from this is that for a class A output stage, we will get the same asymmetrical amplification that we saw in the preamp stage. However, because the complimentary class B and class AB tubes are fed phase inverted signals, each tube amplifies its half-signal with the same nonlinearity so that the resulting amplification is symmetrical as depicted in Figure 19.1c or 19.1d. In both cases (class A and B), the output transformer sews the two out-of-phase output signals back together properly.

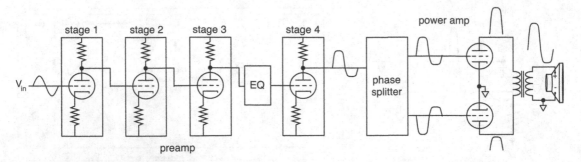

Figure 19.4: In this block diagram of a tube amplifier with a class A preamp and class B power amp, the transformer flips the amplified negative half at the bottom and sews it together with the top half to produce the final signal for the loudspeaker.

A complete block diagram for a tube based musical instrument amplifier is shown in Figure 19.4. The output is designed as class B and each of the output tubes amplifies only the positive half of its input. The transformer flips the phase of the bottom half before recombining it with the top portion to form the output signal. Figure 19.5 shows the time, frequency, and harmonic distortion plots for the class A 12AX7 preamp stage and class B and class AB power amp outputs. Note that these are only simulations and not actual measurements. For a 12AX7 tube preamplifier simulation in Figure 19.5a we see the effect of increasing the input voltage as the output starts clipping at the top with increased input levels. In Figure 19.5b we observe the low frequency shelf produced with cathode self-biasing which is very common in preamp designs. The preamplifier's frequency domain plots in Figure 19.5c and d show the increase in harmonic distortion components as the input is increased from 0.25 V in Figure 19.5c to 1.0 V in Figure 19.5d. For power amplifer stages, Figure 19.5e shows the time domain plots for Class B and Class AB designs; note the crossover distortion present in the Class B version. Finally, Figure 19.5f shows the harmonic distortion components present in the simulated Class AB output stage driven into saturation.

We now have a good starting point for a tube simulation with respect to the frequency response and nonlinear harmonic generation. For the frequency response, we can use a 1st order HPF to both remove the DC offset and emulate the existing HPF between preamp stages, and we can use a 1st order low shelving filter to simulate the effects of the cathode bypass cap. Of course, inverting the output of each stage is simple. Harmonically, we know that the asymmetrical amplification in class A stages will produce both even and odd harmonics, while the symmetrical nature of the class B output produces odd harmonics—of course, these are generalizations and the actual devices (which are far from perfect) will produce variations. We can then seek nonlinear signal processing functions that create matching, or at least similar, harmonic envelopes for the circuits we wish to emulate.

We won't bother with simulating the transformers in this text—the subject of a lot of current academic research—and we can simulate the cabinet enclosure and loudspeaker(s) by convolving the output with the impulse response of an actual cabinet using the fast convolution techniques in Chapter 20. There is plenty of material available—including papers, books, and software—if you want to investigate this topic further, including the passive EQ circuits often found in musical instrument amplifiers.

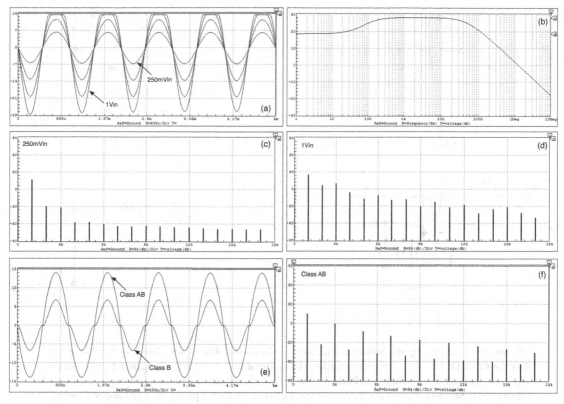

Figure 19.5: Circuit simulation results for a single 12AX7 preamp stage driven into saturation; (a) the time domain plot of the output with input voltages ranging from 0.25V to 1.0V, (b) the frequency response of the 12AX7 stage using a cathode bypass capacitor, (c) the harmonic distortion plot with V_{in} = 250 mV and the same plot with (d) V_{in} = 1V, (e) time domain plots off the outputs of a class B and class AB power amp stage and (f) the harmonic distortion plot of the class AB power amp.

19.3 Solid State Distortion

The original solid state distortion devices that use transistors or diodes to perform the clipping operation were not designed to emulate tubes, but rather to be a unique effect on their own. Later, more devices appeared that attempted to sound tube-like. The fuzz effect is a more drastic electronic distortion sound that produces high gain square waves while lower gain devices like the Ibanez Tube Screamer® and Boss Super Overdrive® use diodes to perform a softer clipping. The Tube Screamer uses two diodes to create symmetrical clipping while the Super Overdrive uses three to create asymmetrical clipping. Most devices consist of one distortion stage, with a notable exception being the original Electro-Harmonix Big Muff Pi® distortion box, which used a cascading series of distortion stages, similar to a tube preamp's topology. The filter from the Korg MS-20® synthesizer includes a built-in diode clipper that is very similar in design with the Tube Screamer topology in Figure 19.6c. As with the tube, the clipper and fuzz circuits are nonlinear devices that add harmonics to the input signal. Figure 19.6 shows some common distortion circuits. The choice

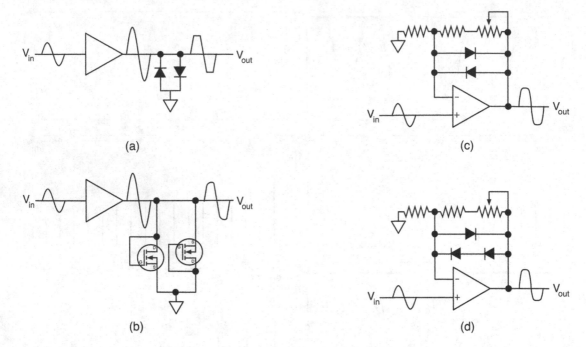

Figure 19.6: Some popular solid state distortion circuit topologies are based on the following: (a) the MXR Distortion+ ® clipping diodes, (b) Fulltone OCD® MOSFET clippers (note that the original circuit uses a germanium diode on one MOSFET branch to create asymmetrical clipping), (c) the Tube Screamer (note the back-to-back diodes in feedback path), and (d) Super Overdrive (note the back-to-back diodes with asymmetrical clipping).

of topology, diode/transistor type and quantity, and gain all influence the harmonic envelope of the resulting distorted outputs.

Figure 19.7 shows simulation results in both the time and frequency domain for several different distortion circuits. Figures 19.7a and b show the time domain and spectrum of symmetrical soft diode clipping while 19.7c and d show the asymmetrical variation. Figures 19.7e and f show the simulation of an NPN Fuzz Face using the same sinusoidal stimulus as the others.

19.4 Bit Crushers

"Bit crushing" is the fun name given to the process of re-quantizing a discrete waveform. The number of bits used to encode the analog waveform dictates the number of quantization levels or stair-steps in the discrete waveform. A 16-bit sampled audio waveform consists of a set of discrete samples at one of $2^{16} = 65,536$ quantization levels. Bit crushers re-quantize the signal at another (lower) bit depth. The equation for re-quantizing a signal is as old as the sampling theorem itself. For a bipolar signal on the range of $[-1.0, +1.0]$, the algorithm is shown in Equation 19.1. For unipolar signals, replace the digit 2 with 1 in the numerator of the QL equation. The smooth 16-bit signal in Figure 19.8a is bit crushed to 2, 4, and 6 bits in Figures 19.8b–d. It should be obvious that these distorted waveforms will

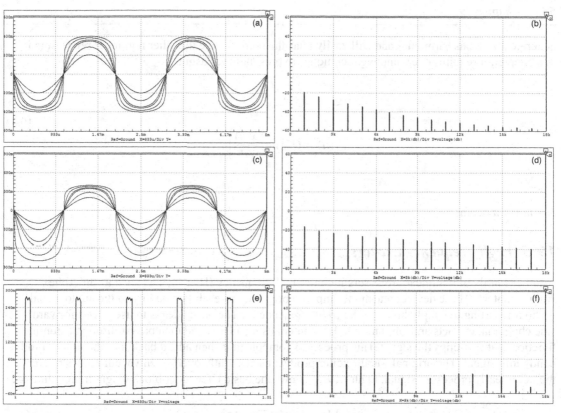

Figure 19.7: (a) The output of a symmetrical diode clipper based on the Tube Screamer circuit for various settings of the distortion knob and (b) its output spectrum at the middle (medium) distortion level. (c) The output of an asymmetrical diode clipper based on the Super Overdrive circuit for various settings of the distortion knob and (d) its output spectrum at the maximum distortion level. (e) The output of an NPN Fuzz Face circuit with sinusoidal input and (f) its output spectrum at the maximum distortion level.

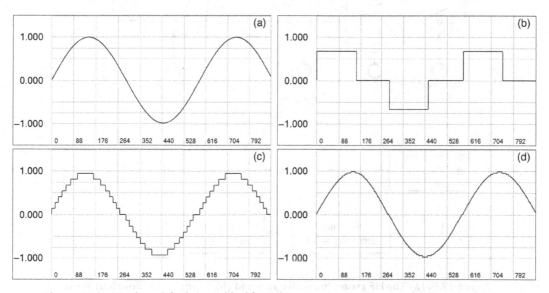

Figure 19.8: (a) The original sinusoid with 16-bit quantization. Re-quantized versions with (b) 2-bit, (c) 4-bit, and (d) 6-bit crushing.

generate numerous harmonics and will easily alias the output—like the other nonlinear processors it this chapter, it requires oversampling techniques to prevent the aliasing.

$$QL = \frac{2}{\left(2^N - 1\right)}$$

$$y(n) = QL\left(\text{int}\left(\frac{x(n)}{QL}\right)\right)$$

(19.1)

where

N = new bit depth

19.5 High Frequency Exciters

Another family of nonlinear processors includes high frequency (HF) exciters such as the Aphex Aural Exciter®. These devices attempt to impart a brighter, more sibilant quality to the audio signal without resorting to simply boosting the treble frequencies. The basic idea is to use a feed-forward side chain that extracts and amplifies the HF content using a HPF, and then pass it through a nonlinear function that soft-clips the signal. This processed component is then added back into the original full bandwidth signal. The original Aural Exciter product used slightly overdriven vacuum tubes to create the nonlinearly processed signal. Figure 19.9a shows the basic HF exciter block diagram including the

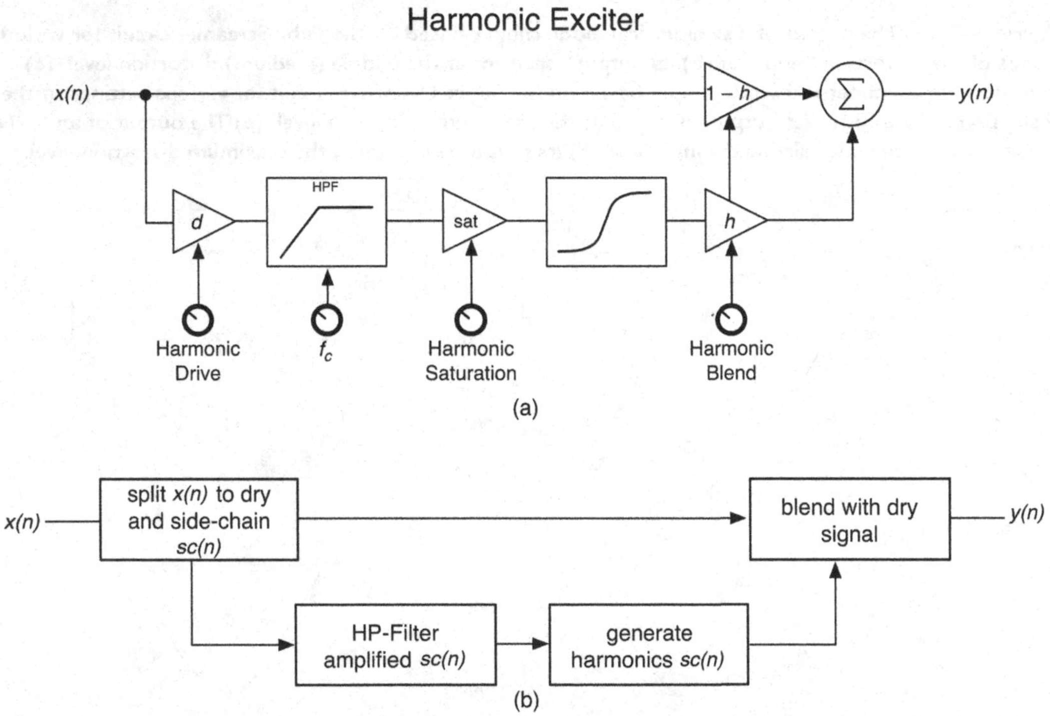

Figure 19.9: (a) The HF exciter block diagram and (b) its equivalent flowchart form.

user controls (knobs) and the blocks they act upon. Figure 19.9b shows the block diagram formatted as a flowchart showing the input signal and side-chain signal directions with arrows. The nonlinear processing block is shown with the "S" shaped curve inside of it. Another variation that does not involve nonlinear processing is to use the same side chain to detect high frequencies and then use that signal to control a dynamic high shelf filter.

19.6 Virtual Bass

Another use for nonlinear signal processing is in Virtual Bass (VB) algorithms that use psychoacoustic effects to trick the ear into perceiving bass frequencies that are not actually present in the signal, or are greatly attenuated. These algorithms first gained popularity in small computer speaker designs, where physical dimensions limit the size of the woofers, and therefore the low frequency capabilities. VB may be implemented in either the time or frequency domain, where harmonics are artificially added to a signal. The time domain version uses a nonlinear processor to add the harmonics. The basic idea is this: if your auditory system detects a set of mathematically related harmonics, it will synthesize the fundamental frequency that is generating those harmonics, producing the auditory illusion of a frequency that isn't really there. There have been several variations on this concept. The block diagram in Figure 9.10 shows the original concept (Larsen & Aarts, 2002, 2004). There are multiple parameters to adjust, including the filter cutoffs, type, and intensity of nonlinear processing, and the NLP gain value G.

19.7 Ring Modulation

Ring modulation is included in this chapter because it adds mathematically related harmonics to an input signal. However, it does not use a nonlinear processing equation. Instead, ring modulation is accomplished by multiplying the input signal against a sinusoid called the *modulator,* as shown in

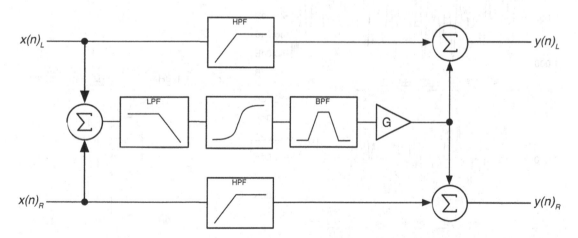

Figure 19.10: A diagram of the basic stereo Virtual Bass algorithm. The mono-ized signal is filtered and nonlinearly processed before being band-pass filtered and amplified; the resulting nonlinear signal is added back to the high-pass filtered channel signals.

Figure 19.11a. This is also known as *frequency mixing*. The resulting signal contains sum and difference pairs of each frequency present in the signal. For example, if the modulator is 500 Hz and the input signal is 1 kHz, the output will consist of two frequencies (1 kHz − 550 Hz = 500 Hz and 1 kHz + 500 Hz = 1.5 kHz) while the original input signal at 1 kHz is eliminated, as shown in Figure 19.12a and b. Double ring modulation as shown in Figure 19.11b multiplies the output again with a second oscillator producing another sum and difference pair, based on the modulation frequencies. In Figure 19.12c and d, the input signal is 1 kHz and the two modulators are 500 Hz and 250 Hz. The first ring modulation produces the sum and difference pair at (500 Hz, 1.5 kHz) and the second adds and subtracts the 250 Hz modulator from the output pair, creating a second pair at (250 Hz, 1.75 kHz). In this case, for each input frequency component, four new frequency components are generated, and the original is destroyed.

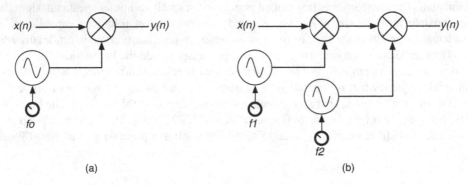

Figure 19.11: (a) A ring modulator block diagram. (b) Double ring modulation adds a second oscillator and multiplier, as shown here.

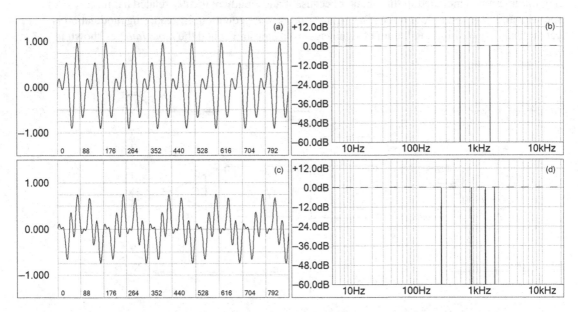

Figure 19.12: Ring modulator time and frequency domain outputs for normal (a) and (b) ring modulation, and double (c) and (d) ring modulation. The input signal is a 1 kHz sinusoid: the modulators are 500 Hz (single) and 500 Hz, 250 Hz (double).

For complex audio signals, this process happens for every frequency component that the signal contains. The resulting output can sound nothing like the input. Drums seem to become more like pitched signals, and full-spectrum signals become sonically obliterated, especially with the double ring modulation. Because of the sum and difference pairing, aliasing will occur when the sum-signal goes outside Nyquist. As with the other nonlinear processors, we can use the oversampling techniques of Chapter 22 to mitigate the aliasing. Lastly, it is important that the sinusoid generator creates as pure of a waveform as possible to prevent other sidebands from appearing. For this plugin, we've created a special direct form sinusoid oscillator that simply places a conjugate pair of poles directly on the unit circle.

19.8 Nonlinear Processing Functions

There are a couple of fundamental ways to generate the nonlinear amplitude processing. For many types of distortion, overdrive, and tube emulation, these curves are similarly shaped and resemble the letter "S." Functions that generate these curves are called *sigmoid functions*. You can find dozens of different functions $y = f(x)$ for producing this soft clipping. The algorithm takes an audio sample as its input x, then runs it through an equation (or several equations) to produce a nonlinearly amplified output y. There is no need to store samples, as there are no delayed sample values required. This also means there are no $H(z)$ transfer functions either. We call these algorithms *waveshapers,* and the functions are called *waveshaping* functions. These waveshapers almost always require either polynomials (squaring, cubing, etc. of the input signal) or trigonometric functions, especially our old friend e. For very complicated waveshaper algorithms, another option is to create the complex curve, sample it, and store it in a lookup table. We will simply use the waveshaper functions directly. If you examine the nonlinear sigmoid functions in Figure 19.1a and c–e, you will see that their maximum input and outputs are exactly 1.0. These are called normalized waveshaper functions. Figures 19.1b and f show un-normalized functions whose input and/or output ranges are smaller than the [−1.0, +1.0] range. Tables 19.1 and 19.2 list a total of 16 waveshaper functions for you to try. Table 19.1 lists soft clipper sigmoid functions, while Table 19.2 includes some more exotic functions for strange distortion effects (use at your own risk!). The bulk of these, and their acronyms are from Woon-Seng Gan and Hawksford (2011).

Most of the functions in Table 19.1 include a variable k, which is a gain multiplier; you use k to amplify the input sample x before applying it to the function. The k term appears in the denominators as well as part of the normalization process. You can un-normalize these functions by removing the k variable from the denominators but keeping it in the numerators. When $k = 1$, you have the classic waveshaper functions as they appear in literature. The Arraya and Sigmoid2 functions are only barely nonlinear and so they do not add much in the way of upper harmonics. The Araya function is interesting because Yamaha owns a patent on a device that uses it as the waveshaper in a vacuum tube emulator. This waveshaper works very well when multiple stages are cascaded in sequence. It is also important to limit the input size of x to the range [−1.0, +1.0] in order to keep the function sigmoid in nature. If you allow x to exceed the normalized range, the Arraya waveshaper will turn into an exotic distortion function instead.

With regards to the exotic waveshapers, the full-wave rectifier has a special place in audio effect algorithms dating back to the octave-up guitar distortion boxes popularized in the 1960s and early 1970s. A full-wave rectified signal has its negative components inverted. When the DC offset is removed, the number of zero-crossings doubles. This effectively doubles the pitch of the signal producing the octave-up effect. The octave-up signal may then be amplified and mixed back into the original signal. Note: overdriven class-A tube amplifiers produce an abundance of the second

Table 19.1: Soft clip sigmoid waveshaper functions ($sgn(x) = -1$ for x < 0 and $sgn(x) = +1$ for $x \geq 0$)

Name	Acronym	Equation (x = in, y = out, k = saturation)	Notes
Arraya	ARRY	$y = \dfrac{3x}{2}\left(1 - \dfrac{x^2}{3}\right)$	No saturation, very mild
Sigmoid	SIG	$y = 2\dfrac{1}{1+e^{-kx}} - 1$	
Sigmoid2	SIG2	$y = \dfrac{(e^x - 1)(e+1)}{(e^x + 1)(e-1)}$	No saturation, very mild
Hyperbolic Tangent	TANH	$y = \dfrac{\tanh(kx)}{\tanh(k)}$	Good for diode simulation
Arctangent	ATAN	$y = \dfrac{\arctan(kx)}{\arctan(k)}$	
Fuzz Exponential 1	FEXP1	$y = sgn(x)\dfrac{1 - e^{-|kx|}}{1 - e^{-k}}$	

Table 19.2: Exotic distortion waveshaper functions ($sgn(x) = -1$ for x < 0 and $sgn(x) = +1$ for $x \geq 0$)

Name	Acronym	Equation (x = in, y = out, k = saturation)
Fuzz Exponential 2	FEXP2	$y = sgn(-x)\dfrac{1 - e^{|x|}}{e - 1}$
Exponential 2	EXP2	$y = \dfrac{e - e^{(1-x)}}{e - 1}$
Arctangent Square Root	ATSR	$y = 2.5\tan^{-1}(0.9x) + 2.5\sqrt{1 - (0.9x)^2} - 2.5$
Square Sign	SQS	$y = x^2\, sgn(x)$
Cube	CUBE	$y = x^3$
Hard Clipper	HCLIP	$y = \begin{cases} 0.5\,sgn(x) & |x| > 0.5 \\ x & \textit{otherwise} \end{cases}$
Half Wave Rectifier	HWR	$y = 0.5\left(x + |x|\right)$
Full-Wave Rectifier	FWR	$y = |x|$
Square Law	SQR	$y = x^2$
Absolute Square Root	ASQRT	$y = \sqrt{|x|}$

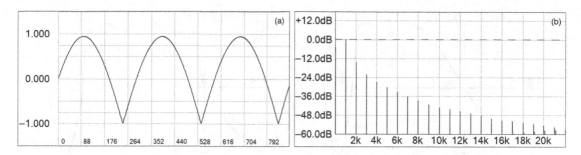

Figure 19.13: (a) A full-wave rectified 500 Hz sinusoid with DC offset removed and (b) its spectrum. The second harmonic is present at 1 kHz.

harmonic, which is one octave above the fundamental. Figure 19.13a shows the time domain response of a full-wave rectified 500 Hz sinusoid. Its spectrum in Figure 19.13b reveals a new fundamental frequency at 1 kHz. These plots were done without oversampling and a tiny bit of aliasing is visible just above 18 kHz as small spikes next to the harmonics.

19.8.1 Asymmetrical Waveshaping

There are several approaches for developing asymmetrical waveshapers, most of which rely on analyzing the input and applying one kind of processing to the positive portions of the waveform and another kind to the negative portions. This might involve different nonlinear equations or (more often) applying a different amount of gain (k) to the input signal depending on whether it is a positive or negative value. Both of these methods will work well as long as the processing of each half is not too drastically different. However, you can wind up with problems near the origin where the two halves connect together. Figure 19.14a shows an example of asymmetrical waveshaping where the top and bottom halves of the input are distorted only slightly differently. Figure 19.14b and c show the problem when the two nonlinear functions are very different. The kink in the curve where the two meet forms a kind of discontinuity and will introduce its own high frequency content—and in many cases this isn't the kind of distortion you want. So it is important when working with the nonlinear waveshapers to understand the limitations. The waveshaper in Figure 19.14d produces asymmetric distortion, but does not attempt to apply different gain values to each half.

For example, to create a similar asymmetric gain curve as Figure 19.14d using the arctangent waveshaper, we might use Equation 9.2 where k is the input gain and g is the negative half saturation limit (in Figure 19.14d, $g = 0.4$).

$$y = \begin{cases} \dfrac{\tan^{-1}(kx)}{\tan^{-1}(k)} & x \geq 0 \\[3ex] g\dfrac{\tan^{-1}\left(\dfrac{k}{g}x\right)}{\tan^{-1}\left(\dfrac{k}{g}\right)} & x < 0 \end{cases} \tag{19.2}$$

$$k \geq 1$$
$$0 < g \leq 1$$

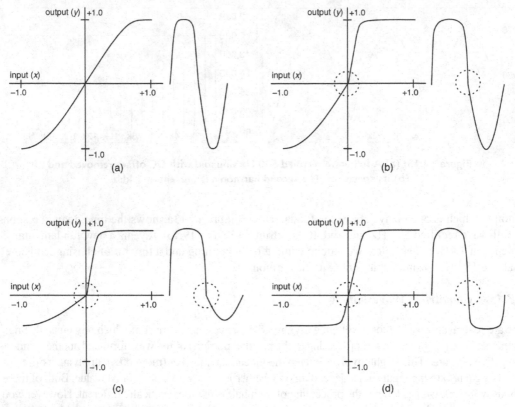

Figure 19.14: (a) The mildly asymmetrical nonlinear waveshaping works well. When the positive and negative half gains are drastically different, a kink forms where the two meet (b) and (c). In (d) the amplification is asymmetrical but only as far as the output limits are concerned.

19.9 *C++ DSP Object:* BitCrusher

The *BitCrusher* object is tiny and simple, implementing the basic bit-crushing equation in Equation 19.1. It is implemented as an object to allow you to extend its functionality as well as to make a minor efficiency improvement by only calculating the *QL* value when the user has altered the bit-depth value.

19.9.1 BitCrusher: *Enumerations and Data Structure*

The object does not have any special enumerations and its custom data structure contains only one member variable to store the bit depth for the re-quantization. Note that this member is a *double* data type and so you mayalso obtain fractional bit-depth crushing (e.g. 8.3 bits).

```
struct BitCrusherParameters
{
        BitCrusherParameters() {}
        BitCrusherParameters& operator=(. . .){ . . . }

        double quantizedBitDepth = 4.0;
};
```

19.9.2 BitCrusher: *Members*

Tables 19.3 and 19.4 list the *BitCrusher* member variables and member functions. It is about a simple as our DSP objects gets.

19.9.3 BitCrusher: *Programming Notes*

The *BitCrusher* object is so simple that there are only two procedures that need to be discussed: setting the *QL* parameter and processing the audio signal, both of which are shown here.

```
void setParameters(const BitCrusherParameters& params)
{
    // --- calculate and store
    if(params.quantizedBitDepth != parameters.quantizedBitDepth)
        QL = 2.0 / (pow(2.0, params.quantizedBitDepth) -1.0);

    parameters = params;
}

virtual double processAudioSample(double xn)
{
    return QL*(int(xn / QL));
}
```

Table 19.3: *BitCrusher* **member variables**

BitCrusher Member Variables		
Type	Name	Description
BitCrusherParameters	*parameters*	Custom data structure
double	*QL*	The discrete quantization level

Table 19.4: *BitCrusher* **member functions**

BitCrusher Member Functions		
Returns	Name	Description
BitCrusherParameters	*getParameters*	Get all parameters at once
void	*setParameters* Parameters: —*BitCrusherParameters* _parameters	Set all parameters at once
bool	*reset* Parameters: —*double* sampleRate	Reset all filters and LFO
double	*processAudioSample* Parameters: —*double* xn	Process input *xn* through the phaser

19.10 C++ DSP Object: DFOscillator

The *DFOscillator* object implements a direct form sinusoid oscillator. To implement this oscillator you take a *reson* filter and place its poles directly on the unit circle. Any input will cause the filter to self-oscillate. For purely sinusoid operation you pre-load the output z^{-1} state storage registers with pre-calculated values representing the $y(n-1)$ and $y(n-2)$ samples that would have occurred before oscillation started. This object is provided for the ring modulator, which needs a very pure sinusoid to operate correctly. You may use the *DFOscillator* wherever you need a pitched, high accuracy sinusoid.

19.10.1 DFOscillator: *Enumerations and Data Structure*

The *DFOscillator* object uses two enumerations to make coding easier to read while accessing its state storage registers. This is the same paradigm we used for the *Biquad* object.

```
enum DFOscillatorCoeffs { df_b1, df_b2, numDFOCoeffs };
enum DFOscillatorStates { df_yz1, df_yz2, numDFOStates };
```

The object is updated via the custom structure *OscillatorParameters* that we used for the *LFO* object in Chapter 13, so it will not be reprinted here. As with the *LFO*, the *DFOscillator* inherits from *IAudioSignalGenerator* and places its output into a *SignalGenData* structure. It outputs both normal and inverted versions of the sinusoid.

19.10.2 DFOscillator: *Members*

Tables 19.5 and 19.6 list the *DFOscillator* member variables and member functions. It implements the output half of a direct form biquad structure.

Table 19.5: *DFOscillator* member variables

DFOscillator Member Variables		
Type	**Name**	**Description**
OscillatorParameters	*parameters*	Custom data structure
double	*stateArray[numDFOStates]*	z^{-1} storage units
double	*coeffArray[numDFOCoeffs]*	Storage for coefficients
double	*sampleRate*	Current sample rate

Table 19.6: *DFOscillator* member functions

DFOscillator Member Functions		
Returns	**Name**	**Description**
OscillatorParameters	*getParameters*	Get all parameters at once
void	*setParameters* Parameters: —*OscillatorParameters* _parameters	Set all parameters at once
bool	*reset* Parameters: —*double* sampleRate	Reset all filters and *LFO*
SignalGenData	*renderAudioOutput*	Renders the oscillator normal and inverted outputs
void	*updateDFO*	Recalculates the coefficients and initial conditions when the generator frequency changes

19.10.3 DFOscillator: *Programming Notes*

The *DFOscillator* is simple in that it implements only half of a direct form biquad structure. Using the object is straightforward:

- Reset the object to prepare it for streaming.
- Set the frequency in the *OscillatorParameters* custom data structure; the object ignores the other parameters and implements only the sinusoid.
- Call *setParameters* to update the calculation type.
- Call *renderAudioOutput* to generate the sinusoid; the output is in a *SignalGenData* structure.

Function: *updateDFO*
Description: When the user adjusts the generator frequency, we back-calculate the previous output values that would have been present under the new oscillator frequency conditions using the *arcsine* function. To properly use the *arcsine* output we need to know whether we are on the increasing or decreasing side of the waveform.

```
void updateDFO()
{
        // --- Oscillation Rate = theta = wT = w/fs
        double wT = (kTwoPi*parameters.frequency_Hz) / sampleRate;

        // --- coefficients to place poles right on unit circle
        coeffArray[df_b1] = -2.0*cos(wT);// < --- a = -2Rcos(theta)
        coeffArray[df_b2] = 1.0;         // < --- R^2 = 1, so R = 1

        // --- update states to reflect the new frequency
        //        re calculate the new initial conditions
        //        arcsine of y(n-1) gives us wnT
        double wnT1 = asin(stateArray[df_yz1]);

        //find n by dividing wnT by wT
        double n = wnT1 / wT;

        // --- re calculate the new initial conditions
        if (stateArray[df_yz1] > stateArray[df_yz2])
            n -= 1;
        else
            n += 1;

        // --- calculate the new (old) sample
        stateArray[df_yz2] = sin((n)*wT);
}
```

Function: *renderAudioOutput*
Description: this simply implements the output half of a biquad structure

```
virtual const SignalGenData renderAudioOutput()
{
        // --- calculates normal and inverted outputs
        SignalGenData output;
```

```
// --- do difference equation y(n) = -b1y(n-1) - b2y(n-2)
output.normalOutput = (-coeffArray[df_b1]*stateArray[df_yz1]-
                        coeffArray[df_b2]*stateArray[df_yz2]);
output.invertedOutput = -output.normalOutput;

// --- update states
stateArray[df_yz2] = stateArray[df_yz1];
stateArray[df_yz1] = output.normalOutput;

return output;
}
```

19.11 C++ DSP Functions: **Waveshapers**

The waveshapers themselves are too simple to be packaged in C++ objects as they require no memory. We've assembled a set of waveshaper functions for you to use. Table 19.7 lists these functions. These are simple to implement, and you are encouraged to add more of your own.

19.12 C++ DSP Object: **TriodeClassA**

The *TriodeClassA* object implements a block diagram that simulates the class-A triode circuit in Figure 19.3. The object does not simulate dynamic compression that is inherent in tube circuits, but it does implement a waveshaper (you may easily customize it), a low shelf filter (to simulate the shelf created with the cathode bybass capacitor), and a high-pass filter (to remove the DC offset of the asymmetrical waveshaping and simulate the coupling capacitor C_c).

19.12.1 TriodeClassA: *Enumerations and Data Structure*

The *TriodeClassA* object uses one strongly typed enumeration to select the waveshaper model (it currently supports three).

```
enum class distortionModel{ kSoftClip, kArcTan, kFuzzAsym };
```

Table 19.7: Waveshaper functions

Function	Implements	Notes
sgn	sgn() function	*sgn* is used in several waveshaper equations; this is a fast implementation
atanWaveshaper	arctangent WS	Includes saturation control, normalized
tanhWaveshaper	hyp-tan WS	Includes saturation control, normalized
softClipWaveshaper	soft clipper WS	Includes saturation control, not normalized
fuzzExp1WaveShaper	fuzz exponential	Includes asymmetry control (Section 19.8.1)

The object is updated via the *TriodeClassAParameters* custom data structure that contains members for adjusting its internal objects and enabling/disabling its filters. The member naming should suffice in making the connection to the underlying object parameters.

```
struct TriodeClassAParameters
{
    TriodeClassAParameters() {}
    TriodeClassAParameters& operator=(. . . .){ . . . . }

    // --- individual parameters
    distortionModel waveshaper = distortionModel::kSoftClip;
    double saturation = 1.0;
    double asymmetry = 0.0; // only for fuzz asymmetric
    double ouputGain = 1.0;
    bool invertOutput = true;
    bool enableHPF = true;
    bool enableLSF = false;
    double hpf_Fc = 1.0;
    double lsf_Fshelf = 80.0;
    double lsf_BoostCut_dB = 0.0;
};
```

19.12.2 TriodeClassA: *Members*

Tables 19.8 and 19.9 list the *TriodeClassA* member variables and member functions. As an aggregate object it is fairly simple.

Table 19.8: *TriodeClassA* member variables

TriodeClassA Member Variables		
Type	Name	Description
TriodeClassAParameters	*parameters*	Custom data structure
AudioFilter	*outputHPF*	The output high-pass filter
AudioFilter	*outputLSF*	The output low shelf filter

Table 19.9: *TriodeClassA* member functions

TriodeClassA Member Functions		
Returns	Name	Description
TriodeClassAParameters	*getParameters*	Get all parameters at once
void	*setParameters* Parameters: —*TriodeClassAParameters* _parameters	Set all parameters at once
bool	*reset* Parameters: —*double* sampleRate	Reset all filters
double	*processAudioSample* Parameters: —*double* xn	Process input xn through the model

19.12.3 TriodeClassA: *Programming Notes*

The *TriodeClassA* object combines the two filters and a waveshaper in a simple C++ package. Note that the output gain is not given in dB, but as a cooked multiplier value because the gain parameter is mimicking a resistor voltage divider only. The saturation control implements the tube's internal gain function. Using the *TriodeClassA* object is straightforward.

1. Reset the object to prepare it for streaming.
2. Set the object parameters in the *PhaseShifterParameters* custom data structure.
3. Call *setParameters* to update the calculation type.
4. Call *processAudioSample*, passing the input value in and receiving the output value as the return variable.

Function: *Constructor* and *setParameter*
Details: The object constructor sets the two filters; the *setParameter* function is nearly identical and simply forwards the f_c values to the two filters, so it is not reprinted here.

```
TriodeClassA() {
     AudioFilterParameters params;
     params.algorithm = filterAlgorithm::kHPF1;
     params.fc = parameters.hpf_Fc;
     outputHPF.setParameters(params);

     params.algorithm = filterAlgorithm::kLowShelf;
     params.fc = parameters.lsf_Fshelf;
     params.boostCut_dB = parameters.lsf_BoostCut_dB;
     outputLSF.setParameters(params);
}
```

Function: *processAudioSample*
Details: This function first performs the waveshaping, then includes the optional filtering operations. Parameters are set in the custom data structure. There are three different waveshapers that are selected with the *waveshaper* parameter. The fuzz asymmetrical waveshaper includes an asymmetry control on the range of [−1.0, +1.0]. It should be noted that this waveshaper's output gain is dependent on the *saturation* control as well as the gains.

```
virtual double processAudioSample(double xn)
{
     // --- perform waveshaping
     if (parameters.waveshaper == distortionModel::kSoftClip)
         output = softClipWaveShaper(xn, parameters.saturation);

     else if (parameters.waveshaper == distortionModel::kArcTan)
         output = atanWaveShaper(xn, parameters.saturation);

     else if (parameters.waveshaper == distortionModel::kFuzzAsym)
         output = fuzzExp1WaveShaper(xn, parameters.saturation,
                                     parameters.asymmetry);
```

```
    // --- inversion, normal for plate of class A triode
    if (parameters.invertOutput)
        output *= -1.0;

    // --- Output (plate) capacitor = HPF, remove DC offset
    if (parameters.enableHPF)
        output = outputHPF.processAudioSample(output);

    // --- if cathode resistor bypass, will create low shelf
    if (parameters.enableLSF)
        output = outputLSF.processAudioSample(output);

    // --- final resistor divider/potentiometer
    output *= parameters.ouputGain;

    return output;
}
```

19.13 C++ *Effect Object:* ClassATubePre

The *ClassATubePre* represents a bare-boned triode preamp that is ripe for extension, alteration and customization. It implements a cascade of four class A triode models and a simple two-band shelving filter as an EQ module, inserted between the third and fourth triode stages as depicted in Figure 19.4.

19.13.1 ClassATubePre: *Enumerations and Data Structure*

The *ClassATubePre* object requires no special enumerations but does declare a constant that sets the number of tube stages, which you may easily change:

```
const unsigned int NUM_TUBES = 4;
```

The object is updated via the *ClassATubePreParameters* structure that holds members to update the four triodes and two-band shelving filter; it also includes members to represent the input and output level controls.

```
struct ClassATubePreParameters
{
    ClassATubePreParameters() {}
    ClassATubePreParameters& operator=(. . .){ . . . . }

    // --- individual parameters
    double inputLevel_dB = 0.0;
    double saturation = 0.0;
    double asymmetry = 0.0;
    double outputLevel_dB = 0.0;
```

```
        // --- shelving filter params
        double lowShelf_fc = 0.0;
        double lowShelfBoostCut_dB = 0.0;
        double highShelf_fc = 0.0;
        double highShelfBoostCut_dB = 0.0;
};
```

19.13.2 ClassATubePre: *Members*

Tables 19.10 and 19.11 list the *ClassATubePre* member variables and member functions. The triode and shelving filter objects do all of the work; the *ClassATubePre* is an aggregate object.

19.13.3 ClassATubePre: *Programming Notes*

The *ClassATubePre* object combines four triodes and one shelving filter that do the bulk of the low-level work. Please be aware that this is a simple tube simulation and does not include many other tube details such as dynamic compression, rectifier choice, and myriad other tube details. Using the *ClassATubePre* object is straightforward.

- Reset the object to prepare it for streaming.
- Set the object parameters in the *ClassATubePreParameters* custom data structure.
- Call *setParameters* to update the calculation type.
- Call *processAudioSample*, passing the input value in and receiving the output value as the return variable.

Table 19.10: *ClassATubePre* member variables

ClassATubePre Member Variables		
Type	**Name**	**Description**
ClassATubePreParameters	*parameters*	Custom data structure
TriodeClassA	*triodes[NUM_TUBES]*	Four triodes
TwoBandShelvingFilter	*shelvingFilter*	Low and high shelf filters
double	*inputLevel*	Cooked version of input level in dB
double	*outputLevel*	Cooked version of output level in dB

Table 19.11: *ClassATubePre* member functions

ClassATubePre Member Functions		
Returns	**Name**	**Description**
ClassATubePreParameters	*getParameters*	Get all parameters at once
void	*setParameters* Parameters: – *ClassATubePreParameters* _parameters	Set all parameters at once
bool	*reset* Parameters: – *double* sampleRate	Resets all triode subcomponents
double	*processAudioSample* Parameters: – *double* xn	Process input xn through the system

Function: *reset*

Details: The *reset* function initializes the sub-components. It also sets some of the triode object members that are hard-coded, including enabling the DC high-pass filter, enabling the low shelf filter, and setting all low shelf cutoff frequencies at 88 Hz (approximate pitch of the low E on a six-string guitar), and a −12 dB low shelf attenuation.

```
virtual bool reset(double _sampleRate)
{
        // --- set triode defaults
        TriodeClassAParameters tubeParams = triodes[0].getParameters();
        tubeParams.invertOutput = true;
        tubeParams.enableHPF = true; // remove DC offsets
        tubeParams.outputGain = 1.0;
        tubeParams.saturation = 1.0;
        tubeParams.asymmetry = 0.0;
        tubeParams.enableLSF = true;
        tubeParams.lsf_Fshelf = 88.0;
        tubeParams.lsf_BoostCut_dB = -12.0;

        for (int i = 0; i < NUM_TUBES; i++)
        {
                triodes[i].reset(_sampleRate);
                triodes[i].setParameters(tubeParams);
        }

        // --- the lone shelving filter
        shelvingFilter.reset(_sampleRate);

        return true;
}
```

Function: *setParameters*

Details: The *setParameters* function distributes the incoming parameters to their respective objects. It also calculates the cooked input and output gain values based on the dB versions we present to the user.

```
void setParameters(const ClassATubePreParameters& params)
{
        // --- check for re-calc
        if (params.inputLevel_dB != parameters.inputLevel_dB)
                inputLevel = pow(10.0, params.inputLevel_dB / 20.0);
        if (params.outputLevel_dB != parameters.outputLevel_dB)
                outputLevel = pow(10.0, params.outputLevel_dB / 20.0);

        // --- store
        parameters = params;
```

```
    // --- shelving filter update
    TwoBandShelvingFilterParameters sfParams =
                            shelvingFilter.getParameters();
    sfParams.lowShelf_fc = parameters.lowShelf_fc;
    sfParams.lowShelfBoostCut_dB = parameters.lowShelfBoostCut_dB;
    sfParams.highShelf_fc = parameters.highShelf_fc;
    sfParams.highShelfBoostCut_dB = parameters.highShelfBoostCut_dB;
    shelvingFilter.setParameters(sfParams);

    // --- triode updates
    TriodeClassAParameters tubeParams = triodes[0].getParameters();
    tubeParams.saturation = parameters.saturation;

    for (int i = 0; i < NUM_TUBES; i++)
        triodes[i].setParameters(tubeParams);
}
```

Function: *processAudioSample*

Details: The *processAudioSample* function implements the four triodes in cascade plus the shelving
   filter EQ inserted between the third and fourth tube stages.

```
virtual double processAudioSample(double xn)
{
    // --- first 3 triodes
    double output1 = triodes[0].processAudioSample(xn*inputLevel);
    double output2 = triodes[1].processAudioSample(output1);
    double output3 = triodes[2].processAudioSample(output2);

    // --- filter stage is between 3 and 4
    double outputEQ = shelvingFilter.processAudioSample(output3);
    double output4 = triodes[3].processAudioSample(outputEQ);

    return output4*outputLevel;
}
```

19.14 *Chapter Plugin:* TubePreamp

The GUI and overall block diagram for the *TubePreamp* plugin is shown in Figure 19.15. It is a
straight-ahead implementation of two of the *ClassATubePre* objects, one for each channel.

19.14.1 TubePreamp: *GUI Parameters*

The GUI parameter table is shown in Table 19.12—use it to declare your GUI parameter interface
for the plugin framework you are using. If you are using ASPiK, remember to first create your
enumeration of control ID values at the top of the *plugincore.h* file, then use automatic variable binding
to connect the linked variables to the GUI parameters.

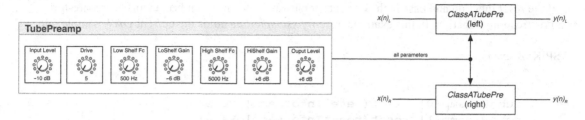

Figure 19.15: The *TubePreamp* GUI and block diagram.

Table 19.12: *TubePreamp* plugin's GUI parameter and linked-variable list

Control Name	Units	Min/max/default or string-list	Taper	Linked Variable	Linked Var Type
Input Level	dB	−60 / +12 / −6	linear	*inputLevel_dB*	*double*
Drive	-	1 / 10 / 2	linear	*saturation*	*double*
Low Shelf Fc	Hz	20 / 2000 / 150	1V/oct	*lowShelf_fc*	*double*
LoShelf Gain	dB	−20 / 20 / 0	linear	*lowShelfBoostCut_dB*	*double*
High Shelf Fc	Hz	1000 / 5000 / 4000	1V/oct	*highShelf_fc*	*double*
HiShelf Gain	dB	−20 / 20 / 0	linear	*highShelfBoostCut_dB*	*double*
Output Level	dB	−60 / +12 / −6	linear	*outputLevel_dB*	*double*

ASPiK: top of *plugincore.h* file

```
enum controlID {
      inputLevel_dB = 0,
      saturation = 1,
      lowShelf_fc = 2,
      lowShelfBoostCut_dB = 3,
      highShelf_fc = 4,
      highShelfBoostCut_dB = 5,
      outputLevel_dB = 6
};
```

19.14.2 TubePreamp: *Object Declarations and Reset*

The *TubePreamp* plugin uses two *ClassATubePre* members to perform the audio processing. It is very simple in design. In your plugin's processor (*PluginCore* for ASPiK) declare two members plus the *updateParameters* method.

```
// --- in your processor object's .h file
#include "fxobjects.h"

// --- in the class definition
protected:
      ClassATubePre tubePreamp[2];
      void updateParameters();
```

Call the *reset* function and pass in the current sample rate in your plugin framework's prepare-for-audio-streaming function. In this example, *resetInfo.sampleRate* holds the current DAW sample rate.

ASPiK: *reset( )*

```
    // --- reset the objects
    tubePreamp[0].reset(resetInfo.sampleRate);
    tubePreamp[1].reset(resetInfo.sampleRate);
```

19.14.3 TubePreamp: *GUI Parameter Update*

Update the left and right channel objects at the top of the buffer processing loop. Gather the values from the GUI parameters and apply them to the custom data structure for each of the filter objects, using the *getParameters* method to get the current parameter set, and then overwrite only the parameters your plugin actually implements. Do this for all channels you support.

ASPiK: *updateParameters( )* function

```
// --- update with GUI parameters
ClassATubePreParameters params = tubePreamp[0].getParameters();
params.inputLevel_dB = inputLevel_dB;
params.outputLevel_dB = outputLevel_dB;
params.saturation = saturation;
params.lowShelf_fc = lowShelf_fc;
params.lowShelfBoostCut_dB = lowShelfBoostCut_dB;
params.highShelf_fc = highShelf_fc;
params.highShelfBoostCut_dB = highShelfBoostCut_dB;

tubePreamp[0].setParameters(params);
tubePreamp[1].setParameters(params);
```

19.14.4 TubePreamp: *Process Audio*

Add the code to process the plugin object's input audio samples through each phase shifter object and write the output samples. The ASPiK code is shown here for grabbing the input samples and writing the output samples; use the method that your plugin framework requires to read and write these values.

ASPiK: *processAudioFrame( )* function

```
// --- read input (adapt to your framework)
double xnL = processFrameInfo.audioInputFrame[0]; //< input sample L
double xnR = processFrameInfo.audioInputFrame[1]; //< input sample R

// --- process the audio to produce output
double ynL = tubePreamp[0].processAudioSample(xnL);
double ynR = tubePreamp[1].processAudioSample(xnR);
```

```
// --- write output (adapt to your framework)
processFrameInfo.audioOutputFrame[0] = ynL; //< output sample L
processFrameInfo.audioOutputFrame[1] = ynR; //< output sample R
```

Rebuild and test the plugin. While you should hear some nice crunchy tones (try with a clean guitar loop), you will also experience severe aliasing, especially as you increase the drive (saturation) control. This plugin, like all nonlinear algorithms, requires oversampling techniques to mitigate the aliasing problem. We will revisit this plugin and perform 4X oversampling to greatly improve its fidelity in Chapter 22.

19.15 Bonus Plugin Projects

Navigate to http://www.willpirkle.com/bonus-projects/ and download two more bonus projects for this chapter, including the *Harmonic Exciter* and *Ring Modulator* plugins. They are already set up for 4X oversampling so you will need to have FFTW installed (see Chapter 20).

19.16 Homework

1. Design a Fuzz Face emulator using the input/output plot of Figure 19.7e, or find the schematics and use a SPICE circuit simulator to investigate this effect further.
2. Add a full-wave rectifier path to the *TubePreamp* plugin and mix in some of its signal to create an octave-up distortion effect.
3. Postulate another method to overcome the kink problem with asymmetrical waveshaping.
4. Implement the VB algorithm in Figure 19.10.
5. Implement a class B tube emulation. What are the fundamental differences you need to observe when compared to a class A version? How would you implement the crossover distortion?

19.17 Bibliography

Boss Super Overdrive Schematic. *Boss SD-1 Super Overdrive Pedal Schematic Diagram*, www.hobby-hour.com/electronics/s/sd1-super-overdrive.php, Accessed August 1, 2018.

Dunlop Fuzz Face Schematic. *Fuzz Face Analysis*, www.electrosmash.com/fuzz-face, Accessed August 1, 2018.

Fulltone OCD Schematic. *Fulltone OCD Guitar Distortion Pedal*, www.youspice.com/spiceprojects/spice-simulation-projects/audio-circuits-spice-simulation-projects/audio-effects-circuits-spice-simulation-projects/fulltone-ocd-guitar-distortion-pedal/, Accessed August 1, 2018.

Ibanez Tube Screamer Schematic, *Tube Screamer Mods*, http://tubescreamermods.blogspot.com/, Accesses August 1, 2018.

Larsen, E. and Aarts, R. 2002. "Reproducing Low-Pitched Signals through Small Loudspeakers." *Journal of the Audio Engineering Society*, Vol. 50, pp. 147–164.

Larsen, E. and Aarts, R. 2004. *Audio Bandwidth Extension: Application of Psychoacoustics*, Signal Processing and Loudspeaker Design. West Sussex: John Wiley & Sons.

MXR Distortion Plus Schematics. *MXR Distortion+ Analysis*, www.electrosmash.com/mxr-distortion-plus-analysis, Accesses August 1, 2018.

Woon N. and Gan, W. 2011. "Perceptually-Motivated Objective Grading of Nonlinear Processing in Virtual-Bass Systems." *Journal of the Audio Engineering Society*, Vol. 59, No. 11.

Pakarinen, J. and Yeh, D. 2009. "A Review of Digital Techniques for Modeling Vacuum-Tube Guitar Amplifiers." *The Computer Music Journal*, Vol. 33, No. 2, pp. 85–100. Cambridge: MIT Press.

Zottola, T. 1995a. *Vacuum Tube Guitar and Bass Amplifier Servicing*, Chap. 4. Westport: Bold Strummer, Inc.

Zottola, T. 1995b. *Vacuum Tube Guitar and Bass Amplifier Theory*, Chap. 3–5. Westport: Bold Strummer, Inc.

CHAPTER 20

FFT Processing: The Phase Vocoder

Jean Baptiste Joseph Fourier was born in 1768 in France. Euler was 61 years old (and would live another 15 years) and Sir Isaac Newton had died only 42 years prior. To put that into a more musical context, Mozart was 12 years old when Fourier arrived, and wrote his seventh and eighth symphonies that same year. Europe was in a period now known as the Age of Enlightenment. Given that we still use Fourier's original theories as the basis for solving problems in engineering and mathematics as well as the other hard sciences like biology and chemistry, we are currently still reaping the rewards of that period of time, and perhaps are still going through some form of enlightenment.

20.1 The Fourier Series

Napolean hired Fourier as a mathematician, historian, and diplomat. In 1807 Fourier published *Théorie analytique de la chaleur* ("The Analytic Theory of Heat") and described a method for analyzing the complex manner in which heat is conducted in solids. The conduction of heat through a solid is not linear but ebbs and flows in a complex manner. Fourier proposed that you could analyze the conduction as a sum of sinusoidal components—sines and cosines—of varying amplitudes and phases. His equation looked something like this:

$$y = \frac{a_0}{2} + \left(a_1 \cos(x) + b_1 \sin(x)\right) + \left(a_2 \cos(2x) + b_2 \sin(2x)\right) + \left(a_3 \cos(3x) + b_3 \sin(3x)\right) + ... \quad (20.1)$$

The number of terms in the summation could be infinite. This didn't sit well with some mathematicians at the time who still tended to be suspicious of infinite series. Nonetheless, Fourier spent the rest of his life working on this topic, expanding it to include more generalized systems and equations. Equation 20.1 is known as the *Fourier series*. The scalar multiplier values a_N and b_N are the coefficients that scale the sizes of the sine and cosine terms. Each of those pairs of terms represents a harmonic of the series. The sine and cosine arguments are *x, 2x, 3x, 4x*, etc., representing the fundamental and harmonics of the signal. The series works correctly when describing smoothly continuous waveforms that are periodic. Looking at Figure 20.1a, it isn't difficult to accept that a periodic function made of complicated wavy curves could be broken down into a set of simple periodic wavy curves. We call this *Fourier decomposition* and we briefly used it back in Section 9.2. In the same manner, it isn't difficult to believe that we can generate a complicated periodic, smooth and wavy signal by adding together a set of very simple smooth and wavy signals—a process called *Fourier synthesis*, which is shown in Figure 20.1b.

If you examine Figure 20.1 you will see that the *x*-axes have double arrow ends and the original signal has ". . ." before and after the waveform. These are there to satisfy the Fourier attorneys, who need us to understand that this whole idea only really works out exactly when the original signal, and all of

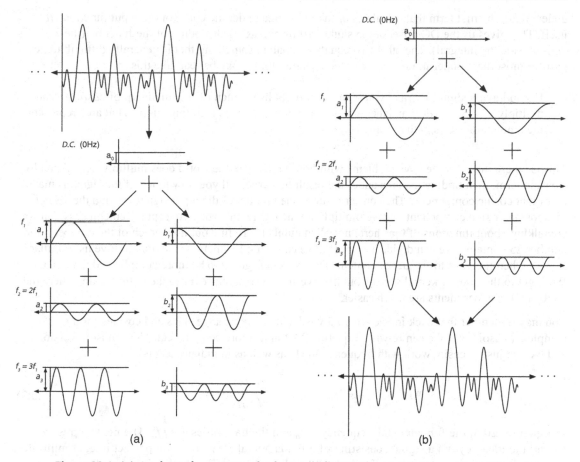

Figure 20.1: (a) As shown here, a complex but periodic waveform can be decomposed into a set of sine an cosine terms with varying amplitudes. (b) The waveform can be re-synthesized from its constituent sinusoidal components.

the components, have existed forever in that past and will exist forever in the future. You can also see that the signal is perfectly periodic and repeats itself in the figure. Figure 20.1 graphically represents the Fourier series in Equation 20.1 and the concepts of decomposition and re-synthesis. The next question is how we calculate the magnitudes of the sine and cosine terms. This is where it suddenly becomes complicated. Equation 20.2 shows the three equations for the a_0 (DC) term plus the a_N and b_N coefficients. Here the value of ω_o is the fundamental frequency in radians/sec. We let n vary on the range $n = \{1, 2, 3 \ldots\}$ to find the harmonics of the fundamental as $n\omega_o$.

$$a_0 = \frac{1}{T}\int_0^T x(t)dt \qquad a_n = \frac{2}{T}\int_0^T x(t)\cos(n\omega_o t)dt \qquad b_n = \frac{2}{T}\int_0^T x(t)\sin(n\omega_o t)dt \qquad (20.2)$$

Do not let the integral signs scare you. The cool thing about DSP is that we don't do integrals—we do discrete summations instead, and we'll get to those versions shortly. The integral signs can be translated as "the area under the curve" and they aren't what makes Fourier math difficult to understand. The first term that calculates a_0 takes the area under the curve of the input function $x(t)$ itself. This gives us the DC offset of the signal $x(t)$ by averaging the values of the function (the $1/T$ term outside the integral). The other two equations contain something that is generally difficult to get past for most that ponder it. For example, the equation for a_n may be read like this:

> If you have a signal $x(t)$ and you want to find out how much of it is made up of $cos(\omega_o t)$ you multiply $x(t)$ by $cos(\omega_o t)$ and then you find the area under the resulting curve. That area represents the amount of $cos(\omega_o t)$ that is in $x(t)$.

This is where most people have problems with Fourier. How in the world does multiplying a signal by a cosine term, then finding the area under the resulting curve, tell you how much of the signal is made up of that cosine component? The Fourier result is the product of the input signal $x(t)$ and the *kernel* (the cosine or sine component that we multiply it with). One important concept to remember is that we are talking about functions of time here, not plain numbers. To find out how much of the number 5 the number 25 contains, we can divide it by 5. But we can't divide functions like that to give us a similar answer when it comes to frequency content. The process is going to be more complicated than that. We'll get to the Fourier kernel shortly, but first we'll convert things over to the digital realm where the math and the experiments are much easier.

You may remember that back in Section 10.3 we decided to exchange sines and cosines for the complex sinusoid $e^{j\omega t}$. We can re-write Equation 20.2 in its more compact complex sinusoid version, and we can just as easily work with frequency in Hz as well as in radians/second.

$$x(t) = \sum_{n=-\infty}^{+\infty} c_n e^{j2\pi n f_o t} \qquad c_n = \frac{1}{T} \int_{-T/2}^{T/2} x(t) e^{-j2\pi n f_o t} dt \qquad (20.3)$$

In Equation 20.3, the fundamental frequency is f_o and the harmonics are nf_o. The next step is to extend the idea beyond a signal consisting of a fundamental frequency and perfect integer multiple harmonics to a signal consisting a fundamental and any harmonic, whether it is a perfect integer multiple or not. This produces a continuous spectrum of frequencies $X(f)$ from a signal in time $x(t)$. The signal $x(t)$ is no longer required to be periodic or existing forever, but it is required to adhere to rules called the *Dirichlet conditions*, which involve the number of discontinuities and the function $x(t)$'s ability to be integrated. This produces the Fourier transform pair of equations. The Fourier transform converts $x(t)$ into a spectrum of frequencies, $X(f)$ while the inverse transform converts it back.

$$X(f) = \int_{-\infty}^{+\infty} x(t) e^{-j2\pi ft} dt \qquad x(t) = \int_{-\infty}^{+\infty} X(f) e^{j2\pi ft} dt \qquad (20.4)$$

Equation 20.4 still includes the integral signs and the annoying infinite integral evaluation limits. Converting the analog Fourier transform into the digital discrete Fourier transform (DFT) replaces the integral sign with a simpler discrete summation, and changes the range of values from an infinite summation into a summation over a finite number of N samples (called the DFT length). The equations become much simpler—you can even perform small DFTs and inverse DFTs in a few minutes with a calculator and pencil. The C++ code becomes fantastically simple and short. The discrete version is shown in Equation 20.5.

$$\text{DFT:} \qquad X(k) = \sum_{n=0}^{N-1} x(n)e^{-j2\pi kn/N}$$

(20.5)

$$\text{Inverse DFT:} \qquad x(n) = \frac{1}{N}\sum_{k=0}^{N-1} X(k)e^{j2\pi kn/N}$$

In the discrete version, the input *x(t)* becomes our old familiar friend, the digitized input *x(n)*. The resulting spectrum is *X(k)*, but you'll notice that the independent variable is *k* and not *f* (Hz) as it is for the analog version. This is the other cool effect that digitizing the Fourier transform has: we no longer care specifically about time in seconds and frequency in Hz. We now care about samples in *n* and *frequency bins* in *k*. The reason we shift away from raw frequencies in Hz is that, in the digitized version, the sample rate now comes into play. The speed of the sample rate will change the number of samples required to encode a various frequency. You will remember that we also stopped thinking about the sample rate in Hz back in Section 10.5, where we only thought about how frequencies relate to Nyquist—½ Nyquist, ¼ Nyquist and so on.

The discrete version of the Fourier transform operates over a set of *N* input values. For mathematical reasons, *N* will always be a power of two. One way to think about the new system is that when we use Fourier in the digital domain, we shift the meaning of 1 Hz. In analog systems, 1 Hz is the frequency of a waveform whose wavelength is 1 second in length. For our DFT, the "bin 1" frequency is the frequency of a waveform that exactly fits inside our analysis window of *N* samples. We reference everything to that—one cycle inside of one block of *N* samples. The "bin 2" frequency contains two cycles in one *N*-sized block of samples—of course, its exact frequency will change with the sample rate. The equation for calculating the bin frequency is shown here.

$$\text{bin\_freq} = \frac{f_s}{N}$$

(20.6)

For a DFT size of 1024 samples and a sample rate of 44.1 kHz, the bin size is 43.066 Hz. For this sample rate and this DFT block size, everything is conceived in relationship to a 43.066 Hz sinusoid. This means that in this case the DFT produces a series of values that correspond to the frequency content of the digital input signal, but each frequency component is a multiple of 43.066 Hz. So the DFT spectrum is labeled *X(k)* where *k* is each bin frequency. For our setup here, the first few values from the DFT will be:

X(k) = { X(0 Hz), X(43.066 Hz), X(86.13 Hz), etc.}

You can see that when we feed the Fourier transform discrete samples *x(n)*, it produces discrete frequency components, *X(k)*. Notice also that the DFT produces exactly the same number of frequency components as we have input samples. A 1024-point DFT will produce 1024 spectral components across the range from DC to Nyquist, then back down to −0 Hz again on the negative set of frequencies (don't worry, we'll show this graphically soon).

If you look in Equation 20.5, you will see that in the kernel we multiply the input signal *x(n)* by a series of $e^{-j2\pi kn}$ terms. There is one $e^{-j2\pi kn}$ term per summation. Each of these terms is actually a test frequency—one of the bins. When *k* = 0, we are testing against the *bin 0* value (DC or 0 Hz) and when *k* = 1, we are testing against the *bin 1* frequency. Notice that the DFT doesn't know or care what the actual sample rate is, or what the actual value of a given bin is in Hz. The DFT cares only about how many cycles will fit into one DFT block of data and everything else is related to that. The outer summation sign is counting bins, and for each possible bin frequency, we test the input signal against that bin by multiplying the input by a sinusoid of the bin frequency. The DFT is nothing more than a massive, brute-force testing algorithm. The input signal is tested repeatedly by multiplying it, sample by sample, against a predetermined bin frequency signal. The summation of all of the points in the product of each of these repeated tests produces each spectral component, $X(k)$. So before we move on, we are going to really tunnel deep into this concept of multiplying a signal by a sinusoid to find out how much of the sinusoid is in the signal. So, let's do a simple experiment to look into that problem first. Please be aware that this is a thought experiment first and foremost and is not a direct proof of the Fourier series, integral, or transforms.

20.2 Understanding How the Fourier Kernel Works

To perform our thought experiment with the Fourier kernel, we are going to generate a digital input signal $x(n)$ and then test it by multiplying it with a sinusoid. To keep things reasonably easy, we are going to go back to old-fashioned cosine signals (though you can repeat the experiment with sines, or even complex sinusoids as well). We are going to naively generate and test the signals to see what happens, and to prove to ourselves that the Fourier kernel frequency-testing idea actually works. The first thing we need is some C++ code to generate a test waveform. We can generate a block of data that contains *numCycles* cycles of a cosine waveform with this simple code (notice that *numCycles* is a double, allowing us to create blocks with fractional numbers of cycles, which we'll do later).

```
// --- DFT block size
const int DFTLEN = 128;

// --- input
double x[DFTLEN]; // --- init to 0.0;
double numCycles = 3;

for (int i = 0; i < DFTLEN; i++)
{
        x[i] = cos(numCycles*(kTwoPi*(double)i) / DFTLEN);
}
```

You can see the resulting array containing three cycles of a cosine waveform in Figure 20.2a. The next thing we are going to do is to test this block by multiplying it, point by point, against an array of data containing the bin 1 test signal, or a single cycle of a cosine wave shown in Figure 20.2b where the two signals are overlaid together while Figure 20.2c shows the product of the two.

Now take a moment to really scrutinize the signal in Figure 20.2c. Specifically, look at the areas of the waveform that are positive (above $y = 0$) and negative (below $y = 0$). The (+) area under the curve exactly equals the (−) area under the curve. The summation of the positive samples and negative samples completely cancel each other out. The result of this test is that there is absolutely none of the bin 1 frequency component in our test signal $x(n)$. This is the crux of the Fourier kernel: multiplying a signal by a sinusoid and then finding the area under the resulting curve by simply averaging the sample values together actually does tell us how much of the test-sinusoid is in the original signal!

This is actually quite remarkable, so you might want to ponder it further. To help, check out Figure 20.3, which shows all of the tests from (a) bin 1 through (d) bin 4 on our input of three cosine cycles. Each

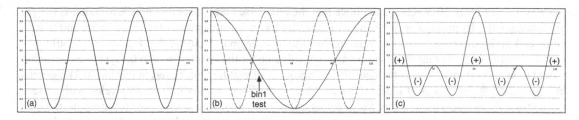

Figure 20.2: (a) An input block of three cosine cycles. (b) The bin 1 test signal with the input signals overlaid. (c) The result of multiplying the input and bin 1 signals point-by-point.

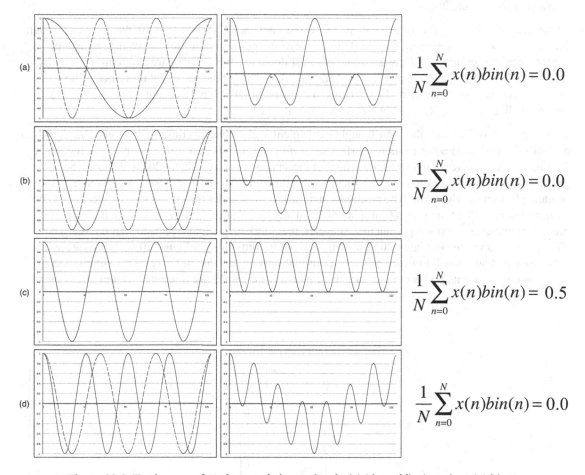

$$\frac{1}{N}\sum_{n=0}^{N} x(n)bin(n) = 0.0$$

$$\frac{1}{N}\sum_{n=0}^{N} x(n)bin(n) = 0.0$$

$$\frac{1}{N}\sum_{n=0}^{N} x(n)bin(n) = 0.5$$

$$\frac{1}{N}\sum_{n=0}^{N} x(n)bin(n) = 0.0$$

Figure 20.3: Testing a perfect three-cycle input signal *x(n)* (dotted line) against (a) bin 1, (b) bin 2, (c) bin 3, and (d) bin 4 signals (solid line) using the Fourier kernel and finding the mathematical area under the curve.

row shows the input signal (dotted) and test bin signal (solid) followed by the resuting point-by-point product of the pair, and finally the mathematical area under the product's curve. Notice we are finding the area under the curve by summing the points and then averaging them by *1/N*, a simple bookkeeping trick to make sense of amplitude values.

Now examine Figure 20.3 again, and notice that the only case that produces a non-zero value is the bin 3 test signal. Its product graph is the only one that has an overall positive average value and that value is 0.5. Suppose we tested our input against the -bin 3 signal, a cosine that is inverted. Would it surprise you to find out that the resulting average value under the curve was also 0.5? The two together would then add up to 1.0—it is the conceptual "amount" of that test bin signal in our input. The graphs in Figure 20.3 speak scores of words to us; this is how the Fourier transform *really* works, with a view from inside of the algorithm. As we keep testing higher and higher frequency bins progressing from Figure 20.3a–d, the resulting product is going to look just like that in the bin 4 case (V-shaped), but with more humps. The summation of the area is always going to be 0.0. In Figure 20.3, we used a DFT block size of 128 points. If we go through all 128 test bins and then plot the results, we will get something like Figure 20.4a.

In the case of our DFT thought experiment, the *x*-axis starts at DC (0 Hz) at $x = 0$, then moves through increasingly positive frequencies up until the center of the plot where it finds ±Nyquist. Above the center, negative frequencies decrease from −Nyquist down to −0 Hz. There are two non-zero results, one at bin 3 and the other at bin 125; both values are 0.5. This is the fundamental essence of how the DFT works. For a sample rate of 44.1 kHz and a 1024-point DFT, bin 3 is about 129 Hz. The DFT in Figure 20.4b shows a spike at +129 Hz and another at −129 Hz.

Before moving on, let's try the same thought experiment again—but this time we will feed the algorithm a block of data that does not contain a perfect set of *n*-cycles. In this case, our input block will contain 3.2 cosine cycles instead, and we'll test it against the same set of bins. Figure 20.5a–d shows the same progression of tests as we did in Figure 20.3, but this time you can see that there is a difference in the product plots on the right side. The positive and negative areas under the curves never cancel out to be zero, and none of the tests give a firm spike of 0.5 at any bin. You can see that when we use an input block that contains a non-integer number of cycles, the Fourier algorithm seems to start breaking down. The firm spikes we see in Figure 20.6a are smeared out to form those in Figure 20.6b, which are wider at the base and not as tall. For convenience, we used the absolute value of the resulting summation as our "amount" of signal present to compare these plots. These results are troubling: for our audio signals,

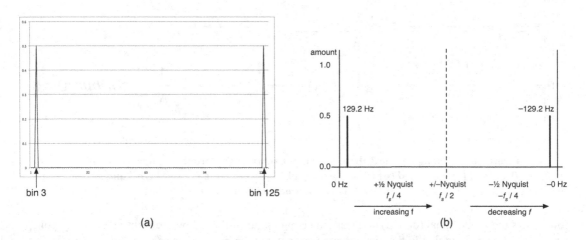

(a)   (b)

Figure 20.4: (a) The graphical results of our Fourier kernel thought experiment.
(b) How to interpret the results.

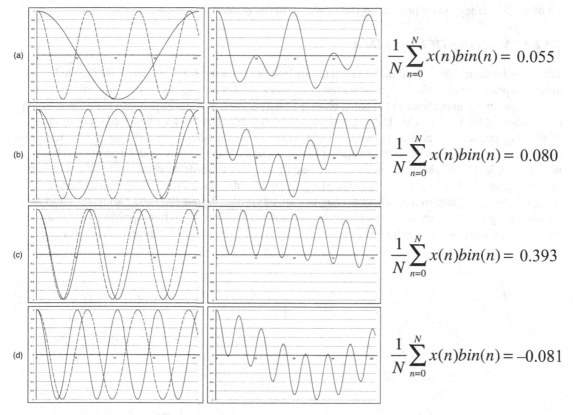

$$\frac{1}{N}\sum_{n=0}^{N}x(n)bin(n) = 0.055$$

$$\frac{1}{N}\sum_{n=0}^{N}x(n)bin(n) = 0.080$$

$$\frac{1}{N}\sum_{n=0}^{N}x(n)bin(n) = 0.393$$

$$\frac{1}{N}\sum_{n=0}^{N}x(n)bin(n) = -0.081$$

Figure 20.5: Testing a input signal *x(n)* (dotted line) with a non-integer number of cycles against (a) bin 1, (b) bin 2, (c) bin 3, and (d) bin 4 signals (solid line) using the Fourier kernel and finding the mathematical area under the curve.

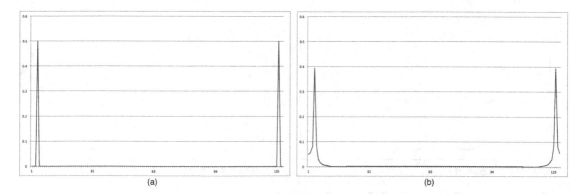

Figure 20.6: (a) The result of testing the Fourier kernel with an input block of exactly three cosine cycles. (b) The result when the input contains 3.2 cosine cycles.

there is no way we can guarantee any particular block of input data is going to always have a matching set of bin frequency cycles present. In fact, it will most likely *never* be the case.

20.2.1 Windowing DFT Input Data

We can make the results of our Fourier kernel testing look better for the 3.2 cycle case if we manipulate the input data prior to multiplying it against the test bins. Specifically, we are going to use a weighting system that will give more importance to the center chunk of the input data, and less importance to the information on the edges of the input blocks. This is called *windowing* the data. There is an established set of common windowing functions, all of which generate weighting that favors the center part of the block. The 3.2 cosine cycle input in Figure 20.7a is multiplied against Figure 20.7b the Hann window and the product is shown in Figure 20.7c. The resulting waveform is more heavily weighted towards the center part of the waveform. The test against the bin 1 frequency is shown in Figure 20.7d and e where the resulting area under the curve is 0.005, a big improvement from the non-windowed version. Figure 20.7f and g show the effect of no windowing versus windowing with clearly better results in the latter. Therefore, applying windowing techniques to our real-world audio input signals is going to be a requirement for meaningful results.

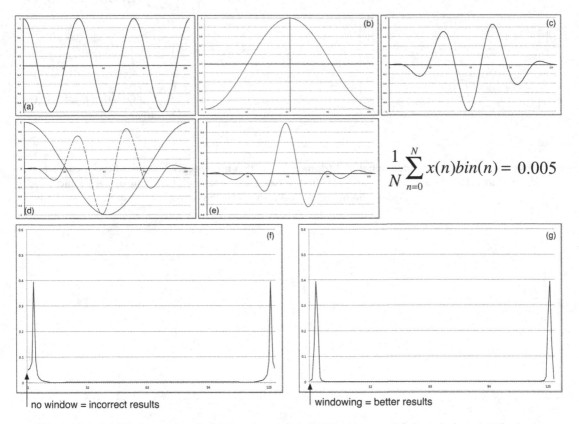

$$\frac{1}{N}\sum_{n=0}^{N} x(n)bin(n) = 0.005$$

no window = incorrect results

windowing = better results

Figure 20.7: (a) The input signal of 3.2 cosine cycles. (b) The Hann weighting window. (c) The input multiplied by the Hann window. (d) The Fourier kernel test with the windowed signal (dotted line) and the first bin signal. (e) The product of the two signals; the area under the curve is 0.005. (f) The results of our Fourier kernel testing with non-windowed input. (g) The result after windowing.

20.3 The Complete DFT

If you've followed our Fourier kernel thought experiment you should have a much better intuitive grasp on how the Fourier transform works, how the testing is done in a brute-force manner, one bin at a time, and how windowing helps remove the error in the Fourier transform response. But the reality is that these experiments used onlybin signals made of cosine waveforms. We did this because we knew the input was a cosine and we were trying to come to grips with the Fourier kernel concept of sinusoid multiplication. Fortunately, we can take the DFT in Equation 20.5 and apply Euler's identity to break the complex sinusoid in the kernel into its real and imaginary parts. This yields the set of equations we use today as the DFT. Equation 20.7 shows the DFT and IDFT pair; note that the independent variable k is the bin number.

$$\text{DFT:} \qquad X(k) = \frac{1}{N} \sum_{n=0}^{N-1} \left[x(n) \frac{\cos(2\pi kn)}{N} - jx(n) \frac{\sin(2\pi kn)}{N} \right]$$

$$(20.7)$$

$$\text{Inverse DFT:} \qquad x(n) = \sum_{k=0}^{N-1} \left[X(k) \frac{\cos(2\pi kn)}{N} + jX(k) \frac{\sin(2\pi kn)}{N} \right]$$

Equation 20.7 shows that for a given input block of N samples $x(n)$, the DFT produces two sets of data: a N-length array of real values and another N-length array of imaginary values (shown as the real and imaginary parts inside of the bracket). The best part is that this is very easy and compact to write in C++. In fact, the complete code to calculate a DFT is shown next. The entire DFT loop, including curly brackets, is only eight lines of code. There are two for loops: the outer loop steps through each bin frequency, and the inner loop performs the multiplication by *cos* (real) and *sin* (imaginary) and then performs the final summations. At the end of the function, we have the real and imaginary parts of the results in two separate arrays. There are *DFTLEN* real values and *DFTLEN* imaginary values that together make up the final DFT output.

```
// --- declare the DFT length, and three arrays for values
const int DFTLEN = 128;
double x[DFTLEN]; // --- init to 0.0;
double im[DFTLEN]; // --- init to 0.0;
double re[DFTLEN]; // --- init to 0.0;

// --- load up the input with a test waveform
double numCycles = 3;
for (int i = 0; i < DFTLEN; i++)
{
    x[i] = cos(numCycles*(kTwoPi*(double)i) / DFTLEN);
}

// --- perform DFT
for (int BIN = 0; BIN < DFTLEN; BIN++) // outer loop: bins to test
{
    // --- inner loop, calculate real and imaginary parts, form summation
```

```
        for (int n = 0; n < DFTLEN; n++)
        {
                re[BIN] = re[BIN] + x[n] * cos (kTwoPi * BIN * n/DFTLEN)/DFTLEN;
                im[BIN] = im[BIN] - x[n] * sin (kTwoPi * BIN * n/DFTLEN)/DFTLEN;
        }
}
```

Before moving on to FFT, we should note that the preceding Fourier kernel thought experiment we performed using bins made of cosine waveforms is actually the first half of the DFT calculation—the part that produces the real portion of the output. In some cases, this is all you need, and this special, real-part-only version of the DFT is called the Discrete Cosine Transform (DCT).

20.4 The FFT

The FFT is really just a more efficient way of calculating the DFT. If you look at the code for the DFT, you can see that there are two for-loops, each iterating over the range of the DFT block size. This immediately squares the number of iterations, and the inner summation loop must execute two trigonometric functions. A 1024-point DFT is going to require more than two million trigonometric function calls in a double-nested loop of 1,048,576 iterations. In 1903, Runge published the paper that described what we now call the FFT, but there have been many improvements to the algorithm over time since then, and you will hear different names given to the various versions. The history of the FFT is actually quite interesting and you should check it out if you are intrigued.

Technically, the FFT results are the same as the DFT, but packaged in a slightly different manner in the real and imaginary output arrays. The FFT algorithms today organize the output arrays in a manner that mimics the DFT output, but there isn't actually a firm standard on it. So when trying out new FFT algorithms, make sure you read the fine print about exactly how the output data are organized, whether the data are normalized or otherwise scaled, etc. The FFT package we will use is called FFTW ("the Fastest FFT in the West") and is a well-known and mature library. It is also wickedly fast and in execution. You will need to download and install it on your system. The procedure is slightly different for MacOS vs. Windows—and in both cases, this is not trivial. You can find lots of information to help with the installation, including www.willpirkle.com. For this chapter we will assume that you have the FFTW package installed and that your compiler (Visual Studio or Xcode) is already set up to easily incorporate the FFTW library code—you can practice with sample projects available all over the Internet first. Once you have a basic FFTW project running, then you may begin working with it in a real-time plugin.

When we use the FFT in a plugin, it typically falls into one of four categories, shown in Figure 20.8. An easy way to use the FFT (see Figure 20.8a) is on the GUI object that displays the FFT magnitude response as a spectrum; we'll look at this in the next chapter. Another way is to use the FFT as an analysis tool, shown in Figure 20.8b. In this case, the FFT is used to extract features from the audio input signal in a side chain and uses information it gleans to alter the plugin's algorithm—a form of adaptive signal processing. Pure frequency domain processing is shown in Figure 20.8c. This is the most computationally expensive way to use the FFT. In this case we transform the audio input data into the frequency domain, manipulate the signal within the FFT by altering the real and/or imaginary components of the various bins, then use the inverse FFT (labeled as FFT^{-1} or IFFT) to convert the data back to a time domain digital signal. In Figure 20.8d the FFT is used twice: once on the input signal

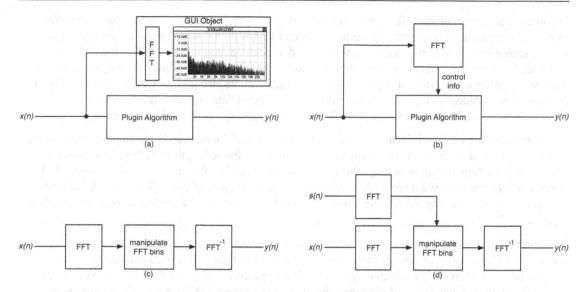

Figure 20.8: Four common ways to use the FFT in a plugin include (a) using the FFT for spectrum visualization; (b) using the FFT to gather information about the signal, then changing the algorithm accordingly; (c) manipulating the audio data in the frequency domain, then converting back to the time domain with the inverse FFT; and (d) using the FFT of another signal (a side-chain signal or a pre-stored filter response) to alter the FFT of the input signal.

and again on another audio signal, or alternatively a pre-stored FFT of something else (a filter, a noise signature, etc.).

Be aware that this is a very deep topic, and entire books have been written on just this frequency domain processing alone. Using the FFT allows us to create plugins that could never exist as analog products. This allows us to step outside the boundary of imitating analog hardware and come up with truly unique plugin ideas.

20.4.1 Overlap/Add Processing

Windowing the input data helps the FFT overcome problems associated with the fact that it only works perfectly when fed buffers that contain exact integer multiples of the bin 1 frequency. However, processing an audio input stream using FFT and IFFT operations while still preserving the original audio data requires more work. For example, in Figure 20.8c, suppose that there is no manipulation of the bins whatsoever and the FFT data was just inverse transformed back into time domain output samples $y(n)$. In this case, we would like the output sequence $y(n)$ to exactly match the input sequence $x(n)$. It is only under this condition that we could say that our FFT processing would be nondestructive to the original signal. One way to do this would be to take one gigantic FFT of the entire input signal from start to finish, manipulate the FFT bins, and then do another massive IFFT operation. Since that is not practical, and owing to the fact that the FFT operation is inherently CPU intensive, we would like to make the FFT blocks as short as possible. In this case, we are using the short term Fourier Transform (STFT) and we operate on a small block of data only, typically 512 to 4096 samples in length. The length of the FFT block size generally depends on the processing algorithm you are using;

in some cases a small block FFT might actually work better than a longer block. In order to further minimize errors, we can overlap FFT frames. Consider Figure 20.9a, which shows a series of input frames gathered from *x(n)* that are windowed, FFT'd, processed, and inverse FFT'd, then overlapped and added to form the output *y(n)*. The upper portion of the algorithm is called the analysis phase, and the lower portion is the synthesis phase. Notice that Figure 20.9a also includes the latency that the FFT processing introduces—we have to wait until the first frame of data is gathered before we can begin outputting meaningful data, so that first IFFT frame is delayed by one frame's worth of time.

The hop size is the number of samples between the frames, and the overlap is the amount of the frame that is reprocessed on the next step. If the hop size equals the frame length then there is no overlap. The smaller the hop, the more overlap and vice versa. The hop size dictates how frequently the FFT operation is performed. What would be the requirements for the size of the FFT frames, the overlap region (or hop size), or the windowing curve used that would guarantee a perfect reconstruction of the input signal?

The criteria for a perfect reconstruction of the signal are called the Constant Over-Lap Add (COLA) constraints. The constraints always come down to the window type that is used and the amount of overlap between frames. The windowing we used in Figure 20.7 during our FFT kernel experiment is the Hann window; this is one of many possible window functions that meet the COLA constraints . You can also see that windowing the output of the IFFT is an optional process; this depends on the algorithm. If no FFT processing occurs, the output windowing is not necessary.

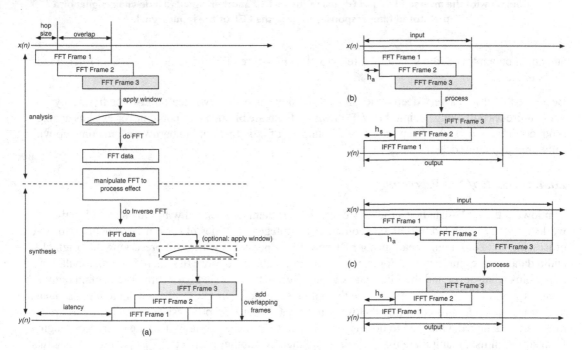

Figure 20.9: (a) A series of windowed frames that are transformed, processed, inverse transformed and overlapped and added together. (b) If the synthesis hop-size is larger than the analysis hop-size, the output grows ever longer (time expansion). (c) If the synthesis hop-size is smaller than the analysis hop-size, the output grows ever shorter (time contraction). Notice that in both scenarios, the length of the FFT never changes.

The system shown in Figure 20.9 is named the *phase vocoder*. This is a generic term given to the overlap-add with FFT bin manipulation that occurs in the algorithm. The phase vocoder originated as a method of telephony at Bell Labs (vocoder stands for VOice enCODER). You will find numerous phase vocoder algorithms to experiment with in the references. One interesting way to use the phase vocoder is to make the analysis and synthesis hop sizes different, as shown in Figure 20.9b and c where the latency has been removed for easier viewing.

When the two hop sizes are different, we create time expansion (stretching) or contraction (shrinking); these are also used in some pitch shifting algorithms. Notice that the size of the FFT never changes— only time between taking a new FFT (the analysis hop size, h_a) and the time between overlapping and adding the IFFT output (the synthesis hop size, h_s) are varied. The ratio of the two hop sizes is called the time stretching factor and is given as follows.

$$\alpha = \frac{h_s}{h_a}$$

stretch: $\alpha > 1$

shrink: $\alpha < 1$

$$(20.8)$$

The *PhaseVocoder* C++ object that we describe in this chapter handles all of the chores of the input and output buffering, the FFT, and overlap adding. Your plugin that uses it needs only to manipulate the FFT bin data, and the rest will be handled automatically. This object is also ripe for extension to include many more effect possibilities. The *PhaseVocoder* object comes prepackaged with three common windowing buffers; you can easily add more of your own windows for experimentation. The three windows are as follows.

- Hanning (called Hann here)
- Blackman-Harris
- Hamming

Because of the similarity of the names Hamming and Hanning, we will always refer to the Hanning window as simply "Hann"; this is our default window for the phase vocoder object. The equations for these windows, packed into buffers of length N—where N is an even number—are shown in Equation 20.9.

$$window_{\text{Hann}}(n) = \begin{cases} 0.5\left[1 - \cos\left(\frac{2\pi n}{N}\right)\right] & 0 \leq n < N \\ 0 & elsewhere \end{cases}$$

$$window_{\text{BlackmanHarris}}(n) = \begin{cases} 0.42323 - 0.4976\cos\left(\frac{2\pi n}{N}\right) + 0.7922\cos\left(\frac{2\pi n}{N}\right) & 0 \leq n < N \\ 0 & elsewhere \end{cases} \quad (20.9)$$

$$window_{\text{Hamming}} = \begin{cases} 0.54 - 0.46\cos\left(\frac{2\pi n}{N}\right) & 0 \leq n < N \\ 0 & elsewhere \end{cases}$$

Table 20.1: Some windows and a few of their common overlap ratios for COLA constraint

Window	COLA Overlap Ratios
Hann	1/2, 2/3, 3/4, 4/5, 5/6, 6/7, 7/8, plus more
Blackman-Harris	3/4, 4/5, 5/6, 6/7, 7/8, plus more
Hamming	1/2, 2/3, 3/4, 4/5, 5/6, 6/7, 7/8, plus more

It should be noted that the strictly legal versions of these windows technically require buffers that have an odd number of points to create a perfectly symmetrical window: that is, N is odd, and in Equation 20.8 you would replace N with $N - 1$. For our even valued buffers and the phase vocoder overlap/add strategy, we use Equation 20.8 directly.

Table 20.1 shows some of the COLA constrained ratios for each window; there are actually many more ratios available.

20.4.2 Window Gain Correction

Any window other than rectangular necessarily reduces the overall power and gain of the signal passing through it. The overlap-added signal will then have a reduced gain compared to the original signal. To accommodate this, you calculate a window gain correction value. To calculate the gain correction value, you sum the individual samples in the window function that normally range from [0.0, 1.0] and then invert the sum. You can see examples in the chapter's *PhaseVocoder* object.

20.4.3 FFT and IFFT Magnitude and Phase

The phase vocoder consists of complimentary FFT and IFFT processing blocks. On the analysis side, the N-point FFT accepts N samples in time $x(n)$ and produces arrays of N real and imaginary components. The large majority of phase vocoder algorithms do not operate on the real and imaginary components directly. Instead, the real and imaginary components are converted to magnitude and phase arrays, each consisting of N components. One way to conceptualize this process that will make understanding the phase vocoder algorithms easier is to imagine the FFT as a bank of ideal, super narrow band-pass filters, each centered on a bin (k) frequency and including a 0 Hz DC detector block, which is simply an average of the samples. This is shown in the left side of Figure 20.10. The real and imaginary output arrays are then mathematically massaged into the N-point magnitude and angle (phase) arrays using the equations from Section 10.2.

$$X = a + jb = \text{Re} + j\,\text{Im}$$
$$|X| = \sqrt{a^2 + b^2} \qquad\qquad \Phi = \tan^{-1}(b\,/\,a) \qquad\qquad (20.10)$$

Each of the conceptual narrow band-pass filters produces a two-part result: the amplitude and starting phase of the frequency that is able to pass through it. Knowing this, we can then move to the IFFT or synthesis phase. On this side, you can imagine the IFFT as a bank of sinusoidal oscillators. The magnitude value sets the amplitude and the phase value sets the starting location (phase) of the oscillator. Then, we let the oscillators each run for N samples, all in parallel. The outputs of the oscillators are summed together to produce the final output sequence $y(n)$.

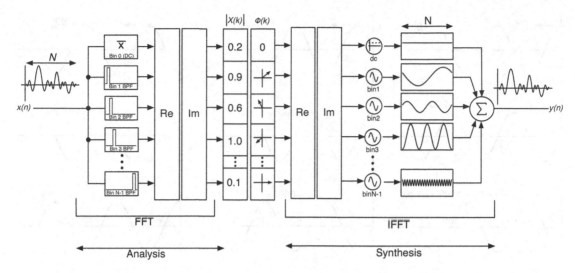

Figure 20.10: As illustrated here, phase vocoder concepts include the fact that the FFT acts as a bank of filters and the IFFT acts as a bank of oscillators. The magnitude and phase information is used to implement the algorithm.

So, N samples at the input produces $N-1$ frequency bins plus the DC component. All of the exact sample information is lost. The phase vocoder algorithm manipulates the magnitude and phase information in some interesting manner. Then, that information is converted back to its native real and imaginary format using the reverse equations shown here, which are simply the standard polar to Cartesian coordinate conversion functions.

$$C = |X| \angle \Phi$$
$$\mathrm{Re}_C = |X| \cos(\Phi) \qquad\qquad \mathrm{Im}_C = |X| \sin(\Phi) \tag{20.11}$$

Finally, the IFFT uses that information to set each oscillator in the bank that synthesize the output $y(n)$ as a sum of sinusoidal oscillators.

20.4.4 Using Phase Information

One question you may have is regarding the meaning of the phase component with respect to the sinusoid. What does a phase of 0.0 degrees mean for a complex sinusoid that is technically neither sine nor cosine? The answer is that a phase of 0.0 represents a sinusoid that starts at the maximum 1.0 level that we usually refer to as a cosine wave. This is shown in the top left of Figure 20.11, where we show the progressive shifting of the sinusoid by 90 degrees to the right. This outlines a fundamental issue: how does the FFT deal with the fact that the multiple phase rotations produce the same phased-sinusoids? The answer is that the phase information we get from the FFT is on the range of −180 to +180 degrees, or −π to +π. If we want to keep track of the total rotation across many FFT frames, then we need to store phase information from one frame to the next. So, many of the phase vocoder algorithms will have some kind of phase "memory" component to them. The value between −180 and +180 degrees, or −π and +π, is sometimes called the *wrapped phase*, and calculating the sinusoid's absolute phase is called *phase unwrapping*.

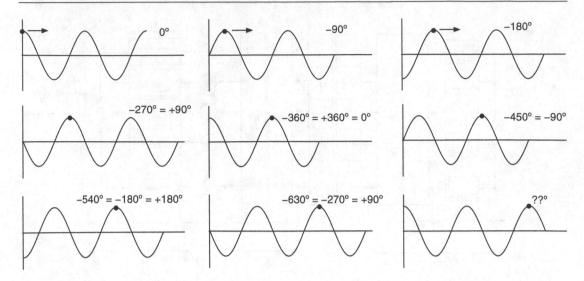

Figure 20.11: Starting with the top row and moving to the right, we shift the sinusoid to the right by 90 degrees. After four shifts we are back to the same sinusoid, but how should we call this last shift?

20.4.5 Phase Deviation

One of the most important aspects of the phase vocoder is the notion of taking sets of overlapped frames of information, processing them, and adding them together. The hop size on the analysis side h_a determines how quickly each new FFT snapshot is taken. If we feed a perfect bin 3 signal into the FFT as shown in Figure 20.12a with the analysis hop size h_a then each new FFT snapshot is going to start on a new phase shown as black dots. If we store the first phase value for the first FFT frame and we know the hop size and bin number, we can predict what the phase will be for subsequent frames. The phase increment we expect to occur on each frame for bin k is ϕ where ω_k is the normalized bin frequency, as in Equation 20.

$$\phi(k)_{inc} = \omega_k h_a$$

(20.12)

In order to use this we'll need to keep track of the previous frame's unwrapped phase value and then store the current phase value for the next frame to use.

If we feed the FFT a signal that is not a perfect bin frequency, then its energy will be spread over several bins around the actual frequency. If we look inside of bin 3, we will get the results in Figure 20.12b where the input signal is lower in frequency, ~ bin 2.3 or so. The amplitude has been left the same because we are comparing phase only. The captured starting phases are shown as white dots and the phase deviation ϕ is the difference between the predicted and actual values. Calculating the phase deviation is found in many phase vocoder algorithms. To use it, you have to convert it back to a wrapped value on the range of $[-\pi, +\pi]$ called the *principal argument,* which is a modulo operation that repeatedly divides out π until you are left with the remainder on that range. The principal argument is expressed as *princArg( ).*

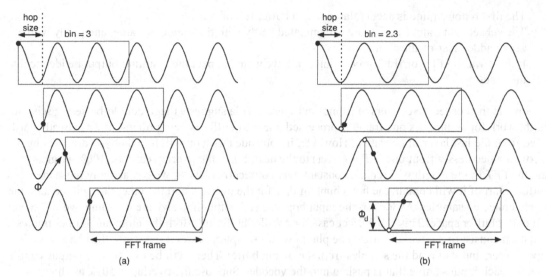

Figure 20.12: (a) Successive frames of a perfect bin 3 signal with starting phase shown as a black dot. (b) Successive frames of a bin 2.3 signal with predicted starting phases shown as black dots and captured starting phases shown as white dots. The phase deviation φ_d is the difference between the predicted and actual values.

The new, actual unwrapped phase increment that includes the phase deviation for the particular bin and FFT frame can then be calculated.

$$\varphi(k)_d = \varphi(k)_{captured} - \varphi(k)_{predicted}$$
$$\varphi_{inc}(k)_{actual} = \omega_k h_a + princArg(\varphi_d(k)) \tag{20.13}$$

In addition to some of the phase vocoder algorithms, the actual unwrapped phase increment may be used to calculate the frequency of the non-perfect-bin-frequency signal that was actually captured—in other words, the bin 2.3 value that allows you to find frequency values that are in-between the bins. The equation for the exact frequency is here.

$$f_{actual}(k) = \frac{\varphi_{inc}(k)_{actual}}{2\pi h_a} f_s \tag{20.14}$$

20.4.6 Phase Vocoder Coding

If you look at Figure 20.9, you can see that the process involves overlapping input frames and the accumulation of the output frames along the *x(n)* and *y(n)* timelines. Figure 20.13 shows details of the overlap/adding phase. There are a few things that we can observe in the overlap and add process.

- The amount of overlap will be between 0 and 99.9%.
- One hundred percent overlap is meaningless: we cannot have more than 100% overlap or less than 0% overlap.

- The first output frame is accumulated with a buffer full of zeros.
- The subsequent output frames are accumulated partly with the previous frame, and partly with a zero-padded rest-of-the-frame.
- If there was no FFT or IFFT processing at all, we would expect the resulting output be identical to the input.

Looking at the consecutive layers of frames in Figure 20.13a suggests that a double-buffering scheme might work for coding. As one frame is processed, the next buffer is being filled—it was already partly filled from the last layer of operation. However, if you code this, you will find that you are copying a lot of data needlessly from one buffer (layer) to the next. Since the overlap can't be 100% or greater and the FFT frame length will remain constant over consecutive layers, we can actually reuse the input buffer—part of it will contain the first chunk of data for the overlap, and the other part will contain the newly collected samples. As long as the input hop size and output hop size are the same, we will never run out of buffer space. This is a perfect case for circular buffering. Both the input and output frames will each consist of a circular buffer. The plugin writes samples, one at a time, into the phase vocoder's input buffer, and will read the samples from the output buffer. There will be exactly one output sample read for each input sample that is pushed into the vocoder. Suppose the overlap is 50%, as shown in Figure 20.13b. Once the first frame is processed, it is added to the output buffer and we begin reading samples out. The strategy here is to read out a sample from the output buffer, then overwrite that value with 0.0, chewing through the data. This is the zero padding that will serve to accumulate the next frame. At the end of the first frame of operation, we reset the next input read location to the center of

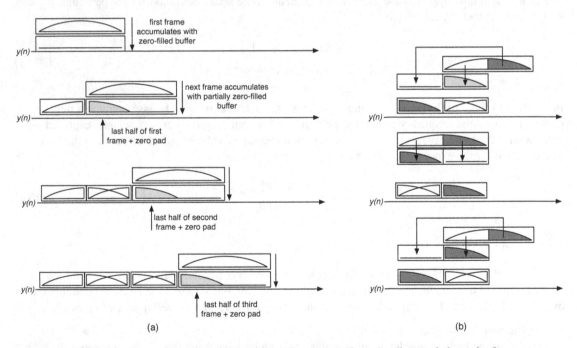

Figure 20.13: (a) The overlap-add mechanism, shown here in finer detail, reveals how the frames are accumulated. At least a portion of the accumulation frame must be zero to accommodate the next chunk; here the overlap is 50%. (b) The circular buffer concept applied to the input and output buffers.

the input buffer, and begin accumulating more input samples into the *front* of the buffer as shown in Figure 20.13b. On the next frame, we begin accessing the circular buffer, reading out the (old) last half of data, wrapping the pointer, then reading the freshly accumulated data in the first half. In a similar manner, the output buffer's read location exactly mirrors this, and it writes and wraps its own read index value accordingly. In this way, each input buffer contains part of the last frame, plus the new data and each output buffer contains part of the last frame, plus zeros written over the old data. The new input frame data are accumulated with the zeros. The good news is that the input and output circular buffers don't necessarily need to be the same length. For the *PhaseVocoder* object, we've set the output buffer to be four times the length of the input buffer to allow many different algorithms that involve stretching or pitch bending. You can easily change this in code—the bookkeeping will all stay intact.

20.5 Some Phase Vocoder Effects

The *PhaseVocoder* object is going produce a new FFT calculation each time the hop size and FFT input counter reset. To make operation as simple as possible, we've set up the *PhaseVocoder* object to contain functions that break the processing operation into pieces. It uses a simple process function that outputs one *y(n)* value for each *x(n)* value you input. If you input a value that triggers a new FFT, the operation occurs automatically and you are notified with a Boolean flag. Your plugin may then access the newly calculated FFT data buffers directly for manipulation. Typically you will convert the real and imaginary parts to magnitude and phase, then do your manipulation, and then convert them back to real and imaginary parts. You may need to store information for the next FFT frame run. You then issue an IFFT command via a function call. After that, your plugin may access the IFFT data arrays if it wishes. Then, the IFFT data are overlapped and added to the last frame. For the processing, you really only need to focus on the FFT data and the manipulation of the bins.

20.5.1 Robot and Simple Noise Reduction

Two very simple algorithms that require minimal processing are robotization and spectral noise reduction. The robotization effect creates a similar effect as half of the ring modulation in that it adds a frequency component to the signal, but not as a pitched sinusoid. For this effect, you reset the phase of each FFT frame right back to 0.0 regardless of its value. The closer the frames are together (the smaller the hop size and/or FFT frame size) the higher the pitch of the added frequency. For voice, this effect makes the human speaker sound like an early computer attempt at voice synthesis. For pitched sounds, it imitates a mild ring modulation. The robotization algorithm steps using the *PhaseVocoder* object are as follows.

1. Obtain real and imaginary parts of FFT result.
2. Convert them to magnitude and phase.
3. Ignore the magnitude values.
4. Reset the phase values to 0.0.
5. Convert magnitude and phase back to real and imaginary components.
6. Issue the IFFT and overlap-add command.

One of the earliest uses of the FFT in DSP was called adaptive noise cancellation (ANC), and this was aimed at removing noise from digital signals (not necessarily audio). A complimentary effect was called active line enhancement (ALE), which attempted to make a narrowband signal "stick out" from a wideband noisy background. The noise-cancelling algorithm uses spectral subtraction. A FFT of the noise signal is subtracted from the input + noise combination. A variation on this in Zölzer (2011) uses a mechanism that reduces the size of small magnitude bin values. In this effect, you multiply each bin's magnitude by a scaling factor, F_{NR}, as shown here:

$$F_{NR} = \frac{|X(k)|}{|X(k)| + R}$$ (20.15)

where

$$R > 0.0$$

The user controls the value of R. As it increases in value, the smaller the magnitude of the bin, and the more it will be scaled downward.

The simple noise reduction algorithm steps using the *PhaseVocoder* object are as follows.

1. Obtain real and imaginary parts of FFT result.
2. Convert them to magnitude and phase.
3. Multiply each magnitude value by the scaling factor F_{NR}.
4. Ignore the phase values.
5. Convert magnitude and phase back to real and imaginary components.
6. Issue the IFFT and overlap-add command.

If the noise is uncorrelated white noise with a magnitude of 1.0, you may also try simply subtracting a constant value from each bin. If the noise can be captured, such as the analog tape noise at the head of a reel of tape, then its FFT may be taken and stored. The FFT spectrum can be subtracted on a frame-by-frame basis. Careful selection of the windowing and overlap is critical.

20.5.2 Time Stretching/Shrinking

The phase vocoder can also be used to stretch or shrink time. The typical way of handling this involves first setting the analysis and synthesis hop sizes to different values, as shown in Figure 20.14a. If the synthesis (IFFT) hop size h_s is larger than the analysis hop size h_a then it means that we are taking a FFT snapshot at a certain time, then pasting the IFFT result to a location *later* in the output timeline, forcing the time to stretch out. The fact that the FFTs are done in chunks allows us to sew the new data to the old data in an overlap-add fashion. A question you might have is "why bother with the FFT—why not just cut and paste the time domain samples directly?" The answer lies with the phase of the transplanted signal: you need to shift its phase to make it line up correctly with the output (this is the crux of the classic phase vocoder algorithms where the two hop sizes are different). This is shown in Figure 20.14b and c where the problem is evident when processing a single sinusoid input. If we don't adjust the phase, the overlap and mixing won't work correctly—there will be phase cancellation or addition in the overlapped regions.

In order to accomplish the phase adjustment, we need to keep track of the running phase of each bin, and estimate the bin frequency according to Equation 20.14. A new phase is calculated for each FFT bin, then saved for the next frame. This maintains the *horizontal phase coherence* of the signal and ensures that successive frames are glued together correctly—but it destroys the *vertical phase coherence*, which is the phase relationship between the sinusoidal components within the FFT frame. Storing the old phases and updating the new ones to maintain horizontal phase coherence is called *phase propagation*. This brings up an interesting issue with time stretching and pitch bending with the FFT: if you try to change the pitch but hold the time constant, you will necessarily need to alter the phase. The same is true if you hold the frequency constant and try to adjust the time. In a sense, this is a game you can't win.

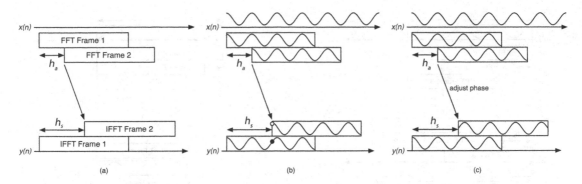

Figure 20.14: (a) Time stretching is done by transplanting the IFFT data to a location later in the timeline. (b) Simply shifting the IFFT result without phase adjustment produces the incorrect output signal (note black and white phase-dot locations). (c) Adjusting the phase of the bin gets it to line up properly with the paste location.

For time shrinking you have to work in the opposite direction, transplanting the IFFT data to locations *earlier* on the timeline, then overlap and mix the signals. A little thought will show that this isn't going to be easy because we are chewing our way through forward time on the analysis timeline. By the time we are ready to make the transplant, the chance to do so has already passed by. To fix that, we would need to pre-buffer the signal by enough time to accommodate the transplant-paste operation. For time stretching, the opposite problem occurs in that we keep pasting information further and further down the timeline in the output buffer, whose size must constantly grow to accept the new data. So we will eventually run out of buffer space, or we will run out of time if we try to perform this in real time. However, there are numerous offline approaches that work well—but of course not in real time.

20.5.3 Pitch Shifting

There are multiple options for pitch shifting. In the classic phase vocoder, you use time stretching or shrinking first to create a longer or shorter output sequences. These are then resampled at a new rate, which corresponds to the pitch shift that is desired. The resampling is done using interpolation to shrink the longer output buffer, or expand the shorter one, to fit back in the original length as shown in Figure 20.15a. This modifies the pitch without changing the tempo. In practice, this can only be done correctly in an offline processing context due to the inherent issues with time stretching. You would stretch the entire piece of music, and then resample it back to the original length in a two-part operation. To perform the process in real time, you can use a trick that results in an equivalent operation, but has some limitations. In this case, you acquire the FFTs and adjust the bin phases as if you were doing a time stretch/shrink. Then, you take the IFFT and resample that time domain signal, stretching or shrinking it according to the desire pitch shift ratio. The resampled signal is windowed and overlap-added into the output timeline as shown in Figure 20.15b at the same hop size as the analysis. This means that you will be overlap-adding buffers that are either shorter (upwards pitch shift) or longer (downwards pitch shift) to the timeline. This places a limit on the amount of pitch shifting you may accomplish. At a minimum, the FFT should be at least 4096 points in length, and a 75% overlap is required for pitch shifting, which sets the hop size at $N/4 - 1024$ points. For upwards pitch shifting, your output buffer will be shrinking; it can't become smaller than $N/4$, which is a pitch shift of two octaves up. For downwards pitch shifting, the output buffer will need to be at least as large as the longest buffer you plan on writing; for a two octave downward shift, that is $4N$ in length.

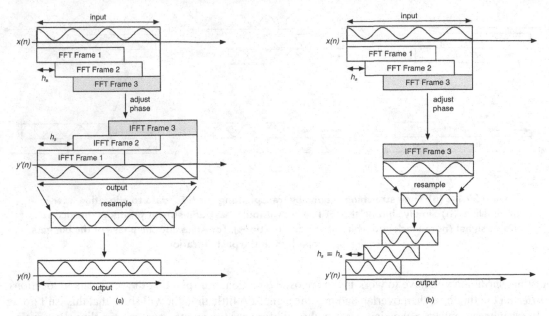

Figure 20.15: (a) Classic pitch shifting starts with a time stretching/shrinking operation, then the output is resampled to the original length. (b) In practice, the phases are adjusted for the time stretch/shrink, and then the IFFT buffer is resampled and overlap-added to the output timeline (the operation here depicts upwards pitch shifting).

Your *fxobjects.h* file contains a phase vocoder based pitch shifter object called *PSMVocoder* that can shift the pitch up or down by two octaves in either direction using the classic phase vocoder algorithm here and the built-in *PhaseVocoder* object. A resampling function is provided in the same file that allows an input array of size N to be resampled into an output array of size M using either linear or Lagrange interpolation. In order to propagate the phase horizontally, you need to calculate and store each phase adjustment value Ψ. This value needs to be on the range of $[-\pi, +\pi]$ so it uses the principal argument calculation. The phase propagation value for the kth bin of the mth FFT frame is calculated using Equation 20.16. The time stretch factor a is equal to the pitch shift ratio; one octave up is a ratio of 2 while one octave down is a ratio of 1/2.

$$\omega_k^m = \frac{2\pi k}{N}$$

$$\phi_{k\,dev}^m = \phi_k^m - \phi_k^{m-1} - \omega_k^m h_a$$

$$\Delta\Phi_k^m = \omega_k^m h_a + princArg\left(\phi_{k\,dev}^m\right)$$

$$\Psi_k^m = princArg\left(\Psi_k^{m-1} + (\alpha)\left(\Delta\Phi_k^m\right)\right) \tag{20.16}$$

$$\alpha = \frac{h_s}{h_a} = \frac{\text{pitch out}}{\text{pitch in}}$$

$$\phi_k^m = \text{current phase of bin k}$$

$$\phi_k^{m-1} = \text{phase of bin k in previous FFT}$$

The C++ code in the *PSMVocoder* object implements these equations directly, saving previous phase values *phi* and running phase correction values *psi* in arrays. Note that PSM_FFT_LEN is the FFT length of 4096 samples:

```
// --- horizontal phase propagation for bin k
// --- first get current phase from FFT data
double phi_k = getPhase(fftData[k][0], fftData[k][1]);
double mag_k = getMagnitude(fftData[i][0], fftData[i][1]);

// --- omega_k = bin frequency(k)
double omega_k = kTwoPi*k / PSM_FFT_LEN;

// --- phase deviation is actual - expected phase
// = phi_k -(phi(last frame) + wk*ha
double phaseDev = phi_k - phi[k] - omega_k*ha;

// --- unwrapped phase increment
double deltaPhi = omega_k*ha + principalArg(phaseDev);

// --- save for next frame
phi[i] = phi_k;

// --- calculate new phase based on stretch factor and save
psi[i] = principalArg(psi[i] + deltaPhi * alphaStretchRatio);
```

The magnitude and new phase value are then applied to the FFT data and converted back to real and imaginary parts:

```
fftData[k][0] = mag_k*cos(psi[k]);
fftData[k][1] = mag_k*sin(psi[k]);
```

If you try out the plugin you will notice that it has a phase-y quality and sometimes sounds like reverb or echoes. In addition, if you apply a signal that has many close frequencies interacting, such as a strummed 12-string guitar loop, you will hear the algorithm fail. You will also hear that transients don't sound natural—they sound sped up or slowed down. These are issues that may be addressed with phase locking.

20.5.4 Phase Locking

Laroche and Dolson (1999) concluded that part of the issue with the classic phase vocoder time stretching and pitch shifting algorithms is the destruction of vertical phase coherence within each FFT frame. They further observed that each FFT bin is treated as if it captures a single bin frequency component, then each sinusoidal phase is adjusted. In reality, a given frequency usually has energy spread across multiple FFT bins. Furthermore, if the input signal contains frequencies that are close together, there may be bins containing energy from multiple frequencies. In this case it is also clear that estimating the frequency inside the bin in order to propagate the horizontal phase adjustment is going to be prone to error as well. Still worse, the error will accumulate over time and the phase adjustment will become less and less accurate. Their solution was to avoid treating every bin as if it contained a pure sinusoid. Instead, they use a peak-picking process to identify peaks in the magnitude of the FFT. The peaks are treated as sinusoidal components and their phases are adjusted as with the normal FFT. However, the bins that surround each peak are treated as non-sinusoidal components. Their phases are also adjusted, but in a way that locks them with that of the peak bin's phase. The area around a peak is called its *region of influence*. The bins in this area have their phases locked to their local peak, which is called *phase locking*. This is an attempt to get the best of both worlds: the peak frequencies maintain horizontal phase coherence, and their neighboring regions of influence are phase locked with them, maintaining at least some type of vertical phase coherence within the FFT frame.

The peak-picking algorithm is going to require a heuristic, and the authors suggest a simple strategy: a bin is considered to be a peak bin if its magnitude is larger than the two bins on either side of it. When the peak bins are located, a line is drawn midway between them, splitting the regions of influence so that the halves nearest to a peak are considered to be in the region of influence (see Figure 20.16a).

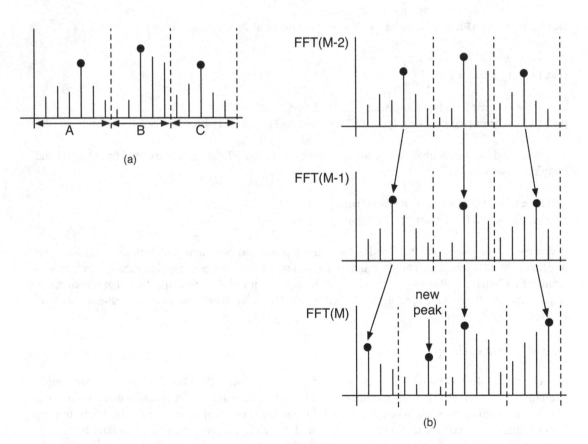

Figure 20.16: (a) A peak-picking algorithm finds three peaks—A, B and C—along with their regions of influence, separated with dotted lines. (b) Over time, peaks may evolve and shift from one bin location to another, die out, or become newly created.

An enhancement to the algorithm assumes that any given peak in a frame may be part of a longer evolution of a moving peak (e.g. a slow glissando) so during the succession of FFT frames peaks may jump from one bin to the next, or they may jump over some number of bins. In this case, you must store the previous set of peaks and their running phase values. For each new FFT frame, you find the nearest peak from the previous frame and use its phase adjustment value to calculate the next value. Figure 20.16b shows the evolution of three moving peaks, plus the addition of a new peak in the current FFT frame. In this case, we are performing peak tracking along with local phase locking. It also requires two passes over the same FFT bins: one to assign peaks and the other to calculate the new phase adjustment value. For a peak bin that has not changed from one frame to the next, the new phase value is calculated as usual using the last phase value, as shown in Equation 20.17. For the bins in the region of influence of a peak bin, the phase is locked to the peak phase as follows.

$$\Psi_k^m = princArg\left(\Psi_{kP}^m - \phi_{kP}^m - \phi_k^m\right)$$

Ψ_{kP}^m = phase update value of peak bin

ϕ_{kP}^m = current phase of peak bin   (20.17)

ϕ_k^m = current phase of bin k

If a peak bin has shifted from the previous FFT frame, its new phase propagation value is calculated from the phase propagation value of the previous peak bin in the previous frame. If the kth bin is detected as a peak, and the nearest peak bin in the previous FFT frame was the gth bin, then the new phase update value is as follows.

$$\Psi_k^m = princArg\left(\Psi_g^{m-1} + (\alpha)(\Delta\Phi_k^m)\right) \qquad (20.18)$$

You may enable both peak-phase locking and peak tracking on the *PSMVocoder* object in *fxobjects.h* by setting two Boolean flags. It will then perform both phase locking and peak tracking using the local maximum search algorithm along with storing of previous peak and phase adjustment information. You will find that the phase locking and peak tracking improves the pitch-shifted quality of some signals, but may perform worse for others. Phase locking and peak tracking both involve heuristics and are separated in the *PSMVocoder* object so that you may implement your own strategies. There have been numerous improvements and modifications to these basic phase vocoder algorithms. One reason that it remains popular in academia is that for real-time processing, it represents a problem that is fundamentally a heuristics issue—there is no perfect solution.

In 1999, Laroche and Dolson proposed a different mechanism for pitch shifting that does not involve time stretching or resampling. In this method, you move the bins around in the FFT by cutting and pasting them to new locations. Pitch shifting upwards involves moving the FFT data to higher bins, while downward shifts are accomplished by moving data into lower bins. The phases of the bins are altered for their new pitches after they have been moved. The Laroche and Dolson (or "modern") approach makes more sense at a base level: why not just move the data around in the FFT bins? But there are numerous issues involved, especially regarding musical pitch shifts, because the bins of the FFT are arranged linearly. This method involves the same peak-picking algorithm and region of influence concept: the bins in the region of influence are cut and pasted along with their local peak and are phase adjusted accordingly into new locations above or below their original bin positions, producing the pitch shift. With some searching you find many more enhancements, including preservation of formants (peak locations relative to one another) and offline time stretching/pitch bending algorithms.

20.6 Fast Convolution

We can also use the DFT or FFT to perform convolution. In Chapter 16, Equation 16.3, we saw that convolution in the time domain is the same as multiplication in the frequency domain. This allows us to multiply the output of one DFT or FFT with another DFT or FFT directly in the frequency domain. This requires a complex multiply operation since you have to multiply both the real and imaginary parts of the FFTs and your *fxobject.h* file defines the function *complexMultiply( )* to implement the operation. In Chapter 16 we noted that there was some fine print regarding the convolution/ multiplication duality. In order to multiply FFTs we need to handle this issue. The problem is that we want to perform linear convolution between two signals and multiplying FFTs performs circular convolution instead. To understand the difference, remember that the output of a convolution of two *N*-point sequences *a(n)* and *b(n)* is *2N − 1* points in length. Go back and examine Figure 16.3 that shows the convolution of two four-point sequences. The *2N − 1* results of that convolution are shown here.

$$
\begin{aligned}
y(0) &= a(0)b(0) & y(4) &= a(1)b(3) + a(2)b(2) + a(3)b(1) \\
y(1) &= a(0)b(1) + a(1)b(0) & y(5) &= a(2)b(3) + a(3)b(2) \\
y(2) &= a(0)b(2) + a(1)b(1) + a(2)b(0) & y(6) &= a(3)b(3) \\
y(3) &= a(0)b(3) + a(1)b(2) + a(2)b(1) + a(3)b(0)
\end{aligned}
\tag{20.19}
$$

An *N*-point sequence of samples *a(n)* will be transformed into an *N*-point DFT, *A(k)*, which is of the same length and where *k* are the frequency bins. Multiplying two *N*-point DFTs *A(k)B(k)* will produce another *N*-point DFT result, *Y(k)*. Taking the inverse DFT will transform *Y(k)* back into time domain samples *y(n)* of the same length *N*. The convolution that occurs when multiplying DFTs is clearly not the same as the result of linear convolution. To understand why and how to fix this requires some math and requires going back to Equation 20.5. If we take the same pair of signals *a(n)* and *b(n)* from Figure 16.3 and transform them into *A(k)* and *B(k)* then multiplying to get *Y(k)*, we get the following.

$$
\begin{aligned}
A(k) &= \sum_{n=0}^{N-1} a(n)e^{-j2\pi kn/N} \\
B(k) &= \sum_{n=0}^{N-1} b(n)e^{-j2\pi kn/N} \\
Y(k) &= \left[\sum_{n=0}^{N-1} a(n)e^{-j2\pi kn/N}\right]\left[\sum_{n=0}^{N-1} b(n)e^{-j2\pi kn/N}\right]
\end{aligned}
\tag{20.20}
$$

Now, running *Y(k)* through the inverse DFT produces Equation 20.12, which looks like a mathematical mess, but there is some good news buried in the result. Notice that we needed to change the indexing variables to *m* and *r* because *n* is used in the IDFT result, as shown here.

$$y(n) = \frac{1}{N} \sum_{k=0}^{N-1} Y(k) e^{2\pi kn/N}$$

$$= \frac{1}{N} \sum_{k=0}^{N-1} \left[\sum_{m=0}^{N-1} a(m) e^{-j2\pi km/N} \right] \left[\sum_{r=0}^{N-1} b(r) e^{-j2\pi kr/N} \right] e^{2\pi kn/N} \qquad (20.21)$$

$$= \frac{1}{N} \left[\sum_{m=0}^{N-1} a(m) e^{-j2\pi km/N} \right] \left[\sum_{r=0}^{N-1} b(r) e^{-j2\pi kr/N} \right] \left[\sum_{k=0}^{N-1} e^{j2\pi(n-m-r)/N} \right]$$

The good news is that the last term in Equation 20.21 evaluates out to be either 0.0 or N depending on the values of n, m and r. This means that the last term may be rewritten as Equation 20.22.

$$\sum_{k=0}^{N-1} e^{j2\pi(n-m-r)/N} = \begin{cases} N & n-m-r = \ldots-2N, -N, 0, N, 2N, \ldots \\ 0 & otherwise \end{cases} \qquad (20.22)$$

This causes the summations in Equation 20.20 to collapse down to a single summation in modulo form, as shown here.

$$y(n) = \sum_{n=0}^{N-1} a(m) b(n-m) \big|_{\text{mod } N} \quad n = 0, 1, 2, \ldots, N-1 \qquad (20.23)$$

This is the same as the linear convolution case in Equation 16.2, except that the second signal sequence *b(n)* here repeats itself. For the $N = 4$ point sequences, Equation 20.23 will produce four results (as expected), *y(0)* to *y(3)* after which the sequence repeats. This is called *circular convolution*. To help understand the difference between linear and circular convolution, examine Figure 20.17, the linear convolution of two sequences *a(n)* and *b(n)*, broken into iterations on each row where the *b(n)* signal slides to the left by one sample interval each time a summation is calculated. Figure 20.18a graphically depicts the circular convolution that Equation 20.23 describes, where the *b(n)* sequence is repeated in a modulo 4 fashion. Notice how the *b(n)* sequence circulates, wrapping back on itself in the repeated patterns.

If you look at the four outputs *y(0)* through *y(3)*, you can see that each output contains at least part of the original linear convolution. Compare the four outputs here to the seven outputs in Equation 20.24 and you will see that there is overlap in their shared terms, shown in bold.

$$\begin{aligned} y(0) &= \mathbf{a(0)b(0)} + a(1)b(3) + a(2)b(2) + a(3)b(1) \\ y(1) &= \mathbf{a(0)b(1)} + \mathbf{a(1)b(0)} + a(2)b(3) + a(3)b(2) \\ y(2) &= \mathbf{a(0)b(2)} + \mathbf{a(1)b(1)} + \mathbf{a(2)b(0)} + a(3)b(3) \\ y(3) &= \mathbf{a(0)b(3)} + \mathbf{a(1)b(2)} + \mathbf{a(2)b(1)} + \mathbf{a(3)b(0)} \end{aligned} \qquad (20.24)$$

In order to force the circular convolution into linear convolution, we need to isolate one copy of the *b(n)* signal by extending it, then padding it with zeros as shown in Figure 20.18b. This is a conceptual diagram because the actual convolution is occurring *during* the inverse DFT process, after multiplying the *A(k)* and *B(k)* DFTs together. The zero insertion must be performed back at the start of the process, before the first DFTs are taken. In order to get things to work properly after the DFT, we need only zero-pad the ends of each of the sequences *a(n)* and *b(n)* to extend their lengths to at least $2N - 1$. In practice we can just double the length of each sequence, padding the last half with zeros.

Linear Convolution

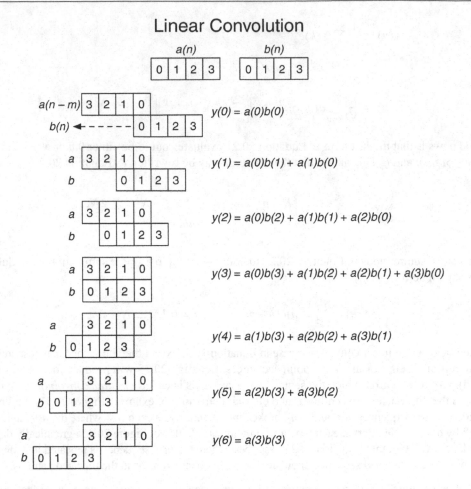

**Figure 20.17: The step-by-step discrete convolution of the four-point sequences *a(n)*
and *b(n)* reveals seven outputs, *y(0)* through *y(7)*.**

The last problem to deal with is the fact that the number of output samples for the convolution needs
to be *2N − 1* and we are performing the DFTs in chunks of *N* original samples. There are two strategies
for this: overlap-save and overlap-add. In both cases we extend the lengths of the two sequences to
2N by zero padding the last halves of each and perform the DFT as a *2N* point DFT. However, each
original *N*-point input sequence needs to result in *N* output points, so a 50% overlap is used. The
multiplication of the two DFTs must be done as a complex multiply.

In overlap-save, you take advantage of the fact that a portion of the circular convolution is identical
to that of the linear convolution, then you save the similar parts and discard the others. Overlap-
add works in a similar manner where the two halves of the successive DFT outputs are mixed
and in each case we get output chunks of *N* points again. This process is called *fast convolution*
because it takes advantage of the fact that both FFT and inverse FFT operations may be done
with fewer CPU instructions than normal linear convolution, even though the lengths are *2N* after

Circular Convolution

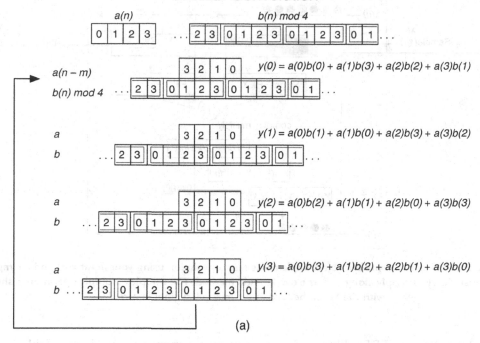

(a)

Circular Convolution as Linear Convolution

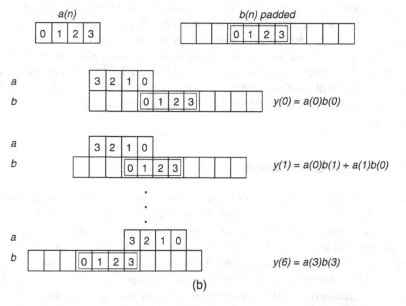

(b)

Figure 20.18: (a) The circular convolution of the four-point sequence *a(n)* and the repeated (modulo 4) *b(n)* sequence results in four outputs *y(0)* through *y(3)*, after which the output pattern repeats and **(b)** zero-padding the *b(n)* sequence allows us to use circular convolution to implement linear convolution.

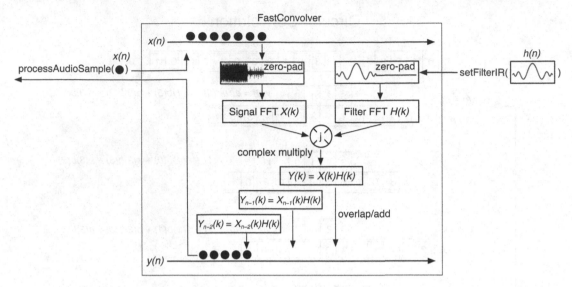

Figure 20.19: The *FastConvolver* object implements fast convolution using your input *x(n)* and an impulse response that you supply along with the overlap-add method. The complex multiply operation is shown with the "*j*" embedded in the multiply symbol in a circle.

zero padding. With the FFTW library, we can perform these operations amazingly quickly. The *PhaseVocoder* object is already set up to perform the overlap-add paradigm, so we may use it as a basis for a new object called *FastConvolver*. This object is a specialized phase vocoder with a hardwired 50% overlap-add mechanism and no windowing. You supply the object with either a pre-calculated FFT or an impulse response *h(n)*, which it will convert to *H(k)* for you. You then simply insert input points *x(n)* and receive outputs *y(n)* that are the fast convolution of the input with the FFT or impulse response you supply. There is no need to break into the FFT results, so the operation is simple. The object does all of the zero padding, FFT, IFFT and overlap-add for you. You might try doing the homework example and converting the *PhaseVocoder* object into your own implementation of a fast convolution object. The block diagram for the *FastConvolver* object is shown in Figure 20.19.

20.7 Gardner's Fast Convolution

In the previous section, we performed the fast convolution by breaking the input into blocks of N samples that we zero-pad to length $2N$. Gardner wanted to perform the convolution with very long impulse responses such as a reverb IR. In this case, the IR may be many seconds in length, corresponding to hundreds of thousands of samples—and far too long to perform zero padding and the DFT. Gardner's fast convolution breaks both the input *and* the impulse response into blocks and performs overlap-add fast convolution in multiple sets of overlapping chunks. The impulse response is broken into pieces of gradually increasing length. This process and the scheduling of the FFTs requires finesse. The very first part of the impulse response (called the *head*) is directly convolved while the remaining pieces use overlap-add fast convolution. Figure 20.20a shows the basic idea. At the top,

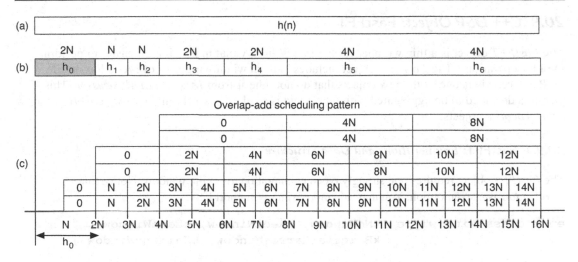

Figure 20.20: Gardner's fast convolution takes the long impulse response in (a) and breaks it into segments consisting of (b) a head h_0 block plus some number of tail segment blocks while (c) shows the FFT and overlap-add scheduling pattern; the first block h_0 shown in grey is linearly convolved with the input while the rest use the overlap-add mechanism.

a long IR $h(n)$ is segmented into blocks based on the FFT length N. The first block h_0 is convolved directly with the input. The successive IR segments double in length in pairs and are shown as h_1–h_M in Figure 20.20b. The FFT and overlap-add scheduling for the tail segments is shown in Figure 20.20c, where the numbers in the blocks are the N-sized block periods, and the output during some block PN is a sum of successive output blocks. Assuming that you have a fast FFT device, the main complexity is in the scheduling of the interblock convolutions.

You can download a specialized object called *TwoStageFFTConvolver* from http://www.willpirkle. com/app-notes/ that is provided by a third party and implements a type of fast convolution by breaking the impulse response into two parts: the head and tail. It uses its own FFT object (though you may enable and link it to FFTW if you have it installed) and performs very fast convolutions on long impulse responses.

20.8 Chapter Objects and Plugins

This chapter introduces multiple DSP and effect objects that are quite complex, due in part to their FFTW requirements, but also because they almost all use the overlap-add reconstruction technique. Owing to this high complexity, we cannot print every code detail, so it is imperative that you download the sample code so that you may access, study, and modify the low-level code. You must have FFTW installed and ready to link to your project in order to use all of these objects. In the *fxobjects.h* file, you need to un-comment the *#define* that will enable FFTW. You can get more help on installing, linking, and enabling the FFTW objects in the ASPiK SDK. You will find the FFT objects near the bottom of *fxobjects.h*.

20.9 *C++ DSP Object:* FastFFT

The *FastFFT* object is a thin wrapper for the FFTW library that makes it a bit easier to perform the FFT operations. The *FastFFT* object includes built-in windowing (or not). As a low-level DSP object, this is one of the few objects that do not inherit from *IAudioSignalProcessor*. This object is designed to be aggregated with others and is used extensively in the *FastConvolver* and *PhaseVocoder* objects.

20.9.1 FastFFT: *Enumerations and Data Structure*

The *FastFFT* object requires one strongly typed enumeration to set the window. Notice that *kNoWindow* is also an option. You may add more window support here.

```
enum class windowType {kNoWindow, kRectWindow, kHannWindow,
                       kBlackmanHarrisWindow, kHammingWindow };
```

20.9.2 FastFFT: *Members*

Tables 20.2 and 20.3 list the *FastFFT* member variables and member functions. The bulk of these members are requires for the FFTW package. You can check the code for all of the FFTW details.

20.9.3 FastFFT: *Programming Notes*

The *FastFFT* object simply wraps the FFTW algorithm in an easy-to-use C++ object.

Function: *initialize*
Description: Initializes a new FFTW session, destroying the old session if it exists. This function also creates and initializes the windowing buffer using the windowing functions from this chapter. See the downloadable code for details.

Table 20.2: *FastFFT* member variables

FastFFT Member Variables		
Type	**Name**	**Description**
*double**	*windowBuffer*	Buffer to hold sampled windowing function
double	*windowGainCorrection*	Gain correction value for selected window
windowType	*window*	Typed enumeration for window
unsigned int	*frameLength*	The FFT length
*fftw_complex**	*fft_input*	Complex FFT input buffers (re,im)*
*fftw_complex**	*fft_result*	Complex FFT output buffers (re,im)
*fftw_complex**	*ifft_input*	Complex IFFT input buffers (re,im)
*fftw_complex**	*ifft_result*	Complex IFFT input buffers (re,im)
fftw_plan	*plan_forward*	FFTW FFT plan
fftw_plan	*plan_backward*	IFFTW FFT plan

Note(*) re = real part, im = imaginary part

Table 20.3: *FastFFT* member functions

FastFFT Member Functions		
Returns	**Name**	**Description**
void	*initialize* Parameters: —*unsigned int* _frameLength —*windowType* _window	Initialize the FFT object with a FFT length and window; may be called more than once
void	*destroyFFTW*	De-allocate all dynamically allocated resources
fftw_complex*	*doFFT* Parameters: —*double** inputReal —*double** inputImag Returns: —the real and imaginary FFT output arrays	Perform FFT on input composed of real and imaginary parts; OK if input is purely real
fftw_complex*	*doInverseFFT* Parameters: —*double** inputReal —*double** inputImag Returns: —the real and imaginary IFFT output arrays	Perform inverse FFT on input composed of real and imaginary parts; OK if input is purely real
unsigned int	*getFrameLength*	Return the current FFT length

Functions: *doFFT* and *doInverseFFT*

Description: These functions load up the FFTW input arrays and process either the FFT or IFFT, using the code from FFTW.

```
fftw_complex* doFFT(double* inputReal, double* inputImag)
{
      // ------ load up the FFT input array
      for (int i = 0; i < frameLength; i++)
      {
            // --- copy array
      }


      // --- do the FFT
      fftw_execute(plan_forward);

      return fft_result;
}


// --- do the inverse FFT
fftw_complex* doInverseFFT(double* inputReal, double* inputImag)
{
      // ------ load up the iFFT input array
      for (int i = 0; i < frameLength; i++)
      {
            // --- copy array
      }
```

```
       // --- do the IFFT
       fftw_execute(plan_backward);

       return ifft_result;
}
```

20.10 C++ DSP Object: PhaseVocoder

The *PhaseVocoder* object implements the hop-based FFT framing, windowing, inverse FFT, and overlap-add operations. It is designed for maximum flexibility as well as simplicity. You have access to the FFT and IFFT data and the ability to manipulate, re-sample, and re-window the output data. The simple algorithms like robotization and noise reduction may be done directly within the object (sub-class it first). For more complex examples, we also offer the *FastConvolver* and *PitchShifter* C++ effect objects that demonstrate the objects advanced capabilities. As per the discussion in Section 20.4.6, the *PhaseVocoder* uses circular buffers to implement the transversal input and output timelines. When you initialize the *PhaseVocoder*, you set up several operating parameters at once:

• the FFT frame length, which sets the input circular buffer size
• the output buffer size, which is set to be four times the FFT frame length to accommodate pitch shifting down by two octaves
• the input time domain windowing
• the synthesis and analysis hop sizes, which are identical

20.10.1 PhaseVocoder: *Enumerations and Data Structure*

The *PhaseVocoder* object uses the same strongly typed enumeration for windowing as the *FastFFT* object. As a low-level DSP object, the *PhaseVocoder* does not use a custom data structure for parameter updates, nor does it inherit from *IAudioSignalProcessor*, implementing its own custom functions that allow the user to break into the *PhaseVocoder* process at every level for the purpose of manipulating the incoming FFT data.

20.10.2 PhaseVocoder: *Members*

Tables 20.4 and 20.5 list the most important *PhaseVocoder* member variables and all member functions. Please refer to the downloadable code to scrutinize the other members as you like. The *PhaseVocoder* is open for modification, especially with regards to the analysis and synthesis hop sizes; if you want to experiment with time stretching, then you will want to sub-class it and override some of the methods.

20.10.3 PhaseVocoder: *Programming Notes*

The *PhaseVocoder* object is very flexible and may be used for simple or quite complex algorithm development. The idea is to set up the FFT framing, inverse FFT, and overlap-add mechanisms, and then allow the programmer to access the data at the different points in the process. Figure 20.21 shows how the *PhaseVocoder* methods work within the block diagram; the methods are listed on the right.

Table 20.4: Important *PhaseVocoder* member variables

PhaseVocoder Member Variables (important)		
Type	**Name**	**Description**
windowType	*window*	The selected window function; the default and suggested window is Hann
double*	*windowBuffer*	Linear buffer to store the window
double*	*inputBuffer*	Circular buffer to maintain the input timeline
double*	*outputBuffer*	Circular buffer to maintain the output timeline
unsigned int	*hopSize*	Hop size in samples
bool	*overlapAddOnly*	Flag for container objects that want to use only the framing and overlap-add capability of the object
fftw_complex*	*fft_input, fft_result, ifft_result*	FFTW arrays for storing and manipulating real and imaginary parts of FFT
fftw_plan	*plan_forward, plan_backward*	FFTW plans for FFT and IFFT

Table 20.5: *PhaseVocoder* member functions

PhaseVocoder Member Functions (all)		
Returns	**Name**	**Description**
void	*initialize* Parameters: —*unsigned int* _frameLength —*unsigned int* _hop_size —*windowType* _window	Initialize the PhaseVocoder with FFT frame length, hop size (in samples), and input windowing choice; this function may be called more than once
void	*destroyFFT*	De-allocate all FFTW resources
bool	*addZeroPad* Parameters: —*unsigned int* count	Zero-pad the input array to accommodate fast convolution or other zero-padding algorithms
double	*processAudioSample* Parameters: —*double* input —*bool&* fftReady	Process input *xn* through the *PhaseVocoder* and produce an output (return variable). If the newly added input triggered an FFT operation, the *fftReady* flag will be set—you can then manipulate the FFT data
fft_complex*	*getFFTData*	Get the newly calculated FFT data real and imaginary parts
fft_complex*	*getIFFTData*	Get the newly calculated IFFT data
void	*advanceAndCheckFFT*	Internal function that increments FFT counter and signals when new FFT needs to be taken
void	*doInverseFFT*	Perform the inverse FFT operation; called internally or externally
void	*doOverlapAdd* Parameters: —*double** outputData —*int* length	The overlap-add mechanism is internal but you may call this function to implement your own; we use this in the *PitchShifter* object
unsigned int	*getFrameLength*	Returns length of FFT
unsigned int	*getHopSize*	Returns hop size in samples
double	*getOverlap*	Returns the hop size/FFT length, or the overlap as a percentage of the total FFT length

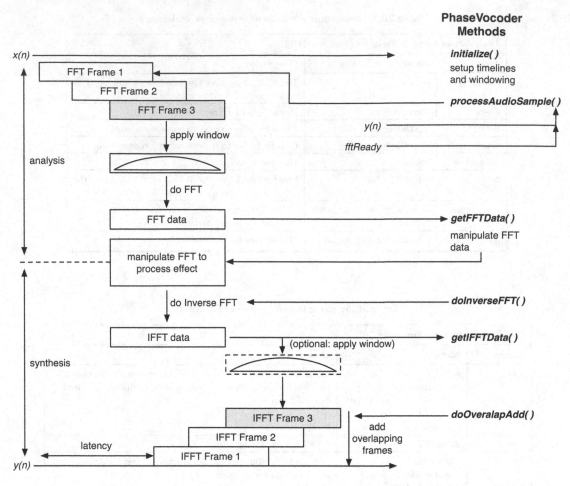

Figure 20.21: The *PhaseVocoder* methods and the block diagram.

Using the *PhaseVocoder* is relatively simple—at the minimum you only need to make two function calls, *processAudioSample* and *getFFTData*. The complete sequence is to first call *processAudioSample* to push an audio sample *x(n)* into the object, and return the current output *y(n)* from the output buffer timeline. Next, monitor the *fftReady* argument that is set in the *processAudioSample* method; when this flag is true, a FFT has just been taken as a result of the *Nth* audio input sample being applied where *N* is the FFT frame length. When the *fftReady* flag is set, you may then call *getFFTData( )* to have direct pointer access to the *PhaseVocoder's* FFT buffers; you manipulate that data directly via the return pointer. At this point, you have two options. You may either do nothing—after your call to *getFFTData( )* returns, the IFFT will automatically be taken on the FFTW buffers and the result will automatically be overlap-added to the output timeline, or you may implement the following sequence to handle these steps yourself and implement more complex algorithms:

1. call *doInverseFFT( )* to manually force the IFFT operation
2. then call *getIFFTData( )* to access the IFFT output (for example, to resample it for pitch shifting),

3. and then call *doOverlapAdd( )* to add your manipulated IFFT data back to the output timeline; if you want to do output windowing, then you must implement it yourself prior to this call—this is normal for pitch shifting where the windowing size is dependent on the pitch scaling factor, which the *PhaseVocoder* knows nothing about.

The details of these functions are available in the downloadable code, are heavily commented, and are not reprinted here. If you plan on modifying this object, then you should scrutinize this code—you may always post questions at www.willpirkle.com/forum/ for help from our community. Note that the FFTW plays a significant role in operations so you may also want to study its documentation.

20.10.3.1 PhaseVocoder *Example:* Robotization

To demonstrate the simple *PhaseVocoder* operation, the robotization code is provided next. Note that you for every input sample *x(n)* you push into the *PhaseVocoder*, you always receive one output sample *y(n)* regardless of whether you manipulate the FFT data. If you ignore the FFT data altogether, the *PhaseVocoder* will still take the IFFT and perform the overlap-add operation. This is an acid test of the object: if you use a COLA-approved hop size and Hann windowing but do no other processing, the *PhaseVocoder* will output a perfect reconstruction of the input signal. Notice the use of helper functions *getMagnitude( )* and *getPhase( )*, which are declared in the *fxobjects.h* file. Notice also that we first convert real and imaginary to magnitude and phase, manipulate the phase, then convert back to real and imaginary. There is no need to call the *doInverseFFT( )* or other methods—they will happen automatically if not called directly. In this example, *xnL* is the left channel audio input sample. First, initialize the *PhaseVocoder* with the FFT length, hop size, and windowing method. For robotization, the phase is reset for each FFT frame, imparting a periodicity onto the signal that creates the non-human effect. In this example, we'll set the FFT for 1024 points and a 50% overlap (hop size = 512).

```
// --- declare
PhaseVocoder phaseVocoder;

// --- initialize
phaseVocoder.initialize(1024, 512, windowType::kHannWindow);

double output = 0.0;
bool fftReady = false;
output = phaseVocoder.processAudioSample(xnL, fftReady);

// --- manipulate FFT bins
if (fftReady)
{
    fftw_complex* fftData = phaseVocoder.getFFTData();

    // --- iterate over FFT frame, bin by bin
    for (int i = 0; i < phaseVocoder.getFrameLength(); i++)
    {
        // --- convert to mag/phase
        double mag_k = getMagnitude(fftData[i][0], fftData[i][1]);
        double phi_k = getPhase(fftData[i][0], fftData[i][1]);
```

```
            // --- robot: set phase to 0.0
            phi_k = 0.0;

            // --- now convert back to real/imaginary
            fftData[i][0] = mag_k*cos(phi_k); // real
            fftData[i][1] = mag_k*sin(phi_k); // real
    }
}
```

20.10.3.2 PhaseVocoder *Example:* NoiseReduction

Another easy algorithm is the noise reduction method in Section 20.5.1 from Zölzer. The code is identical to the code in the last section, except for the FFT manipulation, which operates on the magnitude rather than phase. In this example, the *NR* value comes from a user GUI manipulation and is on the range of [0.0, +*N*] where +*N* is 1.0 or greater. As with the robotization example, there are no other functions to call and the IFFT and overlap-add will happen automatically.

```
for (int i = 0; i < phaseVocoder.getFrameLength(); i++)
{
    // --- convert to mag/phase
    double mag_k = getMagnitude(fftData[i][0], fftData[i][1]);
    double phi_k = getPhase(fftData[i][0], fftData[i][1]);

    // --- NR: manipulate magnitude
    double nrCoeff = mag_k/(mag_k + NR);
    mag_k *= nrCoeff;

    // --- now convert back to real/imaginary
    fftData[i][0] = mag_k*cos(phi_k); // real
    fftData[i][1] = mag_k*sin(phi_k); // real
}
```

20.11 C++ DSP Object: FastConvolver

The *FastConvolver* object implements the fast convolution algorithm in Figure 10.18 directly (note that this figure also references *FastConvolver* function calls). The *FastConvolver* uses a *PhaseVocoder* object to handle the FFT, IFFT, and overlap-add operations. In this algorithm, windowing is not required or performed. It also uses a *FastFFT* object to take the FFT of the incoming impulse response that the user supplies; the incoming IR is automatically zero-padded. The processing function performs the input zero padding when the number of input samples is half the size of the desired FFT. When the FFT is ready, the processing function performs the complex multiplication of the input FFT and the stored filter FFT. The IFFT and overlap-add then happen automatically.

20.11.1 FastConvolver: *Members*

Tables 20.6 and 20.7 list *FastConvolver* member variables and member functions. This object is your first example of using the *PhaseVocoder's* FFT and overlap-add processes in an external object.

Table 20.6: *FastConvolver* **member variables**

FastConvolver Member Variables		
Type	**Name**	**Description**
PhaseVocoder	*vocoder*	Vocoder for FFT, IFFT and overlap-add
FastFFT	*filterFastFFT*	Fast FFT object to calculate the FFT from the user-supplied impulse response
fft_complex*	*filterFFT*	Real and imaginary arrays of the filter FFT
double*	*filterIR*	Array to hold the incoming filter impulse response, prior to FFT
unsigned int	*inputCount*	Counter to trigger the zero-pad and FFT operations
unsigned int	*filterImpulseLength*	Filter IR length, which is half of the FFT size (due to zero-padding)

Table 20.7: *FastConvolver* **member functions**

FastConvolver Member Functions		
Returns	**Name**	**Description**
void	*initialize* Parameters: —unsigned int _ filterImpulseLength	Initialize the *FastConvolver* with the desired filter IR length—it dictates the FFT length (2X)
void	*setFilterIR* Parameters: —double* irBuffer	Set a new filter IR with an array of *doubles*; the FFT is performed and the results are stored for complex multiplication
double	*processAudioSample* Parameters: —double input	Process one input sample $x(n)$, perform the convolution, and output one sample $y(n)$
unsigned int	*getFrameLength*	Returns length of FFT
unsigned int	*getFilterIRLength*	Returns the IR length (which is one-half of the FFT length)

20.11.2 FastConvolver: *Programming Notes*

You can see from the sparse number of member functions that the majority of the operation of the *FastConvolver* is hidden from the user. To see examples of the *FastConvolver* in use, see the *Interpolator* and *Decimator* objects in Chapter 22. The operation is fairly simple.

1. Initialize the object with the desired filter IR length (typically you hold this value constant and you apply IRs that are all the same length).
2. Call the *setFilterIR* function to send a filter IR into the object and take its FFT snapshot that is then stored; note that the name implies a filter IR, but in reality it may be any impulse response that you'd like to fast-convolve with the input.
3. Call *processAudioSample* to push input samples into the object, and extract output samples as the return.

Function: *setFilterIR*

Description: You apply the (filter) impulse response with the *setFilterIR,* which performs the zero-padding and FFT operations. You may acquire the IR from whatever source you like; see Chapter 16 for several IR generation functions.

```
void setFilterIR(double* irBuffer)
{
        if (!irBuffer) return;

        // --- use memset to zero-pad the buffer
        memset(&filterIR[0], 0, filterImpulseLength * 2 *
                sizeof(double));

        // --- copy first half; filterIR len = filterImpulseLength * 2
        int m = 0;
        for (unsigned int i = 0; i < filterImpulseLength; i++)
                filterIR[i] = irBuffer[i];

        // --- take FFT of the h(n)
        fftw_complex* fftOfFilter = filterFastFFT.doFFT(&filterIR[0]);

        // --- copy the FFT into our local buffer for storage
        for (int i = 0; i < 2; i++)
        {
                for (unsigned int j = 0; j < filterImpulseLength * 2; j++)
                {
                        filterFFT[j][i] = fftOfFilter[j][i];
                }
        }
}
```

Function: *processAudioSample*

Description: All of the work happens in the *processAudioSample* function. Note the use of more helper objects and methods from *fxobjects.h* including the *ComplexNumber* structure and *complexMultiply* function. For convenience the function is broken into pieces here. In the first part, we check the last count of the input samples—when we hit the filter IR length, we stop and zero-pad the vocoder input. This will immediately trigger the FFT as long as the user has properly initialized the object.

```
double processAudioSample(double input)
{
        bool fftReady = false;
        double output = 0.0;

        if (inputCount == filterImpulseLength)
        {
                fftReady = vocoder.addZeroPad(filterImpulseLength);

                if (fftReady) // should happen on time
                {
```

With the FFT now ready, we can get to work and perform the complex multiplication; the *ComplexNumber* structure simply holds a real and imaginary value. The *complexMultiply* function performs the textbook complex multiply operation.

```
// --- multiply our filter IR with the vocoder FFT
fftw_complex* signalFFT = vocoder.getFFTData();
if (signalFFT)
{
        // --- complex multiply with FFT of IR
        for (unsigned int i = 0; i < filterImpulseLength * 2; i++)
        {
                // --- get real/imag parts of each FFT
                ComplexNumber signal(signalFFT[i][0], signalFFT[i][1]);
                ComplexNumber filter(filterFFT[i][0], filterFFT[i][1]);

                // --- complex multiply; this convolves in the time domain
                ComplexNumber product = complexMultiply(signal, filter);

                // --- overwrite the FFT bins
                signalFFT[i][0] = product.real;
                signalFFT[i][1] = product.imag;
        }
}
```

If we get to this point, we reset the input counter. Otherwise, we process the audio sample and wait for our input counter variable to trigger another zero-pad/FFT operation.

```
        }
        // --- reset counter
        inputCount = 0;
    }

// --- process next sample
output = vocoder.processAudioSample(input, fftReady);
inputCount++;

return output;
}
```

20.12 C++ *Effect Object:* PSMVocoder

The *PSMVocoder* object not only performs both the classic and phase-locked pitch scale modification algorithms, but it is also an example of using the *PhaseVocoder's* optional functions to manipulate the IFFT data, window it, and then overlap-add it to the output timeline. If you want to extend or modify the *PhaseVocoder*, this is a great object to study. The *PSMVocoder* makes use of another helper function named *resample( )* that resamples any input array of length N into an output array of length *M*, stretching or compressing the array as needed. You may choose linear or 4th order Lagrange interpolation for the resampling function. As with the *PhaseVocoder* object, the *PSMVocoder* is a quite complex C++ construction; be sure to obtain the downloadable code for your own scrutiny of every detail.

20.12.1 PSMVocoder: *Enumerations and Data Structure*

The *PSMVocoder* object has no custom enumerations but does declare a constant for the FFT length, set to 4096 points as detailed in the literature. Feel free to adjust this value. The *PSMVocoder* always uses a 75% overlap and will adjust its hop size according to this value.

```
const unsigned int PSM_FFT_LEN = 4096;
```

The *PSMVocoder* object uses a *PSMVocoderParameters* custom data structure to hold its parameter information. You can see that phase locking and peak tracking are independently set, but remember that peak tracking works only if phase locking is also enabled.

```
struct PSMVocoderParameters
{
        PSMVocoderParameters() {}
        PSMVocoderParameters& operator=(. . .){ . . . }

        // --- params
        double pitchShiftSemitones = 0.0; // half-steps
        bool enablePeakPhaseLocking = false;
        bool enablePeakTracking = false;
};
```

A second custom data structure is defined for the peak searching, phase locking, and peak tracking portions of the object. There is one *BinData* structure that holds the current information for each FFT bin, and another that hold the previous FFT frame's information. This is required for the peak tracking operation that must remember past peaks.

```
struct BinData
{
            BinData() {}
            BinData& operator=(. . .){ . . . }

            void reset() { . . . }

            // --- members
            bool isPeak = false;
            double magnitude = 0.0; // mag
            double phi = 0.0; // angle
            double psi = 0.0; // new (adjusted) angle
            unsigned int localPeakBin = 0;
            int previousPeakBin = -1; // -1 = no peak tracking
            double updatedPhase = 0.0;
};
```

20.12.2 PSMVocoder: *Members*

Tables 20.8 and 20.9 list the *PSMVocoder* member variables and member functions. As an aggregate object, it is actually simpler than it appears.

20.12.3 PSMVocoder: *Programming Notes*

The *PSMVocoder* sets up its member *PhaseVocoder* without help from the programmer. The object implements the *IAudioSignalProcessor* interface and therefore is easy to use. The steps for using the *PSMVocoder* are:

1. Reset the object to prepare it for streaming.
2. Set the object parameters in the *PSMVocoderParameters* custom data structure.
3. Call *setParameters* to update the calculation type.
4. Call *processAudioSample*, passing the input value in and receiving the output value as the return variable.

Table 20.8: Selected important *PSMVocoder* member variables

PSMVocoder Member Variables (selected, important)		
Type	**Name**	**Description**
PSMVocoderParameters	*parameters*	Custom data structure
PhaseVocoder	*vocoder*	Vocoder object
double	*ha, hs*	The analysis and synthesis hop sizes 75% overlap
double	*phi[PSM_FFT_LEN]*	The bin angles from the previous frame (non-phase locking)
double	*psi[PSM_FFT_LEN]*	The adjusted bin angles from the previous frame (non-phase locking)
BinData	*binData[PSM_FFT_LEN]*	*BinData* structures for the current frame (phase locking)
BinData	*binDataPrevious[PSM_FFT_LEN]*	*BinData* structures for the previous frame (peak tracking)
int	*peakBins[PSM_FFT_LEN]*	Peak bins for current frame (peak tracking)
int	*peakBinsPrevious[PSM_FFT_LEN]*	Peak bins for previous frame (peak tracking)

Table 20.9: *PSMVocoder* member functions

PSMVocoder Member Functions		
Returns	**Name**	**Description**
PSMVocoderParameters	*getParameters*	Get all parameters at once
void	*setParameters* Parameters: —*PSMVocoderParameters* _parameters	Set all parameters at once
bool	*reset* Parameters: —*double* sampleRate	Reset all filters and LFO
double	*processAudioSample* Parameters: —*double* xn	Process input *xn* through the phaser
void	*setPitchShift* *double* semitones	Create the time scaled buffers, and windowing
int	*findNearestPreviousPeak*	Search for nearest peak within ±25% of the FFT length
void	*findPeaksAndRegionsOfInfluence*	Heuristic search for peaks; regions of influence are assigned after peaks are found

The *PSMVocoder* is another effect object that is primed for modification. It handles the low-level chores of setting up the phase vocoder and storing the phase information. The phase locking and peak tracking strategies are straight out of the literature, with the exception of a leak limit value that we added to the peak tracking mechanism—a previous nearby peak is considered valid only if within a certain range (in bins) above or below the new peak. You should certainly consider altering that heuristic as well as the others.

Function: Constructor
Description: We need only initialize the *PhaseVocoder* object to the FFT length, 75% overlap, and Hann windowing in the constructor.

```
PSMVocoder() {
vocoder.initialize(PSM_FFT_LEN, PSM_FFT_LEN/4, windowType::kHannWindow
}
```

Function: reset
Description: The reset operation is crucial for clearing out the old phase information to prepare for a new run of audio streaming. The function simply clears out the old arrays and resets the *BinData* structures.

Function: setPitchShift
Description: This function prepares the object for a new pitch shifting modification (distance). It is set to receive semitones as doubles, for exotic in-between semitone shifts, but usually it will receive integer shifts from −24 to +24 semitones (±2 octaves). The function prepares a new output buffer that is scaled in size to hold the resampled audio data, and the new window buffer that must accompany it during the overlap-add process. Only the important parts of the function are shown here.

```
void setPitchShift(double semitones)
{
        // --- this is costly so only update when things changed
        double newAlpha = pow(2.0, semitones / 12.0);
        double newOutputBufferLength =
                        round((1.0/newAlpha)*(double)PSM_FFT_LEN);

        // --- check for change
        if (newOutputBufferLength == outputBufferLength)
            return;

        // --- new stuff
        alphaStretchRatio = newAlpha;
        ha = hs / alphaStretchRatio;

        // --- set output resample buffer
        outputBufferLength = newOutputBufferLength;

        // --- create Hann window
        // --- create output buffer
        <SNIP SNIP SNIP>
}
```

Function: processAudioSample

Description: This function performs the pitch shifting operation. If phase locking is not enabled, it uses the same code as printed in Section 20.5.3. With phase locking and peak tracking enabled, the function uses the identity phase-locking algorithm: peak bins are phase adjusted as normal and region-of-influence bins are adjusted so that their phases are locked to their nearby peaks. For both phase locked and non-phase locked algorithms, the *processAudioSample* function performs the following tasks.

- pushes input sample into *PhaseVocoder* object, get output sample.
- when FFT is ready, gets the FFT arrays for manipulation.
- adjusts phases of bins, either old school or with phase locking.
- calls *PhaseVocoder* to take IFFT and give us the results.
- resamples the IFFT result into a new array that is smaller (pitch shift up) or larger (pitch shift down) than the original FFT frame.
- calls *PhaseVocoder* method to overlap-add the new resampled buffer into the output timeline.

The pertinent code that shows this skeleton follows.

```
bool fftReady = false;
double output = 0.0;

// --- normal processing
output = vocoder.processAudioSample(input, fftReady);

// --- if FFT is here, GO!
if (fftReady)
{

// --- get the FFT data
fftw_complex* fftData = vocoder.getFFTData();

// --- phase/peak locking
if (parameters.enablePeakPhaseLocking)
{
        // --- get the magnitudes for searching
        <SNIP SNIP SNIP>

        // --- do the search for peaks and regions
        findPeaksAndRegionsOfInfluence();

        // --- now propagate phases accordingly
        < SNIP SNIP SNIP>

}// end if peak locking

else // --- > old school
{
```

```
        // --- now propagate phases accordingly
        < SNIP SNIP SNIP>
}

// --- manually so the IFFT (OPTIONAL)
vocoder.doInverseFFT();

// --- can get the iFFT buffers
fftw_complex* inv_fftData = vocoder.getIFFTData();

// --- make copy (can speed this up)
double ifft[PSM_FFT_LEN] = { 0.0 };
for (int i = 0; i < PSM_FFT_LEN; i++)
        ifft[i] = inv_fftData[i][0];

// --- resample the audio as if it were stretched
resample(&ifft[0], outputBuff, PSM_FFT_LEN, outputBufferLength,
          interpolation::kLinear, windowCorrection, windowBuff);

// --- overlap-add the interpolated buffer to complete the operation
vocoder.doOverlapAdd(&outputBuff[0], outputBufferLength);
}

return output;
```

20.13 Chapter Plugin: PitchShifter

The GUI and overall block diagram for the *PitchShifter* plugin is shown in Figure 20.22. With FFTW installed, you should have no problem implementing two *PSMVocoder* objects on a decent CPU. Remember to build your project in Release mode for best efficiency.

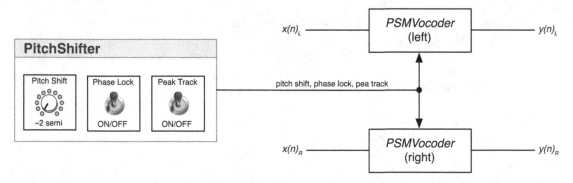

Figure 20.22: The *PitchShifter* GUI and block diagram.

Table 20.10: The *PitchShifter* plugin's GUI parameter and linked-variable list

Control Name	Units	Min/max/default or string-list	Taper	Linked Variable	Linked Var Type
Pitch Shift	semis	−12 / +12 / 0	linear	*lfoRate_Hz*	*double*
Phase Lock	-	"OFF, ON"	-	*enablePeakPhaseLocking*	*int*
Peak Track	-	"OFF, ON"	-	*enablePeakTracking*	*int*

20.13.1 PitchShifter: *GUI Parameters*

The GUI parameter table is shown in Table 20.10—use it to declare your GUI parameter interface for the plugin framework you are using. If you are using ASPiK, remember to first create your enumeration of control ID values at the top of the *plugincore.h* file and use automatic variable binding to connect the linked variables to the GUI parameters.

ASPiK: top of *plugincore.h* file

```
enum controlID {
     pitchShiftSemitones = 0,
     enablePeakPhaseLocking = 2,
     enablePeakTracking = 3
};
```

20.13.2 PitchShifter: *Object Declarations and Reset*

The *PitchShifter* plugin uses two *PSMVocoder* members to perform the audio processing. In your plugin's processor (*PluginCore* for ASPiK) declare two members plus the *updateParameters* method.

```
// --- in your processor object's .h file
#include "fxobjects.h"

// --- in the class definition
protected:
     PSMVocoder psmVocoder[2];
     void updateParameters();
```

Call the *reset* function and pass in the current sample rate in your plugin framework's prepare-for-audio-streaming function. In this example, *resetInfo.sampleRate* holds the current DAW sample rate.

ASPiK: *reset()*

```
     // --- reset the filters
     psmVocoder[0].reset(resetInfo.sampleRate);
     psmVocoder[1].reset(resetInfo.sampleRate);
```

20.13.3 PitchShifter: *GUI Parameter Update*

Update the left and right channel objects at the top of the buffer processing loop. Gather the values from the GUI parameters and apply them to the custom data structure for each of the shifter objects using the *getParameters* method to get the current parameter set, and then overwrite only the parameters your plugin actually implements. Do this for all channels you support.

ASPiK: *updateParameters( )* function

```
// --- update with GUI parameters
PSMVocoderParameters params = psmVocoder[0].getParameters();
params.pitchShiftSemitones = pitchShiftSemitones;
params.enablePeakPhaseLocking = enablePeakPhaseLocking == 1;
params.enablePeakTracking = enablePeakTracking == 1;
psmVocoder[0].setParameters(params);
psmVocoder[1].setParameters(params);
```

20.13.4 PitchShifter: *Process Audio*

Add the code to process the plugin object's input audio samples through each phase shifter object and write the output samples. The ASPiK code is shown here for grabbing the input samples and writing the output samples; use the method that your plugin framework requires to read and write these values.

ASPiK: *processAudioFrame( )* function.

```
// --- read input (adapt to your framework)
double xnL = processFrameInfo.audioInputFrame[0]; //< input sample L
double xnR = processFrameInfo.audioInputFrame[1]; //< input sample R

// --- process the audio to produce output
double ynL = psmVocoder[0].processAudioSample(xnL);
double ynR = psmVocoder[1].processAudioSample(xnR);

// --- write output (adapt to your framework)
processFrameInfo.audioOutputFrame[0] = ynL; //< output sample L
processFrameInfo.audioOutputFrame[1] = ynR; //< output sample R
```

Rebuild and test the plugin. What are the differences you hear with phase locking and peak tracking engaged? Try a variety of audio input files./ What kinds of files work best and worst with the phase vocoder algorithms here?

20.14 Third Party C++ DSP Object: TwoStageFFTConvolver

The third party object *TwoStageFFTConvolver* is included with your ASPiK project files and may be used for fast convolution with long impulse responses, as in convolutional reverb. This object may be used with or without FFTW installed and produces excellent results with a minimum of overhead. You can get an App Note on how to include and use this object in your own plugins at www.willpirkle.com/app-notes/.

20.15 Homework

1. Convert the *PhaseVocoder* object into a fast convolver without looking a the *FastConvolver* code.
2. Implement the modern Laroche and Dolson pitch shifting algorithm found in the *New Phase-Vocoder Techniques for Real-Time Pitch-Shifting, Chorusing, Harmonizing, and Other Exotic Audio Modifications* paper. You can use the *PhaseVocoder* to do the low-level FFT and overlap-add, and you can borrow the PSMVocoder's phase locking and peak detection code, both of which are needed. This algorithm implements pitch shifting by moving data from one FFT bin location to another. Will this method take care of the "chipmunk" problem of non-preservation of formants?

20.16 Bibliography

Barry, D., Dorran, D. and Coyle, E. A. 2008. *Time and Pitch Scale Modification: A Real-Time Framework and Tutorial*. Proc. 11th Intl. Conference on Digital Audio Effects, Vol. DAFx-08.

Clarkson, P. 1993. *Optimal and Adaptive Signal Processing*, Chap. 6. Boca Raton: CRC Press.

Driedger, J. 2011. *Time-Scale Modification Algorithms for Music Audio Signals*. Master's Thesis, Saarland University Faculty of Natural Sciences and Technology I Department of Computer Science. Saarbrücken.

Driedger, J. and Müller, M. 2016. "A Review of Time-Scale Modification of Music Signals." *Applied Sciences*, Vol. 6. No. 57.

Karrer, E. and Bochers, J. *PhaVoRIT: A Phase Vocoder for Real-Time Interactive Time-Stretching*, https://hci.rwth-aachen.de/phavorit, Accessed August 1, 2018.

Laroche, J. and Dolson, M. 1997. *Phase Vocoder: About This Phasiness Business*, Proc. IEEE Workshop on Applications of Signal Processing to Audio and Acoustics, New Paltz, NY.

Laroche, J. and Dolson, M. 1999. *Phase Vocoder: Improved Phase Vocoder Time-Scale Modification of Audio*, IEEE Transactions on Speech and Audio Processing, Vol. 7 pp. 323–332.

Laroche, J. and Dolson, M. 1999, 2000. *New Phase-Vocoder Techniques for Real-Time Pitch-Shifting, Chorusing, Harmonizing, and Other Exotic Audio Modifications*, Proc. IEEE Workshop on Applications of Signal Processing to Audio and Acoustics, WASPAA'99 (Cat. No.99TH8452).

Quatieri, T. and McAulay, R. *Speech Transformation Based on a Sinusoidal Representation*. Technical Report 717. Lincoln Library, Massachusetts Institute of Technology, Lexington.

Ramirez, R. 1985. *The FFT Fundamentals and Concepts*, Chap. 1–5. Englewood Cliffs: Prentice-Hall.

Reiss, J. and McPherson, A. 2015. *Audio Effects Theory, Implementation and Application*, Chap. 8, 11, Boca Raton: CRC Press.

Zölzer, U. 2011. *DAFX (Digital Audio Effects), 2nd Ed.*, Chap. 7. New York: Wiley.

CHAPTER 21

Displaying Custom Waveforms and FFTs

The subject of displaying graphical information for the user comes up from time to time on the Forum at www.willpirkle.com. For ASPiK users, we've created several custom view objects that you can use in your plugins to display audio histograms and FFT results. These are included in your ASPiK projects in the files *waveformfiew.h* and *waveformview.cpp*. The custom graphic code in these objects is heavily dependent on VSTGUI4. However, the basic ideas are universal across plugin frameworks. So, only these basic concepts are presented here.

You can get much more information about the advanced GUI features in ASPiK projects at www.willpirkle.com. These include the following.

- creating a tabbed GUI with multiple internal pages (views)
- creating a custom view by sub-classing VSTGUI4 objects
- implementing advanced behavior between controls with sub-controllers

21.1 Custom Views for Plotting Data

We can already send data back to the GUI (albeit in a primitive form) during the post-buffer processing phase when the metering information is copied to the outbound parameters. But metering is often not enough to convey the information you would like for the user. One popular custom view displays a rolling histogram of the audio waveform that scrolls from one side of its window to another. A more complex, but highly desirable, view will plot the spectrum of the audio signal using the FFT. These are shown in Figure 21.1a and b. Both plots consist of two parts: the graticule, which is the background grid of lines plus the axis labels, and the actual plot of data. In the time domain

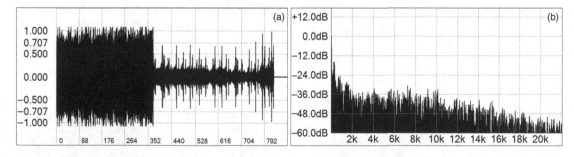

Figure 21.1: (a) An audio histogram of Blink-182's "Feeling This" intro in the sample domain.
(b) A linear spectrum graph view reveals the frequency content.

plot, each vertical line represents the maximum amplitude value of the audio data that was received since the last update. In the spectrum plot, each vertical line is the normalized magnitude of each FFT bin.

For an example plugin, we will be generating the data plot portion and you may then add the graticule and change the graphing options for homework. There are several issues to address that are equally applicable no matter what graphics platform you are using:

- The rectangle that we draw our waveform data into is a control called a *view*, and it is embedded somewhere in the surface of the plugin's GUI; we need to send data to that control only and not to the rest of the plugin object or other controls.
- The plugin GUI is refreshed and usually based on a timer. Each time the timer resets, the GUI redraws any items that have been marked as "dirty" (needing a redraw) during the last timer cycle.
- The plugin needs to send bulk data to the GUI for display; this data transfer must happen during each GUI refresh cycle.
- The data that needs to be sent is raw audio data streaming though the buffer processing cycle that is part of the audio processing thread.
- The audio and GUI threads are asynchronous.
- The GUI thread is a low-priority thread and its refreshing may be delayed due to other higher priority events occurring, so we never know exactly how much data we will be blasting out to the GUI for display.
- For the spectrum view, we should do the FFT processing on the GUI thread so as to not use excess CPU cycles during audio processing.
- We must observe thread safety when transferring data to the GUI.

21.1.1 ASPiK: The GUI Lifecycle and Messaging

The ASPiK *PluginCore* object handles the audio signal processing for the plugin. But, as a monolithic object, it also handles the parameter update process. In addition, it implements a message system for receiving notifications from the GUI thread, or from the plugin shell. If you are using another framework, then a similar system likely exists. For the *PluginCore*, this function is:

```
bool PluginCore::processMessage(MessageInfo& messageInfo)
```

The argument is a *MessageInfo* structure that contains the message itself, along with other information, depending on the message. The first few message enumerations and *MessageInfo* structure are shown here. We'll concentrate on the bold items first.

```
enum messageType{
    PLUGINGUI_DIDOPEN,         /* called after window is created */
    PLUGINGUI_WILLCLOSE,       /* called before window is destroyed */
    PLUGINGUI_TIMERPING,       /* timer ping for custom views */
    PLUGINGUI_REGISTER_CUSTOMVIEW,    /* register a custom view */
    PLUGINGUI_REGISTER_SUBCONTROLLER, /* register a subcontroller */
    // and others

struct MessageInfo
{
```

```
    // --- message: see enumeration
    uint32_t message = 0;

    std::string inMessageString;
    std::string outMessageString;

    void* inMessageData = nullptr;
    void* outMessageData = nullptr;
};
```

The user may open and close the plugin GUI many times during operation. However, the GUI lifecycle is well established as it relates to the *PluginCore* and the plugin shell. Figure 21.2a shows the relationship between the components, while Figure 21.2b shows the connection between the GUI and plugin shell.

When the user requests to see the plugin GUI, the plugin shell creates and opens the GUI object—the shell is the only entity qualified to do this since it receives a pointer to the GUI window from the DAW. After the GUI's *open* function is called and the GUI has successfully opened and returned a success flag, the plugin shell then notifies the *PluginCore* that the GUI did open by sending it the *PLUGINGUI_DIDOPEN* message. It is during this message that your plugin may do any special initialization to prepare for delivering data to the GUI. After the GUI opens, it begins sending the plugin shell notifications that the GUI is preparing to enter the redrawing cycle. The shell then forwards those messages to the *PluginCore* by sending *PLUGINGUI_TIMERPING* messages. For the *PluginGUI* object, the refresh timer will ping every 50 mSec (note: you can control this value if you would like to speed up or slow down GUI refreshes; the constant is declared in *guiconstants.h* as METER_UPDATE_INTERVAL_MSEC).

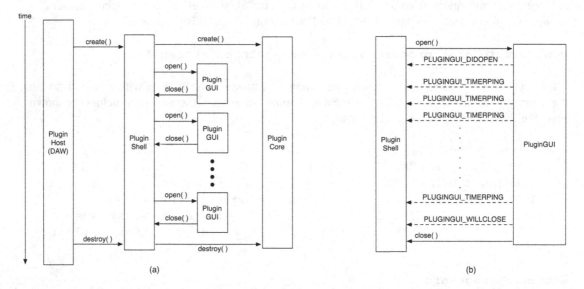

Figure 21.2: (a) The plugin shell, plugin GUI and plugin core lifecycles. (b) The GUI details.

During each *PLUGINGUI_TIMERPING* message handler, the *PluginCore* will pump audio data into the GUI. From that point on, it is up to the GUI to massage the data into whatever format it needs to display for the user. The GUI refresh cycle continues until the user closes down the GUI, or the session is closed, or the DAW itself is closed. In all these cases, the closing of the GUI windows is well defined for the *PluginGUI* paradigm. Just prior to closing, the GUI issues one last message to the shell: PLUGINGUI_WILLCLOSE. This message is always sent prior to any GUI object destruction. The shell forwards the message to the *PluginCore* to notify it that the GUI will close and the core object does what is needed for that. The GUI refresh timer is shut off after the PLUGINGUI_WILLCLOSE message is sent and handled. Notice that the GUI object never queries the shell for information; the message flow is one-way only, from the GUI.

21.1.2 Multi-threading: The Lock-Free Ring Buffer

We discussed multi-threading in Chapter 2— although you haven't needed to invest any time in this, because it is all handled for you between the plugin shell and *PluginKernel*. Now it is time to deal with the ramifications of the two-thread plugin. The audio data are streaming into the plugin object during the buffer processing cycle and the GUI timer ping messages are arriving asynchronously on the GUI thread. We need a mechanism to allow data transfer between the two threads in a safe manner, but also in a manner that does not require thread synchronization devices—which will attempt to force one thread to wait while the other thread finishes its operation. It would be disastrous for the GUI thread to force the audio thread to stop its operation while we wait for a data transfer to complete. To deal with this problem, we will need to use a *lock-free ring buffer*. A ring buffer is just another name for a circular buffer or queue. A lock-free ring buffer is specifically designed to have more than one thread accessing its data in a safe manner without using a synchronization mechanism that requires one thread to wait for the other (called *locking*). The theory behind lock-free ring buffers is deep and the implementations are clever, so we won't go into details here. But they are all based on one of at least two operation paradigms: single-producer–single-consumer and single-producer–multiple-consumer. These are also known as *queues,* and they involve buffers within buffers (or queues of queues) for operation. The lock-free ring buffer also solves another problem: we never know exactly how much data we will be writing or reading at any time due to the asynchronous nature of the two threads. A lock-free ring buffer is designed to grow itself if needed, and the really good implementations will do this in a highly efficient manner.

In our setup, the audio thread will be the producer, writing (also called *pushing*) raw audio data into the ring buffer, once per sample period. The GUI thread is the consumer; during the timer ping messages, we must pull data from the ring buffer and deliver it to the GUI. All of this is done in one object, the *PluginCore* but there are two different threads accessing the ring buffer. Because we only have one producer thread and one consumer thread, we may use the single-producer-single-consumer ring buffer. This is shown on the left side of Figure 21.3. The audio thread calls *processAudioFrame* that pushes data into the producer end of the ring buffer. The GUI thread sends the timer ping via *processMessage* and data are pulled from the consumer end of the buffer. Note that if you are not using ASPiK, then your framework will have a different mechanism for pinging the plugin for data, or you may be required to implement it yourself. This is usually not too difficult; it requires setting up an asynchronous clock to fire during the update interval. Refer to your framework documentation for more information.

Now at this point, there are a couple of options, depending on the architecture. For the *PluginKernel*, we know that the GUI timer ping messages are part of the redrawing cycle on the GUI thread. This

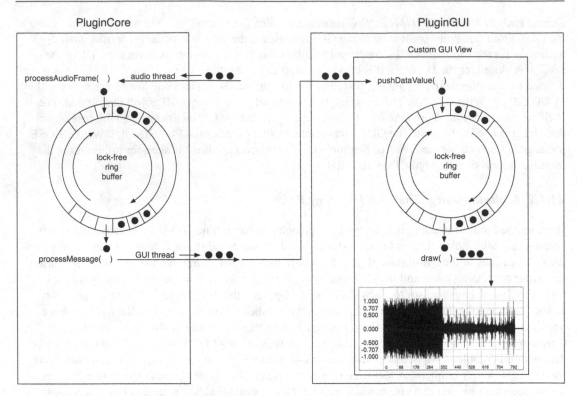

Figure 21.3: For complete thread safety, two lock-free ring buffers may be used to transfer data between the plugin and the custom data view, as shown here.

means that the drawing function on the custom view is called only *after* data are pushed into the GUI. However, this might not be the same way that your plugin framework implements the GUI. We would like to create a more generic and safer way to handle the incoming data so that if the drawing is being handled on yet another thread, there will be no data corruption. This method should also accommodate our spectrum view, which is going to require taking a FFT of the data. One solution is to provide another lock-free ring buffer on the GUI side. Incoming data are pushed into the buffer on the producer side, and then the drawing function pulls data from the consumer side and graphs it, as shown on the right side of Figure 21.3. The good news is that there are many good open-source and free lock-free ring buffers to choose from. They have similar behavior: you initialize the buffer's length (the number of data cells), then the producer pushes data into the queue using an en-queuing function and the consumer pulls data from the queue with a de-queuing function. In our case, the parent objects own the queues, and the data are moved inside of different member functions that have access to the same queue.

The lock-free ring buffer chosen for our plugin projects is called *ReaderWriterQueue* and it is a simple single-producer-single-consumer ring buffer. You can get the queue at https://github.com/cameron314/ readerwriterqueue, and you can get the full theory of its design at http://moodycamel.com/blog/2013/a-fast-lock-free-queue-for-c++. The ring buffer is namespaced as "moodycamel," so you will see definitions with that namespace prefix.

21.1.3 A Waveform Histogram Viewer

The waveform histogram (shown in Figure 21.1a) shows the amplitude of the waveform as vertical line segments. There is one vertical line segment for each column of pixels in the view's rectangle. The line segment reflects the absolute value of the audio sample at that point in time, normalized on a range of [0.0, +1.0]. In order to make the waveform scroll across the view at a reasonable speed, we need to consider the update interval for the GUI. For our GUIs, the refresh time is 50 mSec, therefore the *PluginCore* will be pinged approximately every 50 mSec. It will dump all of the audio it has accumulated since the last refresh cycle into the custom view's input queue. For $f_s = 44.1$ kHz this amounts to 2,205 samples per channel (in this example, we will plot only one channel's worth of data). After pulling the data from its ring buffer and placing them into the view's queue, the *PluginCore* will notify the view that the data dump is complete and that the view needs to refresh itself. It is important to issue the refresh command once every timer ping to make the waveform march across the view.

When the view gets its update command, it goes through all of the audio data that were pushed into it and finds the largest value. This value is the next vertical line that will appear on the left side of the window. It is placed next to the previous maximum value that arrived on the previous refresh interval. To simplify programming, we will use a circular buffer to push audio samples into and to make life even easier, we will force our circular buffer to be exactly the same length (M) as the view rectangle is wide, shown in Figure 21.4a. Each slot in the circular buffer corresponds to one column of pixels in the view's rectangle.

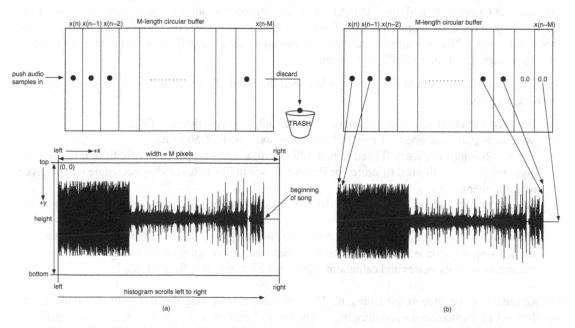

Figure 21.4: (a) A circular buffer provides the mechanism for easy painting of the histogram. (b) As the samples move through the circular buffer, they are painted as vertical lines, one at a time.

In Figure 21.4b you can see that we are at a point where the circular buffer has almost filled with the first M * 50 mSec worth of audio maxima. The very front (oldest) samples in the buffer are still 0.0s, and these form the blank spot at the head of the waveform. Once they are gone, the circular buffer will always be filled with valid data. By selecting the maximum from each refresh frame of data and pushing a new vertical line onto the left side of the graphic, the waveform scrolls smoothly and at a reasonable rate.

Notice in Figure 21.4a how the coordinate system works when drawing to the custom view's rectangular drawing space. The origin of the system at $x = 0$ and $y = 0$ is the upper left corner of the rectangle; positive direction in the x-axis moves to the right as normal. However, positive y-axis values move in the downward direction, so a point in the center of the rectangle would be at $p = (+width/2, +height/2)$, noting that this system forces all coordinates within the rectangle to have positive values. When we draw the vertical lines on the view, we will draw them using this coordinate system, so we need to make sure the polarity of the y-coordinates is correct.

21.1.4 An Audio Spectrum Analyzer View

To display the spectrum of an audio waveform, we will need to take the FFT of the audio data. This section assumes that you are familiar with the FFT information in Chapter 20, that you have FFTW installed on your system, and that you've already got a FFTW project running properly. FFTW will handle 100% of the FFT math for us, and it will do so very efficiently. For our project, we will use a FFT length of 512 points, but FFTW is so fast that you may easily increase that if you wish. We will need to calculate the magnitude response of each FFT buffer and display the magnitude. The magnitude array is calculated by finding the magnitude of the real and imaginary components as we discussed in Chapter 20, and the data in this array are mirrored about the center, so we will plot only the first half of the data, representing the frequency bins between 0 Hz and Nyquist. For our FFT size, this will become 128 data points. Figure 21.5 shows how the FFT calculation results in a vertical line (filled) or segmented line (unfilled) spectrum.

There are several fundamental differences between this kind of viewer and the previous audio histogram viewer.

- We must buffer up audio samples until we have enough to calculate the FFT.
- We can display only one FFT per GUI refresh period (one FFT/50 mSec).
- For a 512-point FFT, we will need to plot 128 magnitude points spaced evenly from 0 Hz to Nyquist, and we will need to stretch or shrink the data to fit whatever view rectangle we are given to graph them.
- We don't need, and can't use, the circular buffer idea from the histogram, so we will need another way to deal with graphing data.
- If we assume the most generalized case of multi-threading with the GUI then this means that the drawing function will possibly need to display one FFT magnitude result while we are accumulating data points and calculating the *next* FFT magnitude for it to display.

So, we seem to have three major issues: the FFT calculation, stretching the data to fit the window, and interleaving FFT calculations and drawing safely. We will save the data plotting details for later in the chapter and concentrate on the FFT problems to solve.

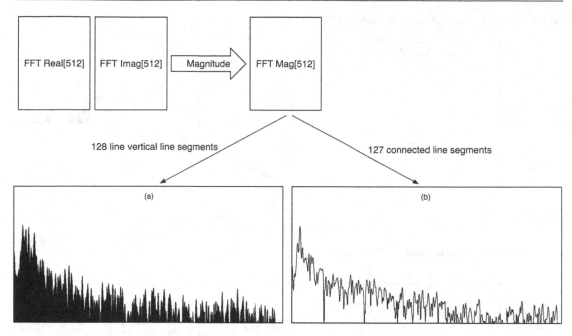

Figure 21.5: The audio spectrum is a plot of the magnitude of the FFT; we can use a set of vertical line segments in plot (a) to create a filled graph or (b) a series of connected line segments to plot the unfilled version.

There are multiple ways to handle the requirement that we display information from one FFT calculation while buffering and calculating the next one. One way would involve double buffers of magnitude arrays. While the drawing function is plotting from one array, the FFT function is writing into the other array. When the next FFT magnitude buffer is done, then we need to swap the old drawing array for the newly calculated FFT array. Once again we have a threading issue, assuming the most generic threading scenarios. To handle this problem, we will use a pair of queues (lock-free ring buffers) that will handle the shuffling. One queue will be for empty magnitude arrays, and the other will be for arrays that are calculated and ready to display. The two queues may have as many slots as you like to hold pointers to the magnitude arrays. Our system will use a double buffer—that is, two separate magnitude arrays. The FFT function and the drawing function will trade them back and forth. Figure 21.6 shows the behavior graphically. In Figure 21.6a, when the very first FFT data buffer is acquired, the FFT function pulls an empty magnitude array from the empty-queue and fills it with the freshly calculated FFT data, then pushes it into the ready-queue. The drawing function monitors the ready-queue and, when the first array arrives, it plots the new array of data while the next batch of data is being acquired. In Figure 21.6b when the second FFT has been calculated, the FFT function pulls another empty magnitude array from the empty-queue and fills it with the new FFT data, then pushes it into the ready-queue. The drawing function monitors the ready-queue and, when the next array arrives, it sends the old one back to the empty-queue to be refilled again, plots the new FFT data, and the cycle repeats, swapping out old data for new. With the double buffering system one buffer's data is being rendered to the graphic while the other is being filled with the next chunk. Pointers are used as the mechanism for shuffling the arrays around. You can experiment with the system by adding more

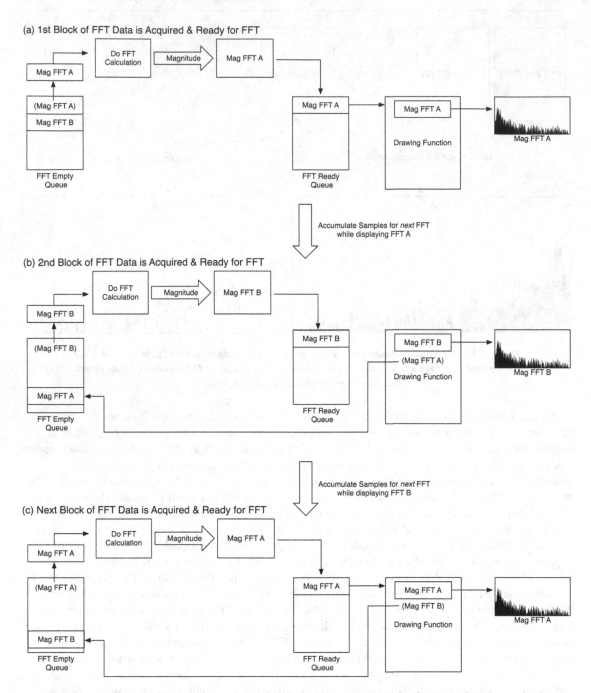

Figure 21.6: The spectrum analyzer view's double buffer system: (a) the first FFT data is acquired, calculated and displayed while the second batch of data is being collected, then (b) when the next FFT is ready, the drawing function sends its current buffer back to the empty-queue for refilling while plotting the next one and (c) the process repeats until the GUI window is closed.

magnitude array buffers (quad or octal buffering is possible) and you can also easily change the size of the queues. This setup is somewhat similar to that of Apple's *CoreAudio* API in which you are required enqueue fresh audio buffers and dequeue emptied buffers.

The last two sections have outlined the general strategies for implementing the audio histogram and spectrum custom views. If you are not using ASPiK, then you may begin working in your framework of choice. If you are using the *PluginKernel* or want to get deeper into how this is all implemented with ASPiK, then you may find the CustomViews project in the ASPiK SDK.

CHAPTER 22

Sample Rate Conversion

Sample rate conversion may be used to convert a recording done at one sample rate to a new sample rate that may be either higher or lower in frequency. An interpolator increases the sample rate of a signal by inserting new points into the existing audio stream. A decimator lowers the sample rate of a signal by removing samples. In audio FX and synth plugins, we may use the interpolator to raise the sample rate of an input signal or a synthesized signal, pushing its Nyquist frequency up to a higher value. Then, we might do some kind of nonlinear processing that inherently generates high frequency harmonics. If these harmonics do not exceed the new Nyquist rate, or do exceed it but only at extremely low amplitudes, then we may use the decimator to return the signal to its original sample rate, removing the added harmonics and eliminating (or at least reducing) the aliasing. Figure 22.1a shows harmonic components that have been generated above the Nyquist frequency and that fold back into the audio range as incorrect frequency components in Figure 22.1b—they are aliased or "in disguise."

Figure 22.1c shows the output of a soft clipper waveshaper and the spectrum Figure 22.1d that reveals aliasing. The bottom of the spectrum analyzer plot is −60 dB; lowering it to the 16-bit digital noise floor of −96 dB Figure 22.1e reveals much more aliasing. Using 4X oversampling and decimation, the aliasing is reduced nearly to the noise floor in Figure 22.1f. If the spectrum analyzer lower limit were to be raised back to −60 dB, the aliased components would disappear.

Of course, there has to be some sort of legal fine print with the up-sampling (interpolation) and down-sampling (decimation) processes: very steep anti-aliasing low-pass filters must be applied during both the interpolation and decimation, and for the best results possible, these filters need to be linear phase. Typically the linear phase Optimal (Parks-McClellan) method filters are used with lengths from 256 to 1024 points or more. If we use linear convolution from Chapter 16, then we will spend the majority of the CPU processing on the filtering alone, not to mention the other processing we wish to introduce. On top of that, we'll need to process the signal multiple times when oversampling. For example, with 4X oversampling, for each sample $x(n)$ that comes into the process function, we will need to interpolate it out to four samples, then process each of those as if they had arrived in the normal input stream. Only then may we decimate the signal back to the original sample rate, recovering a single output $y(n)$ that will be used for that sample period. In addition we must remember that any processing that occurs during the oversampling operation must be designed at that new rate: for example, if we are filtering, we must design the digital filter at the new, higher sample rate for it to work properly.

22.1 Interpolation: Overview

The interpolator consists of an up-sampler that inserts each input sample $x(n)$ plus $L - 1$ zero valued samples into the audio stream $w(n)$ followed by a low-pass filter whose cutoff frequency is set to the original Nyquist frequency. However, this filter will be processing L times as many points, so it is

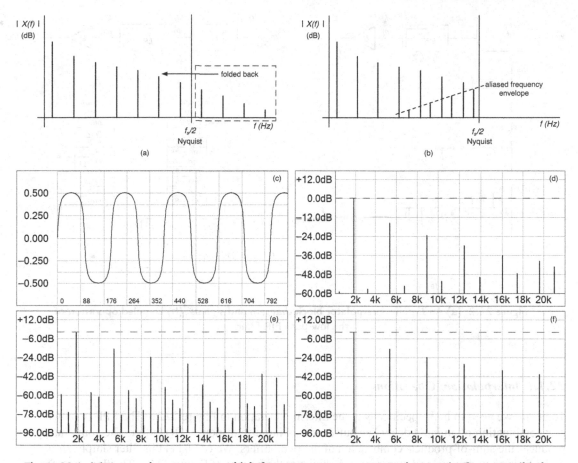

Figure 22.1: (a) A waveshaper generates high-frequency components past the Nyquist frequency (b) that are folded back into the audio spectrum as aliased components. (c) The time domain response of a soft clipping waveshaper. (d) The spectrum reveals aliased components. (e) Lowering the analysis magnitude limit to −96 dB reveals much more aliasing. (f) This spectrum using 4X oversampling and decimating around the nonlinear process shows that the aliasing is virtually eliminated.

running at a sample rate of Lf_s. This filter must be designed at that higher sample rate. The low-pass filter effectively interpolates the missing points during its convolution with the input. The multiplier L is required at the end to amplify the signal. This is due to the fact that the convolution spreads the impulse response energy across L samples; the multiplier makes up that loss. Figure 22.2a shows the block diagram for a L-point interpolator. The interpolator outputs L samples $p(n)$ for each input sample.

Interpolator:

- outputs L samples for every sample that enters
- uses a steep LPF to perform interpolation
- LPF needs to be linear phase, e.g. Optimal (Parks-McClellan) design
- any processing algorithms at the new sample rate Lf_s must be designed to operate at that new rate

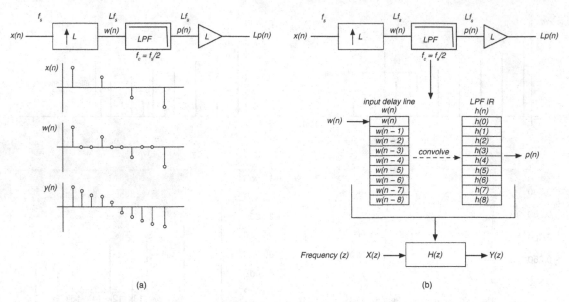

Figure 22.2: (a) An *L*-point interpolator block diagram and signals. (b) Detailed operation of the low-pass FIR filter.

22.1.1 Interpolation: Operations

To begin, we need to look specifically at the filtering operation in Figure 22.2b. The LPF is shown as a combination of an input delay line buffer and an impulse response (IR) buffer. The convolution produces the sum-of-products combination of the two buffers. We can label this filter simply *H(z)* in the frequency domain and represent it as shown at the bottom of Figure 22.2b. The flowchart for the *L*-point interpolator is shown in Figure 22.3. You can see that the two operations (zero insertion and convolution) are combined together. One input sample produces three output samples, so this is a three-point interpolator. The operation involves inserting either the input sample *x(n)* the first time through the loop, or zeros on any other loop into the delay line. A convolution is performed after each input value is applied, which results in that output value *p(n)*.

As an example, consider a three-point interpolator. For each input sample, it must produce three output samples. Let's look closely at the zero insertion and filtering operations. The operation is easiest to consider when the FIR filter is a multiple of the oversampling rate. In this case, our FIR filter will have nine taps or delay registers. To simplify the notation and stay with the book's convention, we are going to stick with *x(n)* as the input and *y(n)* as the filter output. Figure 22.4 shows the details of the operation. The non-zero input samples are shaded to show how the input signal moves through the delay line with the zero insertions.

The operations in Figure 22.4 are straightforward, assuming you've designed the FIR filter properly. There are *L* convolutions, one on each of the oversampled periods. But, you can see that the delay line is full of zeroes and convolving with all these zeros is inefficient: they have no effect on the final output sum. To optimize this process and ensure that we don't unnecessarily convolve with zeros, we need to change the strategy and use *polyphase decomposition*.

L-point interpolator flowchart combine zero insertion and filtering

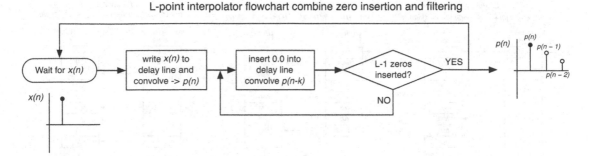

Figure 22.3: *L*-point interpolator flowchart.

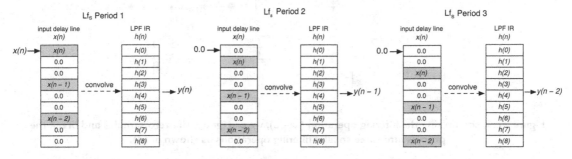

Figure 22.4: As shown here, the three output samples are produced on each of the oversampled Lf_s periods. The input $x(n)$ is pushed into the delay line on the first oversampled period, which is followed by two zero insertions on the next two subsequent periods. A convolution occurs on each interval, producing an output $y(n)$ and three outputs in total.

22.1.2 Interpolation: Polyphase Decomposition

Polyphase decomposition sounds imposing (and it could make a good title for a Steve Morse song), but it is really just a way to make the convolution more efficient by eliminating the convolution with zero samples. To understand it, you need to look once more at the filtering with a bit more detail. What we really want is to do the low-pass filtering first (at the slower sample rate) and then insert the zeros. It's more efficient to run the filter at the slower rate; it won't have the zeros in it. But, how do we deal with inserting the zeros later without destroying the signal? If you look at the first convolution on the first oversampled period Lf_s in Figure 22.5a and ignore the 0.0 values, you can see the output $y(n)$ is a sum of L products. It's really a mini-convolution that uses only $1/L$ the number of coefficient/data pairs. Examining the next two oversampled periods in Figure 22.5b and c, you can see there is another set of mini-convolutions with the each next set of three $h(n)$ coefficients. The most important observation is that the same input sequence $\{ x(n), x(n-1), x(n-2) \}$ is applied three times to produce the three outputs. Only the filter coefficients change during this operation.

This means that you can conceptually crack the original FIR filter into L pieces, each containing every Lth coefficient, as shown in Figure 22.5d. This produces L mini-FIR filters, each with an impulse response h and a frequency transfer function $H(z)$. Don't worry if the indexing of the filters looks "backwards"—that notation will become clear shortly. These little impulse responses in the mini-filters

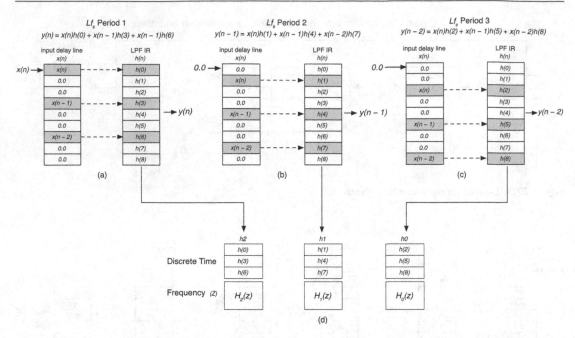

Figure 22.5: scrutinizing the filtering operation (a)–(c), we can ignore the zero samples and break the process into three smaller filtering operations as shown in (d).

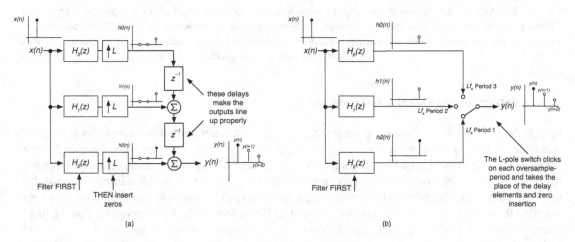

Figure 22.6: Two different polyphase interpolator block diagrams. (a) In the classic version, delays are used to make the outputs line up. (b) In the alternate version, a symbolic multi-pole switch is used. Note: This figure is adapted from (Porat, 1997).

are made up of pieces of the original; none of them are, within themselves, a complete impulse response of a LPF. They are called *sub-band filters*, *polyphase filters,* or *sub-band polyphase filters*. So you can see that you don't really design a polyphase filter: you take an existing filter and decompose it into the polyphase sub-filters. Decomposition follows a pretty simple rule about skipping through the original impulse response $h(n)$ by a factor of L. The decomposition of the original nine-tap FIR filter results in three three-tap polyphase (mini) filters. The last three steps to complete the algorithm are as follows, and are also shown in the classic polyphase interpolator block diagram in Figure 22.6a. An alternate version is shown in Figure 22.6b, where a switch takes the place of the delays.

1. The input delay line is now shortened to the same length as the polyphase filters.
2. Perform the filtering first, then insert the zeros (0s).
3. Use delays to offset the outputs of the three mini-filters so their samples line up properly; these "delays" are not actually storage locations or registers but are actually just trips through a *for( )* loop.

The same input is applied to each filter, just as we saw in the original sequence when we optimized it by skipping the zeros. The zeros are indeed inserted *after* the filtering operation—but there are delays and summers in place that cause the output samples to line up properly. The first sample out is output from $H_2(z)$, followed by $H_1(z)$, and then $H_0(z)$ two sample periods later. Notice the zero insertion appears to be backwards compared to the original diagram, thus the backwards indexing of the $H(z)$ mini-filters.

The delay elements are actually implemented as iterations through our oversampling for-loop.

22.2 Decimation: Overview

To decimate or down-sample a signal, we first must low-pass the signal with a cutoff of the *target* Nyquist frequency as shown in Figure 22.7a. This is to band-limit the input in the same way you low-pass filter the analog input of an ordinary digital audio device. We need this filter to eliminate frequencies above the target Nyquist frequency. After filtering, $M - 1$ samples are discarded, resulting in the signal $y(n)$ at the lower sample rate. The low-pass filter must be designed at that higher sample rate since it is running at the higher rate and processing M samples for each output. Figure 22.7b shows the same operation as (a) with DSP block details, using an FIR LPF convolved with the input $p(n)$ and it's digital $H(z)$ equivalent.

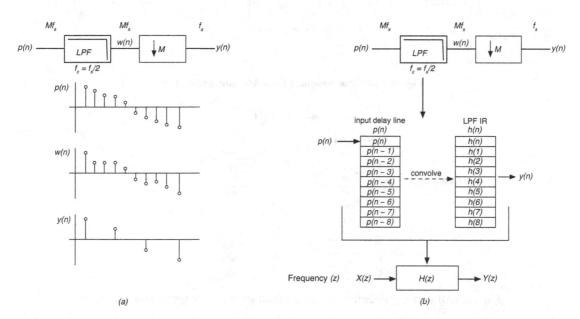

Figure 22.7: (a) An *M*-point decimator block diagram and signals.
(b) Detailed operation of the low-pass FIR filter.

Decimator

- outputs one sample for every *M*-sample that enters
- uses a steep LPF to prevent aliasing from the higher-rate input signal
- LPF needs to be linear phase—for example, Optimal (Parks-McClellan) design

22.2.1 Decimation: Operations

For decimation or down sampling, we also start with the LPF, shown as a combination of an input delay line buffer and an impulse response (IR) buffer. The convolution produces the sum-of-products combination of the two buffers. We can label this filter simply *H(z)* in the frequency domain and represent it as shown at the bottom. This is identical to the interpolation case. The flowchart for the *M*-point decimator is shown in Figure 22.8.

We can make the process more efficient and break apart the filter the same way that we did for the interpolation operation. On the down-sampling side, the key is to identify that we are doing at least one full convolution and break that one apart. The result is that we will decimate before we filter, running our polyphase filters at the original (slower) sample rate. For these examples, we will load up the input delay line a bit differently, showing sets of *L* (3) samples labeled *a,b,c* (etc.). Each set of three samples will get converted into a single output as a result of the decimation process. Figure 22.9 shows the decimator during some sample period *N*. Each input consists of a set of three samples, which are segmented in the delay line as a set of three current input *a* samples, then the older *b* and *c* groups that

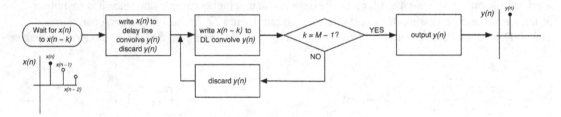

Figure 22.8: The flowchart for a *M*-point decimator.

Mf_s Period N

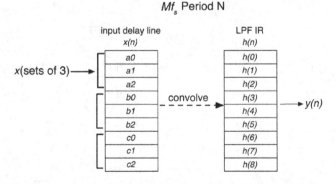

$$y(n) = a0h(0) + a1h(1) + a2h(2) + b0h(3) + b1h(4) + b2h(5) + c0h(6) + c1h(7) + c2h(8)$$

Figure 22.9: The decimator filter state when the next group of three input samples *a0*, *a1* and *a2* arrive at the input. The previous sets of three are shown as *b* and *c* groups.

follow. The final difference equation is the standard linear convolution equation and is shown at the bottom of the figure.

Now we rearrange the difference equation so that it still produces the same result but with a different ordering of the sets-of-three samples. To do this, we decimate the input sequence (take every Mth sample) and decimate the filter (take every Mth coefficient) to produce a rearranged version in Equation 22.1.

$$
\begin{aligned}
y(n) = a0h(0) &+ b0h(3) = c0h(6) \\
&+ a1h(1) + b1h(4) \_ c1h(7) \\
&+ a2h(2) + b2h(5) + c2h(8)
\end{aligned}
\tag{22.1}
$$

We can then group the sets of three as follows.

$$
y(n) = y0(n) + y1(n) + y2(n)
$$

where

$$
\begin{aligned}
y0(n) &= a0h(0) + b0h(3) + c0h(6) \\
y1(n) &= a1h(1) + b1h(4) + c1h(7) \\
y2(n) &= a2h(2) + b2h(5) + c(0)h(8
\end{aligned}
\tag{22.2}
$$

22.2.2 Decimation: Polyphase Decomposition

Notice that the final difference equation (Equation 22.2) didn't change—we just reordered the sub-components. With that in mind, revisit the filtering operation once more in detail as shown in Figure 22.10. The succession of operations during each Mf_s sample period in Figure 22.1.a–c shows

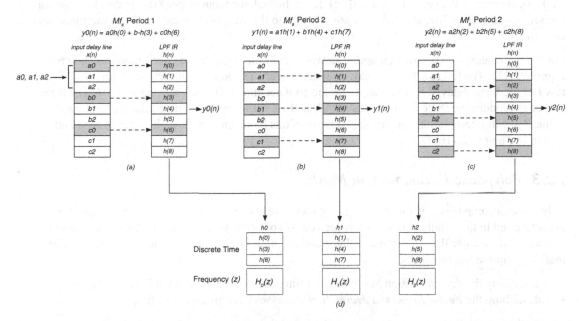

Figure 22.10: Scrutinizing the filtering operation (a)–(c) we can reorganize the convolution process into three smaller filtering operations as shown in (d).

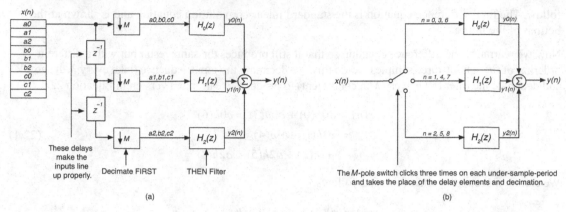

Figure 22.11: Two different polyphase decimator block diagrams. (a) In the classic version, delays are used to make the outputs line up. (b) In the alternate version, a symbolic multi-pole switch is used. Note: this figure is adapted from (Porat, 1997).

that you can conceptually crack the original FIR filter into pieces, each containing every Mth (third) coefficient. This produces $M = 3$ mini-FIR filters, each with an impulse response h and a frequency transfer function $H(z)$ shown in Figure 22.10d. In this case the indexing looks "normal"—again, it is of no real consequence to our software operation but it is important when put in context of the block diagrams. Now that we have three sub-band polyphase filters, we need to think about how we decimate the input before filtering. If you look at Figure 22.10, you can see that the inputs to each mini-filter are unique; in fact the inputs are decimated-by-three versions of the original audio stream. The first is { a0, b0, c0 } while the second is { a1, b1, c1 }, etc. Indeed, the input as been decimated by three for each filter—and each filter's coefficients are identical to the interpolation case as far as their indexes are concerned.

For the decimator, we also use conceptual delays to shift the signal before decimation as shown in Figure 22.11a. The top row has no delay, so it takes every Mth input value starting with $a0$. The second row has the input delayed by one sample period so it decimates M times starting with $a1$. The three branches produce the three mini-outputs $y0$, $y1$, and $y2$, then they are summed for the final output value. As with interpolation, an alternate version of this block diagram is a bit easier to read and is shown in Figure 22.11b.

22.3 Polyphase Decomposition Math

Polyphase decomposition is found all over the literature concerning sample rate conversion. If you are interested in the details, this section is for you. If you wish, you may ignore it and move on to Section 22.4, and use the built-in interpolator and decimator objects in your plugins. The polyphase math description has two parts:

- describing the decomposition of the original filter $H(z)$ where we crack it into sub-filters
- describing the *interpolation* and *decimation* processes through these sub-filters

The mathematical descriptions of the interpolator and decimator use the two different forms of decomposition in their proofs. The decimator uses type 1 decomposition while the interpolator uses type 2 decomposition. This is why the indexing on the *H(z)* filters looked backwards for the interpolator but forwards for the decimator.

22.3.1 Type-1 Decomposition

Decomposing the polyphase filter is fairly easy to prove to yourself. The equation for type 1 polyphase decomposition is here:

$$H(z) = \sum_{p=0}^{L-1} H_p(z^L) z^{-p} \tag{22.3}$$

The original *H(z)* becomes a set of *Hₚ(z)* filters. Each of these *Hₚ(z)* filters is shown in Equation 22.4.

$$H_p(z) = \sum_{m=0}^{(K+1)/L} h[p+mL] z^{-m} \tag{22.4}$$

where

$\quad K$ is the order of the original filter.

This produces the "skipping through the impulse response by *L*" that we saw visually earlier. For example, if *L* = 3:

$$
\begin{aligned}
H_0 &= h[0+0(3)]z^0 + h[0+1(3)]z^{-1} + h[0+2(3)]z^{-2} + \ldots \\
&= h[0] + h[3]z^{-1} + h[6]z^{-2} + \ldots \\
H_1 &= h[1+0(3)]z^0 + h[1+1(3)]z^{-1} + h[1+2(3)]z^{-2} + \ldots \\
&= h[1] + h[4]z^{-1} + h[7]z^{-2} + \ldots \\
H_2 &= h[2+0(3)]z^0 + h[2+1(3)]z^{-1} + h[2+2(3)]z^{-2} + \ldots \\
&= h[2] + h[5]z^{-1} + h[8]z^{-2} + \ldots
\end{aligned}
\tag{22.5}
$$

The consequence is that each filter's order becomes reduced by a factor of *p*. For example, suppose *L* = 3 and the original *H(z)* is an 8th order filter (a nine-tap FIR).

$$H(z) = 1 + 2z^{-1} + 3z^{-2} + 4z^{-3} + 5z^{-4} + 6z^{-5} + 7z^{-6} + 8z^{-7} + 9z^{-8} \tag{22.6}$$

The three sub-filters would then be as follows.

$$
\begin{aligned}
H_0(z) &= 1 + 4z^{-1} + 7z^{-2} \\
H_1(z) &= 2 + 5z^{-1} + 8z^{-2} \\
H_2(z) &= 3 + 6z^{-1} + 9z^{-2}
\end{aligned}
\tag{22.7}
$$

You can see that the original 8th order filter has been decomposed into three 2nd order filters. To get these transfer functions to combine together into the 8th order version, convert the filters from *H(z)* to *H(zᴸ)*, then multiply by *z⁻ᵖ* as follows.

$$H(z) = H_0(z^3)z^0 + H_1(z^3)z^{-1} + H_2(z^3)z^{-2} \tag{22.8}$$

Converting the filters from $H(z)$ to $H(z^L)$ is done by replacing z^q with z^{qL} as shown here.

$$H_0(z) = 1 + 4z^{-1} + 7z^{-2}$$
$$H_0(z^3) = 1 + 4z^{-3} + 7z^{-6}$$

$$H_1(z) = 2 + 5z^{-1} + 8z^{-2}$$
$$H_1(z^3) = 2 + 5z^{-3} + 8z^{-6} \tag{22.9}$$

$$H_2(z) = 3 + 6z^{-1} + 9z^{-2}$$
$$H_2(z^3) = 3 + 6z^{-3} + 9z^{-6}$$

Finally, multiply each filter by z^{-p} and reorder the terms in the proper sequence to get the final result.

$$\begin{aligned}
H(z) &= H_0(z^3)z^0 + H_1(z^3)z^{-1} + H_2(z^3)z^{-2} \\
&= (1 + 4z^{-3} + 7z^{-6})z^0 + (2 + 5z^{-3} + 8z^{-6})z^{-1} + (3 + 6z^{-3} + 9z^{-6})z^{-2} \\
&= 1 + 4z^{-3} + 7z^{-6} + 2z^{-1} + 5z^{-4} + 8z^{-7} + 3z^{-2} + 6z^{-5} + 9z^{-8} \\
&= 1 + 2z^{-1} + 3z^{-2} + 4z^{-3} + 5z^{-4} + 6z^{-5} + 7z^{-6} + 8z^{-7} + 9z^{-8}
\end{aligned} \tag{22.10}$$

Type 1 decomposition is used for the decimator. In Figure 22.12 you can see the relationship between the block diagram and decomposition equations.

22.3.2 Type-2 Decomposition

The polyphase decomposition equation can also be shown in a different form where the indexes run backwards. This is called the type 2 polyphase decomposition:

$$H(z) = \sum_{p=0}^{L-1} H_p(z^L)z^{-(L-1-p)} \tag{22.11}$$

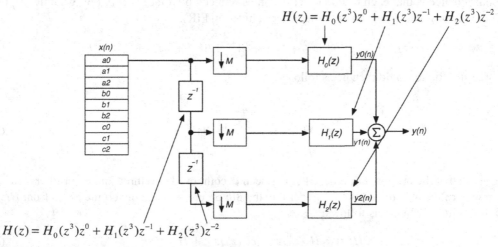

$$H(z) = H_0(z^3)z^0 + H_1(z^3)z^{-1} + H_2(z^3)z^{-2}$$

Figure 22.12: The decimation operation shown as a type-1 polyphase decomposition.

Now, each filter is defined by Equation 22.12.

$$H_p(z) = \sum_{m=0}^{(K+1)/L} h[(L-1-p)+mL]z^{-m}$$ (22.12)

where

K is the order of the original filter.

For example (L = 3):

$$
\begin{aligned}
H_0 &= h[3-1-0+0(3)]z^0 + h[3-1-0+1(3)]z^{-1} + h[3-1-0+2(3)]z^{-2} + \ldots \\
&= h[2] + h[5]z^{-1} + h[8]z^{-2} + \ldots \\
H_1 &= h[3-1-1+0(3)]z^0 + h[3-1-1+1(3)]z^{-1} + h[3-1-1+2(3)]z^{-2} + \ldots \\
&= h[1] + h[4]z^{-1} + h[7]z^{-2} + \ldots \\
H_2 &= h[0+0(3)]z^0 + h[0+1(3)]z^{-1} + h[0+2(3)]z^{-2} + \ldots \\
&= h[0] + h[3]z^{-1} + h[6]z^{-2} + \ldots
\end{aligned}
$$ (22.13)

The combination is now reversed:

$$H(z) = H_0(z^3)z^{-2} + H_1(z^3)z^{-1} + H_2(z^3)z^0$$ (22.14)

Using the original example, we get this:

$$H(z) = 1 + 2z^{-1} + 3z^{-2} + 4z^{-3} + 5z^{-4} + 6z^{-5} + 7z^{-6} + 8z^{-7} + 9z^{-8}$$ (22.15)

The three sub-filters would then be as follows:

$$
\begin{aligned}
H_2(z) &= 1 + 4z^{-1} + 7z^{-2} \\
H_1(z) &= 2 + 5z^{-1} + 8z^{-2} \\
H_0(z) &= 3 + 6z^{-1} + 9z^{-2}
\end{aligned}
$$ (22.16)

Notice that only the indexing of the filters has changed, compared to the first case. Finally, we get Equation 22.17—which produces the exact same result!

$$
\begin{aligned}
H(z) &= H_0(z^3)z^{-2} + H_1(z^3)z^{-1} + H_2(z^3)z^0 \\
&= (3 + 6z^{-3} + 9z^{-6})z^{-2} + (2 + 5z^{-3} + 8z^{-6})z^{-1} + (1 + 4z^{-3} + 7z^{-6})z^0
\end{aligned}
$$ (22.17)

Type 2 decomposition is used for the interpolator. In Figure 22.13 you can see the relationship between the block diagram and decomposition equation. The reason for the reverse in indexing has to do with which side of the parallel branches the z^{-1} elements are located (compare with Figure 22.12).

22.3.2.1 What Is the Meaning of "Polyphase?"

The Type 1 decomposition equation shows that each mini-filter is as follows:

$$H_p(z) = \sum_{m=0}^{(K+1)/L} h[p+mL]z^{-m}$$ (22.18)

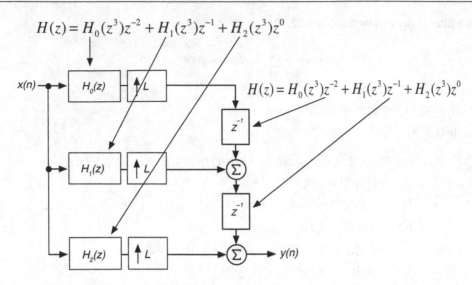

$$H(z) = H_0(z^3)z^{-2} + H_1(z^3)z^{-1} + H_2(z^3)z^0$$

Figure 22.13: The interpolation operation shown as a type-2 polyphase decomposition.

The original filter $H(z)$ is decomposed into $(K+1)/L$ mini-filters. The impulse response of each of those filters is a set of $h[n+L]$ terms.

$$h[p + mL]$$

or more generally   (22.19)

$$h[n + L]$$

The p value in Equation 22.18 is a counter that produces the sequences, and the L value skips through the original impulse response. So, for example when $p = 0$, it produces an impulse response $h = \{ h(0),\ h(3),\ h(6)$ etc. $\}$; and when $p = 1$ we get the next set $h = \{ h(1),\ h(4),\ h(7) \ldots \}$.

Manipulating the impulse response so that is shifted by L, producing the $h[n+L]$ version, is called *time-shifting by L* in the time domain. In the frequency domain, this produces a phase shift.

$$h[n + L] \leftrightarrow e^{j\theta L} H(\theta)$$

$e^{j\theta L}$ represents a phase-shift of θL (22.20)

The term *polyphase* comes from this phase shift component (the $e^{j\theta L}$ term). Each sub-band filter is a phase-shifted sub-component of the original impulse response. The interpolator and decimator filter branches use delays to line up the shifted/filtered components properly.

22.4 C++ DSP Objects: Interpolator and Decimator

The two DSP objects *Interpolator* and *Decimator* perform the sample rate conversion operations and are included in your *fxobjects.h* file. You need to have FFTW installed and linked to your project. These objects require FIR filter coefficients to perform the filtering required for either up-sampling or down-sampling. Your projects come with a file named *filters.h* that contains a set of pre-computed FIR coefficient arrays for the steep low-pass filters at a variety of sampling rates and array lengths. All of these filters were designed with the Optimal/Parks-McClellan method using the free software at http://t-filter.engineerjs.com that produces C++ arrays that you may cut and paste as you like. In the *filters.h* file, you will find definitions for filters for the following up/down sampling factors, sample rates, and FIR filter lengths (note that polyphase implementation may be enabled or disabled:

- 2X and 4X up/down sampling
- 44.1 kHz and 48 kHz sampling rates
- 128, 256, 512, and 1024 point FIR filters

The *fxobjects.h* file includes a helper function to find and load the FIR filter array from the *filters.h* file, as shown here.

```
inline double* getFilterIRTable(unsigned int FIRLength,
                                rateConversionRatio ratio,
                                unsigned int sampleRate)
```

It also includes a helper to decompose a FIR into an array of polyphase mini-filters:

```
// --- polyphase decomposition on a big FIR into a set of sub-band FIRs
inline double** decomposeFilter(double* filterIR,
                                unsigned int FIRLength,
                                unsigned int ratio)
```

Those two objects use the *FastConvolution* object to perform the fast FIR filtering needed for the sample rate conversion, so you should check them out to see how to use the *FastConvolution* object in your own designs. The following support functions should be modified if you want to add your own ratios beyond 2X or 4X, or you want to implement FFTs that are smaller than 128 points or larger than 1024.

```
// --- strongly typed enum for ratio
enum class rateConversionRatio { k2x, k4x };

// --- 4X max, you can easily change this
const unsigned int maxSamplingRatio = 4;

// --- returns ratio as a number, from the strongly typed enum
inline unsigned int countForRatio(rateConversionRatio ratio)

// --- get table pointer for built-in anti-aliasing LPFs
inline double* getFilterIRTable(unsigned int FIRLength,
                                rateConversionRatio ratio,
                                unsigned int sampleRate)
```

22.4.1 C++ DSP Object: Interpolator

The *Interpolator* object performs up-sampling or interpolation and is coded to implement 2X or 4X oversampling. You may easily add more ratios. but you'll need to design the appropriate FIR filter as well.

22.4.2 Interpolator: *Enumerations and Data Structure*

This object requires the strongly typed enumeration to define ratios that is listed in Section 22.4. The object is outputs its oversampled data in an *InterpolatorOutput* structure. For each point that enters the *Interpolator*, either two or four points will be rendered; those points and their count are stored in this structure.

```
struct InterpolatorOutput
{
        InterpolatorOutput() {}
        double audioData[maxSamplingRatio] = { 0.0 };
        unsigned int count = maxSamplingRatio;
};
```

22.4.3 Interpolator: *Members*

Tables 22.1 and 22.2 list the *Interpolator* member variables and member functions. This object is highly specialized, so it does not inherit the *IAudioSignalProcessor* interface.

Table 22.1: *Interpolator* **member variables**

Interpolator Member Variables		
Type	Name	Description
FastConvolver	*convolver*	Fast convolver for anti-aliasing filter
double	*sampleRate*	Current sample rate
double	*firLength*	Filter length
rateConversionRatio	*ratio*	Conversion ratio enum
bool	*polyphase*	Flag for polyphase operation
FastConvolver	*polyPhaseConvolvers[maxSamplingRatio]*	Set of fast convolvers for polyphase sub-filters

Table 22.2: *Interpolator* **member functions**

Interpolator Member Functions		
Returns	Name	Description
void	*initialize* Parameters: —*unsigned int* _FIRLength —*rateConversionRatio* _ratio —*unsigned int* _sampleRate —*bool* polyphase	Initialize the *Interpolator* with the filter length, conversion ratio, sample rate and polyphase flag
InterpolatorOutput	*interpolateAudio* Parameters: —*double* xn	Interpolate the input sample and return the new samples in the output structure

22.4.4 Interpolator: *Programming Notes*

As you can see from Table 22.2, this object implements only two functions. To use the object, you initialize it with the FIR length (which sets up the *FastConvolver*), ratio, sample rate, and the *polyphase* flag. Note that due to the speed of FFTW, the polyphase operation will save you about 5% to 10% of the overall CPU task. To process audio, you send one sample into the object and get either two or four samples back. Its initialization and use are shown here:

```
// --- initialize 512 point FIR, 4X, 44.1 kHz, polyphase
interpolator.initialize(512, rateConversionRatio::k4x, 44100, true);

// --- do interpolation
InterpolatorOutput interpOut = interpolator.interpolateAudio(xnL);

// --- process the four samples
for (int i = 0; i < interpOut.count; i++)
{
    // --- process each sample in audioData
    filter.processAudioSample(interpOut.audioData[i]);
}
```

Function: *interpolateAudio*
Description: The *interpolateAudio* function works with or without polyphase filters. In the standard configuration it inserts the input sample into the fast convolver, followed by $L - 1$ zero samples where L is the oversample ratio, in a for-loop.

```
for (unsigned int i = 0; i < count; i++)
{
    output.audioData[i] = i == 0 ?
                ampCorrection*convolver.processAudioSample(xn) :
                ampCorrection*convolver.processAudioSample(0.0);
}
```

With polyphase enabled, it takes the output of the L polyphase filters in succession, as shown in Figure 22.6b. The filters are packed into an array called *polyPhaseConvolvers[]*. In both cases, the amp correction factor (which equals L) is applied, as shown in Figure 22.2. The counter using the index m variable counts backwards due to the reverse notation used with indexing the polyphase filters:

```
for (unsigned int i = 0; i < count; i++)
{
    output.audioData[i] = ampCorrection*polyPhaseConvolvers[m-]
                                    .processAudioSample(xn);
}
```

22.4.5 *C++ DSP Object:* Decimator

The *Decimator* object performs down sampling or decimation and is coded to implement 2X or 4X down sampling. You may easily add more ratios but you'll need to design the appropriate FIR filter as well.

22.4.6 Decimator: *Enumerations and Data Structure*

This object requires the strongly typed enumeration to define ratio as listed in Section 22.4. For every *M* points that enters the *Decimator*, only one point is produced. The multiple samples arrive in a *DecimatorInput* structure.

```
struct DecimatorInput
{
    DecimatorInput() {}
    double audioData[maxSamplingRatio] = { 0.0 };
    unsigned int count = maxSamplingRatio;
};
```

22.4.7 Decimator: *Members*

Tables 22.3 and 22.4 list the *Decimator* member variables and member functions. This object is highly specialized, so it does not inherit the *IAudioSignalProcessor* interface.

22.4.8 Decimator: *Programming Notes*

As you can see from Table 22.4 this object only implements two functions. To use the object, you initialize it with the FIR length (which sets up the *FastConvolver*), ratio, sample rate,

Table 22.3: *Decimator* **member variables**

Decimator Member Variables		
Type	**Name**	**Description**
FastConvolver	*convolver*	Fast convolver for anti-aliasing filter
double	*sampleRate*	Current sample rate
double	*firLength*	Filter length
rateConversionRatio	*ratio*	Conversion ratio enum
bool	*polyphase*	Flag for polyphase operation
FastConvolver	*polyPhaseConvolvers[maxSamplingRatio]*	Set of fast convolvers for polyphase sub-filters

Table 22.4: Decimator member functions

Decimator Member Functions		
Returns	**Name**	**Description**
void	*initialize* Parameters: —*unsigned int* _FIRLength —*rateConversionRatio* _ratio —*unsigned int* _sampleRate —*bool* polyphase	Initialize the *Interpolator* with the filter length, conversion ratio, sample rate and polyphase flag
double	*decimateAudio* Parameters: —*DecimatorInput* data	Decimates the *M* input sample and returns a single sample output

and the *polyphase* flag. Note that due to the speed of FFTW, the polyphase operation will save you about 5% to 10% of the overall CPU task. To process audio, fill the *DecimatorInput* structure's *audioData* array with the *M* data points to be decimated, then call the *decimateAudio* method.

```
// --- initialize 512 point FIR, 4X, 44.1 kHz, polyphase
decimator.initialize(512, rateConversionRatio::k4x, 44100, true);

// --- do decimation
DecimatorInput decimatorIn;
decimatorIn.audioData[0] = // <-sample 1
decimatorIn.audioData[1] = // <-sample 2
decimatorIn.audioData[2] = // <-sample 3
decimatorIn.audioData[3] = // <-sample 4

// --- process the four samples down to one
double yn = decimator.decimateAudio(decimatorIn);
```

Function: *decimateAudio*

Description: The *decimateAudio* function works with or without polyphase filters. In the standard configuration, it calls the convolver to process the *M* input samples. The results of the first *M − 1* are discarded and the last is kept, via overwriting the same variable.

```
double output = 0.0;
for (unsigned int i = 0; i < count; i++)
{
    output = convolver.processAudioSample(data.audioData[i]);
}
```

With polyphase enabled, the object sums the outputs of the *M* polyphase filters in succession as shown in Figure 22.11b. The filters are packed into an array called *polyPhaseConvolvers[]*; note this is verbose for the book print only.

```
for (unsigned int i = 0; i < count; i++)
{
    double sample = data.audioData[i];
    output += polyPhaseConvolvers[i].processAudioSample(sample);
}
```

22.5 Chapter Plugin: TubePreamp *Revisited*

The *TubePreamp* plugin in Chapter 19 easily created significant aliasing with even modest saturation levels—and high frequencies exacerbated the problem. We can mitigate the aliasing with oversampling. First we use the *Interpolator* to increase the sample rate to 4X the original value, placing the new Nyquist frequency at 88.2 kHz (four times the original Nyquist rate). We then process the four new samples through the *ClassATubePre* object, one after the other, gathering each output and placing it in a *DecimatorInput* structure. When the last sample is processed, we form the output *y(n)* by calling the *decimateAudio( )* function. There are some things to note here.

- In an oversampled plugin, every component that is part of the oversampling operation will be running at the new and higher sample rate; this means that any object that uses the sample rate in its calculations needs to be given the new, higher rate.
- With FFTW and a Release mode build, the sample rate conversion is extremely fast and efficient.
- You may limit the oversampling portion of your process function to the nonlinear parts of your algorithm, and run the other objects at the normal rate.

For the *TubePreamp*, you'll need to enable FFTW in the compiler and the *fxobjects.h* file. Then, it is just a matter of preparing the *Interpolator* and *Decimator* objects for use and resetting the components at the new sample rate. You'll need two sets of objects for stereo operation.

Declarations:

```
Interpolator interpolator[2];
Decimator decimator[2];
```

Reset:

```
// --- interp/decimate, with polyphase
interpolator[0].initialize(1024, rateConversionRatio::k4x,
                           resetInfo.sampleRate, true);
interpolator[1].initialize(1024, rateConversionRatio::k4x,
                           resetInfo.sampleRate, true);

decimator[0].initialize(1024, rateConversionRatio::k4x,
                        resetInfo.sampleRate, true);
decimator[1].initialize(1024, rateConversionRatio::k4x,
                        resetInfo.sampleRate, true);

// --- two preamps running at 4X sample rate
tubePreamp[0].reset(4.0*resetInfo.sampleRate);
tubePreamp[1].reset(4.0*resetInfo.sampleRate);
```

Process audio (left channel only):

```
InterpolatorOutput interpOut;
DecimatorInput decimatorIn;

double xnL = // < from your framework Left channel

// --- up sample
interpOut = interpolator[0].interpolateAudio(xnL);

// --- process through preamp into decimator structure
for (int i = 0; i < interpOut.count; i++)
{
    double val = interpOut.audioData[i];
    decimatorIn.audioData[i] = tubePreamp[0].processAudioSample(val);
}

// --- get downsampled output
double ynL = decimator[0].decimateAudio(decimatorIn);
```

22.6 Homework

The quality of the anti-aliasing in the sample rate conversion comes down to two factors:

- How nonlinear is your processing?
- How well are your filters designed? The longer the FIR the better, in general.

Experiment with the sample rate conversion objects by trying longer FIRs or different FIR filtering criteria. You should know that the criteria were varied to produce the set of coefficients in the *filters.h* file; you may very well be able to get better results with re-designs of the filters. You can copy and paste the C++ arrays into other software to examine the stop-band attenuations. The website in the text is highly recommended for quickly generating coefficients that are packaged in C++ arrays. You can find it at http://t-filter.engineerjs.com and you simply enter the sample rate and filter design specifications in the table at the lower left of the page and click the "Design Filter" button. You may then cut and paste the C++ code that appears in the output window on the right side of the page.

22.7 Bibliography

Ifeachor, E. and Jervis, B. 1993. *Digital Signal Processing, A Practical Approach*, Chap. 4, 6. Menlo Park: Addison Wesley.

Porat, B. 1997. *A Course in Digital Signal Processing*. New York: John Wiley and Sons.

Stanford University. *N-Channel Polyphase Decomposition*, https://ccrma.stanford.edu/~jos/sasp/N_Channel_Polyphase_Decomposition.html, Accessed August 1, 2018.

Turek, D. 2004. *Design of Efficient Digital Interpolation Filters for Integer Upsampling*, Master's Thesis, Department of Electrical Engineering and Computer Science Massachusetts Institute of Technology., Massachusetts.

Index

Printed in the United States
by Baker & Taylor Publisher Services